STRATEGIES AND APPLICATIONS IN QUANTUM CHEMISTRY

TOPICS IN
MOLECULAR ORGANIZATION AND ENGINEERING

Volume 14

The titles published in this series are listed at the end of this volume.

Strategies and Applications in Quantum Chemistry

From Molecular Astrophysics
to Molecular Engineering

edited by

Y. ELLINGER

Ecole Normale Supérieure
& Observatoire de Paris,
Paris, France

and

M. DEFRANCESCHI

Commissariat à l'Energie Atomique,
Saclay, France

Springer Science+Business Media, B.V.

Library of Congress Cataloging-in-Publication Data

Strategies and applications in quantum chemistry : from molecular
  astrophysics to molecular engineering / edited by Y. Ellinger and M.
  Defranceschi.
      p.   cm. -- (Topics in molecular organization and engineering ;
  v. 14)
   Includes index.
   ISBN 0-7923-3837-5 (hb : alk. paper)
   1. Quantum chemistry.   I. Ellinger, Y. II. Defranceschi,
Mireille, 1955-   . III. Series.
QD462.5.S73 1996
541.2'8--dc20   95-45654

ISBN 978-94-017-3786-9

ISBN 978-94-017-3786-9   ISBN 978-0-306-46930-5 (eBook)
DOI 10.1007/978-0-306-46930-5

"The logo on the front cover represents the generative hyperstructure of alkanes", printed with permission from J.E. Dubois, Institut de Topologie et de Dynamique des Systèmes, Paris, France.

Printed on acid-free paper

Introduction to the Series

The Series 'Topics in Molecular Organization and Engineering' was initiated by the Symposium 'Molecules in Physics, Chemistry, and Biology', which was held in Paris in 1986. Appropriately dedicated to Professor Raymond Daudel, the symposium was both broad in its scope and penetrating in its detail. The sections of the symposium were: 1. The Concept of a Molecule; 2. Statics and Dynamics of Isolated Molecules; 3. Molecular Interactions, Aggregates and Materials; 4. Molecules in the Biological Sciences, and 5. Molecules in Neurobiology and Sociobiology. There were invited lectures, poster sessions and, at the end, a wide-ranging general discussion, appropriate to Professor Daudel's long and distinguished career in science and his interests in philosophy and the arts.

These proceedings have been arranged into eighteen chapters which make up the first four volumes of this series: Volume I, 'General Introduction to Molecular Sciences'; Volume II, 'Physical Aspects of Molecular Systems'; Volume III, 'Electronic Structure and Chemical Reactivity'; and Volume IV, 'Molecular Phenomena in Biological Sciences'. The molecular concept includes the logical basis for geometrical and electronic structures, thermodynamic and kinetic properties, states of aggregation, physical and chemical transformations, specificity of biologically important interactions, and experimental and theoretical methods for studies of these properties. The scientific subjects range therefore through the fundamentals of physics, solid-state properties, all branches of chemistry, biochemistry, and molecular biology. In some of the essays, the authors consider relationships to more philosophic or artistic matters.

In Science, every concept, question, conclusion, experimental result, method, theory or relationship is always open to reexamination. Molecules do exist! Nevertheless, there are serious questions about precise definition. Some of these questions lie at the foundations of modern physics, and some involve states of aggregation or extreme conditions such as intense radiation fields or the region of the continuum. There are some molecular properties that are definable only within limits, for example, the geometrical structure of non-rigid molecules, properties consistent with the uncertainty principle, or those limited by the neglect of quantum-field, relativistic or other effects. And there are properties which depend specifically on a state of aggregation, such as superconductivity, ferroelectric (and anti), ferromagnetic (and anti), superfluidity, excitons, polarons, etc. Thus, any molecular definition may need to be extended in a more complex situation.

Chemistry, more than any other science, creates most of its new materials. At least so far, synthesis of new molecules is not represented in this series, although the principles of chemical reactivity and the statistical mechanical aspects are included. Similarly, it is the more physico-chemical aspects of biochemistry, molecular biology and biology itself that are addressed by the examination of questions related to molecular recognition, immunological specificity, molecular pathology, photochemical effects, and molecular communication within the living organism.

Many of these questions, and others, are to be considered in the Series 'Topics in Molecular Organization and Engineering'. In the first four volumes a central core is presented, partly with some emphasis on Theoretical and Physical Chemistry. In later volumes, sets of related papers as well as single monographs are to be expected; these may arise from proceedings of symposia, invitations for papers on specific topics, initiatives from authors, or translations. Given the very rapid development of the scope of molecular sciences, both within disciplines and across disciplinary lines, it will be interesting to see how the topics of later volumes of this series expand our knowledge and ideas.

WILLIAM N. LIPSCOMB

Table of Contents

Prefaces

Why this Book ?

Starting from pure academic knowledge, Quantum Chemistry has gained the rank of full partner in most chemical research carried out today, from organic chemistry to solid state chemistry, from biology to material sciences, from astrophysics to chemical engineering.

Evolution has been rapid.

After a promising start in the mid twenties, Quantum Chemistry then faced the first technological block: the numerical barrier.
The consequence has been a flourishing of concepts still found in the present literature and in the common langage of all physico-chemists when it comes to understanding the basic phenomena.

Then appeared the time of computers. Quantum chemists developed semi-empirical codes that rapidly evolved into ab-initio complex systems of programs. According to their optimistic or pessimistic views, colleagues have seen this period either as that of semi -quantitative or of semi-qualitative theoretical chemistry. Very recently came the age of super computers, and a generation of quantum chemists have seen their dream come true: at last, the quality of the calculation is in harmony with the quality of the concepts.

At a time when an increasing number of chemists are being dangerously attracted by the fascination of supposedly easy computing, it seemed an appropriate opportunity to dedicate a volume to Gaston Berthier. Born in 1923, a year in which the scientific community celebrated the 10th anniversary of Bohr's quantum theory of the atom, but also the year when de Broglie published the fundamental idea that the orbit of an electron is linked to a stationary condition on the associated wave, Gaston Berthier took part in all the evolution and unavoidable conflicts between the *anciens* and the *modernes* at each stage of development of the discipline. Often in advance of his time, but never rejecting the past, Gaston Berthier has inspired generations of young, and less young, scientists in almost every branch of theoretical chemistry and its applications to experiment.

The title "Strategies and Applications in Quantum Chemistry" was chosen to illustrate his dual philosophy. The response of former collaborators of Berthier, often former students, always friends, has been overwhelming, as is evident from the size of this book.

Most of the signatures come from the community of the Theoretical Chemists of Latin Expression, a melting pot conceived by Berthier and a few others. The contributions range from the prefaces with a personal assessment by the first witnesses of Berthier's beginning through a series of articles covering basic developments in MCSCF theory, perturbation theory, basis sets, charge densities, wave function instabilities, correlation effects, momentum space theory, through contributions to understanding EPR spectroscopy, magnetic properties, electric field effects, electronic spectra, interaction between radiation and molecular structures, puzzling astrophysical systems, clusters and, as an opening to a different world, muonium chemistry.

Although the broad spectrum of Quantum Chemistry represented by the thirty or so articles contained in this volume only partially reflects the variety and richness of Berthier's preoccupations, it should convey to the coming generation of quantum chemists and to all readers the useful lesson of an outstanding chemist maintaining wide interests and resisting the drift of fashion.

It is a pleasure to thank all the contributors to this volume for their gracious and enthusiastic cooperation, with a special mention to J.P. Flament for his kind assistance. This book is simply a measure of universal regard and affection and the date is just what it is, no more, since we all know that G.B. will never retire.

Y. Ellinger   M. Defranceschi

At the Dawn of Quantum Chemistry:
The Role of Gaston Berthier

The active involvement of french scientists in the development of quantum chemistry started rather late. Strangely, as is may seem to be, the native country of Louis de Broglie was for a long time rather insensitive, if not resistant, to the possible significance and usefulness of quantum theories for the development of chemical knowledge. The strangest opponents were found, as can be guessed, among... the chemists. When we were students, the highly popular series of courses leading to the Certificate of General Chemistry, which implied a whole year of studies, hardly mentioned the existence of an electronic structure in atoms or molecules. The situation was somewhat better in the Certificate of Physical Chemistry in which the foundations of the quantum theory were studied in more details but even there the practical incidence of the theory was limited to the case of the H_2 molecule. Nobody seemed to have considered seriously that quantum theory could be of practical use in contributing to the solution of chemical problems involving larger molecules.

It is on this background of indifference, if not of open animosity, that a group of young research workers have undertaken, in the forties, the courageous, but rather risky for their career, task of promoting the development of quantum chemistry in France, with the well-conceived goal of exploring its capacity for studying realistic problems related to the exploration of molecular properties, without any a priori limitation as to the size of the molecules involved. Gaston Berthier was one of the earliest members of these local pioneers.

A crucial event which greatly helped to stimulate the interest of the french scientific community in the potentialities of quantum chemistry was the holding in Paris, in 1948, of an international symposium on the methods, achievements and status of quantum chemistry, which was attended by the most eminent specialists in this field. Suffice it to mention the presence of Linus Pauling and Robert Mulliken. It is that year that Gaston Berthier joined our group.

This happened at an interesting moment in our activities. These were concentrated at that period on the structure and properties of polynuclear aromatic hydrocarbons, in particular but far from exclusively, in relation with their carcinogenic activity. For purely historical reasons, due to a large extent to the international prestige of Linus Pauling, the method which we have been using for this sake, was the valence bond method. By 1948 we have clearly realized the practical limitations of this procedure for the exploration of large molecular systems and turned our attention to the molecular orbital method much more suitable for such an endeavour. Berthier joined our laboratory just at this methodologically turning point. He was thus immediately associated with what were the first works and publications ever performed in France by this method. They dealt with the electronic structure of aromatic hydrocarbons composed of four and five benzene rings, a tremendous task at that time.

We may say that from that period on, Berthier became one of the best experts in France of the molecular orbital method and an acknowledged pioneer in its application and development.

In the few years which followed his first steps in the field he succeeded in investigating by the relatively simple Hückel approximation a large number of new and fascinating problems in organic chemistry of conjugated systems. This involved, in particular an extremely vast exploration of non-aromatic benzenoid compounds, starting from the small fulvene and azulene up to rather very complicated thermochromic ethylenes, a study which led to the discovery of a number of unpredicted and surprising properties, in particular in the field of dipole moments and ultra-violet spectra, which contradicted a number of rules "established" by the resonance theory. A special mention must be given to his contribution to the theoretical exploration of the diamagnetic properties of polycyclic hydrocarbons and of the electronic structure of free radicals and biradicals. This very prolific activity was carried out in the early fifties.

Parallel to this use of relatively simple approximations of the molecular orbital theory to the study of complex molecules Berthier has investigated the possible utilization of more refined molecular orbital procedures in the study of necessarily smaller molecules. We owe him the first application of the SCF method to the study of fulvene and azulene and also a pioneering extension, presented in 1953, of the SCF method to the study of molecules with incomplete electronic shells.

This was altogether a most important period in the history of quantum chemistry in France, when slowly but surely the electrons gained the right of citizenship in chemistry. Berthier is largely responsible for this success both through his scientific contributions and through the influence which he has exerted on an number of young students and the enthusiasm which he has distilled in them. He had the chance of becoming a good friend of another of our research collaborators in these early years, Madame Serre who was to become later the Director of the Ecole Normale Supérieure de Jeunes Filles. Berthier used to divide his time between our laboratory and that of Madame Serre; the latter enabled him to have the greatest number of female students among all the quantum chemists in the world. This is, may be, why he remained a bachelor.

We are referring here only to the early years of Berthier's activity in the field of quantum chemical theory. These were the decisive ones and from some point of view the most difficult but also exciting ones. Needless to say, everybody knows it, that he has continued since and continues still, to contribute in a most efficient way to the development and propagation of quantum chemistry and, at this time, not only in France but on the world scene.

A. Pullman   B. Pullman

Quantum Chemistry: The New Frontiers

J. TOMASI
Dipartimento di Chimica e Chimica Industriale. Università di Pisa
Via Risorgimento 35, 56126 Pisa, Italy

1. Introduction

The members of the scientific community are accustomed to work within a frame of rules, laws protocols, which constitutes the accepted paradigm of a specific discipline. Moreover, the paradigms of the various disciplines, or the scientific programs, if one prefer a different terminology [1], are interrelated and connected in a wide and at the same time tight set of general truths and criteria which constitutes the basic layout of science. Innovation means to modify protocols, to question truths, to introduce new models and ultimately to infringe rules, if necessary, but all these innovations are accurately examined before presentation to the community, planned and justified according to considerations inherent to the specific protocol and of the general layout of correct scientific methodology.
Things are different when a scientist has to give a overview of the future trends of his discipline. Prediction is an art, more than a science, and also the more modest goal of a critical appraisal of the trends of evolution, and of selection of a set of themes for which progress is expected or hoped cannot be performed with the same instruments used in the everyday research.

I will thus rely on my tastes, my biases, with an attempt of tempering them with considerations on the past.

This modest essay will be inserted in a book in honour of Gaston Berthier. He is decidedly more qualified than myself to undertake this task, having a far larger experience and clearer ideas on what is good or less good in the theoretical chemistry production. It is a detriment for the book that Berthier is not the author of these pages. The collection of authors gathered here to honour Berthier, and the titles of the contributions they are providing for the book lead me to suspect, and to hope, that the reader will derive a clearer and better idea of the future trends in quantum chemistry by the global appreciation of the whole book rather than by the lecture of these few pages.

2. The "modern" quantum chemistry of the past 30 years

Theoretical chemistry may be considered an old discipline: the first steps of "modern" chemistry (in the 18th century!) are imbued of theoretical considerations, and in the course of the past century theoretical arguments and approaches have grown into a wide body of methods and concepts which can be collected under the heading "Theoretical Chemistry". In the last decades there has been a remarkable shift, making more precise and restricted

1

Y. Ellinger and M. Defranceschi (eds.), Strategies and Applications in Quantum Chemistry, 1–17.
© 1996 *Kluwer Academic Publishers.*

the meaning of theoretical chemistry which now may be defined as the discipline studying molecules with quantum mechanical methods.

The progress in science does not proceed with a steady pace. Periods of quantitative growths, often rich of results, begin, and end, with sudden changes which gives rise to a quantitative turn in the research methods.
In my opinion, the last qualitative change in theoretical chemistry corresponds to the introduction of computers in chemistry. A conventional date for the beginning of this last period may be indicated in the Boulder Conference of 1959 [2], i.e. more than thirty years ago. The use of more and more large and efficient computers has shifted the attention of theoretical chemists to an extensive use of quantum mechanical calculations.

Quantum mechanics was the dominant theory in chemistry even before the advent of electronic computers. The conventional date for the beginning of this period may be fixed at 1927 with the publications of the Heitler and London paper on hydrogen molecule [3]. The growth of theoretical chemistry (or better, theoretical quantum chemistry) between 1930 and 1960 (thirty years, again, as for the last period) has followed a research programme different from that accepted in the most recent period.

We shall return later on this difference of approach. Before the advent of quantum mechanics theoretical chemistry was influenced by the lack of a comprehensive theory for matter at the microscopic level. In the preceding thirty years, i.e. from the beginning of this century, there has been an evolution of the main line of research, based on the adoption of approaches (paradigms) derived from physics with a progressive shift from an alternative approach, based on chemical concepts, elaborated during the last part of the preceding century by structural chemists. The physical approach has given much emphasis to the molecule, considered as a physical entity, the properties of which are sufficient to interpret, and to predict, the chemical behaviour of matter.

According to this partisan view of the evolution of theoretical chemistry we draw the impression of a choice, in which the single molecules represent the basic unit of investigation, the quantum theory provide the theoretical basis, and computer calculations the final step. The three periods of growth are, in reality related, and the "sudden" changes in between do not corresponds to "revolutions" in according to the meaning this word has in the Kuhn's analysis [4].
Just at the closing speech of the event we have chosen as indicative of the beginning of the last period, the Boulder Conference of 1959, C.A. Coulson [5] expressed the preoccupation that the new era of theoretical chemistry, so bright of exciting promises, would also lead to a splitting of the discipline into two (or to be more precise, three) separate domains, each having its own set of paradigms, and not paying much attention to the evolution of the other domains.
According to the Coulson words, the exponents of group I were committed to "in-depth computing" and "prepared to abandon all the chemical concepts and simple pictorial quality in their results" "in order to achieve complete accuracy"; while "the exponent of group II argue that chemistry is an experimental subject, whose results are built into a pattern around quite elementary concepts". The third group was at that moment (1959) more a hope than a reality; the "spreading of quantum chemistry to biology": "Group I exponents will throw up their hands in horror at such attempts", "group II members will mistrust the complete neglect of many terms which are known to be large", but "the prizes are immense": "there is much experience possessed by professional biologists which could be linked with the deeper levels of interpretation associated with quantum chemistry", even if "biological systems are much more perverse than any laboratory chemical system".

Thirty years later we may compare these indications of possible trends in the evolution of theoretical chemistry with the present status of the discipline. The separation into three separate disciplines has not been happened. Grace to the efforts of a small number of persons aware that the primary objective of theoretical chemists is to interpret chemical phenomena, and that for this interpretation the semi-quantal interpretative tools elaborated in the preceding stage of the discipline were nothing more that imperfect, provisional instruments, the links between the first two approaches have been reinforced. The handful of persons I refer here - the names of whom are familiar to every theoreticians and among whom I would here remember Berthier - took the correct position that the most efficient way of improving interpretative tools was to work, personally or through the younger men under their control, on both sides, in the implementation of new computational and theoretical methodologies, and in the exploitation of these progresses for the definition and refinement of the interpretative tools.

An analogous role has been played by other scientists in strengthening the ties between quantum chemistry of type I (and type II) with the area corresponding to biochemistry (or complex molecular systems in general), a task made more difficult by the explosive growth of structural and functional information about biomolecular systems. It is worth to remark here that such a fruitful use of quantum chemical concepts in biology has requested the extension of the methods to approaches different from quantum molecular theory in the strict sense introduced before. We shall come back to this remark later.

I think that the majority of my colleagues will agree with this statement: Coulson's worries are not become a reality, theoretical chemistry has survived to the impact with computerized quantum mechanics and has grown in a complex discipline, rich of different facets, and with an increasing weight in chemistry.
It rests however that many aspects of the Coulson analysis were correct: he rightly singled out the three most important directions of progress in theoretical chemistry (we may add now a fourth group). There has not been a complete divorce among the three groups but a sizeable number of scientists continues to work with enthusiasm and success (we shall distinguish later between enthusiasm and success) on the development of formal theories and on their translation into computational codes, without bothering much about chemistry, while another active group produces model about complex structures and functionalities in large complex systems, without paying much attention to the congruence with the basic paradigms of the discipline which remain those of quantum mechanics.

As anticipated before there is now another group, called in the following pages group III, those of persons making computations on chemical systems (and on chemical phenomena) with the goal of getting specific information about structure, energetics, observables, as derived from the calculations, without any methodical attempt of "understanding" the phenomenon

The presence of these strong differences in the community of theoretical chemists is mainly due to the explosive growth of the discipline. Theoretical chemistry has kept the pace with science in general in the ever increasing rate of development, and this remark, which could be substantiate with many indicators, like the number of journals, the number of new of sub-disciplines, of new scientific societies, etc., points out that theoretical chemistry enjoys good health
We have to acknowledge the situation sketched here, a discipline which has got over the peril of complete fragmentation during a stage of sudden change, and the presence of at least four subgroups with well differentiated interests. This will be starting point for our attempt of indicating the most probable, and the most hoped, lines of evolution.

3. Models and interpretation in quantum chemistry

To put things in a clearer perspective it is convenient to introduce - in a compendious form - few methodological concepts.

Theoretical chemistry works on models. My point of view on models in chemistry - and quantum chemistry in particular - has been expressed elsewhere [6]; this view closely corresponds to that expressed by other colleagues [7-11]. I suggested a partition of a quantum chemical model into three components, and in my scientific practice I have always taken into consideration the presence and interplay of these three components. The consideration of the evolution of the whole quantum chemistry suggests me now the introduction of a fourth component of the models. My revised partition of quantum chemical models may be put in the following form.

1) The material composition of the model (material model) which states what portion of matter is explicitly considered in the model. This portion of matter may be described in a "realistic" way, or reduced to a simplified description (e.g. a set of coupled oscillators).
2) The physical aspects of the model (physical model) which collects the selection of physical interactions considered in the model. It may be convenient to introduce a distinction between interactions involving the components of the material model alone and interactions involving the exterior.
3) The mathematical aspects of the model (mathematical model). Methods and approximations used to study the selected physical interactions in the given material model.
4) The interpretative aspects of the model (interpretative model). The collection of chemical "concepts" (according to the definition given by Coulson) or other interpretative tools selected to "understand" the output of a model.

The introduction of the last components in quantum chemical models makes easier the analysis of the second methodological point I will consider here.

What is the ultimate goal of theoretical investigation in chemistry? There could be a difference of opinions on this point and I will not express here my point of view. There will be however unanimity on the statement that this ultimate goal - whatever its nature is - will be achieved in a safer way if there is a good understanding of the results obtained with the model.
Scientific inquiry, which requires the definition of a model, the examination of the results, then, if necessary, the elaboration of another model, in a sequence of steps, is a complex taste in which interpretation (or description) plays a crucial role. I hope that the introduction of three levels, or steps, in the process of interpretation will be of some help for our task.

1) Report. The outcome of the model must be collected, and selected, to put in evidence the results of interest for the desired scope. The first step collects the relevant empirical evidence provided by the model. In many cases the report is sufficient to reject a given model (e.g. for reasons due to its mathematical component) but usually it provides the material for the following steps.

2) Interpretation (or description). The aspects of the phenomenon brought in evidence by the report are related to a set of "chemical concepts" which introduces a rationale in the empirical evidence. This is the realm of chemical (or structural) "theories" which

are not complete and often in competition (A given phenomenon may be described in different ways, using different concepts, and invoking different "causes").

3) <u>Explanation</u>. This last step aims at reaching a fuller comprehension of the phenomenon. Contrasting descriptions must find here a synthesis. A satisfying explanation cannot be reached by examining the descriptions of the report of a single model, but must consider a whole set of models. As members of the chemical community we also require that the explanation so obtained applies to the "objects" of the real word of which the model is a schematic representation.

A partition of the process of understanding a complex phenomenon into three steps has been supported and justified by Runcimann [12] for social sciences. The definitions done by Runcimann cannot be directly translated into our field, nor the names he selected for the sequence of levels, but his scheme, presented here in a modified form, gives a contribution to appreciate the strategy and the impact of specific works of research.

4. The different facets of quantum chemistry

4.1.   GROUP I

The definitions given by Coulson to quantum chemists belonging to this group (electronic computors, or ab initio-ists) is surely outdated. Every quantum chemists is now an "electronic computor" and the difference between ab-initioists and non ab-initioists is rather feeble.
There is a large variety of motivations and strategies for persons and works collected here under this heading. The effort of making more efficient the computational algorithms, extending thus the area of material and physical models for which the report becomes satisfactory and quite exhaustive, has produced results of paramount importance.

The good success of these efforts has greatly improved the status of quantum chemistry in the scientific community.Quantum chemistry is now one respectable branch of chemistry, like organic synthesis or molecular spectroscopy, because their practitioners have shown that high-level quantum calculations are not confined to models composed by 2-10 electrons, and that the information thus gained is valuable and comparable to that obtainable with the aid of other methods. This achievement could be considered of secondary interest ("well, there is another technique which confirms our evidences"), but actually has had a great impact on the evolution of chemical thinking and teaching, suffice to compare textbooks of chemistry ante 1960 with the present ones.

The future evolution of chemistry will be more and more based on theoretical concepts, and we have to ascribe to "in-depth computors" the merit of this evolution, even if quite probably the most significant progresses will not directly derive from very accurate calculations.

This line of research has not lost his momentum. One of the reasons is the continuing progress in the computer hardware and software. Methods and algorithms are, and will be, continuously updated to exploit new features made available by computer science, as for example the parallel architectures, or the neuronal networks, to mention things at present of widespread interest, or even conceptually less significant improvements, as the increase of fast memory in commercial computers. Computer quantum chemistry is not a mere recipient of progresses in computer science. Many progresses in the software comes from quantum chemists, and also the stimulus of our discipline on the progress of hardware is not negligible, as the example of Clementi tells.

Efficient computational algorithms for "in-depth calculations" produces, as output - apparently this is a truism - accurate calculations. Let me consider now more in detail what this means.

At the basis there is the tacit assumption of a reductionistic ideal. Quantum mechanics in the current version is the correct theory, and the process of extracting from the whole universe the molecule subjected to accurate calculations does not create problems. I do not object this assumption, being however aware that there are objections, mainly for the process of abstraction of one molecule from the whole universe (see, e.g. Primas [13]) and for the definition of an isolated molecule (see, e.g. Wolley,[14],Claverie [15], Sutcliffe [16]).

The material model is just a bit of matter - a molecule -, all the physical interactions are in principle considered (even if some terms are discarded in actual calculations), the modelization is thus reduced to the mathematical part. In addition, the report has the characteristics of an explanation. Making reference to a celebrated sentence opining the textbook on Quantum Chemistry by Eyring, Walter, Kimball [17]: *"In so far as quantum mechanics is correct, chemical questions are problems in applied mathemathics"* ; it may be said that this program is a realization of that sentence.

This research program is far from being devoid of practical interest. Numerous are the problems in which the interpretation of experimental evidence is dubious. (Theoreticians often forget that rarely experimental evidence is directly amenable to the properties of more general interest for the progress of our knowledge of the material work; at the molecular level at least theory and computations have the advantage of getting directly the property of interest.) In case of doubt, as in the choice between two different values of a molecular parameter, both fitting the experimental data, the computed value often has univocally provided the right answer.

In many other cases the experimental approach is unable to give the desired answer, because the material system is not available in laboratory, as it is the case for many astrophysical problems. or because the experiment is too complex and delicate. I am unable to find now the exact source of the remark performed some years ago by a well known experimentalist that the determination of the octupole moment of a molecule requires a stay of the molecule in the measuring device for a time of the order of a week, while a computer gives an equivalent answer within few hours. I am not expert of the progresses in measuring octupole moments, but surely the quality of the quantum mechanical results in the meantime greatly ameliorated, the computer time reduced, the size of the molecules for which this calculation is possible is noticeably enlarged (We may also add that the use of this kind of program may be now left to a medium-grade technician, while the experimental determination requires a good technical skill. This remark is not essential here, but has its weight in the general economy of the research).

One of the fields in which the competition of ab-initio methods is evident is that of molecular vibrations. The experimental technique is relatively easy - transient species apart- but the ab-initio methods give now results of comparable quality and a wealth of additional information [18]. It is reasonable to forecast that the number of experimental measurements which will find in accurate ab-initio calculations a serious competitor will increase.

It is worth to remark that the opposite also happens. There is an evolution in the experimental techniques too, and in some cases this progress makes possible (or competitive) the measurement of a quantity formerly available via computations only. One example is the detailed measurement of the electronic density of a molecule, and of the related molecular electrostatic potential. The determination of these two observables has been for many years a task feasible only by quantum-mechanical methods, now the progresses in the elaboration of diffraction technique measurements makes possible a direct determination.

I have reported this last example not for the sake of completeness in our discussion, but to underline a different point. Quantum chemistry, in the work of group I and even more in the work of group II, put the emphasis on some properties which by tradition are not object of direct experimental determination. Electron charge distribution and MEP are just two examples. The use of these quantities by theoreticians has spurred the elaboration of experimental methods able to measure them. This positive feedback between theory and experiment is an indication that quantum and experimental chemistry do not live in separate worlds.

The competition between theory and experiments may be expressed in another way. Is quantum chemistry able to predict special properties unknown to the experimentalists, or the existence of compounds not yet synthesized? We are here considering the activity of group I and the question thus regard a definitive demonstration of the existence (or non-existence) of a given property or of a given compound; the question must be put in a different way when addressed to persons belonging to group II. In the present case the answer is partially positive.

There are several examples in the literature of recent years of convincing numerical demonstrations that a compound not yet observed has a stable structure. It must be remarked that these studies usually regard compounds of marginal chemical interest, and that for innovative problems the quantum approach has always been late with respect to the experiments. This delay decreases, but it is unlikely to expect that the leadership in the search of new compounds will be assumed by in-depth calculations.
To substantiate this statement I will quote three examples. In the early sixties the discovery of noble gases compounds came after the elaboration of the first codes for the ab-initio calculation of polyatomic molecules: it was not possible however to give at that moment a serious demonstration of the existence of XeF_4 or related compounds because the technical means for in-depth calculations were not sufficient. Ten years later there has been a great fuss about a presumed form of "polymeric water". Several theoreticians tried to corroborate (or to disprove) that claim. The computational theory was at that time sufficient to give a reliable description of the water dimer, but completely inadequate to disprove the existence of that particular state of aggregation. Twenty years later the experimental discover of a new stable form of carbon, C_{60}, aroused first a sceptical reaction, then a widespread interest. A definitive prediction of the stability of C_{60} via quantum mechanics was at that time within the possibilities of an efficient computational centre, and it was not necessary to wait until 1991 to discover the existence of carbon nanotubes. Other more complex carbon structures are good candidates as potential carriers of new properties, but there is no indication of resolute efforts of group I in this direction.

Another aspect of quantum chemical activity which we connect with group I is the formal elaboration of new approaches. At the beginning of the "computational era" (i.e. 30 years ago) there has been a blossoming of new formulations and new approaches which have given origin to to computer algorithms constituting the basic structure of today in-depth calculations. The names of Roothaan, Boys, Nesbet, McWeeny, Löwdin, Shavitt are few examples randomly selected to make more clear what kind of activity I am considering here. This activity has continued in the following years, and the formal framework of quantum chemical theory has greatly changed from 1960 to now.

It is my impression, however, that the momentum has decreased, and not for the lack of enthusiasm. Several interesting approaches have been formulated ten, or more, years ago and still wait the final step necessary to pass from an exploratory stage to efficient tools to be used for chemical applications. Formal quantum chemistry is a mature discipline and the progresses occur now at a slower pace. The potentialities are not fully exploited, however, and further efforts must be encouraged, the reward being now larger than 30 years ago, because the larger impact quantum theory has now in chemistry. An example is given by the density functional theories, which after many years of induction, are now amply paying for the efforts of elaboration. The future of quantum chemistry is also in the hands of the persons struggling with unconventional approaches.

4.2.   GROUP II

The definition given by Coulson for this second branch of quantum chemistry has not been a well selected choice. Coulson spoke of "non-electronic computors", when the use of computers to examine "chemical concepts" and to elaborate new interpretation tools was already initiated. Within few years from the Coulson analysis the computer become the main instrument for interpretation (or description).

What is the main difference in the use of quantum calculations between group I and II ?

Group I relies, as said before, on the reductionistic ideal that everything, in the field of chemistry, is amenable to the first principles and that a correct applications of the principles, accompanied by the necessary computational effort, will give the answer one is searching. It is a rigourous approach, based on quantum mechanical principles, in which the elements of the computation have no cognitive status, unless when employed to get numerical values of physical observables or of other quantities having a well defined status in the theory.

Group II accepts the basic postulates of group I, performs molecular calculations as group I but with a different philosophy which may be appreciated by contrasting two quotations. The opening sentence of Quantum Chemistry by Eyring et al. I have already quoted is often considered as a shortened re-formulation of another famous saying by Dirac [19] which deserves to be reported here because its second part is often omitted: *"The underlying physical laws necessary for the mathematical theory of a large part of physics and the whole of chemistry are thus completely known, and the difficulty is only that the exact application of these laws leads to equations much too complicated to be soluble. It therefore becomes desiderable that approximate practical methods of applying quantum mechanics should be developed, which can lead to an explanation of the main features of complex atomic systems without too much computation"*. The members of group II address their attention to the last part of this sentence, with the emphasis put on the "explanation". It is a quite different philosophy, in which the extreme variety of chemical phenomena plays the essential role in assessing the strategy.

To achieve their objective the members of group II are compelled to complement the quantities with a correct formal status in quantum mechanics (e.g the observables) with others which, according to the Primas definition [13] are g̲r̲a̲c̲e̲l̲e̲s̲s̲. These graceless quantities are defined and selected with the scope of "understanding", i.e. in our terminology with the scope of giving an i̲n̲t̲e̲r̲p̲r̲e̲t̲a̲t̲i̲o̲n̲ of the chemical phenomena.

The "nature of the chemical bond" , the "chemical group effects" are examples of "concepts" accepted by group II as objects of theoretical investigation. To perform these studies it is allowed to introduce other "concepts" and "quantities" which have a questionable status in the formal theory.

There is a large number of concepts and quantities of this kind used in actual investigations, and a large rate of increase of new formulations. This abundance of interpretative tools could lead somebody to suspect that we have lost any control on the growth and on the use of of these instruments and that interpretation in chemistry is becoming an exercise in which is possible to reach the conclusions one desires by an appropriate selection of the tools.
This is not my opinion, and I will try to explain why.

Interpretation, as it has been already said, is not univocal. There is competition among different interpretations, and the concept of "generality" of an interpretation (i.e. its range of applicability) should be - and in fact is - an important criterion to eliminate ad hoc descriptions. In other words a good chemical concept must be r̲o̲b̲u̲s̲t̲ (the adjective is taken again from the Primas book [13]). Gracelessness and robustness must be balanced. As example we may consider the natural orbitals. These quantities have been proposed by group I: they have "grace" and little effectiveness in the interpretation. Their use in the Weinhold's formulation of natural bond analysis (NBOA) [20] makes them graceless but effective; it rests to verify if this formulation is robust enough.

The definition of "concepts" must be accompanied by explicit recipes for computing them is actual cases. There is no more space in theoretical chemistry for "driving forces", "effects', etc. not accompanied by specific rules for their quantification. The impact of a new "concept' will be greater if the rules of quantifications are not restricted to ad hoc methods, but related to methods of general use in molecular quantum mechanics. A concept based exclusively on some specific features of a given method, e.g. the extended Hückel method, is less robust than a concurrent concept which may be quantified also using other levels of the theory.

The "chemical concepts" represent a part of the model and must share with the entire model other requirements, in particular simplicity, falsicability, and agreement with the general laws of physics [6]. These additional criteria make possible to keep under control the growth of methodological proposals.
The elaboration of "concepts" often requires the partition of the molecule into smaller subunits. This partition is not supported by formal theories. and it is thus at a good extent arbitrary. The consideration of the above mentioned criteria introduces strong limitations in the choice of submolecular units. In fact there are only three basic choices: the constituent atoms, the molecular orbitals and the partition of the charge distribution into localized units. Each choice presents advantages and disadvantages which is not convenient to analyze here.
The selection of a type of basic subunit is the first step in the elaboration of interpretative tools. An analysis of this work, which represent the essence of the innovative activity of group II is not possible here. It sufficient to remark that during this process of elaboration there has been important "admixtures" of concepts having their origin in different choices of the basic subunits.

The concept of atom, for example, must be accompanied by a definition of "valence states", based on the definition of hybrid orbitals, which play a prominent role in the definition of VB structures and also constitute an important component in models based on localized orbitals (LO). Localized orbitals represent a bridge between partitions of a molecule based on molecular orbitals and partitions based on a dissection of the charge distribution into local subunits. LO give an alternative view of electron correlation holes, and in a different context offer a good starting point for semiclassical segmental partitions of the charge distribution. The alternative concept of charge partitioning according to the values of some physical quantities (e.g. the Bader's approach) challenge the traditional partition into atoms or into LOs, offering new views of some concepts and permitting new admixtures of tools.

I have selected few examples taken from the theory of molecular bonding, but this aspect of competition and transference of concepts is present in all the fields of chemistry -and they are quite numerous- in which quantum theory has given contributions.
Another important point is the connection with in-depth calculations. We have already remarked that this connection is extremely important for the elaboration of interpretations. There are two main lines, in my opinion.

The first could be called a constructive (synthetic) approach. Pragmatic and theoretical reasons suggest to start computations appropriate for the model one has selected at a low level of the theory; the goal is not to obtain the most complete report on the problem under scrutiny but to reach a satisfactory degree of confidence in the interpretation obtained by analyzing the results, and a set of reports at increasing levels of the mathematical model may be of noticeable help in assessing the quality of the description. The constructive approach may be also employed with the aim of getting accurate reports. In this case an interpretative model is (tentatively) adopted, and the mathematical model is formulated accordingly. The goal now is to get accurate results at a lower computational cost than using brute force methods. Many formulations of this strategy have been elaborated, some addressed to specific problems, others of more general character. The quotation of an example always make more clear what a general statement means; to this end we recall the recent generalized multi-structural wavefunction method (GMS) [21]. This approach remind me the suggestion expressed by Dirac in the last lines of the above reported quotation. It is worth to remark that the modern reformulations of the VB theory play an important role in this field [22-23]. On the whole, the constructive approach is the most potent and versatile method to connect chemical concepts to in-depth calculations,

The second approach is addressed to elaborate methods able to derive from accurate calculations the points of interest for the interpretation The strategy, in general, consists in the adoption of a simpler model (the mathematical aspects of the model are again concerned) and the task consists in reducing the information coming from the full in-depth calculation (not the the numerical values of observables and other statutory quantities alone) to the level of the simpler model. For example accurate calculations may be reduced at the level of a simple VB theory (Robb, Hiberty) or of a simple MO perturbation scheme (Bernardi) making more transparent the interpretation.

Both approaches are subjected to criticisms and to errors. The stipulation of the model constitutes a bias in favour of one among several possible alternatives, and the conclusions will suffer if the choice was not appropriate. All the models, by definition, are subjected to failures of this kind, and it rests with the users to exert their acumen to decide if the model is applicable to the case under examination.

I have touched few items selected in the varied activity of group II not sufficient to give a balanced appraisal of the evolution and of the prospects of the quantum molecular theory addressed to interpret chemical facts, but sufficient, I hope, to show that there is here, after more than thirty years of activity, a noticeable momentum, and that in the foreseeable future there will be other important progresses.

As a last point I would like to reconsider again a question already examined in relation to the activities of group I. Is this kind of quantum chemistry able to predict properties, or molecular species, unknown to experimentalists?

The question is now different from that asked before, because there is no more the demand of a definitive or fully convincing demonstration. Coming back to the three examples considered before, it may be said that the discovery of xenon fluorides could have predicted with theoretical arguments (and in fact this has been partially done, because these compounds have been synthesized not by serendipity, but on the basis of theoretical considerations). The relatively high stability of C_{60}, and the stability of carbon nanotubes , as well as of other more complex structures not yet synthesized, involving knots of different topology and pseudo-3D lamellar structures, has been already predicted , on the basis of simple, not definitive, models. The example of polywater shows, on the contrary, a weakness of the approach. There had been models supporting and describing the properties of polywater. This activity came at an abrupt end when it was provided experimental evidence that polywater is a myth. The weakness of interpretative models put in evidence by this example will be even more critical when put in the context of the activities of group IV which we shall examine later.

This cautionary remark expressed, we may conclude this section giving a positive answer to our question. Quantum chemistry, in the version cultivated by group II, represents an important factor in the growth of chemistry, and constitute one of the cornerstones of molecular engineering, or similar activities addressed to plan, and to produce, new substances, new materials endowed with special properties.

4.3.   GROUP III

We supplement here the classification proposed by Coulson. The success of quantum chemistry has given in fact origin to another group, hardly foreseeable in 1959. I am collecting into group III persons not interested in producing new techniques for the improvement of in-depth calculations), nor interested in elaborating and checking interpretative tools, but simply interested in performing and using molecular calculations.

It is a reasonable activity for persons belonging to groups I and II to use, also for extensive applications, the tools they have elaborated. This activity does not be confused, in my opinion, with that of members of group III.

There is a variety of motivations for using molecular calculations, some of which are of interest for the future evolution of theoretical chemistry.

When quantum calculations, at the ab initio and at the semiempirical level, gained foot in the realm of chemistry, a steadily increasing number of experimentalists began to use quantum calculations as a supplement in the exposition of their findings. In many case this was - and still is - nothing more that an ornament, like decorations on a cake. This use of quantum chemistry has been, in general, harmless, because results in contrast with experimental evidence have been rarely published, and this production may be considered now as a sort of advertising for the new-born computational chemistry. A more serious use of the facilities offered by the computational techniques is done by scientists provided of the adequate training for a sound appreciation of the limits of the methods they were using. Theoretical chemistry, as the other domains of experimental sciences, is not restricted to the persons who have elaborated or improved methods, but open to all the researchers with the necessary background. A large portion of the good work done in theoretical applied chemistry can be classified here.

The dissemination of computers, and the diffusion of complex computational packages, has given origin to another type of members of group III, the "molecular computers". The links with the underlying theory become feeble. The computors are omnivorous: observables or other statutory quantities attract their interest as well as molecular indexes or other interpretative tools, these last often considered at the same degree of "realism" as the physical observables. The selection of the level of the theory and of the method is based on the criterion of "continuity", i.e. by looking at the methods used by other computors and trying to do something better. A criterion to decide what is better is simply the cost of the method: ab initio methods are better that the semiempirical ones which in turn are better than the semiclassical ones. A larger basis set is better than a smaller one, and so on.

I have here purposely drawn a caricature instead that a faithful portrait of a relatively large portion of our community, composed, on average, by young enthusiastic people. The reason is that the future of quantum chemistry depends, at a good extent, on the evolution of this group, and a flattening description overlooking deficiencies and questionable trends of evolution does not help to address the progress along the most fruitful direction.

A crisis similar to that felt by Coulson in 1959 is probably impending now. Our fathers have been able to close the gap between groups I and II offering on the one side methods, computational tools, and confidence in numerical quantum chemistry and on the other side accepting this offering and exhibiting the capability to re-formulate in a new form concepts and approaches.

The present generation is on the verge of a splitting between persons devoted to the theory and persons devoted to the practice of computation. Thirty years ago the young people was mostly on the side of in-depth calculations, and this people gave a quite important contribution to the evolution of chemistry; now the youngsters are on the computers side and they also have contributions to give to the future evolution of theoretical chemistry (in drawing my caricature of this group I have omitted the positive points). This group does not benefit by a privileged situation akin to that enjoyed by group I at its beginning: a good, reliable theory and the opportunity offered by new technical means of making concrete dreams coveted for many years. Group III has no well established traditions, and its driving force is the desire of enlarging our knowledge of the complex realm of matter at the molecular level. This driving force has produced valuable methodological results (I am not interested here to examine practical results, which are not negligible), in particular the attitude of combine methods and approaches of different theoretical level into a unified strategy. Molecular graphics is largely exploited, classical and quantum methods are used in sequence or in parallel, information theory is exploited to enlarge the field of application of the results, etc.

The oldest core of quantum chemistry (i.e. groups I and II) must give its help to strengthen this trend of evolution. Theoreticians already satisfy the increasing demand of integrated computational packages, easy to use. The next generation of programs should make easier not the use only, but the modification of the procedures, the implementation of new algorithms, the establishment of new connections among the several subunits of the computational stock. At the same time the authors should provide more information about the limits of the calculations performed with these programs.

Computors of group III (let me use again this disparaging definition; it would be clear now that my personal position is far from being disparaging) are shifting their interests to problems of ever increasing complexity, because this is the evolution of chemistry, and are now affording problems hardly treatable with canonical procedures elaborated for molecules containing a moderate number of atoms. These problems represent a new challenge to the theory, and this is the field of investigation of the last group in our classification.

4.4.   GROUP IV

Coulson signalled the possible formation of a separate group related to "the spreading of quantum chemistry into biology". This prediction is now a reality and Quantum Biology is an important branch of Quantum Chemistry [27], cultivated by members of all the preceding groups. The contribution of group I via the elaboration of new formalisms as well as via the elaboration of more powerful computational techniques constitutes the basic layout; concepts and interpretations provided by group II find here an exciting field of application (and a challenge to refine and to extend the methods); the computational enthusiasm of group III with its combination of different approaches is especially addressed to these problems.

I prefer do not consider scientists working in quantum biology as a separate group, but rather to collect a sizable part of their activity into a more general group, characterized by the presence in their problems of a large number of degrees of freedom. We could collect here all the problems regarding matter in condensed phases, from real gases to perfect crystals. In this very large body of systems - and of phenomena - many are not sensibly affected by the increase of the degrees of freedom, and the traditional approaches are still sufficient.

More interesting is the consideration of cases in which the traditional approach is ill at ease. The theory of chemical bonding is not profoundly affected (special cases apart) from the extension of the number of degrees of freedom. Clementi rightly pointed out that from the point of view of quantum mechanical calculations there are no "too large" systems: the portion of space including the matter exhibiting a non vanishing interaction with a localized subunits (e.g. an atom or a bond) may be defined in terms of a sphere, with a radius R_{max} not extremely large. Nowadays our computational tools are able to fill almost completely this sphere with interacting matter (electrons, nuclei) and to describe the interactions at a reasonable (and steadily increasing) level of accuracy. This concept may be introduced into our definition of models for quantum chemistry: there will be an overlap of material sub-models, with defined physical interactions, and the whole problem is then reduced to the specification of the mathematical model able to deal with the couplings among subunits. A formidable problem is thus reduced to a more manageable form.

A report on the electronic structure of a large molecule at a given geometry is however the first in a long sequence of steps. Even the next step, the recognition of the features of the potential energy hypersurface presents formidable problems, well known to members of group III who study conformational properties of large molecules. There are now expedient ways to overcome (in part) the difficulties of this specific problem, but analogous questions rise again at a higher level of investigation, when the "large molecules" are involved in chemical reactions. This last problem is present, and perhaps more evident, in the study of chemical reactions involving "small" molecules in condensed media.

The number of detailed studies on these last systems is nowadays sufficiently large to generalize the results, and to project the conclusion to more complex (o "perverse", according to Coulson) systems. The traditional view of a reaction occurring on a well defined surface, with a flux of representative points passing the transition state region is untenable. The separation between static and dynamic aspects of a problem, so often exploited for studies an isolated molecule must be reconsidered.

There is a deluge of papers, as well as of methods and of approaches, addressed to these problems. It is significant that in this blossoming of studies there is space for very simple models (as regards the material composition of the model) as well as for very complex models with a high degree of realism in the chemical composition.

A combination of different approaches is at present the most convenient strategy. Most of the work done an complex material models adopts a classical formalism, disregarding for the moment quantum aspects, while there are significant progresses in quantum description of simple models [26].

I have briefly touched here two examples, structure of large molecules and reactions in condensed media. The number of examples could be by far larger, from isolated molecules again (the dynamics of excited polyatomic molecules, the study of their roto-vibrational levels) to man-made materials with their specific properties (ceramics, polymers, incommensurable phase systems, dispersed mesosystems) to materials of natural origin (mainly, but not exclusively, of biological nature).

Numerous additions to our collection of methodological remarks could derive from the consideration of other examples. The picture drawn here is extremely incomplete, but sufficient to express some remarks and to draw some conclusions.
The various attempts, in the different fields, can be viewed as an effort to combine methods and experience of two disciplines which have reached since longtime their maturity: quantum mechanics for isolated systems and statistical mechanics. This effort of combination produces important results, and the progress in this area is indisputable.

There is however the need of a qualitative jump. The resulting theory should not be called quantum chemistry again: this now is an old and glorious name, corresponding to a very active research domain, promising new progresses and important results; the more generic name of theoretical chemistry is more suitable. Specific suggestions for the elaboration of this theory, not supported by detailed analysis and discussion, could be considered with scepticism or criticized for many reasons (partiality, inconsistency, errors, etc.). For this reason I will refrain from suggestions, but I am unable to resist temptation of adding a few concise remarks. Temperature is not a statutory quantity in quantum mechanics of isolated systems and it is introduced here via a classical picture. A quantum definition of T, e.g. via the fluctuations theory, could be an important supplement to a reformulated and generalized theory. Time has a special status in quantum mechanics [27] but it should be reconsidered when passing to complex systems arranged in hierarchical order [28,29]. We have thus far assumed that all the activities in this domain are "covered" by the usual quantum theory. The proviso expressed by Dirac just before the sentence we have quoted has been until now superfluous. There are no convincing evidences of limits of the quantum theory in the fields covered by groups I to IV. There is however a widespread dissatisfaction with some basic aspects of the theory. If there will be something to change (and a change at this level means the formulation of a new theory, encompassing the old one) the clue should come from the realm of complexity, rather than from a reconsideration of simple gedanken experiments [30]. The future activity of group IV will be the forefront for further attempts to amend and to extend the quantum theory.

5. Conclusions

My attempt of depicting the new frontiers for Quantum Chemistry has no produced exhaustive and detailed indications. It is almost impossible to present in a few pages indications of this type, of questionable validity even if expressed as final report of a panel of experts, after a hard collective work on this theme. Quantum Chemistry is in fact one of the cornerstones of Chemistry, enjoying good health as the other branches of Chemistry, and there are ample and varied perspectives of progress. A selection of some themes would means to indulge too much to personal tastes.

I have tried to sketch a partition of the various approaches in Quantum Chemistry into four groups. This taxonomy is open to criticism and does not imply, in my intention, an exclusive assignation of each quantum chemist to one of the four groups.

The main message of this short undertaking is that Quantum Chemistry in the different facets it displays, still is an unique discipline, and the activities of a single researcher may be often assigned to different groups.

Quantum Chemistry is a mature discipline: the roots are very far in the past, and during his life, more than sixty years, it has been the subject of a "scientific revolution", and a second important change (or "revolution") is on the verge. This change will perhaps modify the relative importance of the various approaches, which I have denoted as groups, and surely will present new challenges to the discipline. Some details and some suggestions have been given in the preceding pages; here I limit myself to few conclusive remarks. Molecular quantum chemistry in its computational version has to merge in a more intimate and effective way with other branches of quantum mechanics and other disciplines or techniques. The main lines of future evolution will be done by the adoption of complex strategies involving several techniques, the molecular quantum chemistry, which embodies the basic understanding of molecular structures and properties, quantum statistics, at the equilibrium and out of the equilibrium, and many ancillary techniques, from information theory to computer graphics, etc. The dynamic methods, and all the aspects involving time, should make more efficient, the temperature should have a better defined status in the theory.

A period of re-formulation of the theory, similar under some aspects to that which has characterized quantum chemistry in the years 1930-1960, but projected toward more complex objectives, should be opened now.
One of the main avenues open to quantum chemistry is that of the complex, very complex, systems. There are the basic premise to reach these goals. I end thus this overview with a note of optimism.

References

1.   I. Lakatos and I. Musgrave "Criticism and the Growth of Knowledge" , Cambridge Univ. Press, London (1970).

2.   Papers from the Conference on Molecular Quantum Mechanics held at the
        University   of Colorado, Boulder, June 21-27 (1959), *Rev. Mod. Phys.*
        32,169 (1960).
3.   W. Heitler and F. London, Z. Physik, **44, 455** (1927).
4.   T.S. Kuhn in "Foundations of the Unity of Science" vol 2, O. Neurath,
        R.Carnap, C. Morris Eds. Chicago Univ. Press, Chicago (1970).
5.   C.A. Coulson, Rev. Mod. Phys. **32,**171 (1960).
6.   J. Tomasi, *Int. J. Mol Struct. (Theochem)* **179**, 273 (1988).
7.   C. Trindle, *Croat. Chim. Acta* **57, 1231** (1984).
8.   Z. Maksic, "Theoretical Models of Chemical Bonding" vol 1, Springer, Berlin
        (1991)
9.   G. del Re, *Adv. Quant. Chem.* **8**, 95 (1974).
10.   P. Durand and J.P. Malrieu, *Adv, Chem. Phys.* **67,**321 (1987).
11.   A. Amman and W. Gans, *Angew. Chem. Int Engl. Ed.* **28**, 268 (1988).
12.   W.G. Runcimann, "A Treatise on Social Theory" , Cambridge Univ. Press,
        London (1983).
13.   H. Primas, "Chemistry, Quantum Mechanics and Reductionism" , Springer, Berlin
        (1983).
14.   R. G. Wolley, *Adv. Phys.* **25**, 27 (1976).
        R.G. Wolley, *Structure and Bonding* **52**,1 (1982).
15.   P. Claverie and S. Diner, *Israel J. Chem.* **19**, 54 (1980)
        P. Claverie, in "Symmetries and Properties of Non-rigid Molecules", J. Maruani
        and J. Serre Eds. Elsevier, Amsterdam (1983).
16.   B. Sutcliffe and Z. Maksic "Theoretical Models of Chemical Bonding" , vol 1,
        Springer, Berlin (1991).
17.   M. Eyring, J. Walter and G.E. Kimball "Quantum Chemistry", Wiley, N. York
        (1944).
18.   G. Fogarasi and P. Pulay , *Ann. Rev. Phys. Chem.* **35**, 191 (1984).
        J. Boggs, in "Theoretical Models of Chemical Bonding", vol 3, Z. Maksic Ed.
        Springer, Berlin (1991).
19.   P.A.M. Dirac, *Proc. Roy. Soc.* **A123**, 714 (1929).
20.   A.E. Reed, L.A. Curtiss and F. Weinhold, *Chem. Rev.* **88,** 899 (1988).
21.   E. Hollauer and M.A.C. Nascimento, *Chem. Phys. Letters*, **184**, 470 (1991).
22.   D.L. Cooper, J. Gerratt and M. Raimondi, *Adv. Chem. Phys.* **65**, 319 (1987).
23.   R. Mc Weeny, *Int.J. Quant. Chem.* **34**. 25 (1988).
24.   B. Pullman , *Int J. Quant. Chem. Q.Biol S.* **17**, 81 (1990).
25.   E. Clementi, *J. Phys. Chem.* **89**, 4426 (1985).
26.   H.J. Kim and J.T. Hynes, *J. Chem. Phys.* **96**, 5088 (1992).
27.   J.T. Fraser, F.C. Haber and G.M. Müller "The Study of Time", Springer, Berlin
        (1972; J.T. Fraser and N. Lawrence"The Study of Time II", Springer, Berlin
        (1972)
28.   H. Primas, ref [13], chapter 6.
29.   I. Prigogine, in "Ecological Physical Chemistry", C. Rossi and E. Tiezzi Eds.
        Elsevier, Amsterdam (1991).
30.   P. Grigolini, "Quantum Irreversibility and Measurement", University press (1993).

*Toute tentative de faire rentrer les questions chimiques
dans le domaine des doctrines mathématiques doit être
réputée jusqu'ici, et sans doute à jamais,
profondément irrationnelle, comme étant antipathique à
la nature des phénomènes : elle ne pourrait découler
que d'hypothèses vagues et radicalement arbitraires
sur la constitution intime des corps, ainsi que j'ai eu
occasion de l'indiquer dans les prolégomènes de cet
ouvrage.*

A. Comte
Cours de Philosophie Positive
Tome Troisième, Trente-cinquième Leçon - 1838

Strategies and Formalisms

Some 150 years after ....

Theory of Orbital Optimisation in SCF and MCSCF Calculations

C. CHAVY, J. RIDARD and B.LEVY
Groupe de Chimie Quantique, Laboratoire de Physico-Chimie des Rayonnements,
(UA CNRS 75), bât. 337, Université Paris Sud, 91405, Orsay Cedex, France

The aim of the present article is to present a qualitative description of the 'optimised' orbitals of molecular systems *i.e.* of the orbitals resulting from SCF calculations or from MCSCF calculations involving a valence CI : we do not present here a new formal development (although some formalism is necessary), nor a new computational method, nor an actual calculation of an observable quantity ... but merely the description of the orbitals.

In fact, it turns out that the orbitals resulting from SCF or valence MCSCF calculations in molecules can be described in extremely simple terms by comparing them with the RHF orbitals of the separated atoms.

In the case of a valence MCSCF calculation the difference between the optimised orbitals and these atomic RHF orbitals simply represents the way in which the atoms are distorted by the molecular environment. Thus, this difference is closely related to the idea of 'atoms in molecules'(1). However, here, the atoms are represented only at the RHF level, and the difference concerns only the orbitals, not the intra- atomic correlation.

The starting step of the present work is a specific analysis of the solution of the Schrödinger equation for atoms (section 1). The successive steps for the application of this analysis to molecules are presented in the section 2 (description of the optimised orbitals near of the nuclei), 3 (description of the orbitals outside the molecule), and 4 (numerical test in the case of H_2^+). The study of other molecules will be presented elsewhere.

1. The atomic case

We briefly recall here a few basic features of the radial equation for hydrogen-like atoms. Then we discuss the energy dependence of the regular solution of the radial equation near the origin in the case of hydrogen-like as well as polyelectronic atoms. This dependence will turn out to be the most significant aspect of the radial equation for the description of the optimum orbitals in molecules.

Y. Ellinger and M. Defranceschi (eds.), Strategies and Applications in Quantum Chemistry, 19–37.
© *1996 Kluwer Academic Publishers.*

1.1. HYDROGEN-LIKE ATOMS

In the case of hydrogen-like atoms the Schrödinger equation can be written as (in atomic units) :

$$(T - \frac{Z}{r} - e) \, \varphi = 0 \tag{1}$$

where T represents the kinetic energy operator, Z the nuclear charge, $-Z/r$ the Coulomb electron-nuclear attraction, e the energy and φ the orbital.

The solution φ of this equation can be factorised into the product of a radial part and an angular part (spherical harmonic Y_{lm}), where the radial part $f_l(r)$ depends of the quantum number l but not of m (2).

$$\varphi = f_l(r) \, Y_{lm}(\theta, \varphi) \tag{2}$$

Inserting this form of φ into the eq.(1) gives the equation to be satisfied by $f_l(r)$, the so called radial equation :

$$\left(-\frac{1}{2} \frac{1}{r} \frac{\partial 2}{\partial r^2} r + \frac{l(l+1)}{2r^2} - \frac{Z}{r} - e \right) f_l(r) = 0 \tag{3}$$

It can be demonstrated (2) that two linearly independent solutions of this equation can be chosen in general (*i.e.* except for some values of e) in such a way that one of them (the so called 'regular' solution) is continuous at the origin and diverges at infinity, and the other one (the so called 'irregular' solution) diverges at the origin and tends to zero at infinity.

Neither of these two solutions is square summable in general. However for some values of e (the 'eigen values') these two solutions coincide and can be accepted physically for atoms since they both are continuous at the origin and they both tend to zero at infinity.

It should be emphasized that we are not interested here specifically by these particular values of e. On the contrary , what is useful here *i.e.* for the description of optimum orbitals in molecules is to study the variation of the regular solution when e varies continuously.

To solve that problem, we depart here from the development used for instance in (2) and we write $f_l(r)$ in the form :

$$f_l(r) = r^l \, R_l(r) \tag{4}$$

Substituting this form of $f_l(r)$ into the eq.(3) leads to :

$$r \, \frac{\partial^2}{\partial r^2} R_l + 2(l+1) \frac{\partial}{\partial r} R_l + 2(Z + er) R_l = 0 \tag{5}$$

But we are interested here only by the 'regular' solution, and we can write R_l in the form of a power expansion

$$R_l = \sum_{k=0,\infty} a_k \, r^k \tag{6}$$

where the a_k's are numerical coefficients depending of l.

Substituting this form of R_l into the eq.(5) gives a recursion relation which allows to determine all the a_k's for any arbitrary choice of one of them. Choosing $a_0=1$, one gets

$$a_0 = 1 \qquad a_1 = -\frac{Z}{l+1} \qquad a_2 = \frac{Z^2}{(l+1)(2l+3)} - \frac{e}{2l+3}$$

$$a_3 = \frac{-Z^3 + e\,Z\,(3l+4)}{3(l+1)(2l+3)(l+2)} \qquad etc... \tag{7}$$

These expressions of the a_k's will allow us now to discuss the energy dependence of R_l and then to derive some consequences from this dependence.

1.2. THE VALLEY THEOREM

We first note that the choice $a_0=1$ made in deriving the eq.(7) simply consists in a particular norm of R_l (and thus of φ). In fact the standard norm $< \varphi|\varphi >=1$ cannot be used here since for most values of e the orbital φ is not square summable. The choice $a_0=1$ is a convenient alternative for $< \varphi|\varphi >=1$.

Next we consider the value of a_1. It implies the relation :

$$\left(\frac{1}{R_l} \frac{\partial}{\partial r} R_l \right)_{r=0} = -\frac{Z}{l+1} \tag{8}$$

which is the well known 'Cusp' theorem (see $e.g.$ the ref.3).

An other aspect of the eq.(7) concerns the energy dependence of R_l. In fact one deduces from this equation that :

$$\frac{1}{R_l(0)} \frac{\partial}{\partial e} R_l = -\frac{1}{2l+3} r^2 + \frac{Z(3l+4)}{3(l+1)(2l+3)(l+2)} r^3 + ... \tag{9}$$

The meaning of the eq. (9) can be stated as : the energy dependence of R_l vanishes like r^2 near the origin (or even faster than r^2 since there is a partial cancellation between the r^2 and r^3 terms). Therefore the energy dependence of φ vanishes like r^{2+l} or faster.

This statement will be referred to here as the 'Valley' theorem. It constitutes the formal basis of our description of the optimum orbitals in molecular systems.

In fact, the Valley theorem is a simple extension of the Cusp theorem. However, the Cusp theorem provides only a local information (for $r=0$), while the Valley theorem is the extension providing a qualitative information (weak e dependence) valid inside a finite volume. This last aspect (finite volume) is the one that allows the description of the optimum orbitals in molecular systems.

The Cusp and the Valley theorems express the same aspect of the Schrödinger equation, eq.(1) : since $e\varphi$ has no pole for $r=0$, the pole of $(-Z/r)\varphi$ can be compensated only by $T\varphi$; but a pole of $T\varphi$ with a residue equal to $-Z$ implies the Cusp theorem (at the origin) and the Valley theorem (inside a finite volume around the origin).

It should be noted that the weak energy dependence of the orbitals inside a finite volume around the nucleus has already been noted and used in different contexts : the numerical determination of atomic orbitals (4) as well as the scattering of electrons by atoms (5).

1.3. ORBITAL OPTIMISATION IN POLYELECTRONIC SYSTEMS

The equation determining the optimum orbitals of polyelectronic systems in the case of the SCF and MCSCF theories can be written in the form :

$$\sum_j \left(h <i^*j> + \sum_{k,l} V_{kl} <i^*k^*lj> \right) \varphi_j = \sum_j \varphi_j e_{ji} \tag{10}$$

$$e_{ij} = e_{ji} \qquad <\varphi_i|\varphi_j> = \delta_{ij}$$

where

- i^*, k^* are the creation operators corresponding to the orbitals φ_i and φ_k and j, l the anihilation operators for the orbitals φ_j and φ_l

- h is the one electron part of the total Hamiltonian.

- V_{kl} is a local operator :

$$V_{kl}(r) = \int \varphi_k(r')\varphi_l(r') \frac{1}{|r - r'|} d^3r' \tag{11}$$

- the e_{ij} factors are the Lagrange multipliers that take care of the orthonormality constraints.

1.4. POLYELECTRONIC ATOMS

We consider here only the SCF case where the off diagonal $<i^*j>$ factors vanish. In addition, we assume that the orbitals satisfy the usual symmetry constraint $i.e.$ that they are pure s, p, d ... functions (RHF approach). On the other hand, no spin constraint is assumed. Then the eq.(10) is most conveniently written as :

$$\left(T - \frac{Z}{r} + V_i - e_i\right) \varphi_i = G_i \tag{12}$$

$$e_i = e_{ii} / <i^*i> \qquad e_{ji} = e_{ij} \qquad <\varphi_i|\varphi_j> = \delta_{ij}$$

with

$$V_i = \frac{1}{<i^*i>} \sum_{k,l} V_{kl} <i^*k^*li>$$

$$G_i = \frac{1}{<i^*i>} \left(\sum_{j\neq i} \varphi_j \, e_{ji} - \sum_{j\neq i,k,l} V_{kl} <i^*k^*lj> \varphi_j \right)$$

The eq.(12) is similar to the eq.(1) in the sense that it requires a compensation between T and $-Z/r$. The main difference comes from the presence of V_i and G_i that might reduce the range of that compensation. In order to solve the eq.(12) one writes φ_i, V_i and G_i in the form :

$$\varphi_i = f_{ilm}(r) \, Y_{lm}(\theta,\varphi) = r^l \, R_{ilm}(r) \, Y_{lm}(\theta,\varphi)$$

$$V_i = \sum_{l,m} r^l \, V_{ilm}(r) \, Y_{lm}(\theta,\varphi) \qquad (13)$$

$$G_i = \sum_{l,m} r^l \, G_{ilm}(r) \, Y_{lm}(\theta,\varphi)$$

Multiplying now the eq.(12) on left by Y_{lm}^* and integrating over the solid angle Ω gives :

$$r \frac{\partial^2}{\partial r^2} R_{ilm} + 2(l+1) \frac{\partial}{\partial r} R_{ilm} + 2[Z + (e_i - J_{ilm})r] R_{ilm} + 2rG_{ilm} = 0 \qquad (14)$$

$$J_{ilm} = \sum_{l''} r^{l''} V_{il''0} \int Y_{lm}(\theta,\varphi)^* \, Y_{l''0}(\theta,\varphi) \, Y_{lm}(\theta,\varphi) \, d\Omega$$

We now study the e_i dependence of the solution of the eq.(14) using the following scheme :

- we first determine normalised R_{ilm}'s by using some standard program of Quantum Chemistry ;

- using these normalised R_{ilm}'s we determine the functions J_{ilm} and G_{ilm} ;

- then we set up the equation :

$$r \frac{\partial^2}{\partial r^2} f + 2(l+1) \frac{\partial}{\partial r} f + 2[Z + (e - J_{ilm})r] f + 2r\sigma G_{ilm} = 0 \qquad (15)$$

$$\sigma = \frac{f(0)}{R_{ilm}(0)}$$

where f is an unknown function, e is a variable parameter, J_{ilm} and G_{ilm} are the functions evaluated at the preceding step using the normalised R_{ilm}'s and $f(0)$, $R_{ilm}(0)$ are the values of f and R_{ilm} at the origin (note that $f(0)$ is unknown).

- finally we solve the eq.(15) with various values of e but always with the same functions J_{ilm}, G_{ilm} and $R_{ilm}(0)$.

The factor $f(0)/R_{ilm}(0)$ ensures that the solution of the eq.(15) is independent of a multiplicative factor (if f is a solution, then λf is also a solution for any number λ) and that f is proportional to R_{ilm} when $e=e_i$. It turns out that no useful comparison with the molecular case can be made in the absence of this factor.

The eq.(15) can be solved by mean of a power expansion of f, J_{ilm} and of G_{ilm} in the same way as the eq.(5) :

$$f = \sum_{k=0,\infty} a_k r^k \qquad J_{ilm} = \sum_{k=0,\infty} v_k\, r^k \qquad G_{ilm} = \sum_{k=0,\infty} g_k\, r^k \qquad (16)$$

Substituting the eq.(16) in the eq.(15) gives a recursion relation which allows to determine the a_k's. Owing to the σ factor it is possible here to choose $a_0=1$ as done in the eq.(7), so that one gets :

$$a_0 = 1 \qquad a_1 = -\frac{Z}{l+1} \qquad a_2 = \frac{Z^2}{(l+1)(2l+3)} - \frac{e - v_0 + \sigma g_0}{2l+3}$$

$$a_3 = \frac{-Z^3 + Z(e - v_0)(3l+4) + \sigma g_0 Z(l+1) - (\sigma g_1 - v_1)(l+1)(2l+3)}{3(l+1)(2l+3)(l+2)}$$

(17)

etc ...

The main aspect of the eq.(17) is that the orbital energy e occurs only in the coefficients a_k with $k \geq 2$. Therefore we obtain here the same results as the one obtained in the case of hydrogen-like atoms (§1.1 and §1.2) :

- the energy dependence of the RHF orbitals of polyelectronic atoms decrease faster than r^2 in the region close to the nucleus (Valley theorem);

- and the corollary that these orbitals depend very weakly of the orbital energy in a finite volume around the nucleus. The range of that volume, which depends of the magnitude of J_{ilm} and G_{ilm}, will be now determined numerically.

1.5 NUMERICAL ILLUSTRATIONS

We present here numerical results illustrating that the solutions of the radial equations (eq.(5) for the hydrogen-like case and eq.(14) for polyelectronic atoms) are 'weakly' dependent of e in a finite volume.

In the case of polyelectronic atoms we have calculated the J_{ilm} and G_{ilm} parameters as described in the preceding section (see above, the §1.4) i.e. using the normalised orbitals resulting from a RHF calculation of the atom in a gaussian basis (11).

The radial equations was then solved using the Runge-Kutta method (7).

We present in Fig. (1-6) the function $f_l(r)$ defined in the eq.(2) (or $f_{ilm}(r)$ defined in the eq.(13)), in the case of the orbital 1s of Hydrogen (Fig.1), 2s and 2p of Carbon

(Fig. 2 and 4) , 3s and 3p of Silicon (Fig. 3 and 5), and 3d of Scandium (Fig.6). In each case three values of e have been chosen : the RHF value, one value higher by 0.2 H and one value lower by 0.2 H. Thus we can study the deviation δ of the orbitals when e varies by $\pm$ 0.2 H around the RHF value.

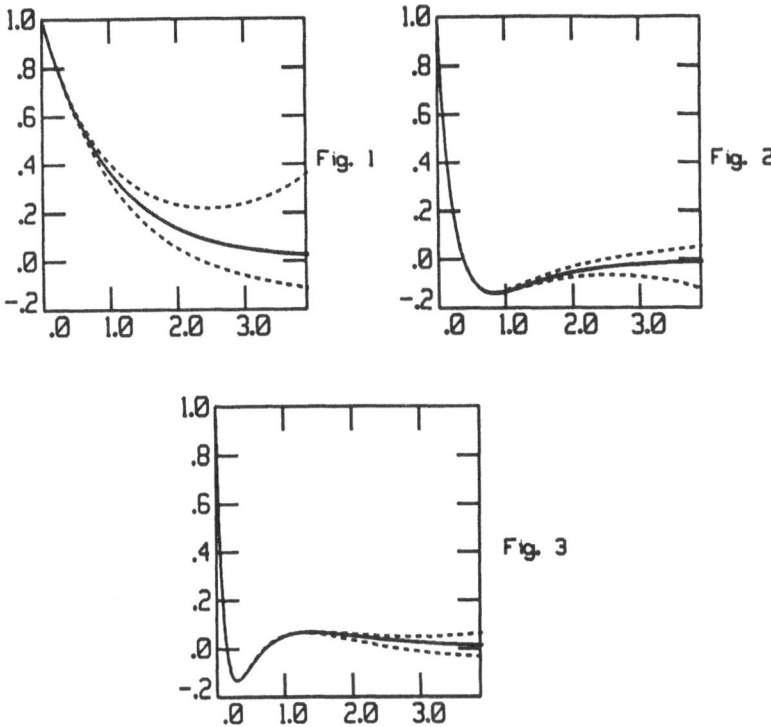

Fig.1. Radial part $f_l(r)$ of three 1s type orbitals ($l = 0$, no node) of the Hydrogen atom corresponding to three different energy values. The full line corresponds to the RHF energy and the other ones to the RHF energy plus or minus 0.2 H . The radius r is given in Bohr units.

Fig.2. Same as fig.1 for 2s type orbitals ($l=0$, 1 node) of the Carbon atom.

Fig.3. Same as fig.1 for 3s type orbitals ($l=0$, 2 nodes) of the Silicon atom.

It is seen on the fig.(1) - 1s orbital of the Hydrogen atom - that this deviation δ is smaller than 5% of the orbital for $r \leq 0.7$ B (close of the covalent radius of the H atom). In the case of 2s(C) and 3s(Si), similar deviations (less than 5%) are observed for r smaller than the position of the last extremum of the function (the one obtained with the largest r) *i.e.* for $r \leq 0.8$ B in the case of Carbon, $r \leq 1.2$ B in the case of Si. These distances are smaller than the covalent radii of these atoms (*ca.* 1.5 B for C and 2 B for Si). But close to the covalent radius, (at 1.4 B for C and 1.8 B for Si) the deviation δ is still smaller than 10% of the value of the orbital at the last extremum.

These results illustrate the fact that the orbital is weakly dependent of the energy e at the peak for $r=0$ (Cusp theorem) but also down in the valley and even on the next hill if any (Valley theorem).

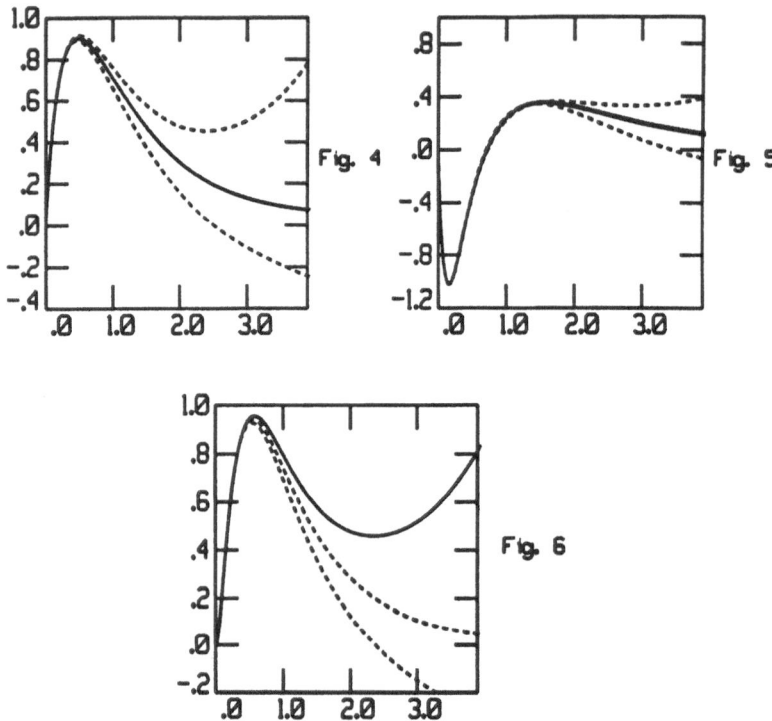

Fig.4. Radial part $f_l(r)$ of three 2p type orbitals ($l=1$, one radial node) of the Carbon atom corresponding to three different energy values. The full line corresponds to the RHF energy and the other ones to the RHF energy plus or minus 0.2 H. The radius r is given in Bohr units.

Fig.5. Same as fig.4 for 3p type orbitals ($l=1$, two radial nodes) of the Silicon atom.

Fig.6. Same as fig.4 for 3d type orbitals ($l=2$, one radial node) of the Scandium atom.

In the case of the p and d orbitals (fig. 4-6) the deviations δ are larger than the deviations obtained with s orbitals. This is simply because the magnitudes of the p and d orbitals are larger than those of the s orbitals for r *ca.* 1.0 B due to the particular norm chosen here. In fact, it can be checked that deviations δ smaller than 10% of the value of the last extremum, are obtained for r values up to a limit close to the covalent radius of the atom in all three cases ($r \leq 1.4$ B for the 2p(C) orbital, $r \leq 1.8$ B for the 3p(Si) orbital and $r \leq 1.5$ B for the 3d(Sc) orbital).

We conclude from these numerical examples that it is possible to give a quantitative and probably rather general expression of the Valley theorem (weak e dependence of the orbital in a finite volume around the nucleus) : a variation of the energy of $ca.$ 0.2 H results in a variation of the function $f_l(r)$ smaller than 10% of the last extremum of $f_l(r)$ until a distance of the nucleus equal to $ca.$ 90% of the covalent radius of the corresponding atom.

2. Molecular systems

We arrive now at the main purpose of the present work : to find a qualitative description of the optimum orbitals (obtained by SCF or MCSCF calculations) of molecular systems.

To that end, we will start with the same equation as the one used above in the case of polyelectronic atoms, $viz.$ the eq.(10), and we will try to use the equivalent of the compensation between the kinetic energy and the nuclear attraction (T and $-Z/r$) found in the atomic case.

In fact, it turns out that the compensation between the kinetic energy and the nuclear attraction does lead to a qualitative description of the optimum orbitals in molecular systems, but only in the frame of the following restrictive conditions.

i) *Global versus local description.* In the case of molecular systems, the one electron part of the electronic hamiltonian includes a sum over the electron-nuclear attraction of all the nuclei:

$$h = T - \sum_a \frac{Z_\alpha}{r_\alpha} \tag{18}$$

Therefore it appears that the above mentionned compensation takes place separately in the vicinity of each atom. We can arrive to a description of the optimum orbitals ; however this description is not global, but local in the sense that it concerns separatly the regions around each atom. Thus, we will hereafter consider only the region of a single atom, say A, and study the effect of the compensation between T and $-Z_A/r_A$.

ii) *Natural versus non natural orbitals.* The Z_A/r_A factor is always combined in the eq. (10) with the $< i^*j >$ factor according to

$$\sum_j (T - \frac{Z_A}{r_A}) < i^*j > \varphi_j \tag{19}$$

If $< i^*j > \neq 0$ (i$\neq$ j) then T-Z_A/r_A appears in several terms corresponding to different orbitals, and it is difficult to demonstrate directly that the compensation occurs separately for each orbital. Therefore, we will consider here only the cases where $< i^*j >=0$ $(i \neq j)$ *i.e.* we will consider only natural orbitals.

iii) *Strongly versus weakly occupied orbitals.* It is then seen on the expression 19 that $(T-Z_A/r_A)$ appears in the eq.(10) multiplied by $< i^*i >$ when natural orbitals are used. Thus, if $< i^*i >$ is small $(< i^*i >\ll 1)$, then $T\varphi_i$ and $(Z_A/r_A)\varphi_i$ dominate the remaining terms of the eq.(10) only in a very small volume around the nucleus of A. In the remaining part of the volume occupied by the molecular system the description of this orbital cannot be deduced from the Valley theorem. Therefore, we will consider here only strongly occupied orbitals with $< i^*i >\sim 1$ or $< i^*i >\sim 2$. In fact, a simple description of the weakly occupied orbitals resulting from valence MCSCF calculations has already been presented (12) .

iv) *Canonical versus non canonical orbitals.* Let us now consider the right hand side of the eq.(10) which depends of the off diagonal Lagrange multipliers through terms like $\varphi_j\, e_{ji}$ $(i \neq j)$. Such terms may present very steep variations with r_A so that the Valley theorem may lead to no special conclusion. Therefore, we consider here only the cases where one can have $e_{ji}=0$ $(i \neq j)$. A similar restriction has not been made in the atomic case (section 1.4 above) because it turns out that e_{ji} $(i \neq j)$ is very small in all useful cases.

v) *Partial waves versus orbital* . Finally it is worth noting already that the present approach will tell us nothing concerning the orbitals themselves! It will tell us something only on each of the partial wave around A separately : the relative weights of the different partial waves in the total orbital do not result from the local compensation between T and $-Z_A/r_A$. It appears rather as a global property of the molecular system .

Let us note that the two conditions $e_{ji}=0$ and $< i^*j >=0$ $(i \neq j)$ can be satisfied only with canonical SCF orbitals. Thus, in fact, the present theory can be applied only in such cases. However it has been demonstrated (12) that in most systems, the strongly occupied MCSCF orbitals and the SCF orbitals are extremely close one to the others. Therefore, in practice, the present theory also applies to the strongly occupied MCSCF orbitals.

On the all, the limitations coming from the above hypotheses *i-v* are :

- one can find a description of the partial waves of the optimum orbitals near each atom separately , not of the orbitals themselves;

- these descriptions concern only the strongly occupied canonical orbitals, not any type of orbitals.

We now return to the eq.(10). In the frame of the hypotheses *i-v* it writes :

$$(T - \frac{Z_A}{r_A} + V_i - e_i)\, \varphi_i = G_i \qquad\qquad e_i = e_{ii}/ < i^*i > \qquad\qquad (20)$$

with

$$V_i = \frac{1}{<i^*i>} \left(\sum_{\alpha \neq A} \frac{-Z_\alpha}{r_\alpha} + \sum_{k,l} V_{kl} <i^*k^*li> \right)$$

$$G_i = \frac{-1}{<i^*i>} \sum_{j \neq i,k,l} V_{kl} <i^*k^*lj> \varphi_j$$

Let us now introduce the partial waves expansion of φ_i, with the origin on the atom A :

$$\varphi_i = \sum_{l,m} r^l \, R_{ilm}(r) \, Y_{lm}(\theta,\varphi) \tag{21}$$

and the expansions given by the eq.(13) for V_i and G_i.

Thus, the eq.(20) becomes

$$r\frac{\partial^2}{\partial r^2} R_{ilm} + 2(l+1)\frac{\partial}{\partial r} R_{ilm} + 2\left[Z_A + (e_i - J_{ilm})r\right] R_{ilm} + 2r\left(G_{ilm} - U_{ilm}\right) = 0$$

$$J_{ilm} = \sum_{l'} r^{l'} \, V_{il'0} \int Y_{lm}(\theta,\varphi)^* \, Y_{l'0}(\theta,\varphi) \, Y_{lm}(\theta,\varphi) \, d\Omega \tag{22}$$

$$U_{ilm} = \sum_{l',m',l'',m''} r^{l'+l''-l} \, V_{il'm'} \, R_{il''m''} \int Y_{lm}(\theta,\varphi)^* \, Y_{l'm'}(\theta,\varphi) \, Y_{l''m''}(\theta,\varphi) \, d\Omega$$

where the summations over l'' and m'' in the expression of U_{ilm} are restricted by the conditions : $l'' \neq l$ and/or $m'' \neq m$.

The presence of the U_{ilm} term is the only formal difference between the eq.(14), obtained in the atomic case, and the eq.(22). This term comes from the fact that the partial wave expansion of φ_i includes several terms here instead of a single term in the atomic case. In fact J_{ilm} and U_{ilm} are two components of the Coulomb type potential V_i : J_{ilm} is diagonal in the partial wave R_{ilm} while U_{ilm} gives rise to a coupling between different partial waves R_{ilm} and $R_{il'm'}$ ($l,m \neq l',m'$) of the same orbital φ_i.

We now transform the eq.(22) in the same way as done for the eq.(14) : we assume that the (normalised) optimum orbitals φ_i have been determined by some existing Quantum Chemistry program along with the partial waves R_{ilm} and with the potential terms J_{ilm}, U_{ilm} and G_{ilm}. Using these quantities we then set up the equation

$$r\frac{\partial^2}{\partial r^2} f + 2(l+1)\frac{\partial}{\partial r} f + 2\left[Z + (e - J_{ilm})r\right] f + 2r\sigma(G_{ilm} - U_{ilm}) = 0 \tag{23}$$

$$\sigma = \frac{f(0)}{R_{ilm}(0)}$$

This equation is similar to the eq.(15) obtained in the atomic case. Thus one can switch at will between the atomic and the molecular cases : if we give to the parameters J_{ilm}, σG_{ilm} and σU_{ilm} the values determined for the atom as described in the above section 1.4 (this implies $U_{ilm}=0$), then f is proportional to the RHF orbital of the atom A with the quantum numbers l and m and the energy e ; if alternatively we give to the parameters the values obtained for a molecular system as just explained, then f is proportional to the l, m partial wave of the orbital of the molecular system with the energy e.

We now use the Valley theorem : the atomic function f depends weakly of the e parameter in a large region near the nucleus. It can be seen by inspection of the eqs.(15),(16),(17) and (23) that the critical parameter in the molecular case is an effective energy $e_{eff}=e-\Delta v_0+\Delta\sigma g_0-\Delta\sigma u_0$ where Δv_0, $\Delta\sigma g_0$ and $\Delta\sigma u_0$ are the differences between the values of J_{ilm}, σG_{ilm} and σU_{ilm} at the origin in the molecular case and in the atomic case. Therefore, if e_{eff} is not too different from the atomic orbital energy,then the two f functions obtained with the atomic and molecular values of the parameters are extremely close to be proportional one of the other in a finite region near the nucleus. Stated differently : in a finite region near the nucleus of an atom A, the partial waves of the optimum orbitals centered on A are proportional to the corresponding RHF orbitals of the atom A , unless the atomic and molecular parameters are very different from each other (*i.e.* unless the difference is much larger than the variations mentionned in the section 1.5).

3. Asymptotic conditions

The Valley theorem leads to simple conditions for the optimised orbitals near the nuclei. However these conditions are not sufficient to characterize these orbitals : one needs in addition to take the asymptotic form of the equations into account.

In the asymptotic region, an electron approximately experiences a Z'/r potential, where Z' is the charge of the molecule-minus-one-electron ($Z' = 1$ in the case of a neutral molecule) and r the distance between the electron and the center of the charge repartition of the molecule-minus -one-electron. Thus the φ orbital describing the state of that electron must be close to the asymptotic form of the irregular solution of the Schrödinger equation for the hydrogen-like atom with atomic number Z'.

$$\varphi(r) = r^{(Z'/\zeta)-1}exp(-\zeta r) ; \qquad \zeta = \sqrt{-2e} \tag{24}$$

(see for instance the Eq.13.5.2 of Ref.9) where e is the orbital energy. Since e is different in the molecule and in the separated atoms, this asymptotic behaviour cannot be represented properly if the molecular orbital is approximated by a linear combination of the RHF orbitals of the separated atoms.

4. The case of H_2^+

The interest of H_2^+ in the present context is that it provides a good test for the present orbital optimisation theory because one knows the exact solution.

Thus we will use the result of calculations of the wave function of H_2^+ expanded in a gaussian basis to provide numerical tests of the qualitative discussion on the orbital optimisation theory presented in the above sections 2 and 3.

We have calculated several approximations of the energy of H_2^+ (ground electronic state) using various GTO bases (Table 1). In all cases the internuclear distance used was equal to 2 B, close to the experimental equilibrium distance (R_e=1.98 B (10)).

The accuracy of the results obtained here using gaussian bases - and the usefulness of the numerical tests based on these results - can be seen from the values given in the Table 1. It is seen that the dissociation energy De obtained in the largest basis used here is excellent (error equal to 0.01 eV). On the other hand, the error on the value obtained using the minimum basis is as high as 1.35 eV (or 48% in relative value). This proves, if need be, the importance of the orbital optimisation studied in the present article.

It is also useful to note that the major part (77%) of the effect of the orbital optimisation is obtained in the intermediate basis where no polarisation orbital is used.

Table 1: Values of the energy of H_2^+ calculated with different CGTO's basis sets.

Basis set	Total energy (H)	Binding energy De (eV)
Minimum basis (a)	-0.5530	1.44
4s basis (b)	-0.59088	2.47
4s + 2p basis (c)	-0.60233	2.78

Internuclear distance used : R=2 B;
Exact binding energy : De=2.79 eV (10)
Exact total energy : E=-0.602704 H.
(a) This basis set consists of the 1s orbital of of each H atom calculated in a 9s CGTO basis set (Ref.11)
(b) This basis is obtained from the 9s basis set by a (6,3*1) contraction.
(c) Exponents of the 2p functions : 1.0 and 0.3

4.1. OPTIMISATION IN THE VICINITY OF A NUCLEUS

We first consider what happens when comparing directly the optimum orbital of H_2^+, the un-optimised orbital of H_2^+ (*i.e.* the sum of the two 1s orbitals of the H atoms) and the orbitals of the H atom itself. The comparison between the values of these orbitals along the bond axis is presented on the fig.(7).

It is seen that in the inner region (positive values of the abscissae), the atomic orbital is close neither to the optimal orbital nore to the un-optimised orbital. On the contrary, the atomic orbital is very close of the un-optimised orbital but not of the optimised one in the outer region (negative values of the abscissae). The inverse conclusion is obtained in the perpendicular direction presented in the fig.(8) : the atomic orbital is very close to the optimal molecular orbital but not of the un-optimised one. Thus, no clear conclusion can be reached in this way.

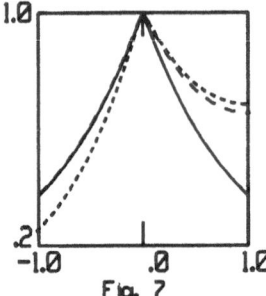 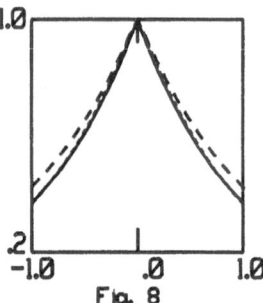

Fig. 7.

Fig. 8.

Fig. 7. Values of different orbitals of H_2^+ and H along the bond axis of H_2^+. The r distance is given in Bohr units ; $r=0$ corresponds to the position of one of the H nuclei, $r > 0$ corresponds to the region between the two nuclei.

dotted line : optimised molecular orbital of H_2^+
dashed line : un-optimised molecular orbital of H_2^+
full line : RHF 1s orbital of Hydrogen

Fig. 8. Same as fig.7 for the values of the orbitals along an axis perpendicular to the bond direction, and containing one of the nuclei.

Let us now consider what happens when comparing the orbital of the H atom, no longer with the orbitals of the H_2^+ system, but with the partial waves of these latter orbitals.

In the case of the s wave ($l = 0$) of the optimised orbital the effective energy defined in the section 2 is given here by $e_{eff} = e + 1/R = -0.602H$ (R is the internuclear distance). According to the analysis of that section it is seen on the fig.(9) that the s wave of the optimum orbital obtained in the gaussian basis is actually very close to the numerical regular atomic s orbital with $e_{eff} = -0.602$ H while the s wave of the un-optimised orbital is significantly different from these two functions.

Here the effective energy is very close to the energy of the genuine atomic orbital (-0.602 H to be compared to -0.5 H). Correspondingly, it can be seen on the fig.(9) that the s wave of the optimum orbital is also very close to the genuine 1s orbital of the H atom. In fact, the difference between these two functions is smaller than 2.4% in all the considered range of r.

A similar conclusion cannot be reached concerning the p waves ($l = 1$). In fact the coupling term between the s and p waves (the U_{ilm} term of the eq.(22)) is not small here and correspondingly the p wave of the molecular system cannot be expected to be close of any atomic-like orbital.

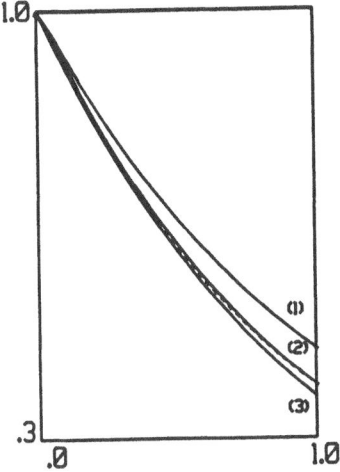

Fig.9. Partial waves ($l=0$) of different orbitals near one of the nuclei.
(1) s wave of the non-optimised orbital of H_2^+.
(2) s wave of the optimised orbital of H_2^+
(3) RHF 1s orbital of Hydrogen atom ($e=-0.5$ H)
dotted line : regular solution of the radial equation of Hydrogen ($l=0$) with an energy equal to $e_{eff}=-0.602$ H.

4.2.   OPTIMISATION IN THE ASYMPTOTIC REGION

When expanding the orbital in partial waves with origin at the midpoint of the molecule (center of charge of the molecule-minus-one-electron) the p wave vanishes, and only the s wave has to be considered. According to Sec.3, this partial wave must be proportionnal to the irregular solution of the hydrogen-like system with atomic number $Z'=2$ and with e equal to the exact orbital energy (-1.102 H).

We present in the Table 2 the ratio of the irregular solution of the hydrogen-like system with the s wave of the optimised orbital, and with the s wave of the unoptimised orbital. It is seen that the irregular numerical solution is actually much closer to be proportional to the s wave of the optimised orbital than to that of the unoptimised orbital.

Table 2 : Ratio of the irregular solution of the hydrogen like system with the s wave of the optimised orbital (1), and with the s wave of the unoptimised orbital (2).

r(B)	1.0	2.0	3.0	4.0	5.0	6.0	7.0
(1)	2.60	3.00	3.25	3.39	3.42	3.32	2.31
(2)	2.40	3.43	4.99	7.45	11.29	17.32	26.84

In fact, the ratio between the numerical and the optimised orbital is nearly constant (relative variation smaller than 11%) for $2 < r < 6$ B, while the ratio with the s wave of the un-optimised orbital is multiplied by *ca.* 5 when r increases from 2 B to 6 B (r=distance to the midpoint of the two nuclei). The decrease of the ratio at larger distances in the case of the optimised orbital just comes from the fall off of gaussian functions at large distances.

4.3.   CONSTRUCTION OF THE ORBITAL OF H_2^+.

In the two preceding sections (4.1 and 4.2) we have presented numerical test of the following description (resulting from the analysis of the sections 2 and 3) of the optimised orbital of H_2^+ :

- near of a nucleus, the s wave (with origin on that nucleus) of the optimised orbital of H_2^+ is proportional to the s regular solution of the radial equation (eq.(2)) with Z=1 and a shifted energy e_{eff} given by $e_{eff} = e + 1/R$ (e=orbital energy, R=internuclear distance);

- outside the molecule, the s wave (with origin at the midpoint of the two nuclei) of the optimised orbital of H_2^+ is proportional to the s irregular solution of the radial equation with Z=2 and the actual energy of the orbital.

We examine now a numerical test of the reciprocal of this description : if a function satisfies this description, then it is the optimised orbital of H_2^+. If both the description and the reciprocal are true we can conclude that the description is complete.

To that end we first introduce the following notations :

- the ' internal zone ' corresponding to the nucleus A is defined by the condition $r_A < R/2$ (in H_2^+ there are two 'internal zones', but , due to the symmetry, only one of them will be considered here) ;

- the 'external zone' is the region outside the molecule ;

- φ is the orbital to be determined in the form of an expansion in a gaussian basis :

$$\varphi = \sum_p \chi_p C_p \tag{25}$$

where χ_p is a gaussian function and C_p the numerical coefficients to be determined;

- φ_I is the s partial wave of φ with origin on the nucleus A;

- φ_E is the s partial wave of φ with origin at the middle of the bond;

- φ_{reg} and φ_{irreg} are the regular and irregular solutions of the two radial equations corresponding to the internal and external zones (assuming e_{eff} and e to be known).

Then φ_I and φ_E are obtained by mean of the partial wave expansion of the gaussian functions :

$$\varphi_I = \sum R_{p,00}(r_A) \; C_p$$

$$R_{p,00}(r_A) = \int Y_{00}^*(\theta,\gamma) \; \chi_p(r_A,\theta,\gamma) \; sin\theta \; d\theta \; d\gamma \tag{26}$$

where Y_{00} is a spherical harmonic centered on the nucleus A and χ_p is a gaussian centered on either of the two nuclei. A similar relation holds for φ_E with r_M in the place of r_A, where M is the midpoint of the two nuclei.

We introduce now two unknown numerical constants α and β and we try to check that the two conditions

$$\varphi_I = \alpha\,\varphi_{reg}, \quad \varphi_E = \beta\,\varphi_{irreg} \tag{27}$$

imply a set of C_p coefficients that are equal (or sufficiently close) to the ones obtained directly by diagonalising the hamiltonian matrix.

To do that, we first guess starting values of α and β ; secondly we determine the C_p coefficients by minimising the quantity Q given by

$$Q = \sum_{r_A} |\sum_p R_{p,00}(r_A)\,C_p - \alpha\,\varphi_{reg}(r_A)|^2 +$$

$$+ \sum_{r_M} |\sum_p R_{p,00}(r_M)\,C_p - \beta\,\varphi_{irreg}(r_M)|^2 \tag{28}$$

where r_A and r_M are two set of points in the internal and external regions respectively; thirdly we evaluate the energy of H_2^+ using the C_p coefficients just determined. The steps two and three are repeated with different values of α and β until the energy is minimised.

It is seen that this process is essentially a least square fit of $\alpha\varphi_{reg}$ and $\beta\varphi_{irreg}$ by φ_I and φ_E, subject to a minimum energy condition which allows to determine α and β. Note that α and β are related by the norm of φ so that there is in fact a single parameter in this minimisation.

This calculation has been made here using the 4s basis set (which includes no polarisation p gaussian orbitals). The energy obtained in this way is very good : it reproduces the energy obtained by diagonalisation (*viz.* -0.59088 H ; cf the Table 1) with an error equal to 0.02 eV.

Concerning the expansion coefficients, the most significant comparison concerns the values of the two orbitals : the one obtained by the fitting process just described and the one obtained by diagonalising the matrix of the hamiltonian in the gaussian basis. In fact we have found that the difference between these two orbitals never exceeds 3% in the internal region as well as in the external region.

We conclude that the description of the orbital by the proportionality between φ_I and φ_{reg} on the one hand and between φ_E and φ_{irreg} on the other hand is supported by the present calculation and that it is indeed complete.

5. Conclusion

We have demonstrated formally that the optimum orbitals of any given molecular system (canonical SCF orbitals or strongly occupied MCSCF orbitals that are closed to the SCF ones) can be described very simply in the regions surrounding each nucleus of the system, approximatly the region extending from the nucleus to the middle of the bonds starting from that nucleus. In that region each partial wave of the optimum orbital is proportional to the atomic orbital with the same value of the quantum number l, unless the molecular potential differs too much from the atomic potential, or unless the coupling term with other waves of the same orbital is too strong (polarisation orbitals).

This description results from the fact that the optimum orbitals are essentially determined in the region surrounding each atom by the compensation between the kinetic energy T of the electron and the Coulomb attraction of the electron by the nucleus of that atom. This compensation implies that the orbital is very weakly dependent of the environment of the atom in the molecular system so that it is essentially determined by atomic conditions (Valley theorem).

A special aspect of this description appears if one starts the orbital optimisation process with orbitals obtained by linear combinations of RHF orbitals of the isolated atoms (LCAO approximation s.str.). Let $\varphi_{n.opt}$ and φ_{opt} be the starting and final orbitals of such a calculation. Then the difference between $\varphi_{n.opt}$ and φ_{opt} in the vicinity of each atom merely consists in a distortion of the atomic orbitals of each atom. This distortion just compensates the contribution of the orbitals of the other atoms to $\varphi_{n.opt}$ in order to restore the proportionality between the partial waves of φ_{opt} and the appropriate atomic orbital.

This description is completed by describing what happens outside the molecule : the partial waves of the optimum orbital are there proportional to the irregular solution of a radial equation involving the actual energy of the orbital .

We have checked, using H_2^+ as a test case, that the description of the optimum orbital of the molecular system is then complete in the sense that it allows (assuming that the orbital energy is known) to construct by a fit process an optimum orbital which is very close to the one obtained by a diagonalisation process in a gaussian basis.

Clearly, several aspects of the orbital optimisation remain to be clarified. Firstly a numerical test using a system more complex than H_2^+ should be made. What happens to π orbitals or strongly hybridized orbitals should be also examined. It would be also interesting to explain how the optimisation - as described here - is related to an energy lowering, as well as the practical use of the present description in actual calculations, etc ... These different aspects will be examined in forthcoming publications.

References

1. W.Moffit, *Proc. Roy. Soc London* **A** 210,224,245 (1951).
   R.Parr and J.Rychlewski, *J.Chem.Phys* **84**,1 (1986).

2. L.D.Landau and E.M.Lifshitz,Quantum Mechanics, Pergamon Press, Oxford, 1977.

3. I.N.Levine, Quantum Chemistry, Allyn and Bacon, Boston, (1093).

4. D.R.Hartree, *Proc. Cambridge Pil. Soc.* **24**, 89,111,426 (1928).

5. Y.N.Demkov, *Sov. Phys. JETP* **19**, 762 (1964).
   J.P.Gauyacq, *J. Phys.* **B 18**, 1859 (1985).

6. E.Clementi and C.Roetti, Atomic Data and Nuclear Data Tables 14, 1974.

7. W.H.Press, B.P.Flannery, S.A.Teukolsky and W.T.Vetterling,
   Numerical Recipes, Cambridge University Press, Cambridge, 1986.

8. P.O.Löwdin, *Phys. Rev.* 97 , 1474 (1955).

9. M.Abramowitz and I.A.Stegun, Handbook of Mathematical Functions, Dover
   Publication, New York, 9th ed., 1970.

10. K.P.Huber et G.Herzberg, Constants of Diatomic Molecules, van Nostrand,
    New York, 1979.

11. F.B. van Duijneveld, Gaussian basis sets for the atoms H-Ne for use
    in molecular calculations, IBM Research, 1971.
    A. Veillard, *Theoret. Chim. Acta* 12,405(1968).
    S.Huzinaga . Gaussian basis sets for molecular calculations, Physical Sciences data 16, Elsevier, 1984.

12. G.Chambaud, M.Gérard-Ain, E.Kassab, B.Lévy and P.Pernot, *Chem. Phys.*
    90, 271 (1984).

13. R.Ahlrichs, *Phys. Rev.* **64** , 2706 (1976)

A Coupled MCSCF-Perturbation Treatment for Electronic Spectra

O. PARISEL and Y. ELLINGER
Laboratoire de Radioastronomie Millimétrique, E.N.S. et Observatoire de Paris,
24 rue Lhomond, F. 75231 Paris Cedex 05, France

1. Introduction

1.1. PURE VARIATION OR PERTURBATION APPROACHES

Despite continuous efforts over many decades, the determination of accurate wavefunctions and energies for polyatomic systems remains a challenging problem. Although carefully-designed implementations of various codes and computational developments allow now for calculations that were unrealistic even a few years ago, the evaluation of correlation energies and highly-correlated wavefunctions, that are necessary to properly describe excited states or potential energy surfaces, remains still in most cases a tremendous task which can hardly be performed routinely and rigorously for large systems.

If we except the Density Functional Theory and Coupled Clusters treatments (see, for example, reference [1] and references therein), the Configuration Interaction (CI) and the Many-Body-Perturbation-Theory (MBPT) [2] approaches are the most widely-used methods to deal with the correlation problem in computational chemistry. The MBPT approach based on an HF-SCF (Hartree-Fock Self-Consistent Field) single reference taking RHF (Restricted Hartree-Fock) [3] or UHF (Unrestricted Hartree-Fock) orbitals [4-6] has been particularly developed, at various order of perturbation n, leading to the widespread MPn or UMPn treatments when a Möller-Plesset (MP) partition of the electronic Hamiltonian is considered [7]. The implementation of such methods in various codes and the large distribution of some of them as black boxes make the MPn theories a common way for the non-specialist to tentatively include, with more or less relevancy, correlation effects in the calculations.

It is however too often forgotten that the usual single-reference MBPT is relevant only for structures that are already well-described by a single determinant: even a second-order perturbation treatment on a closed-shell molecule using RHF orbitals and the SCF determinant as zeroth-order function for the perturbation will be relevant only if this function dominates the exact wavefunction of the system [8]. It follows that using standard MPn approaches for the determination of potential energy surfaces which invoke distorted geometries and breakings of chemical bonds or in the description of molecules involving transition metals should be considered with an extreme critical mind. This point is even more crucial for excited states where appropriate perturbative excitonic treatments are necessary as shown in the pioneering works by Berthier or Pauzat [9-13]. Moreover, there

39

Y. Ellinger and M. Defranceschi (eds.), Strategies and Applications in Quantum Chemistry, 39–53.
© 1996 *Kluwer Academic Publishers.*

is no method currently avalaible to perform efficient MPn calculations on open-shell systems described by a spin-clean single ROHF determinant (Restricted Open Hartree-Fock) [14]; when dealing with unpaired electrons, UHF orbitals are used instead, sometimes leading to the well-known drawbacks of spin contamination (for an extreme example, see reference [15] or poor convergence [16,17] and even to dramatic failures [18-21].

The advantages of MPn perturbation treatments are however clear on both the theoretical and computational points of view. For example, size-consistency is ensured, analytical gradients and Hessians are avalaible, parallelization of the codes is feasable.

Most of the previous advantages are lost in the variational approaches: getting upper-bound energies has to be paid for and despite numerous and ingenious implementations using a large variety of algorithms, large-scale CI are not easily tractable. The cost-effectiveness argument leads either to carefully design a CI space or to truncate it in order to accommodate the storage limitations of modern computers, whatever the method used. The single-reference SDCI (Singles and Doubles Configuration Interaction) approach is an example of such a truncation which is known to give an unbalanced description of the correlation energy between excited states [22]. Even the extension to the SDTQ CI appears to be insufficient [23], especially as soon as the single reference does not dominate the exact wavefunction by a large margin. Also the lack of size-consistency of such dramatically truncated CIs [24-26] makes them too flimsy to accurately deal with correlation problems. Major improvements in variational methods have been reached using MRCI (Multi-Reference CI) [27,28]: however, a careful choice of the reference configurations has to be made in order to avoid both the inflation of the CI expansion and the lack for some potentially important configurations needed for a proper description of the phenomenon under investigation. Even carefully truncated MRCI may lead to deceptive results when one deals with excited states.

It is seen that neither the MBPT nor the CI approaches are the panacea.

1.2.   THE COUPLING OF VARIATION AND PERTURBATION TREATMENTS

The idea of coupling variational and perturbational methods is nowadays gaining wider and wider acceptance in the quantum chemistry community. The background philosophy is to realize the best blend of a well-defined theoretical plateau provided by the application of the variational principle coupled to the computational efficiency of the perturbation techniques.[29-34]. In that sense, the aim of these approaches is to improve a limited Configuration Interaction (CI) wavefunction by a perturbation treatment.

One of the first attempts was done more than 20 years ago and led to the so-called 'CIPSI' method whose basic idea is to progressively include the most important correlation terms in the variational space to be improved by a forthcoming second-order perturbation treatment [35]. The selection of the terms to be included in the variational zeroth-order space is made according to a user-fixed numerical threshold based either on the contribution of these terms to the perturbed wavefunction, as in the original CIPSI approaches [35,36], or on their energetic contribution to the total energy [37,38]. The pitfalls to avoid when using such iterative algorithms are now well-established, although often forgotten: in particular, extreme caution must be taken to ensure an homogeneous treatment of correlation energies along a reaction path or between excited states.

In order to systematically remedy the previous drawbacks, we recently proposed to perform a perturbation treatment, not on a wavefunction built iteratively, but on a wavefunction that already contains every components needed to properly account for the the chemistry of the problem under investigation [34]. In that point of view, we mean that this zeroth-order wavefunction has to be at least qualitatively correct: the quantitative aspects of the problem are expected to be recovered at the perturbation level that will include the remaining correlation effects that were not taken into account in the variational process: any unbalanced error compensations or non-compensations between the correlation recovered for different states is thus avoided contrary to what might happen when using any truncated CIs. In this contribution, we will report the strategy developed along these lines for the determination of accurate electronic spectra and illustrate this process on the formaldehyde molecule H_2CO taken as a benchmark.

2. Theoretical background in the perturbation theory

2.1. PERTURBATIONS AND THE SPECTRAL DECOMPOSITION OF THE HAMILTONIAN

Let suppose $|0>$ is an exact solution to the eigenvalue problem :

$$H°|0> = E_0^0 |0> \tag{1}$$

where $H°$ is an hermitian zeroth order hamiltonian. Considering the perturbation to $|0>$ and E_0^0 induced by the perturbation operator V on $H°$, the first order correction $|1>$ to $|0>$ can be developed on a set of basis functions $\{|j>\}$ $(j=1, ...M)$:

$$|1> = \sum_{j=1}^{M} c_j |j> \tag{2}$$

The Rayleigh-Schrödinger Perturbation Theory (see [2]) leads then to the following system of linear equations for the determination of $\{c_j\}$ $(j=1, ...M)$:

$$\sum_{j=1}^{M} c_j <i|H°- E_0^0 |j> = -<i|V- E_0^1 |0> \qquad (i=1, ...M) \tag{3}$$

where E_0^1 is the first order correction to the zeroth order energy E_0^0.

Let us now define :

$$E_j^0 = <j|H° |j> \qquad (j=1, ...M) \tag{4}$$

If the following relations are both valid:

$$<i|H° |j> = E_i^0 \delta_{ij} \qquad (i,j = 0, ... M) \tag{5}$$

$$<i |j> = \delta_{ij} \qquad (i,j = 0, ... M) \tag{6}$$

then, equation (3) can be simplified, which gives :

$$c_j \left[E_j^0 - E_0^0\right] = -<j |V |0> = - <j |H |0> \qquad (j=1, ... M) \tag{7}$$

which is the usual first order perturbation coefficient for $|j>$ in the first order correction to the initial wavefunction $|0>$. The first- and second-order corrections to the energy are in that case :

$$E_0^1 = <0 |V| 0> \tag{8a}$$

$$E_0^2 = \sum_{j=1}^{M} \frac{|<0 |V| j>|^2}{E_j^0 - E_0^0} = \sum_{j=1}^{M} \frac{|<0 |H| j>|^2}{E_j^0 - E_0^0} \tag{8b}$$

We emphasize that the validity of equations (7) and (8) depends on that of equations (5) and (6) which reflect the fact that expansion (2) is performed on the set of the eigenvectors of $H^°$.

    If, for example, we suppose that $|0>$ is the ground state electronic configuration of interest and $|j>$ are Slater determinants built on a set of orthogonal orbitals $\{\phi_k\}$, then equation (6) is automatically fulfilled.

Furthermore, if $\{\phi_k\}$ are eigenvectors of some one-electron operator $h(v)$ such that :

$$H^° = \sum_{v} h(v) \tag{9}$$

equation (5) becomes also valid. An immediate application of these results is the usual MP2 theory for a set of RHF or UHF orbitals with $H^°$ taken as the Fock operator for the polyelectronic system.

In the CIPSI theory, the reference is the zeroth order space S which consists in a set of determinants $\{|\Phi_I>\}$. Let then P be the perturbation space formed of Slater determinants $|j>$ arising from all the single and double excitations relative to the Slater determinants included in S for the description of $\{|\Phi_I>\}$. We define the zeroth-order wavefunction as :

$$|0> = \sum_{I \in S} c_I^0 |\Phi_I> \tag{10}$$

The expansion coefficients $\{c_I^0\}$ are determined variationally so that $|0>$ is one of the eigenvectors of the restriction of H to the S space with eigenvalue E_0^0 :

$$P_S H P_S |0> = E_0^0 |0> \tag{11}$$

where P_S defines the projection operator onto the S space. We now chose the zeroth-order hamiltonian $H^°$ so that :

$$H^° = \sum_{I \in S} E_I^0 |\Phi_I><\Phi_I| + \sum_{j \in P} E_j^0 |j><j| \tag{12}$$

$H°$ is thus defined through the two sets $\left\{E_I^0\right\}$ and $\left\{E_j^0\right\}$, namely by its matrix elements in the $S \cup P$ space. If the determinants $|j\rangle$ are built on orthogonal orbitals, equation (6) is automatically fulfilled which ensures that equation (5) is also valid due to the definition of $H°$. The matrix elements of $H°$ are then easily calculated :

* for the P-P interaction :

$$\langle i|H°|j\rangle = \delta_{ij}\, E_i^0 \qquad (i \in P, j \in P) \tag{13}$$

* for the S-S interaction :

$$\langle 0|H°|0'\rangle = \delta_{00'}\, E_0^0 = \delta_{00'} \sum_{I \in S} |c_I|^2 E_I^0 \tag{14}$$

* for the S-P interaction :

$$\langle 0|H°|j\rangle = 0 \qquad (j \in P) \tag{15}$$

With this choice for $H°$, equations (7) and (8) are automatically valid for the perturbation. The only restriction is that we have to use orthogonal orbitals and Slater determinants rather than Configuration State Functions (CSFs) as a basis for the perturbation. None of these restrictions is constraining, however.

2.2. THE MOELLER-PLESSET PARTITION

A detailed study of the various possibilities in the choice of the partition to be used in performing the perturbation falls outside the scope of the present contribution (see reference [34]): here we will limit the discussion to the widely used Möller-Plesset partition [7] in which the diagonal matrix elements are defined by :

$$E_j^0 = \langle j\,|F|j\rangle \qquad (j \in P) \tag{16}$$

$$E_I^0 = \langle \Phi_I\,|F|\Phi_I\rangle \qquad (I \in S) \tag{17}$$

where F is the usual Fock operator. For a multireference zeroth-order wavefunction, equation (18) gives the usual expansion of the definition of the zeroth-order energy [35]:

$$E_0^0 = \sum_{I \in S} |c_I|^2 E_I^0 \tag{18}$$

This approach extends the usual MP single-reference approach and will be hereafter referred to as "Barycentric Möller Plesset" (BMP) perturbation theory [35]. If the orbitals used are of RHF or UHF type, a single reference BMP calculation is analogous to a MP2 or UMP2 calculation. However, as emphasized above, we only need to have orthogonal orbitals, which means that the orbitals to be used are not necessarily those that diagonalize the usual Fock operators $F(v)$ for a closed-shell system :

$$F = \sum_v F(v) = h_{core} + \sum_k^{occ.} [2J_k - K_k] \tag{19}$$

F can be advantageously taken as the following effective operator [39] in which orbital occupancies are explicitely considered :

$$F = h_{core} + \sum_k n_k [J_k - K_k/2]$$ (20)

If we define the orbital energies by :

$$\varepsilon_k = <\phi_k|F(v)|\phi_k>$$ (21)

$\{E_I^0\}$ are easily expressed as :

$$E_I^0 = <\Phi_I|F|\Phi_I> = \sum_k n_k \varepsilon_k$$ (22)

More generally, given a set of orbitals and their corresponding energies with respect to some one-electron operator, it is always possible to define a (non local) one-electron operator having these orbitals and energies as eigenvectors and eigenvalues [40]. Such a possibility which amounts to extending relations (12), (19), (20) and (22), will not be developed further here, but allows us to use various level of correlated orbitals in the calculations [34,41,42] and gives the opportunity to circumvent the problem of the invariance of the perturbation energy correction relative to any arbitrary rotations of the orbitals when those are not unambiguously defined . Furthermore, in the implementation used here, and contrary to the CASPT2 approach [22,31,43], the zeroth-order wavefunction is not necessarily supposed to ensure the Generalized Brillouin Theorem [44].

3. The "Chemical" choice of the zeroth-order wavefunction

There is no general way to choose the "best" zeroth-order wavefunction to be used. However, to avoid large variational expansions or to be sure not to miss some important effects by a too drastic truncation, it may be wise to keep some rules in mind.

3.1. DESIGNING A "GOOD" ZEROTH-ORDER WAVEFUNCTION

First of all, the wavefunction has to contain the necessary ingredients to properly describe the phenomenon under investigation: for example, when dealing with electronic spectra, it thus has to contain every CSFs needed to account at least qualitatively for the description of the excited states. The zeroth-order wavefunction has then to include a number of monoexcitations from the ground state occupied orbitals to some virtual orbitals. In that sense, the choice of a Single CI type of wavefunction as proposed by Foresman et al. [45,46] in their treatment of electronic spectra represents the minimum zeroth-order space that can be considered.

However, the restriction of this space to monoexcited configurations wrongly sweeps away the complexity of excited state wavefunctions [22]. In particular, such a truncated space lacks all the CSFs that account for non-dynamical correlation effects. These effects are poorly recovered by any subsequent second-order perturbation while being essential in the description of excited states or potential energy surfaces. In those cases, a wavefunction generated by a specific configuration interaction is necessary. The structure of the corresponding multiconfiguration reference space must however be carefully designed if one does not want to handle large expansions that might include useless CSFs. The "chemical" (or spectroscopic, or quantum chemical...) intuition can help in designing the most relevant CI space as will be shown in the case of H_2CO in the next section. In particular, CAS spaces which are often used to build zeroth-order wavefunctions before performing large-scale CI can be split into products of smaller CAS or GVB [47] spaces without loss of accuracy: the formal completeness of the treatment may be lost, but the computing time saving is considerable.

It is furthermore logical to use some sets of orbitals that are coherent with the zeroth-order space used: the natural MCSCF orbitals issued from an MCSCF treatment using the space defined previously are then attractive candidates for the perturbation.

Finally, in order to ensure an homogeneous treatment of all excited states at the variational level, the MCSCF calculation should be averaged on the states under investigation. The lowest eigenfunctions of the MCSCF Hamiltonian will provide the zeroth-order wavefunctions to build the perturbation on.

As a conclusion, the calculation will be performed using a state-averaged MCSCF treatment in a well-designed active space.

3.2. THE ACTIVE SPACE FOR H_2CO

The space spanning the CSFs used in the calculation is presented in Table 1. Orbitals are distributed into several sets, and ordered by symmetry. They are denoted in terms of localized orbitals (Fig. 1) in order to emphasize their "chemical" significance: Table 1 presents the various distributions of the correlated electrons into these sets, with 'R' standing for Rydberg orbitals or Rydberg states.

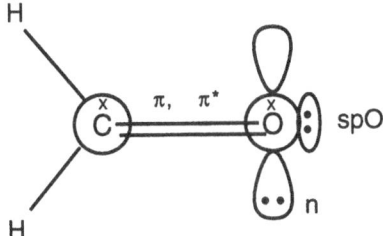

Figure 1 : Lewis structure of formaldehyde

For the description of the vertical spectrum of H_2CO, it is necessary to account for $\pi \rightarrow \pi^*$, $n \rightarrow \pi^*$, $\pi \rightarrow R$ and $n \rightarrow R$ transitions, so that the MCSCF space has been built as a product of smaller MCSCF spaces as follows:

part A : Two electrons in the $(\pi\pi^*)$ set describing both the ground state and the excited
$\pi \rightarrow \pi^*$ states using a CAS space.

part B : Three electrons in the $(\pi\pi^*)$ set and only one in the (n) set describing the excited
$n \rightarrow \pi^*$ states using a MCSCF space.

part C : One electron in the $(\pi\pi^*)$ set and one in the (R) set describing the excited $\pi \rightarrow R$
states using a MCSCF space.

part D : One electron in the (n) set and one in the (R) set describing the excited $n \rightarrow R$ states
using a MCSCF space.

Table 1: Variational space for the formaldehyde molecule (see text for details)

symmetry	Orbital repartition							
	set 1	set 2	set 3	set 4	set 5	set 6	set 7	set 8
a_1	σCO	σCO^*	spO	spO*				10 R
b_1							$\pi\,\pi^*$	4 R
b_2					n	n*		4 R
a_2								2 R
Electronic distributions[a]								
A	2	0	2	0	2	0	2	0
A	0	2	2	0	2	0	2	0
A	2	0	0	2	2	0	2	0
A	2	0	2	0	0	2	2	0
A	1	1	2	0	2	0	2	0
B	2	0	2	0	1	0	3	0
B	0	2	2	0	1	0	3	0
B	2	0	0	2	1	0	3	0
B	1	1	2	0	1	0	3	0
C	2	0	2	0	2	0	1	1
C	0	2	2	0	2	0	1	1
C	1	1	2	0	2	0	1	1
C	2	0	0	2	2	0	1	1
C	2	0	2	0	0	2	1	1
D	2	0	2	0	1	0	2	1
D	0	2	2	0	1	0	2	1
D	2	0	0	2	1	0	2	1
D	1	1	2	0	1	0	2	1

(a) 2 K-shells are doubly occupied in all CSFs

We emphasize that the Rydberg states are included in the variational MCSCF treatment in order not to be undercorrelated relative to the valence states [48,49]. Furthermore, it is seen that all the previous distributions are coupled to diexcitations from the σ_{CO} to the σ_{CO}^* orbital : that way, we account for the non-dynamical repolarization of the polar σ_{CO} bond in a GVB-like approach.
Finally, we also account for an explicit relaxation of both lone pairs of H_2CO by the inclusion of corresponding correlating orbitals [50] and treat them at a GVB-like level.

The final space spanned by these distributions can then roughly be seen as a product of CAS, GVB and MCSCF spaces. It accounts for all effects supposed to be essential to get a good zeroth-order description of the ground and the excited valence and Rydberg states.The dynamical correlation of the 8 electrons included in this variational treatment will be recovered, in the ground state and in the excited states, with the perturbation, and so will be the correlation energy arising from the core and from the σ_{CH} remaining electrons. The perturbation will also account for any coupling not explicitly included in the distributions presented in Table 1.

4. The vertical electronic spectrum of formaldehyde

4.1. SOME HISTORY ON FORMALDEHYDE STUDIES

Initiated by the pioneering work of Burawoy [51], a number of experimental and theoretical studies were performed on the carbonyl group [52-55]. A complete review is beyond the scope of this paper. We will mention only some of them that we consider of particular importance for a comprehensive coverage of the electronic spectrum of formaldehyde for both the theoretical and experimental points of view.

A review of the early experimental works can be found in references [56-58]. More recently, Chutjian recorded the electron-impact excitation spectrum of formaldehyde [59,60] and reported transition energies that are taken as reference values in many other works. So are the experimental values compiled by Robin [61].
A few years ago, Brint et al. [62] focused on the vacuum high-resolution spectrum, pointing out a number of well-defined Rydberg series, of special importance for theoretical benchmarks.

On the theoretical hand, calculations have been performed as soon as in the 50ies [56,63] since formaldehyde represents the smallest member of the carbonyl series. References to early works are avalaible in the compilation by Davidson and McMurchie [64] and in references [56-58,63]. Of particular interest for a comprehensive assignment of the experimental transitions are the very fine and accurate calculations by Harding and Goddard using their GVB-CI method [60,65].

4.2. COMPUTATIONAL DETAILS

The MP2/6-311++G** C_{2v} geometry [45] was used for H_2CO in the present report (CO=1.2122 Å, CH=1.1044 Å, HCO=121.94°). It is very close to the experimental geometry [66]. The molecule is supposed to lie in the yz plane; the z axis corresponds to the C_2 axis, as in Figure 1.

The MCSCF and the subsequent perturbation calculations were done using a 6-31+G* basis set expanded by a set of spd Rydberg functions. Exponents of this additional gaussians were : 0.032 and 0.028 for the s and p shells for the oxygen atom, and 0.023 and 0.021 for the carbon atom. For the d functions, a common value of 0.015 was chosen for both heavy atoms.

The MCSCF calculation was performed using the configuration space described in section 3.2. The state-averaging was done for seven A_1, six B_1, ten B_2 and seven A_2 states for both singlet and triplet multiplicities.

The variational calculations were performed using the Alchemy II package [67] while the further perturbation calculations used a code derived from the original CIPSI module. Proper interfaces between the two programs were developed.

Table 2: Singlet states of formaldehyde

state	this work		experiments	HG[h]		HFW[i]		FHGPF[j]	
X^1A_1	0.00		0.00	0.00		0.00		0.00	
2^1A_1	7.95	$nR3p_y$	7.91[a,b] 7.97[c] 8.05[d] 8.14[e,f] 8.15[g]	8.09	$nR3p_y$	8.47	$nR3p_y$	8.06	$nR3p_y$
3^1A_1	9.11	$nR3d\pi$	9.03[e] 9.07[g]	9.23	$nR3d_{yz}$	8.75	$nR4d_{yz}$		
4^1A_1	9.88	$nR3d\pi$	9.85[d]						
5^1A_1	10.30	$\pi\pi^*$	10.70[g]	10.77	$\pi\pi^*$	9.19	$\pi\pi^*$	9.63	$\pi\pi^*$
6^1A_1	11.60	$\pi R3p_x$	11.60-11.90[g]	12.00	$\pi R3p_x$				
7^1A_1	11.80	$\pi R3p_x$	11.60-11.90[g]						
1^1B_1	9.06	$nR3d\delta$	9.22[n]	9.21	$nR3d_{xy}$				
2^1B_1	9.90	$nR3d\delta$	10.25[d]			9.97	$spO\pi^{*}$[k]	9.97	$spO\pi^{*}$[k]
3^1B_1	10.68	$\pi R3s$	10.7[g]	10.70	$\pi R3s$	10.84	$\pi R3s$		
4^1B_1	11.65	$\pi R3p_z$	11.60-11.90[g] 11.46[p]	11.66	$\pi R3p_z$	11.84	$\pi R3p_z$		
5^1B_1	12.66	$\pi R3d\sigma$	12.50-12.80[g] 12.48[p]	12.58	πRd_{y2}				
6^1B_1	12.78	$\pi R3d\delta$	12.50-12.80[g]	12.70	$\pi R3d$				
1^1B_2	6.90	$nR3s$	7.13[l] 7.11[m] 7.09[c,d,e]	7.16	$nR3s$	6.85	$nR3s$	6.60	$nR3s$
2^1B_2	7.77	$nR3p_z$	7.97[e] 8.00[l] 8.14[c] 8.13[d]	8.08	$nR3p_z$	7.66	$nR3p_z$	7.52	$nR3p_z$
3^1B_2	8.80	$nR3s$	8.88[d,e] 8.92[g]	9.05	$nR3d_{y2}$	8.46	$nR3d$		
4^1B_2	8.95	$nRb2$	8.88[d,e] 8.92[g]			8.94	$nRa1$		
5^1B_2	9.11	$nRb2$	9.26[c,d]	9.17	$nR3d$	8.96	$nR3d$		
6^1B_2	9.39	$nRb2$	9.63[c,d]			9.19	$nR3ds$		
7^1B_2	9.82	$nRb2$	9.85[d]						
8^1B_2	9.84	$nRb2$	9.97[d]						
9^1B_2	10.35	$nRb2$	10.39[d]						
10^1B_2	12.83	$\pi R3d$	12.50-12.80[g]	12.58	$\pi R3d_{xy}$				
1^1A_2	3.83	$n\pi^*$	>3.81[l] 4.0[f] 4.07[m] 4.1[g] 4.2[o]	4.09	$n\pi^*$	4.58	$n\pi^*$	4.58	$n\pi^*$
2^1A_2	8.46	$nR3p_x$	8.37[n]	8.32	$nR3p_x$	7.83	$nR3p_x$		
3^1A_2	8.82	$nRa2$	8.88[d,e] 8.92[g]	9.24	$nR3d_{xz}$				
4^1A_2	9.88	$nRa2$	9.59[d]			10.08	$nR3d$		
5^1A_2	10.44	$nR3p_x$	10.13[d]			10.13	$nR4p_x$		
6^1A_2	11.83	$\pi R3p_y$	11.60-11.90[g]	11.78	$\pi R3p_y$	11.63	$\pi R3p_y$		
7^1A_2	13.01	$\pi R3d$	12.50-12.80[g]	12.88	$\pi R3d_{yz}$				

a) [73], b) [56], c) [58],d) [62], e) [74,75], f) [76], g) cited in [60] as unpublished results by Chutjian, h) [60,65], i) [68], j) [45], k) The spO $\rightarrow \pi^*$ transition was found to lie at 13.69 eV in our work, l) [59], m) [61], n) [69], o) [77], p) [78]

4.3. RESULTS AND DISCUSSION

The analysis of the variational wavefunctions clearly shows admixtures of valence and Rydberg characters in many states, either at the orbital level or at the CI level. We will not discuss this point here, but will focuse on transition energies.

The transition energies from the ground state to the lowest 60 vertical excited states considered in this study are reported in Table 2 (30 singlets) and in Table 3 (30 triplets) where they are compared to the avalaible experimental results and to some previous theoretical calculations [45,60,65,68].
It is immediately seen that the agreement of our computed values with experimental transitions is excellent for both valence and Rydberg states. The discrepancies vary from 0.00 eV to 0.40 eV for the largest of them. An exact value of the deviation is however difficult to obtain due to both the experimental band widths and the fact that many observed transitions are not necessary vertical so that structural effects and vibrational shifts are involved. However, the calculated root-mean-square deviation of the computed values from their experimental assignment is found to be, for the whole spectrum, about 300 cm^{-1}. To our knowledge, there has been no report, whatever might have been the theoretical method used, of such a small deviation between theory and experiment when dealing with so many excited states together.

Within a few exceptions, all singlet states can be correlated to an observed experimental feature. Especially, the high density of states around 11.8 and 12.7 eV is compatible with the observation of unresolved broad peaks in the 11.6-11.9 eV and 12.5-12.8 eV spectral intervals [60]. Unfortunately, the lack of spectroscopic resolution makes any unambiguous one-to-one assignment impossible in these regions.
The situation is more favorable at lower energies: up to about 11 eV, each calculated singlet state correlates unambiguously to a well-resolved experimental line, and the deviation from the experiment does not exceed 0.35 eV which is the largest discrepancy observed. Compared to the calculations by Harding and Goddard [60], the agreement between both methods is excellent. Each state reported by these authors is found in our calculations. In addition, we report some new singlet states of Rydberg character whose description has been made possible essentially because of the larger flexibility of both our MCSCF calculation and one-particle space (basis set including semi-diffuse orbitals that were not in reference [60]). Our calculations provide a clear-cut assignment for the 4^1A_1, 7^1A_1, 2^1B_1, 6^1B_2, 7^1B_2, 8^1B_2, 9^1B_2, 4^1A_2, and 5^1A_2 states which were not reported previously. It is important to notice that most of these new states correlate to the recent experimental results obtained in the study by Brint and Sommer [62] which is devoted to the Rydberg series. It is worth to emphasize that all their lower terms of the ns (3 states), np (6 states) and nd (4 terms) series can be related to a calculated state. Getting a correct description of the higher terms of these series would however require the inclusion of a Rydberg orbital progression in the basis set, so as the consideration of f functions as suggested in reference [62].
The same comments apply to the triplet states, although comparison to experiments is more difficult due to the lack of experimental determinations, even in the low energy region. However, as seen in Table 3, the agreement with avalaible data is excellent, and shows the same quality as for singlet states. So is the correlation with the results by Harding and Goddard [60]. In the triplet manifold, as in the singlet one, the largest flexibility of the present method allows for more states to be found: as an example, we tentatively assign the 6^3B_2 or the 4^3A_2 state, missing in reference [60], to a peak reported at 9.59 eV [69].

Table 3: Triplet states of formaldehyde

state	this work	experiments	HG[h]	HFW[i]	FHGPF[j]
1^3A_1	5.86 $\pi\pi^*$	5.60-6.20[a] 5.86[b] 6.0[a,c]	5.95 $\pi\pi^*$	6.72 $\pi\pi^*$	6.81 $\pi\pi^*$
2^3A_1	7.86 nR3py	7.96[b] 8.11[c]	8.05 nR3py	7.78 nR3py	7.61 nR3py
3^3A_1	9.01 nR3d		9.17 nR3dyz		
4^3A_1	9.83 nR				
5^3A_1	10.29 nR				
6^3A_1	11.91 nR	11.60-11.90[c]	11.77 πR3px		
7^3A_1	12.99 πR	12.50-12.80[c]	12.76 πR3dxz		
1^3B_1	9.03 nR3dδ		9.21 nR3dxy	9.97 spOπ^*d	
2^3B_1	9.87 nR3dδ				
3^3B_1	10.59 πR3s	10.70[c]	10.68 πR3s		
4^3B_1	11.48 πR3pz	11.60-11.90[c]	11.57 πR3pz		
5^3B_1	12.61 πR3dσ	12.50-12.80[c]	12.57 πR3d		
6^3B_1	12.73 πR3dδ	12.50-12.80[c]	12.68 πR3d		
1^3B_2	6.82 nR3s	7.09[a,c] 6.83[b]	7.08 nR3s	6.97 nR3s	6.79 nR3s
2^3B_2	7.72 nR3pz	7.92[c] 7.79[b]	7.99 nR3pz	7.75 nR3pz	7.59 nR3pz
3^3B_2	8.80 nR3s			8.67 nR3d	
4^3B_2	8.97 nRb2		9.01 nR3dy2		
5^3B_2	9.12 nRb2		9.16 nR3d		
6^3B_2	9.37 nRb2	9.59[e]			
7^3B_2	9.85 nRb2				
8^3B_2	9.87 nRb2				
9^3B_2	10.35 nRb2				
10^3B_2	12.88 πR3d	12.50-12.80[c]	12.75 πR3dxy		
1^3A_2	3.41 nπ^*	3.3-3.6[a] 3.2[a] 3.3[a,f,g] 3.5[c,b,f] 3.35[a,k] 3.19[k]	3.68 nπ^*	4.15 nπ^*	4.16 nπ^*
2^3A_2	8.43 nR3px	8.31[c]	8.31 nR3px	8.16 nR3px	
3^3A_2	8.72 nRa2		9.23 nR3dxz	9.12 nR3px	
4^3A_2	9.89 nRa2	9.59[e]			
5^3A_2	10.43 nR3px			10.52 nR3d	
6^3A_2	11.81 πR3py	11.60-11.90[c]	11.63 πR3py		
7^3A_2	12.56 πR3d	12.50-12.80[c]	12.74 πR3dyz		

a) [59], b) [61], c) cited as unpublished results by Chutjian in [60],

d) The spO $\rightarrow \pi^*$ transition was found to lie at 14.31 eV in our work,

e) [69], f) [79], g) [80], h) [60,65], i) [68], j) [45], k) [81]

The comparison to the results obtained using the SCI/MP2 approach [45,68] leads to unquestionable conclusions: not only the SCI/MP2 method does not provide acceptable transition energies for the lowest valence and Rydberg states but it misses some of them and does not provide any good energetical ordering of the excited states. Even if this method presents interesting computational advantages, it can only provide a flimsy quantitative electronic spectrum, as anticipated in section 3.1 and outlined in reference [22].

5. Conclusions and prospects

The present approach is one of the second-generation multireference perturbation treatments first opened by the CIPSI algorithm 20 years ago. Even if the spirit of these new treatments is different, mainly because the reference space is chosen on its completeness rather than on energetical criteria, it remains that the unavoidable problems of disk storage, bottleneck of variational approaches, can now be conveniently transferred to the problem of CPU time which is less restrictive.

The methodology presented here expands the recent CASPT2 approach to more flexible zeroth-order variational spaces for a multireference perturbation, either in the Möller-Plesset scheme or in Epstein-Nesbet approach [70-72]. Furthermore, it allows for the use of a wide set of possible correlated orbitals. These two last points were discussed elsewhere [34].

The reliability of this method for the evaluation of (vertical) electronic spectra has been clearly established in the present work, and further calculations on other molecules (ethylene, vinylydene... for example) have confirmed the very promising potentialities of such an approach that avoids the possible artefacts brought in by any arbitrary truncated CIs when dealing with excited states [49]. We also emphasize that this methodology is able to give reliable splittings between states ranging from 10 kcal/mol to more than 10 eV.

References

1   B.O. Roos, Lecture Notes in Quantum Chemistry II, Vol. 64, Springer Verlag, Berlin, 1994.
2   K.A. Brueckner, The Many-Body Problem, J. Wiley and Sons, Inc., New York, 1959.
3   C.C.J. Roothaan, Rev. Mod. Phys. 23, 69, (1951).
4   G. Berthier, C. R. Acad. Sc. (Paris) 238, 91, (1954).
5   G. Berthier, J. Chim. Phys. 51, 363, (1954).
6   J.A. Pople and R.K. Nesbet, J. Chem. Phys. 22, 571, (1954).
7   C. Möller and M.S. Plesset, Phys. Rev. 46, 618, (1934).
8   A. Masson, B. Lévy and J.-P. Malrieu, Theor. Chim. Acta 18, 193, (1970).
9   G. Berthier, Y.A. Meyer and L. Praud, in : "Aromaticity, Pseudo-Aromaticity, Anti-Aromaticity", The Jerusalem Symposia on Quantum Chemistry and Biochemistry III, Jerusalem, 1971.
10 F. Pauzat, J. Ridard and B. Lévy, Mol. Phys. 23, 1163, (1972).
11 F. Pauzat, J. Ridard and P. Millié, Mol. Phys. 24, 1039, (1972).
12 B. Lévy and J. Ridard, Chem. Phys. Lett. 15, 49, (1972).
13 L. Praud, B. Lévy, P. Millié and G. Berthier, Int. J. Quant. Chem. Symp. 7, 185, (1973).
14 C.C.J. Roothaan, Rev. Mod. Phys. 32, 179, (1960).
15 S. Bell, J. Chem. Soc. Faraday Trans. II 77, 321, (1981).
16 P.M.W. Gill and L. Radom, Chem. Phys. Lett. 132, 16, (1986).

17   M.B. Lepetit, M. Pélissier and J.-P. Malrieu, *J. Chem. Phys.* **89**, 998, (1988).
18   R.H. Nobes, D. Moncrieff, M.W. Wong, L. Radom, P.M.W. Gill and J.A. Pople, *Chem. Phys. Lett.* **182**, 216, (1991).
19   R.P. Messmer and C.H. Patterson, *Chem. Phys. Lett.* **192**, 277, (1992).
20   N.L. Ma, B.J. Smith and L. Radom, *Chem. Phys. Lett.* **193**, 386, (1992).
21   N.L. Ma, S.S. Wong, M.N. Paddon-Row and W.-K. Li, *Chem. Phys. Lett.* **213**, 189, (1993).
22   B.O. Roos, K. Andersson and M.P. Fülscher, *Chem. Phys. Lett.* **192**, 5, (1992).
23   F. Sasaki, *Int. J. Quant. Chem. Symp.* **11**, 125, (1977).
24   R. Ahlrichs, H. Lischka, V. Staemmler and W. Kutzelnigg, *J. Chem. Phys.* **62**, 1225, (1975).
25   A. Meunier, B. Lévy and G. Berthier, *Int. J. Quant. Chem.* **10**, 1061, (1976).
26   W. Kutzelnigg, A. Meunier, B. Lévy and G. Berthier, *Int. J. Quant. Chem.* **12**, 777, (1977).
27   P.S. Bagus, B. Liu, A.D. McLean and M. Yoshimine, "Application of wave mechanics to the electronic structure of molecules through configuration interaction" in : Wave Mechanics : the First Fifty Years, Butterworth, London, 1973.
28   R.J. Buenker, S.D. Peyerimhoff and W. Butscher, *Mol. Phys.* **35**, 771, (1978).
29   R.B. Murphy and R.P. Messmer, *Chem. Phys. Lett.* **183**, 443, (1991).
30   R.B. Murphy and R.P. Messmer, *J. Chem. Phys.* **97**, 4170, (1992).
31   K. Andersson, P.-A. Malmqvist and B.O. Roos, *J. Chem. Phys.* **96**, 1218, (1992).
32   H. Nakano, *J. Chem. Phys.* **99**, 7983, (1993).
33   P.M. Kozlowski and E.R. Davidson, *J. Chem. Phys.* **100**, 3672, (1994).
34   O. Parisel and Y. Ellinger, *Chem. Phys.* **189**, 1, (1994).
35   B. Huron, J.-P. Malrieu and P. Rancurel, *J. Chem. Phys.* **58**, 5745, (1973).
36   S. Evangelisti, J.-P. Daudey and J.-P. Malrieu, *Chem. Phys.* **75**, 91, (1983).
37   Z. Gershgorn and I. Shavitt, *Int. J. Quant. Chem.* **2**, 751, (1968).
38   R.J. Harrison, *J. Chem. Phys.* **94**, 5021, (1991).
39   H.C. Longuet-Higgins and J.A. Pople, *Proc. Roy. Soc. (London)* **A68**, 591, (1955).
40   E. Steiner, *J. Chem. Phys.* **46**, 1727, (1967).
41   G. Berthier, A. Daoudi and J.-P. Flament, *J. Mol. Struct. (Theochem)* **166**, 81, (1988).
42   G. Berthier, A. Daoudi, G. Del Re and J.-P. Flament, *J. Mol. Struct. (Theochem)* **210**, 133, (1990).
43   K. Andersson, P.-A. Malmqvist, B.O. Roos, A.J. Sadlej and K. Wolinski, *J. Phys. Chem.* **94**, 5483, (1990).
44   B. Levy and G. Berthier, *Int. J. Quant. Chem.* **2**, 307, (1968).
45   J.B. Foresman, M. Head-Gordon, J.A. Pople and M.J. Frisch, *J. Phys. Chem.* **96**, 135, (1992).
46   M. Head-Gordon, R.J. Rico, M. Oumi and T.J. Lee, *Chem. Phys. Lett.* **219**, 21, (1994).
47   W.J. Hunt, P.J. Hay and W.A. Goddard, *J. Chem. Phys.* **57**, 738, (1972).
48   L. Serrano-Andrés, M. Merchan, I. Nebot-Gil, R. Lindh and B.O. Roos, *J. Chem. Phys.* **98**, 3151, (1993).
49   O. Parisel and Y. Ellinger, *Chem. Phys.*, submitted for publication, (1994).
50   A.D. McLean, B.H. Lengsfield, J. Pacansky and Y. Ellinger, *J. Chem. Phys.* **83**, 3567, (1985).
51   A. Burawoy, *Chem. Ber.* **63**, 3155, (1930).

52 H.L. McMurry and R.S. Mulliken, *Proc. Natl. Acad. Sci. U.S.A.* **26**, 312, (1940).
53 H.L. McMurry, *J. Chem. Phys.* **9**, 231, (1941).
54 M. Kasha, "Ultraviolet radiation effects : molecular photochemistry" in : Comparative Effects of Radiations, Wiley, New York, 1960.
55 H.M. McConnel, *J. Chem. Phys.* **20**, 700, (1952).
56 M. Sender and G. Berthier, *J. Chim. Phys.* **53**, 384, (1958).
57 G. Berthier and J. Serre, "General and theoretical aspects of the carbonyl group" in The Chemistry of the Carbonyl Group, Interscience, 1966.
58 D.C. Moule and A.D. Walsh, *Chem. Rev.* **75**, 67, (1975).
59 A. Chutjian, *J. Chem. Phys.* **61**, 4279, (1974).
60 L.B. Harding and W.A. Goddard III, *J. Am. Chem. Soc.* **99**, 677, (1977).
61 M.B. Robin, Higher Excited States of Polyatomic Molecules, Academic Press, New York, 1985.
62 P. Brint and K. Sommer, *J. Chem. Soc. Faraday Trans. II* **81**, 1643, (1985).
63 R.S. Mulliken and W.C. Ermler, Polyatomic Molecules. Results of ab initio calculations, Academic Press, New York, 1981.
64 E.R. Davidson and L.E. McMurchie, Excited States, Vol. 5, Academic Press, New York, 1982.
65 L.B. Harding and W.A. Goddard III, *J. Am. Chem. Soc.* **97**, 6293, (1975).
66 K.T. Takagi and T. Oka, *J. Phys. Soc. Jpn.* **18**, 1174, (1963).
67 ALCHEMY II by A.D. McLean, M. Yoshimine, B.H. Lengsfield, P.S. Bagus and B. Liu in MOTECC-90
68 C.M. Hadad, J.B. Foresman and K.B. Wiberg, *J. Phys. Chem.* **97**, 4293, (1993).
69 S. Taylor, D.G. Wilden and J. Comer, *Chem. Phys.* **70**, 291, (1982).
70 P.S. Epstein, *Phys. Rev.* **28**, 695, (1926).
71 R.K. Nesbet, *Proc. Roy. Soc. (London)* **A230**, 312, (1955).
72 P. Claverie, S. Diner and J.-P. Malrieu, *Int. J. Quant. Chem.* **1**, 751, (1967).
73 A.D. Walsh, *J. Chem. Soc.* 2306, (1953).
74 E.P. Gentieu and J.E. Mentall, *Science* **169**, 681, (1970).
75 J.E. Mentall, E.D. Gentieu, M. Krauss and D.J. Neumann, *J. Chem. Phys.* **55**, 5471, (1971).
76 M.J. Weiss, C.E. Kuyatt and S. Mielczarek, *J. Chem. Phys.* **54**, 4147, (1971).
77 J.G. Calvert and J.N. Pitts, Photochemistry, Wiley, New York, 1966.
78 P.M. Guyon, W.A. Chupka and J. Berkowitz, *J. Chem. Phys.* **64**, 1419, (1976).
79 J.C.D. Brand, *J. Chem. Soc. (London)* 858, (1956).
80 A.D. Cohen and C. Reid, *J. Chem. Phys.* **24**, 85, (1956).
81 G.W. Robinson and V.E. DiGiorgo, *Can. J. Chem.* **36**, 31, (1958).

Reduced Density Matrix versus Wave Function: Recent Developments

C. VALDEMORO

Instituto de Ciencia de Materiales, Serrano 123, 28006 Madrid, Spain

1.Introduction

Much of the great interest that the Reduced Density Matrices (RDM) theory has arisen since the pioneer works of Dirac [1], Husimi [2] and Löwdin [3], is due to the simplification they introduce by *averaging* out a set of the variables of the many body system under study. For all practical purposes, the *averaging* with respect to N-1 or N-2 electron variables which is carried out in the 1-RDM or 2-RDM respectively, does not imply any loss of the necessary information. The reason for this is that the operators representing the N-electron observables are sums of operators which depend only on one or two electron variables.

The RDM's are therefore much simpler objects than the N-electron Wave Function (WF) which depends on the variables of N electrons. Unfortunately, the search for the N-representability conditions has not been completed and this has hindered the direct use of the RDM's in Quantum Chemistry. In 1963 A. J. Coleman [4] defined the N-representability conditions as the limitations of an RDM due to the fact that it is derived by contraction from a matrix represented in the N-electron space. In other words, an antisymmetric N-electron WF must exist from which this RDM could have been derived by integrating with respect to a set of electron variables.

The research for finding these conditions, has been intense and fruitful [5-13]. Thus, although an exact procedure for determining directly an N-representable 2-RDM has not been found, many mathematical properties of these matrices are now known and several methods for approximating RDM's and for employing them have been developed [14-19].

To study the electronic structure of small systems within the framework of the RDM formalism is a good strategy to adopt, but where it is of the foremost importance is in the study of the electronic structure of very large systems. In this latter case, to work within the framework of an N-electron WF does not seem the best approach to take even now that large and fast computers are available. It seems clear to me that it would be advantageous to approach the study of these large systems within a theoretical framework having a quantum statistical character. Since the RDM's are statistical objects their formalism would fit in a natural way in such a framework.

The aim of this paper is to review the work done by our group in this direction in the last ten years. The reader wishing to have a broader outlook of this vast and fascinating field of research is referred to the Proceedings of the A. J. Coleman Symposium on Reduced Density Matrices and Density Functionals [20]. In this book the opening contribution is by A. J. Coleman himself, where he masterly describes the history of the Reduced Density Matrix (RDM) research from 1929 up to 1987.

Y. Ellinger and M. Defranceschi (eds.), Strategies and Applications in Quantum Chemistry, 55–75.
© 1996 *Kluwer Academic Publishers.*

This book also collects the contributions of most specialists in the field.

Two different approaches to this problem will be described in this work. They are based in quite different philosophies, but both are aimed at determining the RDM without a previous knowledge of the WF. Another common feature of these two approaches is that they both employ the discrete Matrix representation of the Contraction Mapping (MCM) [17,18]. Applying this MCM is the alternative, in discrete form, to integrating with respect to a set of electron variables and it is a much simpler tool to use.

In this work, we will concentrate on describing the ideas leading to the relevant formulae and only the essential algebraic developments will be described.

2.Notation and Basic Definitions

2.1. CONSTANTS AND STATES

$N =$ number of electrons of the system

$K =$ number of orbitals of the basis set

$S =$ spin quantum number

$\sigma_1, \sigma_2 \ldots$ denote spin variables

$i_1, i_2 \ldots, j_1, j_2 \ldots$ denote orbitals

$I, J \ldots$ denote correlated states

$\Lambda, \Omega, \Gamma \ldots$ denote p-electron configurations with $N \geq p$

$\mathcal{L}, \mathcal{L}' \ldots =$ denote N electron states

$\mathcal{E} =$ energy

2.2. OPERATORS AND EXPECTATION VALUES

2.2.1. *Replacement Operators and Reduced Density Matrices*

Most operators used in this work may be written in terms of the q-order *Replacement Operators* $(q - RO)$ [21,27] which, in our notation, take the form:

$$ {}^q E_{j_1, j_2, \ldots, j_q}^{i_1, i_2, \ldots, i_q} = \sum_{\sigma_1, \sigma_2, \ldots, \sigma_q} b_{i_1 \sigma_1}^+ b_{i_2 \sigma_2}^+ \cdots b_{i_q \sigma_q}^+ b_{j_q \sigma_q} \cdots b_{j_2 \sigma_2} b_{j_1 \sigma_1} \tag{1} $$

where b^+ and b are the usual fermion operators.

The *expectation values* of the q-RO's are the $q - RDM$'s. Thus, the general definition of the $q - RDM$ in this formalism is:

$$ {}^q D_{i_1, i_2, \ldots, i_q; j_1, j_2, \ldots, j_q}^{\mathcal{L}\mathcal{L}'} = \frac{< \mathcal{L} |^q E_{j_1, j_2, \ldots, j_q}^{i_1, i_2, \ldots, i_q} | \mathcal{L}' >}{q!} \tag{2} $$

When $\mathcal{L} \neq \mathcal{L}'$ relation (2) defines a transition $q - RDM$. In what follows, unless it is necessary, the upper indices which indicate the bra and ket states will be omitted since that only the case $\mathcal{L}L = \mathcal{L}L'$ is considered.

Since

$$|I,s> = \sum_{i_1,i_2,\dots,i_q} C^I_{i_1,i_2,\dots,i_q} \; b^+_{i_1\sigma_1} b^+_{i_2\sigma_2} \cdots b^+_{i_q\sigma_q} |0> \tag{3}$$

where the symbol s denotes the set of spin variables, the replacement operators can be further generalised. Thus, in the operator

$$^q E^I_J = \sum_{i_1,i_2,\dots,i_q} \sum_{j_1,j_2,\dots,j_q} C^I_{i_1,i_2,\dots,i_q} \; C^J_{j_1,j_2,\dots,j_q} \; ^q E^{i_1,i_2,\dots,i_q}_{j_1,j_2,\dots,j_q} \tag{4}$$

the sum of annihilators destroy q electrons which are in the state J and the sum of creators create the state I.

Therefore, in this basis one has:

$$^q D_{I;J} = \frac{< \mathcal{L} |^q E^I_J| \mathcal{L} >}{q!} \tag{5}$$

2.2.2. The Hamiltonian Operator and the Energy

The spin free many-body *Hamiltonian Operator* can be written in compact form by employing the 2-$\bar{R}O$

$$\hat{H} = \frac{1}{2} \sum_{i_1,i_2,j_1,j_2} H^0_{i_1 i_2;j_1 j_2} \; ^2 E^{i_1 i_2}_{j_1 j_2} \tag{6}$$

where

$$H^0_{i_1 i_2;j_1 j_2} \equiv (i_1 j_1 | i_2 j_2) + \frac{1}{N-1} \left(\delta_{i_2 j_2} \epsilon_{i_1 j_1} + \delta_{i_1 j_1} \epsilon_{i_2 j_2} \right) \tag{7}$$

In this latter formula, the two electron repulsion integral is written following Mulliken convention and the one electron integrals are grouped in the matrix $\underline{\epsilon}$. In this way, the one-electron terms of the Hamiltonian are grouped together with the two electron ones into a two electron matrix. Here, the matrix $\underline{H}^0$ is used only in order to render a more compact formalism.

An element of the matrix representation of the Hamiltonian in the N-electron space is therefore:

$$\mathcal{H}_{\Lambda\Omega} = \sum_{i_1,i_2,j_1,j_2} H^0_{i_1 i_2;j_1 j_2} \; \frac{< \Lambda |^2 E^{i_1 i_2}_{j_1 j_2}| \Omega >}{2} \tag{8}$$

or equivalently

$$\mathcal{H}_{\Lambda\Omega} = tr \left(\underline{H}^0 \; ^2 \underline{D}^{\Lambda\Omega} \right) \tag{9}$$

A particular case of this is the well known expression of the energy of a normalized state $|\mathcal{L} >$:

$$< \mathcal{L}|\hat{H}|\mathcal{L} > = \mathcal{E}_\mathcal{L} = \sum_{i_1,i_2,j_1,j_2} H^0_{i_1 i_2;j_1 j_2} \; \frac{< \mathcal{L}|^2 E^{i_1 i_2}_{j_1 j_2}| \mathcal{L} >}{2} = tr \left(\underline{H}^0 \; ^2 \underline{D}^\mathcal{L} \right) \tag{10}$$

2.2.3. *The Holes Replacement Operators*

An important operator, complementary to the RO, is the *Holes Replacement Operator* (HRO). The general form of this operator is:

$$^q \bar{E}^{j_1,j_2,...,j_q}_{i_1,i_2,...,i_q} = \sum_{\sigma_1,\sigma_2,...,\sigma_q} b_{j_q \sigma_q} \cdots b_{j_2 \sigma_2} b_{j_1 \sigma_1} b^+_{i_1 \sigma_1} b^+_{i_2 \sigma_2} \cdots b^+_{i_q \sigma_q} \tag{11}$$

The HRO's are *holes density operators* and operate by first filling orbitals with electrons (i.e. they annihilate holes) and then removing electrons from orbitals (i.e. they create holes). These operators generate the *Holes Reduced Density Matrix* ($HRDM$) which in our notation takes the form:

$$^q \bar{D}^{\mathcal{L}}_{j_1,j_2,...,j_q;i_1,i_2,...,i_q} = \frac{1}{q!} < \mathcal{L} |^q \bar{E}^{j_1,j_2,...,j_q}_{i_1,i_2,...,i_q} | \mathcal{L} > \tag{12}$$

I wish to stress that the meaning of the word Hole here is different and far more general than in *Many Body Perturbation Theory*. Indeed, no specific reference state is required in this definition and the difference between the RO's and the HRO's follows exclusively from the different order of the creator operators with respect to the annihilator operators in E and in $\bar{E}$ respectively.

3. The Contraction Mapping in Matrix Form (MCM)

The MCM is at the basis of the two formalisms which are described in this work. Its general form for $p < q < N$ is:

$$^p D^{\mathcal{L}\mathcal{L}'}_{i_1,...,i_p;j_1,...,j_p} = \sum_{\Lambda \Omega} {}^p D^{\Lambda \Omega}_{i_1,...,i_p;j_1,...,j_p} \; {}^q D^{\mathcal{L}\mathcal{L}'}_{\Lambda \Omega} \; \frac{\binom{N}{p}}{\binom{q}{p}\binom{N}{q}} \tag{13}$$

where $\mathcal{L}$ and $\mathcal{L}'$ are N-electron states. The symbols Λ, Ω denote q-electron states which may correspond to simple Slater determinants, to eigenstates of the spin operator $\hat{S}^2$, or to any kind of q-electron correlated states.

It must be underlined that, in (13), while the resulting RDM, $^p D$, may be represented either in an orbital basis or in a spin-orbital one, as in (Ref.17), the symbols Λ, Ω stand for uniquely defined states depending on both space and spin variables.

Relation (13) allows us to contract any $q - RDM$ and what is more, it also allows us to contract any $q - HRDM$ by replacing the number N by the number $(2K - N)$. The derivation of the MCM is based on the important and well known relation

$$\sum_i \frac{b^+_i b_i}{N} = \hat{I} \tag{14}$$

which must be used $q - p$ times in order to obtain relation (13).

In many cases a simpler form of this mapping may be used. Thus, the RDM by itself, when it is not involved in matrix operations it can be contracted by using

(14) directly, without any reference to the intermediate matrix $^pD^{\Lambda\Omega}_{i_1,\dots,i_p;j_1,\dots,j_p}$. The algorithm in this case is:

$$
\begin{aligned}
\sum_{i_1} {}^qD_{i_1,i_2,\dots,i_q;i_1,j_2,\dots,j_q} &= \frac{<\mathcal{L}|\sum_{i_1} {}^qE^{i_1,i_2,\dots,i_q}_{i_1,j_2,\dots,j_q}|\mathcal{L}>}{q!} \\
&= \frac{{}^{q-1}D_{i_2,\dots,i_q;j_2,\dots,j_q}\,(q-1)!\,(N-(q-1))}{q!} \\
&= \frac{{}^{q-1}D_{i_2,\dots,i_q;j_2,\dots,j_q}\,(N-(q-1))}{q}
\end{aligned}
\tag{15}
$$

This direct sum with respect to an index which is common to the creator and the corresponding annihilator cannot always be carried out as will be seen in the following sections, hence the usefulness of the general MCM (relation (13)).

4. The Spin adapted Reduced Hamiltonian (SRH)

The *Spin adapted Reduced Hamiltonian* (SRH) is the contraction to a p-electron space of the matrix representation of the *Hamiltonian Operator*, $\mathcal{H}$, in the N-electron space for a given *Spin Symmetry* [17,18,25,28]. The basis for the matrix representation are the eigenfunctions of the $\hat{S}^2$ operator. The block matrix which is contracted is that which corresponds to the spin symmetry selected; In this way, the spin adaptation of the contracted matrix is insured.
Let us consider the spectral resolution of $\mathcal{H}$

$$
\mathcal{H} = \sum_{\mathcal{L}} \mathcal{E}_{\mathcal{L}} \; {}^N\underline{D}^{\mathcal{L}}
\tag{16}
$$

Now, in order to obtain the $p-SRH$, let us apply the MCM to both sides of this relation,

$$
{}^pH'_{i_1,\dots,i_p;j_1,\dots,j_p} = \sum_{\Lambda\Omega}(\sum_{\mathcal{L}} \mathcal{E}_{\mathcal{L}} \; {}^N D^{\mathcal{L}}_{\Lambda\Omega})\, {}^pD^{\Lambda\Omega}_{i_1,\dots,i_p;j_1,\dots,j_p} \equiv \sum_{\mathcal{L}} \mathcal{E}_{\mathcal{L}} \; {}^pD^{\mathcal{L}}_{i_1,\dots,i_p;j_1,\dots,j_p}
\tag{17}
$$

According to relation (17), the $p-SRH$ matrix is a sum of terms and each term is the product of the energy and the RDM corresponding to an eigen-state of the system. Therefore for $p \geq 2$ the $p-SRH$ matrix has all the relevant information about the eigen-states of our system and is represented in a reduced space which renders it easy to handle.
Note that by generalizing the concept of N-representability it can be said that the $p-SRH$ are both *N-representable* and *S-representable*.

4.1. CONSTRUCTION OF THE $p-SRH$ MATRIX

A very convenient feature is that, in order to construct the $p-SRH$ matrix it is not necessary to evaluate the $\mathcal{H}$ matrix and then apply the MCM since both operations can be combined and carried out simultaneously. Thus, by replacing in relation (17) the element of $\mathcal{H}$ (relation 16), which appears between brackets, by its value in terms of the electronic integrals (relation 8) one gets after some simple algebra

$$^pH'_{i_1,\ldots,i_p;j_1,\ldots,j_p} = \frac{1}{2\,p!} \sum_{k_1,k_2,m_1,m_2} H^0_{k_1k_2;m_1m_2} \sum_{\Lambda} <\Lambda|^2 E^{k_1k_2}_{m_1m_2} \,^pE^{i_1,\ldots,i_p}_{j_1,\ldots,j_p}|\Lambda> \quad (18)$$

At first sight, this looks rather complicated, and in fact, evaluating this general relation is not trivial. On the other hand, the results are simple, closed form expressions. The derivation of the algorithms for different values of p have been described in detail previously [25,26,28]. Here we will just sketch the results obtained.

The main point about the SRH matrix is that its only non zero elements are those where the set of the i indices and the set of the j indices are equal or differ at the most in two indices.

The general form of the three different classes of the matrix elements are [28]:

- The set of i indices is equal to the set of j indices (although any ordering of the indices is allowed).

$$^pH'_{i_1,\ldots,i_p,\hat{P}(i_1,\ldots,i_p)} = \sum_{k,m} \left(A^{0,\hat{P}}_{km} H^0_{km;km} + B^{0,\hat{P}}_{km} H^0_{km;mk} \right) \quad (19)$$

Where the symbol H^0 has the same meaning as previously. In this relation, $\hat{P}$, denotes a permutation of the indices $i_1,\ldots,i_p$, and A and B are coefficients.

- An index of the set i differs from that of an index of the set j, say $i_1 \neq j_1$ (although any ordering of the indices is allowed).

$$^pH'_{i_1,\ldots,i_p,\hat{P}(j_1,i_2,\ldots,i_p)} = \sum_{k} \left(A^{1,\hat{P}}_{k} H^0_{ki_1;kj_1} + B^{1,\hat{P}}_{k} H^0_{ki_1;j_1k} \right) \quad (20)$$

- Two indices of the set i differ from two indices of the set j, say $i_1 \neq j_1$ and $i_2 \neq j_2$ (although any ordering of the indices is allowed).

$$^pH'_{i_1,i_2,\ldots,i_p,\hat{P}(j_1,j_2,i_3,\ldots,i_p)} = A^{2,\hat{P}} H^0_{i_1i_2;j_1j_2} + B^{2,\hat{P}} H^0_{i_1i_2;j_2j_1} \quad (21)$$

A very convenient feature of this formalism is that the values of the A and B coefficients only depend on three numbers: the number of electrons (N), the number of orbitals of the basis (K), and the Spin quantum number (S).

The different orderings of the matrix element indices ($\hat{P}$ superscript in relations 19,20,21) give rise to different values of the A and B coefficients. Finally, the value of these coefficients also depends on whether one or more of the matrix element indices are repeated (appears twice).

For $p = 1, 2, 3$ the values of these coefficients have been explicitly obtained; and for large p, a set of diagramatic rules have been reported [28] in order to determine the coefficients in each case.

To calculate a $p - SRH$ once the electronic integrals are known, is therefore a very simple and rapid task.

4.2. SOME PROPERTIES OF THE $p - SRH$ MATRICES

Let us now derive from the fundamental relation (17), some properties of the $p - SRH$ matrices. Taking the *trace* of this matrix we find that:

$$tr(^p\underline{H}') = \binom{N}{p} tr(\mathcal{H}) \qquad (22)$$

Since the trace of $\mathcal{H}$ is an invariant of the system, relation (22) establishes that the trace of the $p - SRH$ matrix is also an invariant of the system.
The *eigen − values* of the $p - SRH$ matrices have the form:

$$\omega_I = \sum_{\mathcal{L}} \mathcal{E}_{\mathcal{L}}\, ^p D_{II}^{\mathcal{L}} \qquad (23)$$

where the symbol I denotes the corresponding $p - SRH$ eigen-state. This relation shows up the *average* character of the $p - SRH$ eigen-values. However, until now we have not been able to find the relation linking the ω values with the energy observables of the N-electron system.
In what follows we will focus our attention on the $p = 2$ and $p = 1$ cases which are the most useful ones. The eigen-vectors of the $p - SRH$ for these values of p are geminals and orbitals respectively. In order to simplify the interpretation of the geminals and to reduce the size of the matrices involved in the calculations it is convenient to apply to the $2 - SRH$ a linear transformation which factorizes this matrix into two blocks according to the representations of the S_2 Symmetric Group of Permutations. In this way the eigen-geminals of both blocks have a clear physical meaning since those of the symmetric block describe the space part of singlet pair states and those of the antisymmetric part describe that of triplet states.
Although the physical meaning of the 2- and 1- electron eigen-states of the $2 - SRH$ has not been established rigorously we interpret them as describing states of *two/one* electrons which in *average* can be considered independent. This interpretation was justified [29] through the analysis of the asymptotic form of the $2 - SRH$ in the coordinate representation for $K \rightarrow \infty$. In this analysis, Karwowski et al. showed that the eigen-geminals of the asymptotic $2 - SRH$ described *isolated* pairs of electrons. Another important feature of the SRH formalism is that it can be generalized by contracting $\mathcal{H}^n$. Taking now the trace of the product of these generalized SRH's matrices and the $\underline{H}^0$ one gets the $n + 1$ moment of the spectral distribution [30].

4.3. APPLICATIONS OF THE SRH THEORY

An outline of the main applications of the SHR theory is presented in this section. In 6.1 the advantage of using the eigen-vectors the 1-SRH as a basis in CI calculations is discussed. The main application until now of this theory is summarized in the following subsection. Then in 6.3 other applications which have been less developed are mentioned.

4.3.1. *Performance of the eigen-vectors of the 1-SRH as a basis in CI calculations*

A set of calculations [31] was recently carried out in order to compare the performance of the $1 - SRH$ eigen-orbitals with other known and easy to get basis sets when doing

CI calculations. Some of the results reported in (Ref. 31) for the ground state of the Water molecule (minimal basis set) are given in Table 1.

Basis set	Ground	1st excited	2nd excited	3rd excited
Without CI				
SCF	-74.963242			
RH	-74.661325			
1-SRH	-74.634825			
$Core\ H$	-73.234196			
CIS				
SCF	-74.963242	-74.481200	-74.408757	-74.349651
RH	-74.985791	-74.420894	-74.345383	-74.286213
1-SRH	-74.983375	-74.407826	-74.334565	-74.276081
$Core\ H$	-74.353107	-73.617477	-73.575950	-73.575133
CISD				
SCF	-75.012321	-74.535866	-74.450994	-74.401800
RH	-75.008348	-74.549180	-74.466995	-74.411620
1-SRH	-75.008626	-74.550633	-74.468182	-74.412463
$Core\ H$	-74.869874	-74.337375	-74.263645	-74.224523
FCI	-75.013035	-74.558053	-74.474124	-74.417688

Table 1. Configuration Interaction energy calculation with a minimal $STO - 3G$ basis of the four lower singlet states of H_2O.

These results show that convergence is favored when the basis used are either the 1−SRH or the Absar and Coleman [32,33] Reduced Hamiltonian (RH) eigen-orbitals. The other two basis used were the SCF orbitals and the core Hamiltonian eigen-orbitals ($Core\ H$)

4.3.2. The 2 − SRH independent pair model

Since the $^2D^{\mathcal{L}}$ which appear in relation (17) are not orthogonal matrices, the energy or the 2 − RDM of a particular eigen-state of the system cannot be filtered out from the contributions of the other eigen-states. To deal with this difficulty and based on the physical interpretation given above of the eigen-geminals and eigen-orbitals of the SRH, the following working hypothesis was proposed [34] in 1985:

"Let us substitute the study of the N-electron system by that of an ensemble of $\binom{N}{2}$ non interacting pairs of electrons, described by the eigen-geminals of the 2 − SRH". According to this hypothesis, each state I, will have the probability, n_I, of being occupied by a pair. Since in an N-electron state there are $\binom{N}{2}$ possible pairings the condition

$$\sum_I n_I = \binom{N}{2} \tag{24}$$

should be fulfilled.
Now, when a pair is described by a state I, its Density Matrix (DM) is

$$D^I_{i_1 i_2 ; j_1 j_2} = \frac{< I | E^{i_1 i_2}_{j_1 j_2} | I >}{2} \qquad (25)$$

and by applying this working hypothesis, the $2 - RDM$ corresponding to the N-electron eigen-state $\mathcal{L}$ can be approximated by

$$^2 \underline{D}^{\mathcal{L}} = \sum_I n^{\mathcal{L}}_I \, \underline{D}^I \qquad (26)$$

It should be noted that this relation is formally identical to the spectral resolution of the $2 - RDM$. That is, in this model, all happens as if the eigen-vectors of the 2-SRH were natural geminals.
In order to determine the n_I, we proposed two main approximations:

- Let us start by assuming that in our N-electron state $\mathcal{L}$ there is a $\hat{S}^2$ and $\hat{S}_z$ eigen-state, Λ, having a coefficient of a much higher absolute value than all the rest i.e., Λ is the dominant configuration in $\mathcal{L}$. Then the n_I are approximated as follows:

$$n^{\mathcal{L}}_I =< \Lambda | E^I_I | \Lambda > \qquad (27)$$

This is called [35] Mixed Pair State approximation (MPS). The name, which probably is not the best one, refers to the fact that the n_I, barring exceptions, has a value smaller than one, which means that the electron pair is not in the pure state I.

- A variant of relation (27) was initially proposed [34] where the n_I were determined as follows. By definition:

$$n^{\Lambda}_\lambda =< \Lambda | E^\lambda_\lambda | \Lambda > \qquad (28)$$

where Λ has the same meaning as before and λ is the orbital part of the eigenstates of the $\hat{S}_2$ operator. The two electron configurations, having a non zero value are thus selected. Now, the eigen-vector whose highest coefficient (in absolute value) is, $c_{\lambda, I}$, is allocated the occupation number:

$$n^{\mathcal{L}}_I = n^{\Lambda}_\lambda \qquad (29)$$

This approximation was denoted initially by the acronym IQG [34] and later on by IP (Independent Pairs) [35]. It gave satisfactory results in the study of the Beryllium atom and of its isoelectronic series as well as in the BeH system. The drawback of this approximation is that when the eigen-vectors are diffuse, i.e. there is more than one dominant two electron configuration per eigen-vector, the determination of the corresponding n_I is ambiguous. In order to avoid this problem the MPS approximation, which does not have this drawback, was proposed.

A detailed discussion of these and other variants was given in (Ref.35). Attention must be called to the fact that these methods are not variational which causes the energies obtained with them to be lower than those obtained with the FCI method. The counterpart to this deffect is that excited states, open-shell systems, and radicals, can be calculated with as much ease as the ground state and closed-shell systems. Also, the size of the calculation is determined solely by the size of the Hilbert subspace chosen and does not depend in principle on the number of electrons since all happens as if only two electrons were considered.

While from the energy point of view, the correlation effects seem to be overestimated, the RDM's are particularly satisfactory. Thus, when comparing the 2-RDM's obtained with these approximations for the ground state of the Beryllium atom with the corresponding FCI one, the standard deviations are: 0.00208236 and 0.00208338 for the MPS and IP respectively. For this state, which has a dominant four electron configuration of the type, $|1\bar{1}2\bar{2}>$, the more important errors, which nevertheless can be considered small, are given in table 2.

In table 2, the elements which are equal due to symmetry have been omitted.

Approximation	Element	Error
IP	$^2D_{12,13}$	0.0364
	$^2D_{22,23}$	0.0277
MPS	$^2D_{12,13}$	0.0293
	$^2D_{22,23}$	0.0274

Table 2. Highest errors found in the approximated 2-RDM with respect to FCI calculations for the ground state of the Be atom.

It can be seen that, even for this case where no ambiguity exists in applying the IP approximation, the results are slightly better with the MPS variant which seems to favor this latter approximation.

The data given in table 2 have been obtained using a double zeta basis [36] and transforming it to the basis which diagonalizes the 1-SRH matrix which as we saw in subsection 5.3 is a basis both good and simple to determine.

Analysis of the different terms of the energy

Another interesting analysis of this method can be carried out by applying a partitioning of the energy [37] which shows up the role played by the $1-RDM$, the 1-$HRDM$ and the 2-$HRDM$.

Thus, it has been shown that the energy can be partitioned as

$$\mathcal{E} = tr(\underline{F}\,^1\underline{D}) + tr(\underline{G}\,^1\underline{D}) + tr(\underline{V}\,^2\underline{\bar{D}}) - tr\bar{v} \tag{30}$$

where:

$$\bar{v}_{ij} = \sum_k [2(ij|kk) - (ik|jk)] = \sum_k v_{ij,kk} \tag{31}$$

$$F_{ij} = \epsilon_{ij} + \frac{1}{2}\sum_k {}^1D_{kk}v_{kk,ij} \tag{32}$$

$$G_{ij} = \frac{1}{2} \sum_k {}^1\bar{D}_{kk} v_{kk,ij} \tag{33}$$

$$V_{ij,kl} = (ik|jl)$$

The only operation used for obtaining this partitioning is the anticommutation rule of the fermion operators. Note, that by adding the F and G terms one falls into the unitarily invariant Absar and Coleman partitioning [32,33] which was obtained by using a Group theoretical approach.

The interesting point about relation (30) is that each of the terms has a clear physical interpretation. Thus the term involving F is a sum (for N electrons) of generalised Hartree-Fock energy levels and clearly is a one-electron term. The term involving G gives the sum (for N electrons) of the energy of *an electron* in the *average* field of *holes*. The $tr(\underline{V}\,{}^2\bar{D})$ term is clearly the repulsion energy between the *holes* and finally the $tr\bar{v}$ value shifts the zero of the energy.

In my opinion this partitioning is particularly suitable for analysing electronic correlation effects. To illustrate this point a set of calculations for the three lowest singlet states of the Beryllium atom are reported in table 3 (in all cases $-tr(\bar{v}) = -19.72037$ Hartrees).

Let us start the analysis of the results given in table 3 by commenting on the FCI one. It is interesting to note that the one-electron term energy becomes lower as the degree of excitation of the state increases. I find this result rather unexpected, since in principle, the low energy orbitals will become more empty. At any rate the stabilization caused by the $tr(\underline{F}\,{}^1\underline{D})$ term is more than counter-balanced by a large increase of the positive terms of the energy, in particular by $tr(\underline{V}\,{}^2\bar{D})$.

The most stricking features, when comparing the FCI results with the IP and MPS ones are:

- The $tr(\underline{F}\,{}^1\underline{D})$ and the $tr(\underline{V}\,{}^2\bar{D})$ terms vary for the different states in a very similar way to the FCI terms. The values obtained with the IP and MPS approximations for the term $tr(\underline{G}\,{}^1\underline{D})$ for the ground and third state show a similar behaviour to those of the FCI calculation. However while the $tr(\underline{G}\,{}^1\underline{D})$ FCI value is higher in the second state (which has a dominant open shell configuration) than in the other states the opposite happens to the IP and MPS results.

- The lowering of the energy in the ground state with respect to the FCI result is due to the $tr(\underline{V}\,{}^2\bar{D})$ term which is much too low in the two approximations. This error is compensated to a certain extent by errors in the opposite direction of the two other terms.

- In the second state the two terms depending on the 1-and 2-$HRDM$ compensate their errors to a large extent but nevertheless the $hole-electron$ positive energy is too low and a global lowering of this state energy results.

- Finally in the third state the two approximations give very similar energy values, both with higher energy than the FCI one. In each approximation, the error of the different terms compensate each other to a certain extent.

State	Method	$tr(\underline{F}^{\,1}\underline{D})$	$tr(\underline{G}^{\,1}\underline{D})$	$tr(\underline{V}^{\,2}\underline{D})$	$\mathcal{E}$
	FCI	-9.94186	11.36318	3.71190	-14.58716
First	IP	-9.93663	11.41360	3.57374	-14.66965
state	MPS	-9.92444	11.41260	3.58135	-14.65086
	FCI	-10.03603	11.52086	3.93415	-14.30139
Second	IP	-10.05360	11.37709	4.01639	-14.38049
state	MPS	-10.04528	11.37187	4.03279	-14.36099
	FCI	-10.45007	11.48956	4.69583	-13.98506
Third	IP	-10.43778	11.51296	4.69471	-13.95047
state	MPS	-10.40760	11.49535	4.67410	-13.95852

Table 3. Partitioning of the energy of the three lowest singlet states of the Be atom in Hartrees.

These results show that at least for the ground state, the correlation effects described by the term $tr(\underline{V}^{\,2}\underline{D})$ are overestimated in the IP and MPS calculations.

4.3.3. *Other applications*

Before concluding this section, it must be pointed out that there are other fields of application of the SRH formalism. Thus, Karwowski et al. have used it in the study of the statistical theory of spectra [30,38]. Also, the techniques used in developing the p-SRH algorithms have proven to be very useful in other areas such as the nuclear shell theory [39,40].

We think that the physical meaning of the SRH matrices is not yet fully understood and can therefore be considered an open field of research. It is to be expected that a more complete understanding of the SRH matrices will lead to new applications of this formalism.

5. The Contracted Schrödinger Equation

In this section I will outline a new line of research recently initiated by our group. It must be emphasized that the only points in common with the SRH formalism previously described are that no call is made upon the N-electron WF of the electronic system and that its basic formal tools are also the MCM and the RDM's.

5.1. THE NAKATSUJI AND COHEN-FRISHBERG EQUATION

The integration with respect to $N - q$ electron variables of the Schrödinger equation was reported simultaneously by H. Nakatsuji [41] and by L. Cohen and C. Frishberg [42] in 1976. The form of this equation (NCF) for $q = 2$ is:

$$E\rho_2 = \hat{H}_2\rho_2 + 3\int\left(\hat{H}_3 + \sum_{i=1}^{2}\hat{\Omega}_{i,3}\right)\rho_3\ dx_3 + \begin{pmatrix}4\\2\end{pmatrix}\int\hat{\Omega}_{3,4}\ \rho_4\ dx_3\ dx_4 \qquad (34)$$

where

$$\rho_2 = \binom{N}{2} \int \rho_N(x_1, ..., x_2, x_{2+1}, ..., x_N; x_1', ..., x_2', x_{2+1}, ..., x_N) \; dx_{2+1}, ..., dx_N \quad (35)$$

is the 2-RDM written in first quantization language. The symbols $\hat{H}_2$, and $\hat{H}_3$ denote the Hamiltonians of two and three electrons respectively and $\hat{\Omega}_{i,j}$ is the two electron repulsion operator.

Since this integro-differential equation depends not only on ρ_2 but also, through the two integral terms, on ρ_3 and ρ_4, it is indeterminate [43].

An important property of the NCF equation is that in it the variational principle is taken implicitly into account [42,44].

5.2. ORBITAL REPRESENTATION OF THE CONTRACTED SCHRÖDINGER EQUATION ($CSchE$)

The matrix form in a spin-geminal representation ($CSchE$) of equation (34) was obtained [18] in 1985 by applying the MCM.

The interest of contracting the matrix form of the Schrödinger equation by employing the MCM, is that the resulting equation is easy to handle since only matrix operations are involved in it. Thus, when the MCM is employed up to the two electron space, the geminal representation of the $CSchE$ has the form [35]:

$$\mathcal{E}\,{}^2D_{pq,rs} = ({}^0\underline{H}\,{}^2\underline{D})_{pq,rs} + 6 \sum_{i,j,k,l} {}^0H_{ij,kl}\,{}^4D_{rsij,pqkl}$$

$$+3 \sum_{i,l,k} \left({}^0H_{iq,kl}\,{}^3D_{rsi,plk} + {}^0H_{ip,kl}\,{}^3D_{rsi,lqk} \right) \quad (36)$$

where the symbols have the same meaning as in the preceding sections. It must be pointed out, that the contraction can also be carried out, up to the first order and the result is:

$$\mathcal{E}\,{}^1D_{pq} = 2 \sum_{j} ({}^2\underline{D}\,{}^0\underline{H})_{pj,qj} + 3 \sum_{i,j,k,l} {}^0H_{ij,kl}\,{}^3D_{pij,qkl} \quad (37)$$

5.3. ITERATIVE SOLUTION OF THE $CSchE$

It was suggested [35,45] that the indeterminacy of the $CSchE$ could be removed by replacing in it the 3- and the 4-RDM's by their corresponding approximations evaluated within the SRH formalism. After this replacement is performed, the matrix equation can be solved with the help of relation (10) and

$$tr({}^2\underline{D}) = \binom{N}{2} \quad (38)$$

as auxiliary conditions.

Recently, a more powerful approach has been initiated. The different steps involved in the procedure just proposed for solving the $CSchE$ are:

- From an initial 2-*RDM* the corresponding 3- and 4- order *RDM*'s are approximated by using a method which will be described in the following section.

- Then, all the approximated *RDM*'s are replaced in the r.h.s. of equation (36) so that its three terms are added into a matrix, say $\underline{R}$, and relation (36) becomes:

$$\mathcal{E}\,^2 D_{pq,rs} = R_{pq,rs} \qquad (39)$$

- By taking the trace of both sides of equation (39) one obtains $\mathcal{E}$ since

$$\binom{N}{2}\mathcal{E} = tr(\underline{R}) \qquad (40)$$

- The following step is to divide $\underline{R}$ by $\mathcal{E}$ which gives a new 2-*RDM* from which the procedure can start again.

All these steps are built into an iterative procedure whose success pivots on the approximation of the higher order *RDM*'s in terms of the 2-*RDM*. This important part of the method will be addressed in the next section.

5.4. APPROXIMATING AN *RDM* IN TERMS OF THE LOWER ORDER ONES

As has been mentioned, the iterative procedure for solving the 2-*CSchE* will only work if sufficiently precise approximations of the 3- and 4-order *RDM*'s in terms of the 2-*RDM* can be obtained. Since the method is based on the N-representability relations, the subsection 8.1 is dedicated to discuss these fundamental equations. Then in 8.2 the method will be outlined and some examples will be given.

5.4.1. *The N-representability conditions*

The basic relations for studying the properties of the *RDM*'s are the anticommutation/commutation relations of groups of fermion operators since their expectation values give a set of N-representability conditions of the *RDM*'s. Thus,

- *The first order condition*

  From the fundamental rule of anticommutation of an annihilator with a creator operator it follows, in our orbital representation, that:

$$^1\bar{D}_{ji} + {}^1D_{ij} = 2\ \delta_{ij} \qquad (41)$$

  Since both the *RDM*'s and the *HRDM*'s are positive matrices, this relation says that the eigen-value μ_i of the 1-*RDM*, must be $0 < \mu_i < 2$ which is the well known ensemble N-representability condition for the 1-*RDM* [10] represented in an orbital basis (in a spin-orbital representation the upper bound would be 1 instead of 2).

• *The second order condition*

From the commutation of two annihilator with two creator operators follows the also well known Q-condition [9] for the 2-RDM. In our notation, this condition takes the form:

$$
\begin{aligned}
{}^2\bar{D}_{j_1 j_2; i_1 i_2} - {}^2 D_{i_1 i_2; j_1 j_2} &= 2\,\delta_{i_1 j_1}\delta_{i_2 j_2} - \delta_{i_1 j_2}\delta_{i_2 j_1} \\
&\quad - 2\left(\delta_{i_2 j_2}\,{}^1 D_{i_1; j_1} + \delta_{i_1 j_1}\,{}^1 D_{i_2; j_2}\right) \\
&\quad + \left(\delta_{i_1 j_2}\,{}^1 D_{i_2; j_1} + \delta_{i_2 j_1}\,{}^1 D_{i_1; j_2}\right)
\end{aligned}
\tag{42}
$$

and replacing the Krönecker deltas by their value according to relation (41) one finds [46]:

$$
\begin{aligned}
{}^2\bar{D}_{j_1 j_2; i_1 i_2} - {}^2 D_{i_1 i_2; j_1 j_2} &= \left(\frac{{}^1\bar{D}_{i_1; j_1}\,{}^1\bar{D}_{i_2; j_2}}{2} - \frac{{}^1\bar{D}_{i_2; j_1}\,{}^1\bar{D}_{i_1; j_2}}{4}\right) \\
&\quad - \left(\frac{{}^1 D_{i_1; j_1}\,{}^1 D_{i_2; j_2}}{2} - \frac{{}^1 D_{i_2; j_1}\,{}^1 D_{i_1; j_2}}{4}\right)
\end{aligned}
\tag{43}
$$

Note, that in this last relation, the part involving $HRDM$'s and that involving RDM's have the same structure.

• *General N-order condition*

The aim of the following discussion under this heading is not to describe the formalism but merely to outline the ideas on which the method for approximating a p-RDM from the q-RDM's with $q < p$ is based. Nevertheless, in order to avoid using vague or imprecise arguments the essential theoretical background supporting the leading ideas must also be included here. The reader interested in going beyond this sketchy discussion is referred to a recent paper [47] where all the details are reported.

The result of commuting/anticommuting (for N even/odd) N annihilator operators with N creator operators is:

$$
{}^N\bar{D}^{\mathcal{L}}_{\Lambda\Omega} \pm {}^N D^{\mathcal{L}}_{\Omega\Lambda} = \delta_{\Lambda\Omega} - tr({}^1\underline{D}^{\Lambda\Omega}\,{}^1\underline{D}^{\mathcal{L}}) + tr({}^2\underline{D}^{\Lambda\Omega}\,{}^2 D^{\mathcal{L}}) - \cdots \\
\pm tr({}^{N-1}\underline{D}^{\Lambda\Omega}\,{}^{N-1}\underline{D}^{\mathcal{L}})
\tag{44}
$$

where the $\Lambda\Omega$ symbols are N-electron configurations. This relation is very elegant and compact but the following, in the orbital representation (obtained by inference [47]), is more practical for our purpose:

$$
{}^N\bar{D}_{i_1 i_2 \ldots i_N; j_1 j_2 \ldots j_N} \pm {}^N D_{i_1 i_2 \ldots i_N; j_1 j_2 \ldots j_N} = \frac{2^N}{N!}\left[\sum_{i=0}^{N-1}(-1)^i\left(\frac{i!}{2^i}\right)\sum_{C_N}\Gamma_{C_N}\sum_{\hat{P}_{C_N}}{}^N G^i_{\hat{P}_{C_N}}\right]
\tag{45}
$$

where:

- C_N represents the classes of the Symmetric Group of Permutations S_N
- $\hat{P}$ are the permutations (of the indices of the annihilator operators) belonging to a particular class
- p_C represents the parity of the permutations belonging to class C
- The symbol Γ is given by:

$$\Gamma_{C_N} = (-1)^{p_{C_N}} \left(\frac{1}{2^{p_{C_N}}} \right) \qquad (46)$$

- The symbol ${}^N G^i_{\hat{P}_{C_N}}$ describes a sum of terms. Each of these terms is a product of $(N - i)$ Krönecker deltas with a i-RDM element. Now, the terms whose addition is represented by a G symbol are those where the indices are ordered according to the permutation $\hat{P}$ of class C_N
  For instance, for $N = 3$ and $P_{[2]_3} = (12)$, the G symbols are:

$$
\begin{aligned}
{}^3 G^0_{(12)} &= \delta_{i_1 j_2} \delta_{i_2 j_1} \delta_{i_3 j_3} \\
{}^3 G^2_{(12)} &= \delta_{i_1 j_2} {}^2 D_{i_2 i_3 ; j_1 j_3} + \delta_{i_2 j_1} {}^2 D_{i_1 i_3 ; j_2 j_3} + {}^2 D_{i_1 i_2 ; j_2 j_1} \delta_{i_3 j_3} \qquad (47)
\end{aligned}
$$

Relations (44,45) describe the general form of the N-order condition; However, some terms must be eliminated from relation (45) because they do not occur when the anticommutation/commutation operations are carried out explicitly. We call these terms $spin-forbidden$ because in all of them the spin correspondence which should exist between the creator and the annihilators forming the p-RO (which generates the p-RDM) is not maintained. These spin-forbidden terms are those having a transposition of at least two indices in their p-RDM. For instance:

$$ {}^2 D_{i_1 i_2 ; j_2 j_1} \delta_{i_3 j_3} \qquad (48) $$

which is the third term of ${}^3 G^2_{(12)}$, is spin-forbidden and must be eliminated.

An equivalent N-order equation having the $same\ structure$ for the particle part and for the holes part (in a similar way as in (43)), may also be inferred. This equation has the form:

$$
{}^N D_{i_1 i_2 \ldots i_N ; j_1 j_2 \ldots j_N} \pm {}^N \bar{D}_{i_1 i_2 \ldots i_N ; j_1 j_2 \ldots j_N} = (-1)^N \left(\frac{N-1}{N!} \right) \sum_{C_N} \Gamma_{C_N} \sum_{\hat{P}_{C_N}} {}^N \mathcal{F}^0_{\hat{P}_{C_N}}
$$

$$
+ \sum_{i=2}^{N-1} (-1)^{N-i+1} \left(\frac{i!}{N!} \right) \sum_{C_N} \Gamma_{C_N} \sum_{\hat{P}_{C_N}} {}^N \mathcal{F}^i_{\hat{P}_{C_N}} \pm \text{Hole Part}
$$

$$ \qquad (49) $$

where all the symbols have the same meaning as in relation (45) except for ${}^N \mathcal{F}^i_{\hat{P}_{C_N}}$. This new symbol, like G^i, also describes a sum of terms. Each of these terms is a product of $(N - i)$ 1-RDM elements with an i-RDM element. (i.e. the 1-RDM plays a similar role to the Krönecker deltas). Thus, for instance:

$$^3\mathcal{F}^0_{(12)} = {}^1D_{i_1 j_2}{}^1D_{i_2 j_1}{}^1D_{i_3 j_3}$$

$$^3\mathcal{F}^2_{(12)} = {}^1D_{i_1 j_2}{}^2D_{i_2 i_3; j_1 j_3} + {}^1D_{i_2 j_1}{}^2D_{i_1 i_3; j_2 j_3} + {}^2D_{i_1 i_2; j_2 j_1}{}^1D_{i_3 j_3} \qquad (50)$$

As in (47) the third term of $^3F^2_{(12)}$

$$^2D_{i_1 i_2; j_2 j_1}{}^1D_{i_3 j_3} \qquad (51)$$

is spin-forbidden and must be eliminated.

The inference process leading to equations (45) and (49) was carried out with the help of a set of graphs specially suited for operating with RO's. This graphical method has been described in several recent publications [26,27,47] and would excessively lengthen this paper; therefore, I would like to mention its usefulness without elaborating.

The interest of relation (49) lies in that the *holes* and the *particle* parts of the equation, have the same structure.

Equations (45) and (49) stress the direct connexion existing between the elements and classes of the Symmetric Group of Permutations and the terms derived by commuting/anticommuting groups of fermion operators after summing with respect to the spin variables.

Two important facts concerning the set of relations given above are that all the N-representability relations known to us, can be derived from (45) (or (44) in a spin-space representation) by varying the value of N and relation (49) condenses them all.

It is interesting to note that relation (45) guaranties that the N-electron state of reference (whose superindex has been omitted) is antisymmetric since the RO's involved on the l.h.s of these equations operate on N-electron states. Now, by contracting this equation to a p-electron space an N-representable equation is obtained (by construction). In view of this, I hoped that a relation obtained in such a way would be a sufficient N-representability condition or at least more stringent than the (45) equation for $N = p$. Now the contraction of equation (45) gives exactly the same equation where N has been replaced by p. On the other hand the contraction of equation (49) gives a very complicated equation where partial traces of RDM's of orders $(p + 1),(p + 2),....(N - 1)$ appear. This equation although difficult to analyse may prove to be useful and it is being studied at the moment.

5.4.2. *Approximation proposed*

The method for approximating an RDM in terms of the lower order ones is based on equation (49). The working hypothesis which has been put forward [46] is:

"Let us assume that *Holes* and *Particles* are totally different objects. If this assumption were true, equation (49) could be exactly decoupled into two equations, one involving RDM's of different orders and the other, of similar structure, linking $HRDM$'s of different orders".

This hypothesis, given that *Holes* and *Particles* are related through the N-representability conditions, is not true. On the other hand, by taking into account several auxiliary conditions, very good approximations have recently been obtained by using this method.

In order to see an example of how these suplementary conditions are imposed let us consider the approximation of a 3-RDM in terms of the 2-RDM. Since an RDM cannot have any negative diagonal element when such an element occurs it is put equal to zero. Until now the negative diagonal elements found were of the type $M_{iij;iij}$, where M represents the approximated 3-RDM. By comparing the approximated matrix with the exact one it was apparent that the deffect in $M_{iij;iij}$ was compensated very closely by an excess in the element $M_{jji;jji}$ therefore this element was corrected in accordance. After this correction was performed the new elements (M') had the value:

$$M'_{iij;iij} = 0$$
$$M'_{jji;jji} = M_{jji;jji} + M_{iij;iij} \tag{52}$$

It has been shown [48] that the related off-diagonal elements $M_{iij;jji}$ and $M_{jji;iij}$ obey definite symmetry relations which must be maintained after the corrections indicated above have been applied.

System and State	Matrix Element	Error
Be Ground State	$M_{113,113}$	-0.00002
	$M_{134,134}$	0.00002
	$M_{223,223}$	-0.00002
	$M_{223,232}$	0.00001
	$M_{233,233}$	0.00002
	$M_{233,323}$	-0.00001
	$M_{244,244}$	0.00001
Be 1st.Excited State	$M_{112,112}$	-0.00001
	$M_{122,122}$	+0.00002
	$M_{223,223}$	+0.00001
	$M_{244,244}$	-0.00001
H_2O Ground State	$M_{226,226}$	-0.00065
	$M_{236,236}$	-0.00048
	$M_{237,237}$	-0.00049
	$M_{246,246}$	-0.00072
	$M_{346,346}$	-0.00117
	$M_{447,447}$	-0.00114
	$M_{556,556}$	-0.00048

Table 4. Highest errors with respect to the exact FCI in the 3-RDM elements of two states of Be and the ground state of H_2O (minimal basis set).

These symmetry relations impose that the following corrections should also be introduced:

- $M_{iij,iji}$, $M_{iij,ijj}$, $M_{ijj,iij}$, and $M_{iij,jji}$ must also be made equal to zero.

- The value $M_{iij;iij}/2$ should be subtracted from $M_{jji,jij}$.

Under these conditions, the 3-RDM of the three lower states of the Beryllium atom and the two lower ones of the Water molecule were determined [48] by taking as initial data the 2-RDM obtained in a Full Configuration Interaction. In Table 4 some of these results are given and as can be seen they are very satisfactory.

The results for approximating the 4-RDM in terms of the lower order RDM's are slightly inferior but still very good. In consequence, I expect that the iterative procedure proposed in the previous section may prove to be a realistic one.

In spite of the good results obtained we continue our search for simple auxiliary conditions directed at ensuring that the approximated matrix is positive and that its trace has the correct value. This search is mainly focused at improving the quality of the 2-RDM obtained in terms of the 1-RDM, which at the moment is the less precise procedure [46]. When this latter aim is fulfilled we expect that the iterative solution of the 1-order $CSchE$ will also be successful although in this $CSchE$ the information carried by the Hamiltonian only influences the result in an average way which probably will retard the convergence.

6.Conclusion

The two previous sections outline the main formal and applicative results obtained in our search for a theoretical framework where the number of variables which are explicitly taken into account would be as small as the observables allow. This framework should permit the use of different levels of approximation for the Hamiltonian operator and its orbital representation. That is, the size of the basis set and the kind of approximation used for the integrals should not be predetermined by the formalism.

Both lines of research are far from being closed and we are confident that their development will contribute useful results. However, without considering future performances I think that it can already be said that it is a good strategy to project the future Quantum Chemical methodology in such a way that the WF is by-passed and the 2-RDM or (better still but more difficult) the 1-RDM are directly determined.

Acknowledgements

Investigación Científica y Técnica del Ministerio de Educación y Ciencia under project PB90-0092.

References

1. P. A. M. Dirac, *Proc. Cambridge Phil. Soc.* **27**, 240 (1931).
2. K. Husimi, *Proc. Phys. Soc. Japan* **22**, 264 (1940).

3. P. O. Löwdin, *Phys. Rev.* **97**, 1474 (1955).

4. A. J. Coleman, *Rev. Mod. Phys.* **35**, 668 (1963).

5. J. E. Mayer, *Phys. Rev.* **100**, 1579 (1955).

6. R. H. Tredgold, *Phys. Rev.* **105**, 1421 (1957).

7. R. V. Ayres, *Phys. Rev.* **111**, 1453 (1958).

8. C. A. Coulson, *Rev. Mod. Phys.* **32**, 170 (1960).

9. C. Garrod, J. K. Percus, J. Math. Phys. **5**, 1956 (1964).

10. P. O. Löwdin, *J. Phys. Chem.* **61**, 55 (1957).

11. H. Kummer, *J. Math. Phys.* **8**, 2063 (1967).

12. R. Erdahl, in Density Matrices and Density Functionals Proceedings of the A.J. Coleman Symposium, Kingston, Ontario, 1985, edited by R. Erdahl and V. Smith (Reidel, Dordrecht, 1987), p. 51, and references therein.

13. J. E. Harriman, *Phys. Rev. A* **24**, 680 (1981).

14. R. McWeeny, *Rev. Mod. Phys.* **32**, 335 (1960).

15. D. Ter Haar, Rept. Prog. Phys. **24**, 304 (1961).

16. J. Hinze and J.T. Broad, in CI-energy expressions in Terms of the Reduced Density Matrices Elements of a General Reference, vol 22 of Lecture Notes in Chemistry edited by J. Hinze (Springer, Berlin, 1981), p. 332.

17. C. Valdemoro, *Phys. Rev. A* **31**, 2114 (1985).

18. C. Valdemoro, in Density Matrices and Density Functionals, Proceedings of the A.J. Coleman Symposium, Kingston, Ontario, 1985, edited by R. Erdahl and V. Smith (Reidel, Dordrecht, 1987).

19. M. Rosina and B. Golli and R. M. Erdahl, in Density Matrices and Density Functionals Proceedings of the A.J. Coleman Symposium, Kingston, Ontario, 1985, edited by R. Erdahl and V. Smith (Reidel, Dordrecht, 1987) and references therein.

20. Density Matrices and Density Functionals Proceedings of the A.J. Coleman Symposium, Kingston, Ontario, 1985, edited by R. Erdahl and V. Smith (Reidel, Dordrecht, 1987).

21. M. Moshinsky, in Group Theory and the Many-Body Problem (Gordon and Breach, New York, 1968).

22. J. Paldus, in Theoretical Chemistry: Advances and Perspectives, edited by H. Eyring and D.J. Henderson (Academic Press, New York, 1976), p. 131.

23. W. Kutzelnigg, *J. Chem. Phys.* **82**, 4166 (1985).

24. W. Duch and J. Karwowski, *Comput. Phys. Rep.*, **2**, 95 (1985).

25. J. Karwowski, W. Duch, C. Valdemoro, *Phys. Rev. A* **33**, 2254 (1986).

26. C. Valdemoro, L. Lain, A. Torre, in Structure, Interaction and Reactivity, edited by S. Fraga (Elsevier, Amsterdam, 1991).

27. J. Planelles, C. Valdemoro and J. Karwowski, *Phys. Rev. A* **41**, 2391 (1990).

28. J. Planelles, C. Valdemoro, J. Karwowski, *Phys. Rev. A* **43**, 3392 (1991).

29. J. Karwowski, C. Valdemoro and L. Lain, *Phys. Rev. A* **39**, 4967 (1988).

30. J. Karwowski, in Basic Aspects of Quantum Chemistry, edited by R. Carbó (Elsevier, Amsterdam, 1989).

31. F. Colmenero, P. Viciano, J. Planelles, C. Valdemoro, *Anal. Real Soc. Fis.* xx xxx (1992) (Proceedings of the I South European Conference on Atomic and Molecular Physics 1992).

32. A.J. Coleman and I. Absar, *Int. J. Quantum Chem.* **XV**, 1279 (1980).

33. I. Absar, *Int. J. Quantum Chem.* **XIII**, 777 (1978).

34. C. Valdemoro, *Phys. Rev. A* **31**, 2123 (1985).

35. C. Valdemoro, in Structure, Interaction and Reactivity, edited by S. Fraga (Elsevier, Amsterdam, 1991).

36. E. Clementi and C. Roetti, *Atomic Data and Nuclear Data Tables*, **14**, numbers: 3-4 (1974).

37. C. Valdemoro, M. Reguero and L. Lain, *Chem. Phys. Lett.* **147**, 219 (1988).

38. J. Karwowski and M. Bancewicz, *J. Phys. A* **20**, 6309 (1987).

39. M. Nomura, *Phys. Rev. A* **37**, 2709 (1988).

40. J. Karwowski and C. Valdemoro, *Phys. Rev. A* **37**, 2712 (1988).

41. H. Nakasutji, *Phys. Rev. A* **14**, 41 (1976).

42. L. Cohen and C. Frishberg, *Phys. Rev. A* **13**, 927 (1976).

43. J.E. Harriman, *Phys. Rev. A* **19**, 1893 (1979).

44. H. Schlosser, *Phys. Rev. A* **15**, 1349 (1977).

45. C. Valdemoro, in Basic Aspects of Quantum Chemistry, edited by R. Carbó (Elsevier, Amsterdam, 1989).

46. C. Valdemoro, *Phys. Rev. A* **45**, 4462 (1992).

47. F. Colmenero, C. Perez del Valle, C. Valdemoro, *Phys. Rev. A* **47**, 971 (1993).

48. F. Colmenero, C. Valdemoro, *Phys. Rev. A* **47**, 979 (1993).

The Real Generators of the Unitary Group

P. CASSAM-CHENAI
Equipe d'Astrochimie Quantique, Laboratoire de Radioastronomie
E.N.S., 24 rue Lhomond, F-75231 Paris Cedex 05, France

This note is dedicated to G. Berthier who has always emphasized the importance of a rigorous use of the language in scientific papers. I would like to expose here an "abus de langage" regarding "the generators of the unitary group $U(n)$", usually denoted by E_{ij} which dates back to their introduction in quantum chemistry [1]. As a matter of fact, in the original paper, the author concedes that they are not the generators of $U(n)$ but those of the linear group $GL(n,\mathbb{C})$; however, as far as I am aware, none of his followers has ever mentioned this point.

The n^2 generators $(E_{ij})_{i,j}$, which are chosen such that :

$$E_{ij} = E_{ji}^{\dagger} \tag{1}$$

are Hermitian only for $i = j$. They generate, using complex numbers, the Lie algebra of $GL(n,\mathbb{C})$. This algebra contains the Lie algebra of $U(n)$, but it is indeed much larger. The Lie algebra of $U(n)$ can be generated more specifically, using real numbers, with n^2 Hermitian generators denoted $(A_{ij})_{i\leq j}$, $(A'_{ij})_{i<j}$.

The generators $((A_{ij})_{i<j}$, $(A'_{ij})_{i<j}$, $(A_{ij})_{i=j})$, and the generators $((E_{ij})_{i<j}$, $(E_{ji})_{i<j}$, $(E_{ij})_{i=j})$ are related in the same way as the angular moment operators (J_x, J_y, J_z) and (J_+, J_-, J_z) :

$$A_{ij} = \frac{E_{ij} + E_{ji}}{1 + \delta_{ij}} \tag{2a}$$

$$A'_{ij} = i\,(E_{ij} - E_{ji}) \tag{2b}$$

where δ_{ij} is the Kronecker symbol, and $i = \sqrt{-1}$.
It is convenient to extend these relations to all couples (i,j), and to write compactly :

$$E_{ij} = \frac{1}{2}\,[(1+\delta_{ij})\,A_{ij} - i\,A'_{ij}] \tag{3}$$

The structure constants for Hermitian generators are purely imaginary :
$$[A_{ij}, A_{kl}] = -\,i\,\{\delta_{kj}\,A'_{il} + \delta_{ik}\,A'_{jl} + \delta_{lj}\,A'_{ik} + \delta_{il}\,A'_{jk}\} \tag{4a}$$

77

Y. Ellinger and M. Defranceschi (eds.), Strategies and Applications in Quantum Chemistry, 77–78.
© 1996 Kluwer Academic Publishers.

$$[A_{ij}, A'_{kl}] = i\frac{\delta_{kj}(1+\delta_{il})A_{il} + \delta_{ik}(1+\delta_{jl})A_{jl} - \delta_{lj}(1+\delta_{ik})A_{ik} - \delta_{il}(1+\delta_{jk})A_{jk}}{1+\delta_{ij}} \tag{4b}$$

$$[A'_{ij}, A'_{kl}] = i\{\delta_{kj} A'_{il} - \delta_{ik} A'_{jl} - \delta_{lj} A'_{ik} + \delta_{il} A'_{jk}\} \tag{4c}$$

The fundamental representation of the generators as $n \times n$ matrices is easily obtained; the matrix elements have the following expressions :

$$A_{ij}(\alpha,\beta) = \frac{\delta_{\alpha i} \delta_{\beta j} + \delta_{\beta i} \delta_{\alpha j}}{1 + \delta_{ij}} \tag{5a}$$

$$A'_{ij}(\alpha,\beta) = i(\delta_{\alpha i} \delta_{\beta j} - \delta_{\beta i} \delta_{\alpha j}) \tag{5b}$$

The remarkable fact about the Hamiltonian :

$$H = \sum_{i,j} <i|w|j> E_{ij} + \frac{1}{2} \sum_{i,j,k,l} <ij|v|kl> E_{ik} E_{jl} \tag{6a}$$

is that it decomposes on $(A_{ij})_{i \leq j}$, $(A'_{ij})_{i<j}$, with real numbers, even when the integrals are complex (case of an electromagnetic field, of a molecule whose symmetry group has irreducible representations which are not realizable over real numbers...) :

$$H = \sum_{i \leq j} Re(<i|w|j>) A_{ij} + \sum_{i<j} Im(<i|w|j>) A'_{ij} + \frac{1}{2} \{ \sum_{i \leq k,j \leq l} Re(<ij|v|kl>+<kj|v|il>) A_{ik} A_{jl}$$
$$- \sum_{i<k,j<l} Re(<ij|v|kl>-<kj|v|il>) A'_{ik} A'_{jl} + \sum_{i \leq k,j<l} Im(<ij|v|kl>)+<kj|v|il> A_{ik} A'_{jl}) +$$
$$\sum_{i<k,j \leq l} Im(<ij|v|kl>-<kj|v|il>) A'_{ik} A_{jl} \} \tag{6b}$$

with w (respectively v) one-electron (respectively two-electrons) Hermitian operator and $Re(x)$ (respectively $Im(x)$) real part (respectively imaginary part) of the complex number x.

So the genuine generators of the unitary group have original properties and do not deserve to be forgotten. It would seem weird to build the theory of angular momentum using only J_+, J_- with no mention of J_x and J_y. It is equally surprising that only the E_{ij} 's appear in the theory of the unitary group. In short, in the traditional approach, one builds the Lie algebra of the linear group but uses only the Lie subalgebra corresponding to the unitary group. A more satisfactory approach would consist in generating the Lie algebra of the unitary group only, using its real generators, then to define in this algebra with Eq.(3) the rising and lowering operators $(E_{ij})_{i<j}$, $(E_{ji})_{i<j}$.

References

1.   J. Paldus, *J. Chem. Phys.* **61**, 5321 (1974).

Convergence of Expansions in a Gaussian Basis

W. KUTZELNIGG

Lehrstuhl für Theoretische Chemie, Ruhr-Universität Bochum,
Universitätsstr. 150, D-4630 Bochum, Germany

1. Introduction

Few papers have had as much impact on the progress of ab-initio quantum chemistry as that of Boys [1] where he proposed to use Gaussians (GTOs) as basis sets. The great breakthrough of ab-initio theory would never have been possible without the invention of Gaussians. Nevertheless, even nowadays it is difficult to explain to a beginner why one should rely on Gaussians, which have the wrong behaviour both near the nuclei and very far from them. The ease with which two-electron integrals over GTOs can be computed is certainly an argument. However, if one has thought a little bit on the importance of choosing basis sets with the right behaviour at the singularities of the Hamiltonian [2], one cannot but be deeply surprised that expansions in GTOs converge decently well in spite of their failure at the singularities of the Hamiltonian.

To appreciate this point somewhat better it is useful to compare three types of Gaussian basis sets, (a) a set of Gaussians with common orbital exponents (for one l) but a sequence of principle quantum-numbers

$$\tilde{\psi}_{nlm} = N r^{n-1} e^{-\eta_l r^2} Y_l^m(\vartheta, \varphi); \ n > l; \ n - l = 1, 3, 5, 7... \tag{1.1}$$

(We consider here only the case of a single center). (b) the same set (1.1) but with $n - l = 1, 2, 3, 4, ...$, (c) a set of Gaussians with the lowest possible n for each l, but with a sequence of orbital exponents $\eta(l, k)$

$$\psi_{klm} = N r^l e^{-\eta(l,k)r^2} Y_l^m(\vartheta, \varphi) \tag{1.2}$$

Sets of orbital exponents $\eta(l, k)$ have been proposed mainly by Huzinaga [3], van Duijneveldt [4], Pople et al. [5]. A systematic construction of basis sets of arbitrary dimension is possible in terms of the 'even tempered' concept of Ruedenberg et al. [6,7], or of some more sophisticated generalizations [8,9,10]. For a recent comprehensive review on basis sets see Feller and Davidson [11].

It does not make a significant difference that in practice one uses 'cartesian Gaussians' rather than Gaussians with explicit inclusion of spherical harmonics. One

79

Y. Ellinger and M. Defranceschi (eds.), Strategies and Applications in Quantum Chemistry, 79–101.
© 1996 *Kluwer Academic Publishers.*

should also mention that there is a fourth type of basis sets (d), namely that of Gaussian lobes [12,13] i.e. functions of type (1.2) with only $l = m = 0$ but with centers spread over the molecule, not only at the position of the nuclei. These don't differ basically from case (c).

It has been shown [14] for both types of basis sets (1.1) and (1.2) that a given set of dimension n can be regarded as a member B_n of a family of basis sets that in the limit $n \to \infty$ become complete both in the ordinary sense and with respect to a norm in the 1st Sobolev space – which is the condition for the eigenvalues and eigenfunctions of a Hamiltonian to converge to the exact ones. However, as to the speed of convergence the two basis sets (1.1) and (1.2) differ fundamentally.

In a careful study of basis sets of type (1.1) applied to the ground state of the hydrogen atom Klahn and Morgan [15] were able to show that the error of the energy goes as $n^{-3/2}$ (n being the dimension of the basis) for fixed η_0. By optimization of η one can achieve [16] that the error goes as $\sim n^{-2}$. Anyhow this rate of convergence is as bad as one can imagine and it makes basis set (1.1) absolutely useless. Convergence as an inverse-power law with a small exponent generally prevents accurate calculations, as is known from the slow convergence of the partial-wave expansion for the interelectronic coordinate (equivalently the convergence of a CI for an atom with the highest angular equantum number l in the basis set included), where the error goes as $(l + 1)^{-3}$. Inclusion of a single term with the right behaviour at the Coulomb singularity (a 'comparison function' [2]) improves the rate of convergence, such that the error goes as n^{-4} for the expansion of the H-atom ground state in basis (1.1) [16] or as $(l + 1)^{-7}$ for the convergence of a CI [17].

If one includes functions with $n - l$ even in (1.1) (i.e. one uses set b) the basis is formally overcomplete. However the error decreases exponentially with the size of the basis [2,16]. Unfortunately for this type of basis the evaluation of the integrals is practically as difficult as for Slater type basis functions, such that basis sets of type (b) have not been used in practice.

The rate of convergence of expansions in the basis (1.2) has received little attention except for purely numerical studies [3,7,8,9,16] which indicated that the convergence is at least (unlike for bais set of type) *not* frustratingly slow. Rather detailed studies were performed for the even-tempered basis set, i.e. for exponents constructed from two parameters α_l and β_l (for each l)

$$\eta(l, k) = \alpha_l \beta_l^k \qquad (1.3)$$

In a numerical study of basis sets (a), (b) and (c) for the H atom ground state W. Klopper and the present author [16] found that for the basis (c) the error goes as

$$|E_n - E| \sim \exp(-c\sqrt{n}) \qquad (1.4)$$

i.e. the convergence is not exponential (which would be ideal, i.e. generally the case for a basis that describes the singularities correctly) but almost so. This does not only hold for the energy, but for other properties as well. However there are properties for which the limit $n \to \infty$ does not yield the correct result, e.g.

$(\partial\psi/\partial r)_{r=0}$ which is $-\alpha$ for the exact H ground state wave function, but which vanishes for the expansion in (1.2) for all finite n.

Similarly $(\partial^2\psi/\partial r^2)_{r=0}$ is equal to α^2, while this second derivative is negative for any finite expansion with an apparent divergency to $-\infty$ for $n \rightarrow \infty$. Some properties like the density at the nucleus and the variance of the energy converge very slowly to the exact values. These are, nevertheless, relatively minor defects.

Again by adding to the basis at least one function that has the correct behaviour at $r = 0$, e.g.

$$re^{-\gamma r^2} \tag{1.5}$$

the convergence can be speeded up – and the last-mentioned defects can be removed [10,16]. However, the improvement is much less spectacular than for basis (1.1) – unless one is interested in the density at the nucleus or the variance of the energy.

There are hints [9,10,18] that the rate of convergence for basis sets of type (1.2) is even better than (1.4), if one uses better optimized basis sets than those of even tempered type (1.3),

$$|E_n - E| \sim \exp(-an^\gamma); \quad \frac{1}{2} < \gamma < 1 \tag{1.6}$$

and that the same convergence pattern is found for the expansion of $e^{-r}\ln r$ as for e^{-r} [18].

There is no doubt that the convergence behaviour of standard Gaussians is much better than one should have expected in view of their failure at $r \rightarrow 0$.

What is the fundamental difference of basis sets of type (1.1) and (1.2)? Without claiming to give a definite answer we can say that the expansion in the basis (1.1) is closely related to the expansion in terms of Laguerre functions, i.e. in a typical orthogonal basis and that a theory much like that for Fourier series applies. There it generally holds that the singularities of the function to be expanded determine the rate of convergence [19]. An expansion in the basis (1.2) can hardly be traced back to something like a Fourier series. It must rather be viewed as a discretization of the integral representation of an exponential (or another exponential-like) function.

$$e^{-\alpha r} = \frac{\alpha}{2\sqrt{\pi}} \int_0^\infty s^{-3/2} \exp\{-\frac{\alpha^2}{4s} - sr^2\}ds \tag{1.7}$$

and entirely different features determine the error. (As to a direct application of a numerical discretization of the integral transformation (1.7) see ref. 20).

To get analytic results for the convergence behaviour of an expansion in a Gaussian basis we shall proceed in two steps.

1. We replace the integral (1.7) by an integral from s_1 to s_2 rather than from 0 to ∞. The errors due to this restriction of the integration domain – the *cut-off errors* – can easily be estimated.

2. We replace the integral from s_1 to s_2 by a sum over a regular grid. We do this by applying first a variable transformation (to be specified by some criteria) such that after this transformation an equidistant grid can be used. An estimate of the *discretization error* is possible by means of tricky and non-trivial application of analysis. Details on this are given in the appendix, which is a rather important part of this paper.

The integral (1.7), which is the starting point for the expansion of a hydrogen-like 1s function in a Gaussian basis, is rather complicated. There is a much simpler counterpart of (1.7) which is relevant for the expansion of the Coulomb potential $1/r$ in a Gaussian basis, namely

$$\frac{1}{r} = \frac{1}{\sqrt{\pi}} \int_0^\infty s^{-1/2} e^{-sr^2} ds \tag{1.8}$$

It has, in fact, been found in a numerical study [21] that this type of expansion has a very similar convergence behaviours as that of $e^{-\alpha r}$, i.e. that the error also goes as $\exp(-c\sqrt{n})$. The origin of this behaviour is essentially the same for the expansion of the two functions. Since (1.8) is formally much simpler, it is recommended to study the expansion of $1/r$ first.

In fact only the expansion of $1/r$ will be treated here in detail, while a full study of the expansion of $e^{-\alpha r}$ will be published elsewhere.

The key feature is − both for the expansion of $1/r$ or $e^{-\alpha r}$ in terms of 'even-tempered' Gaussians − that, for large n, the cut-off error goes as $\sim \exp(-anh)$ with h the step size and that the discretization errors goes as $\sim \exp(-b/h)$, with a and b constants. While − for fixed n − a small h is good for the discretization error, it is bad for the cut-off error and vice versa. The best compromise is that $h \sim 1/\sqrt{n}$, which implies that the overall error goes as $\sim \exp(-c\sqrt{n})$.

The similarity between $1/r$ and $e^{-\alpha r}$, as far as the expansion in a Gaussian basis is concerned, leads to another interesting aspect. In many-electron quantum mechanics we have in principle to solve both Schrödinger and Poisson equations. We don't realize this usually because the Poisson equations are first solved in closed form − which is not possible for the Schrödinger equation. This procedure destroys the equivalence between the matter field and the electromagnetic field and one may want to consider an approach in which one solves the Poisson equations numerically in a basis of Gaussians rather than solving it exactly. Work on these lines is in progress [21].

2. Expansion of 1/r in a Gaussian basis

We proceed in two steps. Starting point is the identity (1.8) or equivalently

$$\frac{1}{r} = \frac{2}{\sqrt{\pi}} \int_0^\infty e^{-r^2 t^2} dt = \frac{2}{r\sqrt{\pi}} \int_0^\infty e^{-s^2} ds \tag{2.1}$$

We first replace r^{-1} by

$$f(r) = \frac{2}{\sqrt{\pi}} \int_{t_1}^{t_2} e^{-r^2 t^2} dt = \frac{2}{r\sqrt{\pi}} \int_{rt_1}^{rt_2} e^{-s^2} ds \tag{2.2}$$

In doing so we make two 'cut-off' errors

$$f_{c1} = \frac{2}{r\sqrt{\pi}} \int_0^{rt_1} e^{-s^2} ds = \frac{1}{r}\text{erf}\,(rt_1) = g_1 \cdot \frac{1}{r} \qquad (2.3a)$$

$$f_{c2} = \frac{2}{r\sqrt{\pi}} \int_{rt_2}^{\infty} e^{-s^2} ds = \frac{1}{r}\text{erfc}\,(rt_2) = g_2 \cdot \frac{1}{r} \qquad (2.3b)$$

The error function erf$x$ has a power series expansion for small x and an asymptotic expansion for large x

$$\text{erf}x = 1 - \text{erfc}x = \frac{2}{\sqrt{\pi}}\left\{x - \frac{1}{3}x^3 + O(x^5)\right\} \qquad (2.4a)$$

$$\text{erfc}x = 1 - \text{erf}x = \frac{1}{\sqrt{\pi}x}[1 + O(\frac{1}{x^2})]e^{-x^2} \qquad (2.4b)$$

and the following inequalities hold

$$\text{erf}x \le \frac{2}{\sqrt{\pi}}x \qquad (2.5a)$$

$$\text{erfc}x \le \frac{1}{\sqrt{\pi}}x^{-1}e^{-x^2} \qquad (2.5b)$$

which allow us to estimate f_{c1} and f_{c2} in two alternative ways.

$$f_{c1} \le 2t_1/\sqrt{\pi}; \ O(r^2t_1^3) \qquad (2.6a)$$

$$|r^{-1} - f_{c1}| = r^{-1}|1 - g_1| \le \frac{1}{\sqrt{\pi}r^2t_1}e^{-r^2t_1^2}; O(\frac{1}{r^3t_1^2})e^{-r^2t_1^2} \qquad (2.6b)$$

$$f_{c2} \le \frac{1}{\sqrt{\pi}r^2t_2}e^{-r^2t_2^2}; \ O(\frac{1}{r^3t_2^2})e^{-r^2t_2^2} \qquad (2.6c)$$

$$|r^{-1} - f_{c2}| = r^{-1}|1 - g_2| \le 2t_2/\sqrt{\pi}; \ O(r^2t_2^3) \qquad (2.6d)$$

We have indicated the order of errors of these estimates after the semicolons. We see that (2.6a) is a close estimate for f_{c1} if $r \ll t_1^{-1}$, while (2.6c) is a close estimate for f_{c2} if $r \gg t_2^{-1}$. On the other hand the relative error g_1 approaches 1, i.e. 100% for $r \gg t_1^{-1}$ and g_2 for $r \ll t_2^{-1}$. Note that the cut-off error never exceeds 100%. The range of r-value for which $f(r)$ is a good approximation to $1/r$ is

$$t_2^{-1} \ll r \ll t_1^{-1} \qquad (2.7)$$

In this range the total cut-off error $f_c(r) = f_{c1}(r) + f_{c2}(r)$ is determined by the 'lower-cut-off' error (2.6a), with respect to which the 'upper-cut-off error' (2.6c) is negligible. In a wide range of r 'flat' gaussian are more important than 'steep' ones, which only matter for small r.

The next step on the way to an expansion of $1/r$ in a Gaussian basis is to replace the integral (2.2) by a sum. Before we divide the range between t_1 and t_2 into n intervals, we apply a variable transformations, such that after this transformation an equidistant grid can be used.

$$t^2 = p(x) \tag{2.8a}$$

$$f(r) = \frac{1}{\sqrt{\pi}} \int_{p^{-1}(t_1^2)}^{p^{-1}(t_2^2)} e^{-r^2 p(x)} p'(x) [p(x)]^{-1/2} dx \tag{2.8b}$$

Let us define the normalized functions (with respect to square integration over r)

$$g(r,x) = \sqrt{\frac{8p(x)}{\pi}} e^{-r^2 p(x)} \tag{2.9}$$

Then (2.8b) becomes

$$f(r) = \frac{1}{\sqrt{8}} \int_{p^{-1}(t_1^2)}^{p^{-1}(t_2^2)} g(r,x) p'(x) [p(x)]^{-1} dx \tag{2.10}$$

Obviously we must choose $p(x)$ such that the domain between t_1 and t_2 — which have different orders of magnitude — is covered in a balanced way. One may further require that all $g(r,x)$ have about the same weight in the sum. The latter requirement leads to the condition

$$p'(x) = \text{const} \cdot p(x); \text{ i.e. } p(x) = e^{\gamma x} \tag{2.11}$$

Obviously an exponential mapping looks also good in the sense of the first criterion. One sees easily that $f(r)$ is independent of the choice of γ, such that we may as well take $\gamma = 1$. Of course, this is only a plausiblity argument and we need a rigorous criterion for the optimum mapping. We come back to this problem in the conclusions.

We hence have

$$f(r) = \int_{x_1}^{x_2} F(x) dx \tag{2.12a}$$

$$F(x) = \frac{1}{\sqrt{\pi}} \exp\left\{ \frac{x}{2} - r^2 e^x \right\} \tag{2.12b}$$

$$x_j = 2\ln t_j; \; j = 1, 2 \tag{2.12c}$$

We now approximate (2.12a) as a sum (with $f_d(r)$ the discretization error).

$$f(r) = f_a(r) + f_d(r) \tag{2.13a}$$

$$h = \frac{x_2 - x_1}{n} = \frac{2\ln(t_2/t_1)}{n} \tag{2.13b}$$

$$f_a(r) = h\sum_{k=1}^{n} F(q_k) = \frac{2}{\sqrt{\pi}}\frac{\ln(t_2/t_1)}{n}\sum_{k=1}^{n}\exp\left\{\frac{q_k}{2} - r^2 e^{q_k}\right\} \tag{2.13c}$$

$$q_k = x_1 + (k - \frac{1}{2})h \tag{2.13d}$$

Estimates for the discretization error are derived in the appendix. Unlike the estimates (2.6) these are not obtained as strict inequalities, but rather as leading terms of asymptotic expansions. For the integral (2.12a) with the integration limits $-\infty$ to ∞ the discretization error is (for large n and sufficiently small h, see appendix E)

$$f_d(r) \approx \frac{1}{r}\exp(-\frac{\pi^2}{h}) \tag{2.14}$$

To arrive from (E.2) and (E.7b) at (2.14) one must identify α of appendix E with r^2 and realize that (E.2) or equivalently $f(x)$ in (C.1) is normalized to 1. To establish the relation to (2.1) one must multiply (E.2) by $\alpha^{-1/2} = r^{-1}$. The relative discretization error happens to be independent of r (at least as far as its dominant term is concerned). Using the arguments of the appendix one finds for the optimum interval length as function of dimension n of the basis

$$h \approx \pi\sqrt{\frac{2}{n}} \tag{2.15}$$

and for the overall error (for that range of r values for which f_{c1} and f_{c2} are suffiently small).

$$\varepsilon \sim e^{-\pi\sqrt{n/2}} \tag{2.16}$$

3. Estimation of the error of an expectation value of $1/r$

In practice one will — in fact — not be interested in the accuracy of $f(r)$ as a function of r, but rather in the error of matrix elements like that over a hydrogenlike $1s$ function

$$<r^{-1}> = 4\alpha^3\int_0^\infty e^{-2\alpha r}rdr = \alpha \tag{3.1a}$$

as

$$<f> = 4\alpha^3\int_0^\infty e^{-2\alpha r}f(r)r^2 dr \tag{3.1b}$$

To estimate this error we insert (2.2) into (3.1b) and integrate first over r such that

$$<f> = \int_{t_1}^{t_2}\varphi(t)dt \tag{3.2a}$$

$$\varphi(t) = -\frac{4\alpha^4}{\sqrt{\pi}t^4} + \exp(\frac{\alpha^2}{t^2})\mathrm{erfc}\ (\frac{\alpha}{t})[\frac{4\alpha^5}{t^5} + \frac{2\alpha^3}{t^3}] \qquad (3.2b)$$

The analytic expression for this integral is

$$< f > = \int_{t_1}^{t_2} \varphi(t)dt = (\alpha - \frac{2\alpha^3}{t_2^2})\exp(\frac{\alpha^2}{t_2^2})\mathrm{erfc}\ (\frac{\alpha}{t_2}) + \frac{2\alpha^2}{\sqrt{\pi}t_2}$$

$$- (\alpha - \frac{2\alpha^3}{t_1^2})\exp\ (\frac{\alpha^2}{t_1^2})\mathrm{erfc}\ (\frac{\alpha}{t_1}) - \frac{2\alpha^2}{\sqrt{\pi}t_1} \qquad (3.3)$$

The limit $\{t_2 \to \infty, t_1 \to 0\}$ of (3.3) is not obvious. To get it we must expand the first line of (3.3) in powers of t_2^{-1} and insert the asymptotic expansion of erfc in the second line before we collect powers of t_1. We get for the first and second lines of (3.3) respectively

$$\alpha - \varepsilon_{c2} = \alpha - \frac{\alpha^3}{t_2^2} + \frac{8\alpha^4}{3\sqrt{\pi}t_2^3} + O(t_2^{-4}) \qquad (3.4a)$$

$$-\varepsilon_{c1} = -\frac{2t_1}{\alpha^2\sqrt{\pi}} + \frac{2t_1^3}{\sqrt{\pi}} + O(t_1^5) \qquad (3.4b)$$

Of course, ε_{c1} and ε_{c2} as defined by (3.4) are the 'cut-off' errors due to limitation of the integration domain to t_1 to t_2

We next approximate the integral (3.3) by a numerical integration after performing the variable transformation (2.11) with $\gamma = 1$. This means we first replace (3.3) by

$$< f > = \int_{x_1}^{x_2} \chi(x)dx; \quad x_j = 2\ln t_j \qquad (3.5a)$$

$$\chi(x) = -\frac{2\alpha^4}{\sqrt{\pi}}e^{-3x/2} + \exp(\alpha^2 e^{-x})\ \mathrm{erfc}(\alpha e^{-x/2})(2\alpha^5 e^{-2x} + \alpha^3 e^{-x}) \qquad (3.5b)$$

Then we replace (3.5a) by

$$\tilde{f} = h\sum_{k=1}^{n}\chi_k; \quad \chi_k = \chi(x_1 + [k - \frac{1}{2}]h) \qquad (3.5c)$$

Before we study the 'discretization errors' let us look on how the 'cut-off errors' ε_{c2} and ε_{c1} depend on the number of points chosen in (3.5c). In view of (3.5a), (3.4) and (2.13b) we have

$$\varepsilon_{c1} = 2t_1/\sqrt{\pi} + O(t_1^3) = (2/\sqrt{\pi})e^{x_1/2} + O(e^{3x_1/2}) \qquad (3.6a)$$

$$\varepsilon_{c2} = \alpha^3 t_2^{-2} + O(t_2^{-3}) = \alpha^3 e^{-x_2} + O(e^{-3x_2/2})$$
$$= \alpha^3 e^{-x_1-nh} + O(e^{-3(x_1+nh)/2}) \qquad (3.6b)$$

The minimum with respect to x_1 (for nh fixed — and sufficiently large —) is achieved if

$$x_1 = -\frac{2nh}{3} + \ln(\pi^{1/3}\alpha^2) \tag{3.7a}$$

$$x_2 = \frac{nh}{3} + \ln(\pi^{1/3}\alpha^2) \tag{3.7b}$$

$$\varepsilon_c = \varepsilon_{c1} + \varepsilon_{c2} = 3\sqrt{\pi}e^{x_1/2} = 3\alpha\pi^{-1/3}e^{-nh/3} \tag{3.7c}$$

This means that one should choose roughly $x_2 = -x_1/2$ or $t_2 = t_1^{-1/2}$ and that for fixed h the error decreases exponentially with n (or for fixed n exponentially with h).

The estimation of the discretization error is fortunately rather easy, relying on the results of appendix E (which contains the difficult part of the derivation). In fact the discretization error $f_d(r)$ given by (2.14) is simply proportional to $1/r$. Hence

$$|\varepsilon_d| = 4\alpha^3 \int_0^\infty e^{-2\alpha r} f_d(r) r^2 dr = \alpha e^{-\pi^2/h} \tag{3.8}$$

A derivation of the discretization as

$$\varepsilon_d = \int \varphi(t)\cos\frac{4\pi\ln t}{h} dt = \mathrm{Re}\int \varphi(t) t^{4\pi i} dt \tag{3.9}$$

is very lengthy, but leads essentially to the same result, which is not so obvious, since in appendix E we have done the phase-averaging before integrating over r, and phase averaging and integration over r need not commute.

We use again the argument that the minimum of $\varepsilon = \varepsilon_c + \varepsilon_d$ appears close to the value of h for which the arguments of the exponential agree, i.e.

$$nh/3 \approx \pi^2/h; \quad h \approx \pi\sqrt{\frac{3}{n}} \approx 5.441/\sqrt{n} \tag{3.10a}$$

$$\varepsilon \sim e^{-\pi\sqrt{n/3}} \approx e^{-1.814\sqrt{n}} \tag{3.10b}$$

There is one difficulty insofar as (3.8) is only an estimate of the absolute value of the discretization error. It cannot be excluded that (depending on how the limit $x_1 \to -\infty$, $x_2 \to \infty$ is performed, see appendix D) ε_c and ε_d have opposite sign. In this case the minimum absolute error may vanish, while (3109a) is still valid.

Note that h is related to the β_0 of an even-tempered basis (1.3) for the H atom ground state as

$$h = \ln\beta_0; \quad \beta_0 \approx e^{-5.441/\sqrt{n}} \tag{3.11}$$

Let the smallest orbital exponent in the Gaussian basis be α_0, and the largest ω_0. Then for sufficiently large n we have

$$\alpha_0 \approx t_1^2 \approx e^{-2\pi\sqrt{n/3}} \approx e^{-3.63\sqrt{n}} \tag{3.12a}$$

$$\omega_0 = \alpha_0\beta_0^n \approx t_2^2 \approx e^{-\pi\sqrt{n/3}} \approx e^{1.81\sqrt{n}} \tag{3.12b}$$

these results, especially that for β_0 are in good agreement with results from a purely numerical study [21].

4. Conclusions

We were able to show analytically – in an unexpectedly tricky way (the mathematical ingredients of which are in the appendix) – that the error of an expansion of the function $1/r$ in terms of an even-tempered Gaussian basis of dimension n goes as $\sim \exp(-c\sqrt{n})$ provided that the two parameters of the even-tempered basis are optimized.

We have not shown that this is the optimum convergence, in other words whether there are other (two- or more-parameter) basis sets for which the convergence is even faster.

The examples given in the appendix give some indications on the properties which the mapping function has to satisfy that both the cut-off error and the discretization error decrease exponentially (or faster) with nh and $1/h$ respectively and don't depend too strongly on r. Further studies are necessary to settle this problem.

For quantum chemistry the expansion of $e^{-\alpha r}$ in a Gaussian basis is, of course, much more important than that of $1/r$. The formalism is a little more lengthy than for $1/r$, but the essential steps of the derivation are the same. For an even-tempered basis one has a cut-off error $\sim \exp(-nh)$ and a discretization error $\sim \exp(-\gamma/h)$, such that results of the type (2.15) and (2.16) result. Of course, $e^{-\alpha r}$ is not well represented for r very small and r very large. This is even more so for $1/r$, but this wrong behaviour has practically no effect on the rate of convergence of a matrix representation of the Hamiltonian. This is very different for basis set of type (1.1). Details will be published elsewhere.

At this point one can conjecture that the relatively rapid convergence of Gaussian geminals [22]

$$\exp[-\gamma(\vec{r}_1 - \vec{r}_2)^2]$$

to describe the correlation cusp, has a somewhat similar origin as the example studied here, and goes probably also as $\exp[-c\sqrt{n}]$, with n the dimension of the geminal basis.

Acknowledgement

The author thanks Stefan Vogtner for numerical studies of expansions of $1/r$ in a Gaussian basis which have challenged the present analytic investigation. Discussions with Christoph van Wüllen and Wim Klopper on this subject have been very helpful.

This paper is dedicated to Gaston Berthier, from whom I have learned a lot. Although Berthier's publications have mostly dealt with applications of quantum mechanical methods to chemical problems, he never liked black boxes or unjustified approximations even if they appeared to work. The question why the quantum chemical machinery does so well although it often lies on rather weak grounds has concerned him very much. I am therefore convinced that he will appreciate this excursion to applied mathematics.

Appendix

Estimation of the discretization error

A. GENERAL CONSIDERATIONS

We want to approximate the integral $\int_a^b f(x)dx$ by dividing the integration domain into n intervals of the same length h and by approximating $f(x)$ in each interval by its value at the center of the interval. The discretization error is then

$$\varepsilon = \int_a^b f(x)dx - h\sum_{k=1}^n f(x_k); \quad h = (b-a)/n; \quad x_k = a + (k - \frac{1}{2})h. \qquad (A.1)$$

To estimate ε (in a more traditional way) we make a Taylor expansion of $f(x)$ around $f(x_k)$ in the k-th interval. We write (assuming that $f(x)$ is differentiable an infinite number of times, which is the case for the functions that we study here)

$$\int_a^b f(x)\ dx = \sum_{k=1}^n f_k \qquad (A.2a)$$

$$f_k = \int_{-h/2}^{h/2} f(x_k + \xi)\ d\xi = \int_{-h/2}^{h/2} \{f(x_k) + \xi\ f'(x_k) + \frac{1}{2}\xi^2\ f''(x_k) + ...\}d\xi$$

$$= hf(x_k) + \frac{h^3}{24}f''(x_k) + \frac{h^5}{1920}f^{(4)}(x_k) + O(h^7) \qquad (A.2b)$$

We express

$$h\sum_{k=1}^n f''(x_k) = \int_a^b f''(x)dx - \frac{h^3}{24}\sum_{k=1}^n f^{(4)}(x_k) + O(h^5) \qquad (A.3)$$

and proceed similarly with $h\sum_{k=1}^n f^{(4)}(x_k)$ in a next step and so on such that finally

$$\varepsilon = -\frac{h^2}{24}[f'(b) - f'(a)] + \frac{7\ h^4}{5760}[f^{(3)}(b) - f^{(3)}(a)] + O(h^6)$$

$$= -\sum_{k=1}^\infty \frac{B_{2k} \cdot (1 - 2^{1-2k})h^{2k}}{(2k)!}[f^{(2k-1)}(b) - f^{(2k-1)}(a)] \qquad (A.4)$$

The B_{2k} are Bernoulli numbers.

The expansion coefficients in (A.4) are essentially those of $\operatorname{cosech}(x/2)$.

The equality sign in (A.4) only holds if the series converges. Otherwise the series is at least asymptotic in the sense that the sum truncated at some k differs from the exact ε by $O(h^{2k+2})$. This also holds if f is only $(2k-1)$ times differentiable, such that one has has to truncate the expansion anyway.

The discretization studied here is related to that of the Euler-McLaurin method well-known in numerical mathematics (see e.g. [23]). The difference is that in this method one approximates the mean value of $f(x)$ in the interval by the average of the values at the boundaries of the interval, while we approximate it by its value at the center of the interval. This choice is more closely related to the expansion of a function in a basis.

For the Euler-McLaurin discretization an error formula similar to (A.4) holds, namely without the factor $(1 - 2^{1-2k})$, which corresponds to the expansion co-efficients of $\coth(x/2)$.

An equidistant integration grid may not be the best choice. Let us therefore consider that we perform a variable transformation in the integral before we discretize.

$$x = g(y); \int_a^b f(x)dx = \int_{g^{-1}(a)}^{g^{-1}(b)} f[g(y)]g'(y)dy = \int_{g^{-1}(a)}^{g^{-1}(b)} F(y)dy \qquad (A.5)$$

To define the error by (A.1) and to apply the error formula (A.4) we must replace h by $\tilde{h}$, $f(x)$ by $F(y)$, x_k by y_k, b and a by $g^{-1}(b)$ and $g^{-1}(a)$ respectively

$$\tilde{h} = \frac{g^{-1}(b) - g^{-1}(a)}{n}; \quad y_k = g^{-1}(a) + (k - \frac{1}{2})\tilde{h} \qquad (A.6)$$

We are mainly interested in the transformation

$$x = e^{\eta y}; \quad \tilde{h} = \ln(b/a)/(\eta n) \qquad (A.7)$$

Eqn. (A.4) or its counterpart with h replaced by $\tilde{h}$ and $f^{(k)}(x)$ by $F^{(k)}(y)$ allows us to estimate ε for small h (or $\tilde{h}$), it is less convenient for h (or $\tilde{h}$) so large that the Taylor series within an interval converges slowly or diverges.

There is an alternative — and for our purposes more powerful — way to estimate the discretization error, namely in terms of the Fourier expansion of a periodic δ-function. We write $hf(x_k)$, see (A.1), as [24]

$$hf(x_k) = h \int_{a+(k-1)h}^{a+kh} f(x)\delta(x - x_k)dx$$

$$= \int_{a+(k-1)h}^{a+kh} f(x)dx + 2\sum_{l=1}^{\infty}(-1)^l \int_{a+(k-1)h}^{a+kh} f(x)\cos\frac{2l\pi(x - a)}{h}dx \quad (A.8)$$

$$\varepsilon = -2\sum_{l=1}^{\infty}(-1)^l \int_a^b f(x)\cos\frac{2l\pi(x - a)}{h}dx \qquad (A.9)$$

Only the cosine terms contribute, because the sine terms vanish at $x = x_k$.

The larger l and the smaller h the more rapidly oscillating is the cosine factor in (A.9) and the smaller is the contribution to ε. For sufficiently small h usually the term with $l = 1$ dominates in the sum.

A very popular method of numerical integration is that of Gauß [23]. It has the advantage that with n points in a Gauß integration one gets the same accuracy as with $2n$ points on an equidistant grid — provided that the integrand is well approximated as a polynominal of degree n, or is expandable in an orthogonal basis like in Laguerre polynomials. For the examples that we study here this condition is far from beeing satisfied, and therefore the Gauß integration is not supposed to be helpful.

We now study some special examples that are closely related to those that we are interested in.

B. THE EXPONENTIAL FUNCTION WITH AN EQUIDISTANT GRID

For the example

$$f(x) = \alpha e^{-\alpha x}; \quad \int_0^y f(x)dx = 1 - e^{-\alpha y} \qquad (B.1)$$

a closed expression for the truncation error can be obtained

$$f_k = \alpha \int_{-h/2}^{h/2} e^{-\alpha(x_k+\xi)}d\xi = f(x_k)\frac{2}{\alpha}\sinh\frac{\alpha h}{2} \qquad (B.2)$$

$$f(x_k) = \alpha e^{-\alpha x_k} \qquad (B.3)$$

In this case the relative error is the same for all intervals and one gets

$$\varepsilon_d = \{\frac{2}{\alpha h}\sinh\frac{\alpha h}{2} - 1\}h\sum_{k=1}^{n} f(x_k)$$

$$= \{1 - \frac{\alpha h}{2}\operatorname{cosech}\frac{\alpha h}{2}\}\int_0^y f(x)dx \qquad (B.4)$$

We write ε_d to indicate that this is a *discretization* error.

If one expands (B.3) in powers of $\frac{\alpha h}{2}$ one gets the same result as from (A.4) namely

$$\varepsilon_d = \frac{\alpha^2 h^2}{24}[1 - e^{-\alpha y}] + O(\alpha^4 h^4) \qquad (B.5a)$$

noting that

$$f^{(\mu)}(x) = (-\alpha)^\mu f(x) \qquad (B.5b)$$

The series (A.4) has here the radius of convergence $h_c = 2\pi$, but it can be continued analytically beyond its radius of convergence.

Let us now argue that we are actually interested in the integral

$$\int_0^\infty f(x)dx = 1 \tag{B.6}$$

and that the first approximation step is to replace ∞ by y and the second one the discretization, then the total error consists of the cut-off-error

$$\varepsilon_c = \alpha \int_y^\infty e^{-\alpha x}dx = e^{-\alpha y} \tag{B.7}$$

and the discretization error (B.4).

The limit $n \to \infty$ of the discretization error (B.4) is

$$\varepsilon_{d\infty} = 1 - \frac{\alpha h}{2}\text{cosech}\frac{\alpha h}{2} \tag{B.8}$$

while from the Fourier expansion (A.9) we get

$$\varepsilon_{d\infty} = -2\alpha \sum_{l=1}^\infty (-1)^l \int_0^\infty e^{-\alpha x}\cos\frac{2\pi l x}{h}dx = \sum_{l=1}^\infty \frac{2(-1)^{l+1}\alpha^2}{\alpha^2 + 4\pi^2 l^2/h^2} \tag{B.9}$$

The identity between (B.8) and (B.9) is not immediately recognized. One sees at least easily that for small h one gets from (B.9)

$$\varepsilon_{d\infty} = \frac{\alpha^2 h^2}{2\pi^2} \sum_{l=1}^\infty \frac{(-1)^{l+1}}{l^2} + O(h^4) = \frac{\alpha^2 h^2}{24} + O(h^4) \tag{B.10}$$

in agreement with what one gets from the Taylor expansion of (B.8) or immediately from (A.4). The agreement of (B.8) and (B.9) is confirmed in terms of a relation familiar in the theory of the digamma function ψ [25,26]

$$y^2 \sum_{l=1}^\infty (l^2 + y^2)^{-1} = y \text{ Im } \psi(iy) - 1 = \frac{\pi y}{2}\coth\pi y - \frac{1}{2} \tag{B.11}$$

together with

$$\sum_{l=1}^\infty (l^2 + y^2)^{-1}(-1)^{l+1} = \sum_{l=1}^\infty (l^2 + y^2)^{-1} - 2\sum_{l=1}^\infty (4l^2 + y^2)^{-1} \tag{B.12a}$$

and

$$\coth 2x - \coth x = -\operatorname{cosech} 2x \tag{B.12b}$$

which implies

$$y^2 \sum_{l=1}^{\infty} (l^2 + y^2)^{-1}(-1)^{l+1} = -\frac{\pi y}{2}\operatorname{cosech}\pi y + \frac{1}{2} \tag{B.13}$$

from which one is immediately led to the equivalence of (B.8) and (B.9)

If one limits the sum (B.9) to the term with $l = 1$ and expands in powers of h, the coefficient of the leading term in h^2 is $\alpha^2/(2\pi^2) \approx \alpha^2/20$ instead of the correct value $\alpha^2/24$ (see B.10). Convergence with l for small h is pretty (though not extremely) fast.

We want to make the overall error minimal for fixed n. We express the total error in terms of h and n

$$\varepsilon = e^{-\alpha n h} + \{1 - \frac{\alpha h}{2}\operatorname{cosech}\frac{\alpha h}{2}\}(1 - e^{-\alpha n h})$$

$$= 1 - \frac{\alpha h}{2}\operatorname{cosech}\frac{\alpha h}{2}[1 - e^{-\alpha n h}] \tag{B.14}$$

We want to minimize ε as function of h for fixed n. Since the discretization error only depends on h, it is obvious that one should make h as small as possible, in order to minimize it. We can therefore assume that h is so small that

$$\varepsilon = \frac{h^2}{24}\alpha + e^{-\alpha n h} + O(h^4) \tag{B.15a}$$

$$\frac{\partial \varepsilon}{\partial h} = \frac{\alpha h}{12} - \alpha n e^{-\alpha n h} + O(h^3) = 0 \tag{B.15b}$$

Asymptotically for large n the solution of this transcendental equation is

$$h \approx \frac{2}{\alpha}\frac{\ln n}{n} \tag{B.16a}$$

$$\varepsilon \approx \frac{\ln n(1 + \ln n)}{6\alpha n^2} \tag{B.16b}$$

Since $\ln n$ is a slowly varying function of n, the error goes essentially as n^{-2}. This is the typical behaviour of a discretization error for a numerical integration [23], but is atypical for the examples that we want to study.

C. THE GAUSSIAN WITH AN EQUIDISTANT GRID

Our next example is

$$f(x) = 2\sqrt{\frac{\alpha}{\pi}}e^{-\alpha x^2}; \quad \int_0^y f(x)dx = \operatorname{erf}(\sqrt{\alpha}y) \tag{C.1}$$

At first glance this looks similar to (B.1). However, there are two differences between (B.1) and (C.1) that have spectacular consequences.

1. While the function $f(x)$ in (B.1) is convex for all x, the $f(x)$ in (C.1) is concave from $x = 0$ to the inflection point $x_i = 1/\sqrt{2}\ \alpha$ and convex from x_i to ∞. This means that the discretization error is negative for intervals between 0 and x_i and positive between x_i and ∞, such that a partial cancellation of the error is possible.

2. While for $f(x)$ in (B.1) all derivatives at $x = 0$ are non-zero, the odd-order derivatives f^{2k-1} of the $f(x)$ in (C.1) vanish at $x = 0$. Since these enter the error formula (A.4) there is no contribution of the boundary at $x = 0$ to the ε given by (A.4), whereas for $f(x) = \alpha e^{-\alpha r}$ (appendix B) the derivatives at $x = 0$ determine the error.

From (A.4) we conclude that for sufficiently small h

$$\varepsilon_d = -4\alpha\sqrt{\frac{\alpha}{\pi}}\frac{h^2}{24}ye^{-\alpha y^2} + O(h^4) \qquad (C.2)$$

Not only is this error negative, meaning that we overestimate the integral (C.1), but it also appears that the error decreases very rapidly with y, such that one is tempted to conclude that in the limit $n \to \infty$ (and hence $y = nh \to \infty$) ε_d vanishes, independently of h.

In fact for $y = \infty$ the odd-order derivatives of $f(x)$ vanish at either boundary such that (A.4) gives the result zero. Of course (A.4) only holds for h smaller than the radius of convergence h_c of the series. There is no reason why h_c should be independent of y, and we shall, in fact see that $h_c \to 0$ for $n \to \infty$. This makes the estimate (C.2) rather useless because its range of validity is too limited (unlike for the example of appendix B).

The explicit expression for the discretization error is

$$\varepsilon_d = 1 - h\sqrt{\frac{\alpha}{\pi}}\sum_{k=1}^{n}e^{-\alpha\left(\frac{2k-1}{2}\right)^2 h^2} \qquad (C.3)$$

Unlike for the example of appendix B a closed summation is not possible. However, (C.3) allows us to discuss the behaviour of ε_d for large h, where the sum is dominated by the first term

$$\varepsilon_d = 1 - 2h\sqrt{\frac{\alpha}{\pi}}e^{-\alpha h^2/4} + O(e^{-3\alpha h^2/4}) \qquad (C.4)$$

For large h one cannot reduce the error significantly by increasing n. There is obviously a limiting function $\varepsilon_{d\infty}(h)$ for $n \to \infty$, which for large h is given by (C.4). For small h (C.3) is not convenient because it is slowly convergent.

Fortunately the Fourier expansion method helps us for small and intermediate h but large n. We get in the limit $n \to \infty$

$$\varepsilon_{d\infty} = 4\sqrt{\frac{\alpha}{\pi}}\sum_{l=1}^{\infty}(-1)^{l+1}\int_0^\infty e^{-\alpha x^2}\cos\frac{2\pi l x}{h}dx = 2\sum_{l=1}^{\infty}(-1)^{l+1}\exp[-\frac{\pi^2 l^2}{\alpha h^2}] \qquad (C.5a)$$

This is (at variance with C.3) a rapidly converging series for $h \lesssim \pi / \sqrt{\alpha}$.

For h suficiently small the first term with $l = 1$ is a good approximation to the sum (C.5a).

If the upper integration limit in (C.5a) is $y = nh$ rather than ∞, i.e. for finite n, a simple closed expression is not obtained. However, one can estimate the leading term in an expansion in powers of n^{-1}, such that

$$\varepsilon_d = \varepsilon_{d\infty} - \alpha n h^2 \mathrm{cosech}(\alpha n h^2) \mathrm{erfc}\sqrt{\alpha} n h + O(n^{-2})$$

$$= \varepsilon_{d\infty} - \sqrt{\frac{\alpha}{\pi}} h\ \mathrm{cosech}(\alpha n h^2) e^{-\alpha n^2 h^2} + O(n^{-2}) \qquad (C.5b)$$

The asymptotic expansion of (C.5b) in powers of h agrees with (C.2). In fact the first term neglected in (C.5b) starts with $O(h^4)$. In the limit $n \to \infty$, of course, all terms of an expansion in powers of h vanish. $\varepsilon_{d\infty}$ has an essential singularity at $h = 0$.

From this asymptotic expansion in powers of n^{-1} no conclusions on the radius of convergence of $\varepsilon_d(h)$ are possible, but there are some hints that the radius of convergence is that of $\mathrm{cosech}\ (\alpha n h^2)$, i.e. the series (A.4) probably converges for

$$h \le h_c = \sqrt{\frac{\pi}{\alpha n}} \qquad (C.6)$$

This conjecture is consistent with the result that for $n \to \infty$ the radius of convergence reduces to 0.

At $h = h_c$ the arguments of the exponential functions in (C.5a) and (C.5b) agree, which implies that near $h = h_c$, ε_d goes through zero. Between $h = 0$ and $h = h_c$, ε_d is slightly negative and rather well approximated by (C.2), while for $h > h_c$, ε_d increases rapidly and soon approaches 1.

Near $h = h_c$ the cut-off error

$$\varepsilon_c = 2\sqrt{\frac{\alpha}{\pi}} \int_y^\infty e^{-\alpha x^2} = \mathrm{erfc}(\sqrt{\alpha}y) = \frac{1}{2\sqrt{\alpha}nh} \exp(-\alpha n^2 h^2)[1 + O(\frac{1}{n^2 h^2})] \quad (C.6)$$

and the discretization error have the same order of magnitude, hence the minimum of $\varepsilon = \varepsilon_d + \varepsilon_c$ is also close to $h = h_c$. The minimum error therefore goes as

$$\varepsilon_{min} \sim e^{-n\pi} \qquad (C.7)$$

The prefactor of the exponential in (C.7) is less easily obtained. To get it one has to solve the transcendental equation $d\varepsilon(h)/h = 0$ for h and insert this into ε. Numerically one obtains that this factor is close to $1/2$.

The essential message is that the error goes as $\sim e^{-n\pi}$ and the optimum h as $\sim n^{-1/2}$. This means very fast convergence with the number n of intervals, very different from the example of appendix B where the error only decreased as $\sim (\ln n/n)^2$.

In this appendix we have argued that (C.5b) is valid for 'sufficiently small' h. That meant that h should not be significantly larger than $\approx \pi/\sqrt{\alpha}$, which is not very restrictive. However (C.2) only holds for h satisfying (C.6), which limits its validity to extremely small h, in the limit $n \to \infty$ (C.2) becomes even invalid. The two references to 'small' h must be clearly distinguished.

D. THE EXPONENTIAL FUNCTION WITH A LOGARITHMICALLY EQUIDISTANT GRID

We consider again (B.1), but with the transformation

$$x = e^z \tag{D.1}$$

$$\int_0^{y_2} f(x)dx = \alpha \int_{-\infty}^{\ln y_2} \exp(z - \alpha e^z)dz \tag{D.2}$$

The lower integration limit is now changed from 0 to $-\infty$. If we want to discretize, we must also introduce a lower cut-off. I.e. rather than (D.2) we must consider

$$\alpha \int_{-z_1}^{z_2} \exp(z - \alpha e^z)dz; \quad z_2 = \ln y_2 \tag{D.3}$$

The integrand in (D.3) falls off rapidly for $z > 0$, but more slowly for $z < 0$. Therefore the 'lower' cut-off z_1 is more critical than the 'upper' cut-off z_2. We have

$$\varepsilon_{c2} = \alpha \int_{z_2}^{\infty} \exp(z - \alpha e^z)dz = \alpha \int_{\exp(z_2)}^{\infty} e^{-\alpha x}dx = \exp(-\alpha e^{z_2}) = e^{-\alpha y_2} \tag{D.4a}$$

$$\varepsilon_{c1} = \alpha \int_{-\infty}^{-z_1} \exp(z - \alpha e^z)dz = \alpha \int_0^{\exp(-z_1)} e^{-\alpha x}dx = -\exp(-\alpha e^{-z_1}) + 1$$
$$= \alpha e^{-z_1} + O(\alpha^2 e^{-2z_1}) = \alpha y_1 + O(\alpha^2 y_1^2); \quad y_1 = e^{-z_1} \tag{D.4b}$$

We divide the domain $z_2 + z_1$ into n intervals, hence

$$h = (z_2 + z_1)/n = \ln(y_2/y_1)/n \tag{D.5}$$

We minimize the error with respect to y_2 for hn fixed, ignoring terms of $O(\alpha^2 y_1^2)$

$$\varepsilon_c = e^{-\alpha y_2} + \alpha y_1 = e^{-\alpha y_2} + \alpha y_2 e^{-nh} \tag{D.6a}$$

$$\frac{\partial \varepsilon}{\partial y_2} = -\alpha e^{-\alpha y_2} + \alpha e^{-nh} = 0; \Longrightarrow y_2 = nh/\alpha \tag{D.6b}$$

$$\varepsilon_c(opt) = e^{-nh}(1 + nh) \tag{D.6c}$$

and we get for the discretization error

$$\varepsilon_d = 2\alpha \sum_{l=1}^{\infty} (-1)^l \int_{-z_1}^{z_2} \exp(z - \alpha e^z) \cos \frac{2\pi l(z + z_1)}{h} dz$$

$$= 2\alpha \sum_{l=1}^{\infty} (-1)^l \int_{y_1}^{y_2} e^{-\alpha x} \cos \frac{2\pi l(\ln x - y_1)}{h} dx \tag{D.7}$$

We want to take the limit $y_1 \to 0$ and $y_2 \to \infty$ in order to obtain $\varepsilon_{d\infty}$. There is the difficulty that in this limit $\vartheta = \mathrm{Mod}(y_1, 2\pi/l)$ becomes an indefinite phase. Pictorially it is clear what this means.

The intervals near the maximum of the integrand $F(z)$ give the largest contributions. If one changes both integration limits, the intervals close to the maximum are not only changed in length, but also their positions with respect to the maximum are shifted. It makes, especially for large h, a lot of a difference if the 'innermost' interval has its center or a border at the maximum. The limit for the integration from $-\infty$ to ∞ depends somewhat on the position of the innermost interval, especially for large h.

Since the limit $y_1 \to 0$ and $y_2 \to 0$ is not unique, we can either choose a procedure to make it unique, e.g. fix that there is always a border of an interval at $z = 0$, or — what is more realistic — we accept the non-uniqueness and hence an incomplete information and average over the indefinite phase in some consistent way. Leaving the phase unspecified we get

$$\varepsilon_{d\infty} = 2\alpha \sum_{l=1}^{\infty} (-1)^l \int_0^\infty e^{-\alpha x} \cos \frac{2\pi l(\ln x + \vartheta)}{h} dx$$

$$= \sum_{l=1}^{\infty} \exp \frac{2\pi i l(\vartheta - \ln \alpha)}{h} (-1)^l \int_0^\infty e^{-\eta} \eta^{\frac{2\pi i l}{h}} d\eta$$

$$+ \sum_{l=1}^{\infty} \exp \frac{-2\pi i l(\vartheta - \ln \alpha)}{h} (-1)^l \eta^{-\frac{2\pi i l}{h}} \int_0^\infty e^{-\eta} \eta^{-\frac{2\pi i l}{h}} d\eta$$

$$= 2\mathrm{Re} \sum_{l=0}^{\infty} (-1)^l \exp \frac{2\pi i l(\vartheta - \ln \alpha)}{h} \Gamma(1 + \frac{2\pi i l}{h}) \tag{D.8}$$

Since [25,26]

$$|\Gamma(1 + \frac{2\pi i l}{h})| = \left\{ \frac{2\pi^2 l}{h} \mathrm{cosech} \frac{2\pi^2 l}{h} \right\}^{1/2}$$

$$= \frac{2\pi}{\sqrt{h}} \exp(-\frac{\pi^2 l}{h}) + O(\exp[-\frac{3\pi^2 l}{h}]) \tag{D.9}$$

the term with $l = 1$ dominates for sufficiently small h. If we take only this term in (D.8) and form the mean square average over the phase $\vartheta - \ln\alpha$ we get

$$|\varepsilon_{d\infty}| \approx |\Gamma(1 + \frac{2\pi i}{h})| = (\frac{2\pi^2}{h}\text{cosech}\frac{2\pi^2}{h})^{1/2} \approx \frac{2\pi}{\sqrt{h}}\exp(-\frac{\pi^2}{h}) \qquad (D.10)$$

This estimate is independent of α as is the estimate (D.6c) of the cut-off error.

Note that $|\Gamma(1 + 2\pi i/h|$ is a monotonically increasing function of h, while both Re $\{\Gamma(1 + 2\pi i/h)\}$ and Im $\{\Gamma(1 + 2\pi i/h)\}$ oscillate between $|\Gamma(1 + 2\pi i/h)|$ and $-|\Gamma(1 + 2\pi i/h)|$.

The discretization error ε_d for finite integration limits y_1 and y_2 contains in addition to (D.8) two extra terms (under the sum) that contain incomplete Gamma functions. We don't need their explicit form for the estimation of the dominating part of the overall error. Of course, expanding these extra terms in powers of h would lead to the error estimation (A.4), that holds for extremely small h (and sufficiently small l) which is rather irrelevant in the present context.

Somewhat similar to appendix C we have a discretization error that goes as $\sim \exp(-b/h)$ and a cut-off error $\sim \exp(-anh)$. The minimum as function of h is achieved (for large n) if

$$\frac{\pi^2}{h} \approx nh; \ h \approx \pi/\sqrt{n} \qquad (D.11a)$$

$$\varepsilon \sim e^{-\pi\sqrt{n}} \qquad (D.11b)$$

If ε_c and ε_d happen to have opposite sign, the optimum error vanishes, while close to its zero $\varepsilon(h)$ has an inflection point.

The optimum interval length goes as $\sim 1/\sqrt{n}$ and the error as $\exp(-\pi\sqrt{n})$. This is certainly a much faster convergence than for the choice of an equidistant grid for the exponential function as studied in appendix B.

We have not considered the next term in an $1/n$ expansion of ε_d, which would be needed to get the prefactor of ε.

E. A GAUSSIAN WITH A LOGARITHMICALLY EQUIDISTANT GRID

We consider now (C.1) but with the transformation

$$x = e^{z/2}; \ y_1 = e^{-z_1/2}; \ y_2 = e^{z_2/2} \qquad (E.1)$$

Everything is similar to appendix D.

Now (D.3) is replaced by

$$\sqrt{\frac{\alpha}{\pi}} \int_{-z_1}^{z_2} \exp(\frac{z}{2} - \alpha e^z)dz \qquad (E.2)$$

We further get

$$\varepsilon_{c2} = 2\sqrt{\frac{\alpha}{\pi}} \int_{y_2}^{\infty} e^{-\alpha x^2} dx = \text{erfc}(\sqrt{\alpha}y_2) \tag{E.3a}$$

$$\varepsilon_{c1} = 2\sqrt{\frac{\alpha}{\pi}} \int_{0}^{y_1} e^{-\alpha x^2} dx = \text{erf}(\sqrt{\alpha}y_1) = 2\sqrt{\frac{\alpha}{\pi}}y_1 + O(\alpha^2 y_1^2) \tag{E.3b}$$

$$h = (z_2 + z_1)/n = 2\ln(y_2/y_1)/n$$

We minimize $\varepsilon_{c1} + \varepsilon_{c2}$ with respect to y_2 (neglecting y_1^2)

$$\varepsilon_c = \text{erfc}(\sqrt{\alpha}y_2) + 2\sqrt{\frac{\alpha}{\pi}}y_1 \tag{E.4a}$$

$$\frac{d\varepsilon_c}{dy_2} = 2\sqrt{\frac{\alpha}{\pi}}(-e^{\alpha y_2^2} + e^{-hn/2}) = 0 \tag{E.4b}$$

$$y_2 = \sqrt{\frac{hn}{2\alpha}}; \ y_1 = \sqrt{\frac{hn}{2\alpha}}e^{-hn/2} \tag{E.4c}$$

$$\varepsilon_c(opt) = \text{erfc}\sqrt{\frac{hn}{2}} + \sqrt{2hn}e^{-hn/2} \approx \left\{\sqrt{\frac{2}{\pi hn}} + \sqrt{2hn}\right\}e^{-hn/2} \tag{E.4d}$$

For the discretization error we get

$$\varepsilon_d = \sqrt{\frac{\alpha}{\pi}} \int_{z_1}^{z_2} \exp(\frac{z}{2} - \alpha e^z)\cos\frac{2\pi l(z + z_1)}{h}dz$$

$$= 2\sqrt{\frac{\alpha}{\pi}} \int_{y_1}^{y_2} e^{-\alpha x^2} \cos\frac{4\pi l(\ln x + \ln y_1)}{h}dx \tag{E.5}$$

The argument concerning the indefinite phase in the limit $y_1 \to 0$ is similar as in appendix D. The counterpart of (D.8) is

$$\varepsilon_{d\infty} = 2\sqrt{\frac{\alpha}{\pi}} \sum_{l=1}^{\infty}(-1)^l \int_{0}^{\infty} e^{-\alpha x^2} \cos\frac{4\pi l(\ln x + \vartheta)}{h}dx$$

$$= \frac{1}{\sqrt{\pi}} \sum_{l=1}^{\infty}(-1)^l \int_{0}^{\infty} \eta^{-1/2}e^{-\eta}\cos\frac{2\pi l[\ln(\eta/\alpha) + \vartheta]}{h}dx$$

$$= \frac{1}{4\sqrt{\pi}} \sum_{l=1}^{\infty}(-1)^l \left\{e^{\frac{2\pi il}{h}(\vartheta - \ln\alpha)} \int_{0}^{\infty} \eta^{-\frac{1}{2} + \frac{2\pi il}{h}} e^{-\eta}d\eta\right.$$

$$\left. + e^{-\frac{2\pi il}{h}(\vartheta - \ln\alpha)}\right\} \int_{0}^{\infty} \eta^{-\frac{1}{2} - \frac{2\pi il}{h}} e^{-\eta}d\eta$$

$$= \frac{1}{\sqrt{\pi}}\text{Re} \sum_{l=1}^{\infty}(-1)^l \exp\frac{2\pi il(\vartheta - \ln\alpha)}{h}\Gamma(\frac{1}{2} + \frac{2\pi il}{h}) \tag{E.6}$$

Limitation to the term with $l = 1$ (which dominates for sufficiently small h and the same phase averaging as in appendix D leads to [24,25]

$$|\varepsilon_{d\infty}| \approx \frac{1}{2\sqrt{\pi}}|\Gamma(\frac{1}{2} + \frac{2\pi i}{h})| = \frac{1}{2}(\mathrm{sech}\frac{2\pi^2}{h})^{1/2} \qquad (E.7a)$$

For small h this goes as

$$|\varepsilon_{d\infty}| \sim \exp(-\frac{\pi^2}{h}) \qquad (E.7b)$$

The condition analogous to (D.11a) is

$$\frac{\pi^2}{h} \approx \frac{nh}{2} \qquad (E.8a)$$

$$h \approx \pi\sqrt{\frac{2}{n}} \qquad (E.8b)$$

$$\varepsilon \approx \frac{1}{2}e^{-\pi\sqrt{\frac{n}{2}}} \qquad (E.9)$$

Like for the last example the optimum h goes as $\sim 1/\sqrt{n}$ and the error as $\sim \exp(-c\sqrt{n})$. The convergence is slower than for the same function with an equidistant grid, but both h and ε are (on this level of approximation) independent of α, i.e. essentially the same grid can be used for a very steep or a very flat Gaussian. there is only a shift via the α-dependence of y_1 and y_2.

References

1. S.F. Boys, *Proc. Roy. Soc.* **A200**, 542 (1950)
2. R.N. Hill, *J. Chem. Phys.* **83**, 1173 (1985)
3. S. Huzinaga, *J. Chem. Phys.* **42**, 1293 (1965)
4. F.B. v. Duijneveldt, *IBM Tech. Res. Rep.* RJ 945 (1971)
5. W.J. Hehre, R. Stewart and J.A. Pople, *J. Chem. Phys* **51**, 2657 (1969)
6. C.M. Reeves, *J. Chem. Phys.* **39**, 1 (1963)
   K. Rudenberg, R.C. Raffinetti, R.D. Bardo in
   Energy, Structure and Reactivity, Wiley, New York (1973)
   R.C. Raffinetti, *Int. J. Quant. Chem. Sym.* **9**, 289 (1975)
7. M.W. Schmidt and K. Ruedenberg, *J. Chem. Phys.* **71**, 3951 (1979)
8. S. Huzinaga, M. Klobukwoski and H. Tatewaki, *Can. J. Chem.* **63**, 1812 (1985)
9. J.D. Morgan and S. Haywood, *unpublished*, quoted in ref. 19
10. V. Mühlenkamp, *Thesis*, Bochum (1992)
11. D. Feller and E.R. Davidson, in Reviews in Computational Chemistry 1, K.B.
    Lipkowitz and D.B. Boyd Eds., VCH, Weinheim (1990) p.1
12. H. Preuß, *Z. Naturforsch.* **A11**, 823 (1956), *Mol. Phys.* **8**, 157 (1964)
13. J. L. Whitten, *J. Chem. Phys.* **39**, 349 (1963)
14. B. Klahn and W.A. Bingel, *Theor. Chim. Acta* **44**, 2 (1977)
15. B. Klahn and J.D. Morgan, *J. Chem. Phys.* **81**, 410 (1984)
16. W. Klopper and W. Kutzelnigg, *J. Mol. Struct. THEOCHEM.* **135**, 339 (1986)
17. W. Kutzelnigg, *Theoret. Chim. Acta* **68**, 445 (1985)

18. R. Franke and W. Kutzelnigg, *Chem. Phys. Letters* **199**, 561 (1992)
19. J.D. Morgan III, in: Numerical determination of the electronic structure of atoms, diatomic and polyatomic molecules, M. Defranceschi and J. Delhalle Eds., (Kluwer, Dordrecht (1989) p. 49
20. J.R. Mohallem and M. Trsic, *Int. J. Quant. Chem.* **33**, 555 (1988)
21. W. Kutzelnigg and St. Vogtner, to be published
22. K. Szalewicz, B. Jeziorski, H.J. Monkhorst, J.G. Zabolitzky, *J. Chem. Phys.* **78**, 1420 (1983)
23. See e.g. J. Stoer, Einführung in die Numerische Mathematik I, Springer, Berlin (1972)
24. This estimate of the discretization error ought to be known in numerical mathematics. Usually it is easier to derive formulas like this than too look them up in the literature.
25. M. Abramowicz and I.A. Stegun, Handbook of Mathematical Functions, Dover, New York (1965)
26. I.S. Gradsteyn and I.M. Ryzhik, Table of Integrals, Series and Products, Academic Press, New York (1980)

Quantum Chemistry in Front of Symmetry-Breakings

J.P. MALRIEU and J.P. DAUDEY
Laboratoire de Physique Quantique, Université Paul Sabatier
118 route de Narbonne, 31062 Toulouse, France

1. Introduction

Symmetry breaking is a universal phenomenon, from cosmology to the microscopic world, a perfectly familiar and daily experience which should not generate the reluctance that it induces in some domains of Physics, and especially in Quantum Chemistry. In classical physics, the symmetry breaking of an *a-priori* symmetrical problem is sometimes refered to as the lack of symmetry of the initial conditions. But it may be a deeper phenomenon, the symmetry-broken solutions being more stable than the symmetrical one.
Quantum chemistry experiences two types of symmetry breakings.
One is purely formal, it concerns the departure from symmetry of an *approximate* solution of the Schrödinger equation for the electrons (ie within the Born-Oppenheimer approximation). The most famous case is the symmetry-breaking of the solutions of the Hartree-Fock equations[1-4]. The other symmetry-breaking concerns the appearance of non symmetrical conformations of minimum potential energy. This phenomenon of deviation of the molecular structure from symmetry is so familiar, confirmed by a huge amount of physical evidences, of which chirality (i.e. the existence of optical isomers) was the oldest one, that it is well accepted. However, there are many problems where the Hartree-Fock symmetry breaking of the wave function for a symmetrical nuclear conformation and the deformation of the nuclear skeleton are internally related, obeying the same laws. And it is one purpose of the present review to stress on that internal link.

2. Symmetry breakings of the electronic wave function

The Schrödinger equation being linear, H commutes with the symmetry operations of space and spin, and the wave function must be symmetry-adapted. This is the basic doxa which we transmit to our students. If they are critical, they perhaps wonder why the $2p_z$ atomic orbital of the hydrogen atom is an eigenfunction, while symmetry-broken. Actually, we usually do not take time to mention that for degenerate roots, it is the projector on the stable subspace of these degenerate eigenvectors which commutes with the symmetry operators of the problem. But the drama arises when the desired state is non degenerate and when an approximate method delivers a symmetry-broken wave-function. The results is in general considered negatively as spurious, contaminated and irrelevant, despite the fact that meaningfull physics have been introduced in these solutions in a biased way, lowering the energy with respect to the symmetry-adapted description obtained at the same level of sophistication.

Y. Ellinger and M. Defranceschi (eds.), Strategies and Applications in Quantum Chemistry, 103–118.
© 1996 *Kluwer Academic Publishers.*

The most famous case concerns the symmetry breaking in the Hartree-Fock approximation. The phenomenon appeared on elementary problems, such as H_2, when the so-called unrestricted Hartree-Fock algorithms were tried. The unrestricted Hartree-Fock formalism, using different orbitals for α and β electrons, was first proposed by G. Berthier [5] in 1954 (and immediately after by J.A. Pople [6]) for problems where the number of α and β electrons were different. This formulation takes the freedom to deviate from the constraints of being an S^2 eigenfunction.

For $S_z=0$ problems, where the ground state is a singlet state, the use of such a wave function appeared to give significantly lower energies than the orthodox symmetry-adapted solution in many problems, as illustrated below. Later on other types of symmetry breaking have been discovered and Fukutome [7] has given a systematics of the various HF instabilities in a fundamental paper.

2.1.   ATOMIC PHYSICS

In the Be atom, the two valence electrons occupy a 2s, 2p valence shell, the 2s and 2p Atomic Orbitals (AO) having an important "differential overlap" (ie a good coincidence of their spatial extension). The contribution of the 2p AO to the angular correlation of the valence electrons is especially large (the Moller Plesset expansion from $\Phi^0 = \left[core(s\bar{s}) \right]$ being poorly convergent) and the proper valence function should be written

$$\psi = \left[core(\lambda s\bar{s} + \mu(x\bar{x} + y\bar{y} + z\bar{z})) \right]$$

while the RHF approximation is reduced to the $\left[core(s\bar{s}) \right]$ component. One obtains a much lower energy using an UHF function which looses both the space and symmetry constrainsts. The single determinant

$$\Phi_z = (s + \lambda z)(\overline{s - \lambda z}) = s^2 + \lambda (z\bar{s} - s\bar{z}) - \lambda^2 z\bar{z}$$

is lower in energy than the best RHF solution $s\bar{s}$, due to the inclusion of some angular correlation through the $z\bar{z}$ component, despite the contamination by the triplet configuration $(z\bar{s} - s\bar{z})$. This example illustrates wonderfully the physically suggestive potentiality of the symmetry-broken solution. Since it tells us that when the α electrons is on the right side of the nucleus (in an s + λz hybrid), the β electron prefers to move into an s-λz hybrid, ie on the left side of the nucleus. This is the best translation of the angular correlation, and it is clear that superimposing Φ_z and the degenerate non orthogonal solution Φ'_z

$$\Phi'_z = (s - \lambda z)(\overline{s + \lambda z})$$

into

$$\psi_z = \Phi_z + \Phi'_{z'}$$

will restore the singlet character of the wave function by eliminating the triplet contamination but still disobeying the space-symmetry constraint [8].

The space symmetry would only be restored by superposing the degenerate Ψ_x and Ψ_y solutions in $\Psi = \Psi_x + \Psi_y + \Psi_z$

Such phenomena do not occur in heavier alkaline earth atoms due to a poorer differential overlap between the valence s and p orbitals (smaller Ksp integrals) as explained by Kutzelnigg [9].

Another well-known atomic HF symmetry breaking is the O^{2-} problem but it is more artificial since in this unbound state, two electrons leave the atom oppositely in two diffuse orbitals [10].

2.2. THE WEAK SINGLE BOND

The most popular use of the UHF solutions concerned the single bond breaking, since it was rapidly understood that while the RHF solution of H_2

$$\phi^{RHF} = \sigma_g \overline{\sigma_g}$$

with

$$\sigma_g = \frac{a+b}{\sqrt{2(1+s)}}$$

imposed a constant ratio of ionic/neutral VB components whatever the interatomic distance

$$\phi^{RHF} = N\left(a\overline{b} + b\overline{a} + a\overline{a} + b\overline{b}\right)$$

and therefore a spurious asymptote at (IP-EA)/2 above the dissociation into neutral atoms, the UHF solution

$$\phi^{UHF} = \sigma\overline{\sigma'}$$

with

$$\sigma' = \lambda b + \mu a$$
$$\sigma' = \lambda a + \mu b$$

authorized one electron to concentrate on atom A while the second one concentrates an atom B. The detailed conditions for the appearance of the UHF solution have been explicited a long time ago as a special application of the Thouless' relations [2]. This relation analyzes the stability of the symmetry-adapted HF solution, using symmetry-adapted MOs [11]. The transcription of these conditions in Valence Bond terms is easy to derive, [12] and one may show that the symmetry breaking takes place when

$$\left|\frac{\langle a|F|b\rangle}{E_n - E_I}\right| < \frac{1}{2}$$

where $\langle a|F|b\rangle$ in the element of the Fock operator between the valence AOs a and b and $E_n - E_I$ is the energy difference between the neutral and the ionic VB determinants. The solid state physicists would say that

$$\left|\frac{2t}{U}\right| < 1,$$

t being the hopping integral (F_{ab}) and $U = E_n - E_I$ the on-site effective bielectronic repulsion, while the radius of convergence of the Rayleigh Schrodinger perturbation theory from the RHF single determinant is

$$\left|\frac{\langle a|F|b\rangle}{E_n - E_I}\right| = \left|\frac{t}{U}\right| > \frac{1}{4}$$

For $|4t| \langle U$ the relevant perturbation consists in perturbing the covalent (or neutral) VB structures by their interaction with the ionic ones ; this is the strongly correlated or magnetic domain. So that the Hartree Fock symmetry breaking occurs in a zone which covers the whole magnetic domain and a significant part of the "weakly" correlated domain

2.3.   THE MULTIPLE BOND

For more complex problems such as multiple bonds (N_2 for instance [13-14] and Metal-Metal bonds [15-17]) or extended systems (the π system of cyclic polyenes, among others), the symmetry-breakings may take several forms since one may leave different space-and spin-symmetry constraints independently or simultaneously. For C_2 for instance, the RHF symmetry adapted solution is of $x_u^2 y_u^2$ character (π^4 double bond) while one may find at much lower energy a $\sigma_g^2 x_u^2$ solution of closed shell character (a pure singlet) which has broken the symmetry between the x and y π bonds. A UHF solution lies much below, which has a dominant VB character

$$\left|s_a z_a x_a y_a \overline{s_b z_b x_b y_b}\right|$$

at short interatomic distances and a correct asymptotic ($^3P + {}^3P$) content

$$\left|s_a^2 x_a y_a s_b^2 \overline{x_b y_b}\right|$$

at large interatomic distances.

The multiplicity of symmetry breakings have been explored in details in N_2 where they occur near the equilibrium interatomic distance [12].

The fact that symmetry breaking occurs at shorter interatomic distances for multiple bonds than for single bonds may be understood within two different languages. One refers to the instability conditions of the symmetry-adapted solution. In multiple bonds some bonding electron pair are strongly delocalized and would not break the symmetry (for instance the σ bond of N_2) while the p bonds are weaker and enter more rapidly into the :

$$\left|\frac{\langle a|F|b\rangle}{E_n - E_I}\right| < \frac{1}{2}$$

regime (this criterion is only weakly modified for multiple bonds [12]). Of course in a sextuple bond, as Cr_2, the δ bonds are so weak that they induce a strong symmetry breaking [18].

In the other approach one looks at the VB content of the symmetry adapted) wave function, for instance for N_2

$$\Phi_{SA}^{RHF} = \left|\text{core}\left(\sigma_g \sigma_g \overline{x}_u x_u \overline{y}_u y_u\right)\right|$$

and one sees that all VB components in terms of localized orbitals (obtained from the occupied MOs and a proper definition of antibonding valence MOs σ_u, x_g, y_g)

$$z_a = \left(\sigma_g + \sigma_u\right)/\sqrt{2} \qquad z_b = \left(\sigma_g - \sigma_u\right)/\sqrt{2}$$

have equal coefficients, from the neutral ones to the triply ionic ones, which is especially absurd. All effective symmetry breakings lower the energy by reducing the components on the most ionic VB components. For instance in a singlet type symmetry-broken solution of N_2 of the type

$$\phi_{SB}^{RHF} = \left|\sigma_g \overline{\sigma_g} x \overline{x} y \overline{y}\right|$$

x concentrates on atom A while y concentrates on atom B so that the occurence of $N^{++} N^{--}$ and $N^{+++} N^{---}$ VB situations is dramatically reduced. This reduction is even stronger in the UHF solution

$$\phi^{UHF} = \left|\sigma \overline{\sigma} x x' \overline{y} \overline{y'}\right|$$

where the α spin MOs concentrate on atom A and the β spin MOs concentrate on atom B. This increases not only the neutral VB character of the wave function but also the component $N(^4S) N(^4S)$ on the atomic ground states, satisfying the atomic Hund's rules.

As an example of the interest to scrutinise the UHF solution, one may quote the Be_2 problem [19]. The bond is weak but it takes place at short interatomic distance and is definitely not the dispersion well which one might expect from two closed shell atoms (and which occurs in Mg_2 and heavier compounds). Quantum chemical calculations only reproduce this bond when using large basis sets and extensive CI calculations [20]. It is amazing to notice that the UHF solution gives a qualitatively correct behaviour, and suggests a physical interpretation of this bond since in

$$\phi^{UHF} = \left|\sigma \overline{\sigma} \sigma'' \overline{\sigma'}\right| \approx \left|s_a z_a \overline{s_b} \overline{z_b}\right|$$

the two α spin MOs concentrate an atom A and have respectively a 2s and $2p_z$ dominant character (the same being true for the β spin MOs on atom B). The creation of that weak bond would be due to an atomic promotion to the lowest ^3P excited state of the atoms. This suggestion would deserve a verification by defining natural MOs from the full CI wave function and their appropriate localizing transformations.

2.4.   EXTENDED SYSTEMS

HF symmetry breaking for polyatomic molecules have the same origin, i.e. the reduction of the components on highly ionic VB situations. If one considers a strongly localizable electronic system such as a molecule built of covalent bonds like the cubic P_8 (12 P-P single bonds), each bond is weakly correlated but there are equal mixtures of neutral and ionic contents. So that each atom being involved in three bonds, the occurence of P^{3+} and P^{3-} situations is completely exaggerated. Moreover, among the initial situations, the spin distribution is random without any privilege neither for the spread of the three electrons in three p AOs nor for the spin alignment, which would satisfy the atomic Hund's rules. The UHF solution may be written :

$$\phi^{UHF} = \prod_{i=1,2} s^2_{2i-1} x_{2i-1} y_{2i-1} y_{2i-1} s^2_{2i} \overline{x}_{2i} \overline{y}_{2i} \overline{z}_{2i}$$

where the MOs concentrate on different atoms and where the spin distribution is antiferromagnetic (each α spin atom being surrounded by three β spin atoms), and it appears close to the equilibrium interatomic distance [21]. It is clear that it corrects the spurious charge fluctuations on the atoms and satisfy their intrinsic preferences.

Symmetry breakings have been studied for systems with one electron per center such as the π systems of cyclic polyenes [22,23]. One finds here both charge-density-wave RHF solutions , where the bond indexes are alternant (one strong bond (2i, 2i+i) between two weak bonds (2i-i, 2i) and (2i+1, 2i+2) and spin-density-wave UHF solutions where the electrons are spin-alternant (one α electron on atom 2i surrounded by two β electrons on atoms 2i±1). The first one does not "dissociate" properly (when t/u tends to zero), since it remains half neutral and half ionic but it reduces the weight of the most irrelevant VB situations with respect to their importance in the symmetry-adapted solution. The charge-density-wave solution tends to localize the electrons by (α,β) pairs on the "strong bonds", each one supporting a localized MO

$$\phi^{RHF} = \left| \varphi_1 \overline{\varphi}_1 ... \varphi_1 \overline{\varphi}_1 ... \varphi_n \overline{\varphi}_n \right|$$

with small delocalization tails. In such a function the probabilities to find one electron of α spin, one electron of β spin, two αβ electrons or zero electron on each atom remain equal, as it occured in the symmetry-adapted HF function. But the probability to find two adjacent positive or negative charges is now diminished (at least in the "strong bond") and the avoidance of such high energy situation through the charge density wave RHF solution lowers the energy (at least when PPP hamiltonian is prefered to the less realistic Hubbard Hamiltonian which only counts the ionicity of each VB structure). As a consequence of that pairing of electrons in bonds, the probability to find two electrons of the same spin on adjacent atoms is also diminished with respect to its probability of occurence in the symmetry-adapted solution and this reduction is overestimated compared to the exact wave function.

The UHF solution appears when the hopping integral t becomes small and leads to a spin density wave. The localization of the MOs leads to a and b atom centered orbitals, localized around odd and even labelled atoms respectively.

$$\phi^{UHF} = \left| y_1 \overline{y}_2 y_3 \overline{y}_4 ... \overline{y}_{2n} \right|$$

This solution can only be reached in linear or cyclic polyenes for rather unrealistic t/u ratios (i.e. lengthened CC bonds) while it occurs in cyclic ideal Li_{2n} clusters for realistic interatomic distances in ab initio calculations [24]. But in that case another fascinating symmetry breaking takes place, namely a bond-centered spin-density wave, as discovered by Mc Adon and Goddard [24]. This UHF solution is much lower in energy, and it consists in an antiferromagnetic distribution of the electrons, each electron occupying a MO centered midway between adjacent atoms. In this solution the electrons have left the atoms and each of them occupy its own cell, i.e. is delocalized into the largest intersticial zone. This is made possible by the fact that a strong s-p hybridization does not require too much energy. This solution is physically based on a good compromise between the electronic repulsion, which keeps the electrons apart, one per cell, the kinetic energy which is higher than in the delocalized RHF solution but lower than in the atom-centered UHF solution. The benefit of that optimal balance compensates a weak diminution of nuclear attraction.

This discovery, confirmed by GVB [25] and later by CI calculations [26], led McAdon and Goddard to propose a rather revolutionary picture of the metal, the "intersticial picture" [27]. Lepetit *et al.*, [26] have shown that

- for a 2n-electron problems there are C_{2n}^n different $S_z=0$ UHF solutions which differ essentially by the distribution of the spin, the localized UHF MOs of the different UHF being almost identical (except for small tails) and defining an unvariant vectorial space,

- there are similar $S_z \neq 0$ solutions, up to the ferromagnetic one ($S_z=n$), with similar content of the localized MOs,

- the antiferromagnetic solution is the lower in energy, but the hierarchy of the energies obeys the logics of an Heisenberg Hamiltonian. This means that the delocalization between the interstices is small enough to be treated as a perturbation, through effective spin couplings. From that hierarchy of energies of the various UHF solutions one may estimate the amplitude of the spin coupling ; and solving the Heisenberg Hamiltonian for the cluster one obtains an energy quite close to the best CI estimates [26].

This strategy has been successfully applied to infinite periodic 1-D chains of Li atoms [28], through the first symmetry-broken application of the ab-initio UHF version of the Torino's CRYSTAL package [29]. The results of this work and of further treatments of 2-D lattices of Li and even Mg (Lepetit and coworkers, to be published) all confirm the validity of the intersticial picture. This is a case where the symmetry-broken HF solutions have led to a completely new picture of the electronic assembly.

When the symmetry breaking of the wave function represents a biased procedure to decrease the weights of high energy VB structures which were fixed to unrealistic values by the symmetry and single determinant constraints, one may expect that the valence CASSCF wave function will be symmetry-adapted, since this function optimizes the coefficients of all VB forms (the valence CASSCF is variational determination of the best valence space and of the best valence function, i.e. an optimal valence VB picture). In most problems the symmetry breaking should disappear when going to the appropriate MC SCF level. This is not always the case, as shown below.

2.5. SYMMETRY BREAKING IN CASE OF WEAK RESONANCE BETWEEN POLARIZED FORMS

In systems such as $[A....A']^+$ where an electron (or a hole) hesitates or oscillates between two equivalent positions on subsystems A or A', symmetry breakings may occur when the effective transfer integral between the two sites is weak. This will be the case when A and A' are far apart, when they are bridged by an "insulating" ligand, or when the two localized MOs concerned by the electron transfer have a very weak spatial overlap.

Actually in such problems the symmetry-adapted solutions should be

$$\phi_{SA}^{RHF} = \left[core \sigma_g \right] \quad , \quad \phi_{SA}^{RHF} = \left[core \sigma_u \right]$$

The two solutions may be reached independently since they belong to different symmetries. Then one may define localized MOs, on A and A' respectively :

$$a = \left(\sigma_g + \sigma_u\right) / \sqrt{2} \qquad b = \left(\sigma_g - \sigma_u\right) / \sqrt{2}$$

and the amplitude of the transfer integral is given by

$$\langle a|F|b \rangle = \left(\langle \sigma_g|F|\sigma_g \rangle - \langle \sigma_u|F|\sigma_u \rangle\right) / 2$$

This quantity may be very weak. Now starting from the trial function

$$\phi_{SB}^A = [\text{core } a] \quad \text{or} \quad \phi_{SB}^B = [\text{core } .b]$$

one may reach under certain circumstances an HF solution of spin-restricted or spin-unrestricted character (this is not the main point) localizing the hole on site A or on site B

$$\phi_{SB}^{A,HF} = [\text{core}'a']$$
$$\phi_{SB}^{B,HF} = [\text{core}''b'']$$

Now the core functions are adapted to the static electric field of the broken symmetry situation, they are adapted for instance to the A^+....A situation while the symmetry-adapted solution optimized the cores in an $A^{+1/2}$...$A^{+1/2}$ field. We qualitatively understand the physics of the problem :

The symmetry-breaking of the HF function occurs when the resonance between the two localized VB form A^+...A and A...A^+ is weaker than the *electronic relaxation* which one obtains by optimizing the core function in a strong static field instead of keeping it in a weak symmetrical field. If one considers for instance binding MOs between A and A' they do not feel any field in the SA case and a strong one in the SB solution. The orbitals around A^+ concentrate, those around A' become more diffuse than the compromise orbitals of $A^{+/2}$...$A^{+1/2}$, and these optimisations lower the energy of the A^+....A'form. As a counterpart, the energy of the $\phi^A = [\text{core}'b]$ which describes the A...A'$^+$ situations in the field of the core polarized in the A^+-A' situation grows up and the interaction between the two resonant VB forms is now completely underestimated. This phenomenon was first noticed for the ionization of the core levels of homonuclear diatoms (N_2, O_2) ; the ionization potentials from symmetry-adapted HF calculations give correct estimates of the small energy splitting between the σ_u and σ_g ionization potentials but overestimate both of them by several eV while the symmetry-broken solutions gave the correct mean energy and miss the energy splitting. The mechanism was first elucidated by Snyder et al. [30], by Denis et al., [31] in a more general and more correct analysis and later on by Cederbaum et al. (32).

The HF symmetry-breaking also occurs in the valence shell for A_2^+ systems when the overlap between the two electron donating groups is too low. One may quote for instance recent works devoted to molecular architectures designed for Molecular Electronics. An analytic treatment of the symmetry breaking of the HF function for such problems may be found in ref. [33] where it is shown that the symetry breaking is a *bifurcation*.

Another interesting problem concerns the case of a weak resonance between two ionic VB structures as occurs in the singlet $\pi\pi^*$ state of ethylene (and longer polyenes). This state (the allowed ^1B_u state) is of ionic VB content

$$\phi^1_{\pi\pi^*} = \left|\text{core}(\pi\overline{\pi^*} + \pi^*\overline{\pi})\right| / \sqrt{2}$$

$$\phi^1_{\pi\pi^*} = \left|\text{core}(a\overline{a} - b\overline{b})\right| / \sqrt{2}$$

where a and b are the π atomic orbitals of the carbon atoms of C_2H_4. The resonance between the two VB forms $a\overline{a}$ and $b\overline{b}$ is the interatomic exchange integral Kab. The repulsion between the two CH_2 groups tends to rotate the bond, and to put the a and b orbitals in perpendicular orbitals. Then the Kab integral becomes very weak, and we have again a weak resonance between two VB components. But these two forms tend to polarize the σ frame in two opposite directions. The electronic relaxation energy is large and prevails on the resonance. So that for highly twisted ethylene the singlet closed shell HF function will be symmetry-broken and will give two $CH_2^- - CH_2^+$ and $CH_2^+ - CH_2^-$ solutions, polarized. Allowing then the CH_2^- group to pyrimidalize leads to a stable $CH_2^- - CH_2^+$ structure. This phenomenon, discovered by Salem et al. [34], has given raise to a great interest in the early seventies under the name of "sudden polarization" but its suddeness has been questionned when the potential energy surface has been more extensively studied [35].

2.6.   DENSITY FUNCTIONAL AND SYMMETRY BREAKING

As long as it maintains the single determinant picture, the density functional function does not dissociate properly the chemical bonds, and is thus the subject of symmetry breaking at large interatomic distances. We may equivalently say that the correlation potential (i.e. the difference between the exact exchange and the exchange correlation potential) diminishes the electronic repulsion or that it increases the delocalization. Turning to the VB formulation of the HF instability, this implies that the symmetry breaking will occur at larger interatomic distances in DF calculations than in HF ones,

$$r_c^{LDF} > r_c^{HF}$$

Then an interesting question would be : for molecules which present strong HF symmetry breaking at equilibrium distance, such as C_2, Be_2 or cyclic Li_6, (and eventually Cr_2), are the symmetry-adapted LDF solutions stable at these distances ? If they are not, one will face an embarrassing problem ; since the calculated dissociation energies, are already correct when using the SA, solution at short distance and the separated atoms energies what should one think of the lower UHF solutions ? Let remember that most LDF calculations on organometallic systems or metallic clusters are LSD (Local Spin Density) calculations (i.e. performed in the UHF formalism). Why should one accept the symmetry (closed shell) constraint in some cases and not in others ?

2.7.   MC SCF SYMMETRY-BREAKING

The precedings sections concentrated on single determinant variational functions. One may wonder whether going to multiconfigurational SCF functions will restore symmetry and when.

We have attributed the origin of the HF symmetry breaking of homopolar simple and multiple bonds or in symmetric homoatomic clusters to unrealistic constraints on the coefficients of the different VB components. Going to a valence CASSCF (ie an optimal valence CI function) should restore the symmetry.

This will not be the case in the weak resonance systems. Notice that in these problems there are only two dominant VB configurations (for instance $A^+...A$ and $A...A^+$) and that the minimal valence CAS function

$$\psi = \frac{1}{\sqrt{2}}(\phi_{A^+...A} + \phi_{A...A^+}) = \left|\text{core}(a\bar{b}b + a\bar{a}b)\right| / \sqrt{2}$$

reduces to the symmetry-adapted single determinant :

$$\phi_{SA} = \frac{1}{\sqrt{2}}\left|\text{core}\sigma_g^2\sigma_u\right|$$

And actually the two-determinant function will be symmetry-broken for a symmetric configuration when the resonance energy is weaker than the polarization energy. This has been observed first by Ellinger *et al.* [36] in a problem with three electrons in two equivalent σ orbitals on distant oxygen atoms. The authors restored the symmetry by considering a local "antibonding" lone pair (or a 3p type orbital) in the active space in order to reintroduce into the CAS function the instantaneous repolarization of the oxygen orbitals, which are more diffuse when they are occupied by two electrons than by only one. But this is by no means a universal recipe. If the two oxygen atoms were separated by more bonds, the resonance would diminish and the rest of the dynamical polarization (the polarization of the bonds between the oxygen atoms) would be larger than the resonance, inducing a new symmetry-breaking of the enlarged CASSCF function.

The problem has already received a dramatic illustration on the LiF molecule, where the avoided crossing between the ionic and the neutral VB configurations takes place at such a large interatomic distance (~10bohr) that the transfer integral $\langle 2s(Li)|F|2p_z(F)\rangle$ is very small (~0.01ev.). The orbitals of F^- are completely different of those of LiF., in spatial extension and distortion (p-d mixing) and these relaxation phenomena bring much more energy than the interaction between the ionic and the neutral VB structures. So that Bauschlicher et al. [37] have never succeeded in making their CASSCF functions continuous around the avoided crossing (as they should be) despite the enlargement of the active space to all valence electrons in up to 12 active orbitals. As explained in ref. [38] this failure is due to the fact that the part of the dynamical polarization which remains out of the CAS space is still larger than the electron transfer integral.

CASSCF calculations are not a universal solution to symmetry-breaking of the wave functions, and for such weak resonance problems it is far more reliable to start from state average solutions which treat on an equal footing the two configurations which interact weakly.

2.8.   THE DIFFICULTY TO RESTORE SYMMETRY

The major drawback of symmetry-broken solutions is the difficulty to exploit them at a higher level of accuracy. There are three possible attitudes (besides simply refusing symmetry-breaking).

One attitude would consist in restoring symmetry by a symmetric superposition of the degenerate and linearly independent but non orthogonal symmetry-broken solutions, considering the gerade and ungerade combinations of the A^+A' and AA'^+ solutions in the electron transfer problem, or of the ϕ_{ab}^{UHF} and $\phi_{\overline{ab}}^{UHF}$ solutions in the bond breaking. Due to the non orthogonality the calculation of the overlap and of the hamiltonian matrix elements between these solutions is rather difficult (although it is routinely done in GVB programs). This is the first drawback. The second one is that in some cases this combination will not satisfy all symmetry requirements. For instance if one combines spin-polarized UHF solutions the result has no reason to be a spin eigenfunction. Finally one does not see how to go simply beyond this step to treat later on the dynamical correlation effects.

A second attitude consists in projecting the symmetry-broken solution on to the appropriate symmetry-adapted subspace. The exact or approximate projected HF methods have been the subject of an important litterature but the cost of the projections is non negligible compared to a CI and they do not compare efficiently with the traditional avenue which consists in respecting the symmetry from the beginning and performing CI.

The third attitude consists in performing the CI from one symmetry-broken HF solution, using the corresponding MOs. The idea is that if one goes sufficiently close to Full CI (which is independent of the choice of the MOs), the symmetry breaking of the intermediate step will be unimportant. Usual CI codes are written assuming the equivalence between α and β MOs and cannot be used for UHF solutions, but they might be exploited for singlet-type symmetry breakings, in order to study the convergence of the symmetry. Unrestricted Moller-Plesset (UMP) perturbative expansions have been written to the 4th order (essentially for the study of doublet or triplet states in [39] and the convergence of UMPn expansion for an $Sz = 0$ UHF solutions in single bond breaking appears to be fantastically poor, as shown by several authors [40-42]. The reasons for that poor convergence, i.e. of that failure to restore symmetry, have been analysed in details [43] and are twofold. The first one is due to the lack of meaning of the energy denominators in that problem, (a defect which disappears if one uses an Epstein Nesbet zeroth-order Hamiltonian). The other one is the strong coupling between the doubly excited and the singly excited determinants in UHF SB solution, which only plays a role from the 4th order in energy and slows the perturbation convergence. So far the HF symmetry-broken solutions appear as deserving to be searched and analysed, since they tell us very instructive stories about the physical trends acting on the electronic population, but they do not appear as a shortcut towards the exact solution.

3. Symmetry breaking of the nuclear conformation

There is not much to say about this well accepted phenomenon. We would simply like to stress on two peculiar aspects.

3.1.   BEHAVIOUR ON THE CRITICAL REGION

The dominant practice in Quantum chemistry is optimization. If the geometry optimization, for instance through analytic gradients, leads to symmetry-broken conformations, we publish and do not examine the departure from symmetry, the way it goes. This is a pity since symmetry breaking is a catastrophe (in the sense of Thom's theory) and the critical region deserves attention. There are trivial problems (the planar three-fold symmetry conformation of NH_3 is a saddle point between the two pyramidal equilibrium conformations). Other processes appear as bifurcations ; for instance in the electron transfer problem, the energy of the H_4^+ rectangular system $(H_2 + H_2)^+$ as a function of the intersystem distance R and of the relaxation of the intra system coordinate δ from the mean geometry (half-way between those of H_2 and H_2^+) behaves as a typical bifurcation [33]. The potential surface presents a symmetrical delocalized hole at short R and the symmetrical ($\delta=0$) valley for larger R values becomes a symmetrical crest beyond a critical value R_c. Beyond R_c there are two symmetry-broken valleys corresponding to A$^+$...A' and A...A'$^+$ (where A is a H_2 molecule). We have not yet met any problem which would exhibit a multi-stable symmetry breaking where for a certain domain, one would have a co-existence of a symmetrical valley and two symmetry broken valleys. A bistability region has been shown to exist in the Li_2F isocele triangle, between the $Li_2^+ F^-$ and the neutral Li_2 F states. It is likely that the above schematic view of the potential energy surface would be relevant for the mixed valence Donnor Acceptor Donnor (DAD) architectures such as :

$$(r - \delta)_{Li}^{Li}\Big|_{(R)}\cdots F_{(R)}\cdots\Big|_{Li}^{Li}(r + \delta)$$

which would be symmetrical and neutral for larger values and ionic and symmetry-broken for smaller values of R, with a possible domain of multistability

3.2.   ISOMORPHISMS AND INTERFERENCES BETWEEN ELECTRONIC AND NUCLEAR RELAXATIONS

The conformational symmetry breaking in electron transfer problem is governed by the ratio between the nuclear relaxation energy (i.e. energy stabilization when going from the symmetrical to the localized A$^+$...A* equilibrium geometry) and the amplitude of the electron transfer. It is therefore governed by the same inequalities that the HF symmetry breaking for the same problem in the symmetrical conformation, the nuclear relaxation replacing the electronic relaxation [33].
Of course the conformational symmetry breaking may appear or disappear depending on the level of sophistication of the computation, which may unduly favor one term of the crucial (relaxation/resonance) ratio. As an example we would like to mention an open controversy on the geometry of Ar_3^+ [44,45]. Does the surface present a double well Ar_2^+...Ar/Ar...Ar_2^+ or a single symmetric well for a linear [Ar3]$^+$ structure with a delocalized hole ? A simular interference between electronic and geometrical symmetry breaking occured for the allyl radical [46]. It may concerns the weak resonance between two $\pi \rightarrow \pi^*$ excitations [47]. It seems necessary to insist on the possible interference between the electronic symmetry breaking of the approximate wave function and the

treatment of the conformational symmetry breaking. Except for intrinsic degeneracies (leading to conical intersections) the potential energy must have a zero derivative with respect to the symmetry breaking coordinate (in the symmetrical conformation)

$$\left(\frac{\partial E}{\partial \delta}\right)_{\delta=0} = 0$$

and this will be the case for the energy obtained from a symmetry-adapted approximate wave function Φ_{SA}. In case of symmetry breaking of the wave function, there exist two other degenerate solutions Φ_{SB} and Φ'_{SB} of broken symmetry and lower energy

$$E_{SB}^{(\delta=0)} = E_{SB}'^{(\delta=0)} \left\langle E_{SA}^{(\delta=0)} \right.$$

and if the SA solution is unstable (which is not necessarily the case), one cannot build a function $\Phi_{SA}(\delta)$ tending to $\Phi_{SA}(\delta=0)$ when δ tends to 0. The variational procedure falls on $\Phi_{SB}(\delta)$, and one can only find estimates of the potential energy surface around $E_{SA}(\delta=0)$ by stabilization techniques [48].

Thus the potential energy surface E_{SA} does not exist and is reduced to a line for $\delta = 0$. On the contrary the solutions $\Phi_{SB}(\delta)$ and $\Phi'_{SB}(\delta)$ exist whatever δ (at least one of them, the other one may become unstable), but the derivative of the corresponding energy surfaces are different from zero at
$\delta = 0$.

$$\frac{\partial E_{SB}}{\partial \delta} = -\frac{\partial E'_{SB}}{\partial \delta} \neq 0$$
$$\text{for d=0}$$

The shape of the potential energy surface obtained by considering the lowest HF energy is qualitatively wrong near $\delta = 0$, since the resonance between Φ_{SB} and Φ'_{SB} is not treated properly. This is evident at $\delta = 0$, but is necessarily remains a problem for $\delta \neq 0$. If F_{SB} is biased for $\delta = 0$, the bias is necessarily continuous and exists for $\delta \neq 0$. Since we have already mentioned the difficulty to restore symmetry from a symmetry-broken solution, ie to correct the bias of the starting wave function, it is clear that the calculation of the PES even near this potential well is questionable if its starts from Φ_{SB}. This is not an academic comment. One may take as an example the famous problem of bonding alternation in linear polyenes, first rationalized by Longuet-Higgins and Salem [49]. This interpretation in terms of Peierls distorsion was purely monoelectronic and led to some contradiction regarding the r-dependence of the hopping integral and electron correlation must be invoked. A recent ab initio evaluation of the correlation energy for an infinite chain has been reported by Stolhoff [50] starting from an HF determinant, and it is clear that a spurious cusp exists for $\delta = 0$ in this work, which questions its reliability to determine the value of the bond alternation, ie δ_c such that

$$\left(\frac{\partial E}{\partial \delta}\right)_{\delta c} = 0$$

In our opinion it would be better to avoid the HF step, and to start the CI process from any unbiased function, symmetrical for $\delta = 0$, as are the Huckel MOs. We think that it is risky to study the existence and amplitude of a physical symmetry-breaking phenomenon through a computational sequence involving a symmetry-broken wave function at an intermediate step, since the use of this function introduces a prejudice and may result in an overestimation of the geometrical symmetry breaking. In that case the singlet symmetry breaking goes through an overestimation of the pairing of electrons into bonds (bond-centered charge density waves) as previously discussed and this overpairing, evident for $\delta = 0$, necessarily acts for $\delta \neq 0$, constraining the bond alternation. The approximate CI cannot repair the defect of this starting point [51].

4. Final comments

Even if symmetry-broken wave functions are difficult to use for higher levels of computation, their physical content is always instructive about the physical trends of the problem under study and they deserve interest. Their appearance and the more physical geometrical symmetry breaking are internally (but not strictly) related. Since they represent catastrophes on the wave function and/or the energy (or energy derivatives) they should be studied with attention and our ultranumericist discipline has not paid enough attention to these critical behaviours. This neglect is perhaps due to some implicit philosophical "continuism", prevailing in a domain where most instruments are based on variational procedures and optimizations. The use of computers and algorithms as black-boxes, and even the systematic plotting of the results through graphic codes using spline interpolations sometimes lead some quantum chemists of high reputation to miss cusps and intriguing features in their results [13,37]. Since qualitative explanations or pictures may be obtained from symmetry-broken wave functions and since funny behaviours are expected around conformational symmetry breaking, these problems should not be considered as teratological. Pictorial explanations and qualitative problems are both necessary to balance the unavoidable and fruitful research of numerical efficiency.

References

1.   A.W. Overhauser, *Phys. Rev. Letters*, **4**, 415 (1960).
2.   D.J. Thouless, *Nucl. Phys.* **21**, 225 (1960).
3.   W.H. Adams, *Phys. Rev.* **127**, 1650 (1962).
4.   P.O. Lowdin, *Rev. Mod. Physics*, **35**, 496 (1963).
5.   G. Berthier, *J. Chim. Phys.* **52**, 363, (1954).
6.   J.A. Pople and R.K. Nesbet, *J. Chem. Phys.* **22**, 57 (1954).
7.   H. Fukutome, *Intern. J. Quant. Chem.* **20**, 95 (1981).
8.   C.A. Coulson in "Quantum Theory of Atoms, Molecules and the Solid State",
      P.O. Löwdin ed., Acad. Press, N. York (1966) p 601.
      T.A. Kaplan and W.H. Kleiner, *Phys. Rev.* **156**, 1 (1967).
9.   W. Kutzelnigg, *Angew. Chem. Int. Ed. Engl.* **23**, 272 (1984).
10.   R. Prat and G. Delgado Barrio, *Phys. Rev. A*, **12**, 2288 (1975).
11.   for a text-book presentation, see A. Szabo and N. Ostlund, "Modern Quantum Chemistry ",Mc Millam, London (1982) p 221.

12. M.B. Lepetit, J.P. Malrieu and G. Trinquier, *Chem. Phys.* **130**, 229 (1989).
13. W.D. Laidig, P. Saxe and R.J. Bartlett, *J. Chem. Phys.* **86**, 87 (1987).
14. M.B. Lepetit and J.P. Malrieu, *J. Chem. Phys.* **87**, 5937 (1987).
15. K. Yamaguchi, *Chem. Phys. Letters,* **68**, 477 (1979).
16. M. Bénard, *J. Chem. Phys.* **71**, 2546 (1979).
17. R. Wiest and M. Bénard, *Theoret. Chim. Acta*, **66**, 65 (1984).
18. M.M. Goodgame and W.A. Goddard, *Phys. Rev. Letters*, **48**, 135 (1982); *J. Phys. Chem.* **85**, 215 (1981)
19. M.B. Lepetit and J.P. Malrieu, *Chem. Phys. Letters*, **169**, 285 (1990).
20. R.J. Harrison and N.C. Handy, *Chem. Phys. Letters*, **98**, 97 (1983) and references therein.
21. G. Trinquier, private communication.
22. J. Cizek and J. Paldus, *J. Chem. Phys.* **47**, 3976 (1967).
23. J. Paldus and J. Cizek, *Phys. Rev. A* **2**, 2268 (1970).
24. M.H. McAdon and W.A. Goddard, *J. Chem. Phys.* **88**, 277 (1988).
25. M.H. McAdon and W.A. Goddard, *J. Phys. Chem.* **92**, 1352 (1988).
26. M.B. Lepetit, J.P. Malrieu and F. Spiegelmann, *Phys. Rev. B* **41**, 8093 (1990).
27. M.H. McAdon and W.A. Goddard, *Phys. Rev. Letters*, **55**, 2563 (1985).
28. M.B. Lepetit, E. Apra, J.P. Malrieu and R. Dovesi, *Phys. Rev. B*, **46**, 12974 (1992).
29. R. Dovesi, C. Pisani, C. Roetti, M. Causa and V.R. Saunders, Crystal **88**, Program N°577, QCPE, Indiana University, Bloomington, 1989.
30. L.C. Snyder, *J. Chem. Phys.* **55**, 95 (1971). see also, P.S. Bagus and H.F. Schaefer, *J. Chem. Phys.* **56**, 224 (1972).
31. A. Denis, J. Langlet and J.P. Malrieu, *Theoret. Chim. Acta*, **38**, 49 (1975).
32. L.S. Cederbaum and W. Domcke, *J. Chem. Phys.* **66**, 5084 (1977).
33. G. Durand, O.K. Kabajj, M.B. Lepetit, J.P. Malrieu, J. Marti, *J. Phys. Chem.* **96**, 2162 (1992).
34. V. Bonacic-Koutecky, P. Bruckmann, J. Koutecky, C. Leforestier and L. Salem, *Angew. Chem. Inst. Ed. Engl.* **14**, 575, (1975); L. Salem and P. Bruckmann, *Nature (London)* **258**, 526 (1975).
35. For reviews see J.P. Malrieu. *Theoret. Chim. Acta*, **59**, 281 (1981). and J.P. Malrieu, I. Nebot-Gil and J. Sanchez-Marin, *Pure & Appl. Chem.* **56**, 1241 (1984).
36. A.D. McLean, B.H. Lengsfield, J. Pacansky and Y. Ellinger, *J. Chem. Phys.* **83**, 3567 (1985). As other examples one may quote the symmetry-breaking of the CASSCF (4e in 4MO) calculation of the $^1\Pi\Pi^*$ twisted excited state of ethylene (G. Trinquier and

J.P. Malrieu, in : "The structure of Double Bond", Patai ed., John Wiley (1990) p 1, or the symmetry-breaking in electron transfer problems (A. Faradzed, M. Dupuis, E. Clementi and A. Aviram, *J. Amer. Chem. Soc.* **112**, 4206 (1992).
37. C.W. Bauschlicher, and S.R. Langhoff, *J. Chem. Phys.* **89**, 4246 (1988)
38. A. Sanchez de Meras, M.B. Lepetit and J.P. Malrieu, *Chem. Phys. Letters*, **172**, 163 (1990).
39. R. Krishnan, M.J. Frisch and J.A. Pople, *J. Chem. Phys.* **72**, 4244 (1980).
40. M.R. Nyden and, G.A. Peterson, *J. Chem. Phys.* **14**, 6312 (1981).
41. N.C. Handy, P.J. Knowles and K. Somasudram, *Theoret. Chim. Acta*, **68**, 87 (1985).
42. P.M.W. Gill and L. Radom, *Chem. Phys. Letters*, **132**, 16 (1986).
43. M.B. Lepetit, M. Pelissier and J.P. Malrieu, *J. Chem. Phys.* **89**, 998 (1989).

44.   M.U. Bohmer and S.D. Peyerimhoff, *Z. Physik D*, **3**, 195 (1986).
45.   M.T. Bowers, W.E. Palke, K. Robins, C. Roells and S.Walsh, *Chem. Phys. Letters*, **180**, 235 (1991).
46.   J.M. McKelvey and G. Berthier, *Chem. Phys. Lett ers*, **41**, 476 (1976).
47.   D.P. Kleier, R.L. Martin, W.R. Wadt and W. Moomaw, *J. Amer. Chem. Soc.* **104**, 60 (1982) and references herein.
48.   I. Nebot-Gil and J.P. Malrieu, *J. Chem. Phys.* **77**, 2475 (1982).
49.   H.C. Longuet-Higgins and L. Salem, *Proc. Roy. Soc. (London)* A **25**, 172 (1959).
50.   G. König and G. Stolhoff, *Phys. Rev. Letters*, **65**, 1239 (1990.)
51.   For an interesting discussion see E.R. Davidson and W.T. Borden, *J. Phys. Chem.* **87**, 4783 (1983).

Molecular Orbital Electronegativity as Electron Chemical Potential in Semiempirical SCF Schemes

G. DEL RE
Chair of Theoretical Chemistry, Università "Federico II", Via Mezzocannone 4, I-80134 Napoli, Italy

1.Statement of the problem

The identification of the electronegativity of an orbital with the corresponding electron chemical potential – *i.e.* the derivative of the total energy with respect to the orbital occupation – is well known, and was in fact mentioned in Hinze and Jaffé's classical paper on electronegativities [1]. That paper referred to atomic orbitals; as far as we know, the notion of electronegativity of a molecular orbital has not been extensively discussed, although an explicit expression of the electronegativity of a molecular orbital has been given in the context of a theoretical analysis of ground-state charge transfer [2]. That expression closely matches Mulliken's classical expression [3], but does derive from an explicit general equation for the chemical potential of an electron in that orbital. We describe here the derivation of such a general equation with special reference to the semi-empirical methods leading to SCF schemes, which are especially useful nowadays for treating large molecules. Probably the method of that kind that is least charged with unphysical and possibly contradictory assumptions is the BMV method, which G. Berthier developed with his collaborators Millié and Veillard [4] in 1965, and De Brouckère [5] extended to molecules containing transition metals in 1972. It is an all-valence-electron method not involving neglect of differential overlap, in which the the diagonal elements of the Hamiltonian depend of the AO populations and the off-diagonal elements are estimated so as to avoid the drastic simplifications concealed in the Wolfsberg-Helmholtz approximation. Many of the ideas of the present author on SCF schemes and their properties go back to discussions and joint work with Berthier on his method. A late development of those discussions is the question discussed here.

The analytical determination of the derivative $\partial E_{tot}/\partial n_r$ of the total energy E_{tot} with respect to population n_r of the r-th molecular orbital is a very complicated task in the case of methods like the BMV one for three reasons: (a), those methods assume that the atomic orbital (AO) basis is non-orthogonal; (b), they involve non-linear expressions in the AO populations; (c) the latter may have to be determined as Mulliken or Löwdin population, if they must have a physical significance [6]. The rest of this paper is devoted to the presentation of that derivation on a scheme having the essential features of the BMV scheme, but simplified to keep control of the relation between the symbols introduced and their physical significance. Before devoting ourselves to that derivation, however, we with to mention the reason why the MO occupation should be treated in certain problems as a continuous variable.

119

Y. Ellinger and M. Defranceschi (eds.), Strategies and Applications in Quantum Chemistry, 119–126.
© 1996 Kluwer Academic Publishers.

In an ordinary MO scheme, fractional occupation of an orbital can only be accepted as a more or less useful fiction. This is because the whole electronic state is assumed to be correctly described by a single Slater determinant. An improvement which is sometimes indispensable is provided by CI (configuration interaction), which associates different occupation schemes to a given set of orbitals. Now, as is well known, already in the simple case of the linear combination with coefficients c_1 and c_2 of two Slater determinants, the expectation value of the population of an orbital $|j>$ is $c_1^2 n_{j1} + c_2^2 n_{j2}$, the n values denoting the (integral) occupations of that orbital in the two Slater determinants. Thus, as soon as the reference scheme becomes one of configurations over MO's, the expected occupations of the latter must be assumed to be in general fractional. Now, when we juxtapose two molecules D and A acting as a donor-acceptor pair in some redox process, a very reasonable and simple way of treating the situation theoretically, in accordance with Mulliken's original formulation [7], consists in assuming that the two partners are described by (possibly SCF) MO's that are localized on either partner and enter two Slater determinants corresponding to the states $|AD>$ and $|A^- D^+>$ [2]. For a vanishing coupling between the two states, the requirement that the actual situation should be described by a linear combination of those two states corresponding to the lowest energy can be translated into the condition that the chemical potentials of the orbitals differing in occupation in the two states should be equal [8]. This is the foundation for a rigorous derivation of the principle of electronegativity equalization [9].

2. Expression of the variations of the MO's

We consider the equation:

$$\bar{\mathbf{H}}\bar{\mathbf{C}} = \bar{\mathbf{C}}\bar{\mathbf{E}}, \tag{1}$$

where:

$$\bar{\mathbf{H}} = \mathbf{H} + \Delta\mathbf{H}, \tag{2}$$

$$\bar{\mathbf{E}} = \mathbf{E} + \Delta\mathbf{E}, \tag{3}$$

and

$$\bar{\mathbf{C}} = (\mathbf{C} + \Delta\mathbf{C})\mathbf{N}. \tag{4}$$

The Hermitian Hamiltonian matrix $\mathbf{H}$, the diagonal matrix $\mathbf{E}$, and the unitary matrix $\mathbf{C}$ are assumed to satisfy the equation:

$$\mathbf{HC} = \mathbf{CE}. \tag{5}$$

The barred matrices have the same properties as those of eqn 5; in the case of $\bar{\mathbf{C}}$ normalization to unity of the single columns is ensured by an *ad hoc* diagonal matrix $\mathbf{N}$. As will appear below (eqn 6), if terms in $\Delta\mathbf{C}$ of order higher than the first are negligible, $\mathbf{N}$ can be taken equal to the identity matrix. This is what will be assumed in the following.

We now follow the familiar procedure of perturbation theory to extract from eqn 1 the first order expression of the variations indicated by Δ. Let us start from the normalization condition, and denote by $\mathbf{M}_{.j}$ the j-th column of any given matrix $\mathbf{M}$. Since both $\mathbf{C}_{.j}$ and $\bar{\mathbf{C}}_{.j}$ are normalized to unity, to first order in $\Delta\mathbf{C}$, as has been mentioned, $\mathbf{N}$ may be taken to be unity, and we must have:

$$\bar{\mathbf{C}}_{.j}^\dagger \bar{\mathbf{C}}_{.j} = \mathbf{C}_{.j}^\dagger \mathbf{C}_{.j} + \mathbf{C}_{.j}^\dagger \Delta\mathbf{C}_{.j} + \Delta\mathbf{C}_{.j}^\dagger \mathbf{C}_{.j} = 1, \tag{6}$$

whence it appears that $\Delta C_{.j}$ must be orthogonal to $C_{.j}$, *i.e.* a linear combination of all the columns of C except $C_{.j}$ itself:

$$\Delta C_{.j} = C f_{.j}, \qquad \text{with } f_{jj} = 0. \tag{7}$$

We now substitute eqns 2, 3, and 4 into eqn 1, eliminate second order terms in ΔC, multiply on the left by $C^\dagger$, and separate the resulting equation into two as follows. We find, first of all:

$$H \Delta C_{.j} + \Delta H C_{.j} = \Delta E_{jj} C_{.j} + E_{jj} \Delta C_{.j}; \tag{8}$$

Multiplication by the j-th row of $C^\dagger$ on the left gives:

$$C_{.j}^\dagger H \Delta C_{.j} + C_{.j}^\dagger \Delta H C_{.j} = C_{.j}^\dagger \Delta E_{jj} C_{.j} + C_{.j}^\dagger E_{jj} \Delta C_{.j}, \tag{9}$$

which, since $C_{.j}^\dagger H = E_{jj} C_{.j}^\dagger$, yields the familiar expression:

$$\Delta E_{jj} = C_{.j}^\dagger \Delta H C_{.j}. \tag{10}$$

Multiplication by the k-th row ($k \neq j$) of $C^\dagger$ on the left gives:

$$C_{.k}^\dagger H \Delta C_{.j} + C_{.k}^\dagger \Delta H C_{.j} = C_{.k}^\dagger \Delta E_{jj} C_{.j} + C_{.k}^\dagger E_{jj} \Delta C_{.j}, \tag{11}$$
$$\text{for } k \neq j.$$

Since $C_{.k}$ is orthogonal to $C_{.j}$, with the same consideration as has led to eqn 10, eqn 12 becomes:

$$(E_{jj} - E_{kk}) C_{.k}^\dagger \Delta C_{.j} = C_{.k}^\dagger \Delta H C_{.j} \tag{12}$$
$$\text{for } k \neq j.$$

Comparing with eqn 7, we find:

$$f_{kj} = \frac{C_{.k}^\dagger \Delta H C_{.j}}{E_{jj} - E_{kk}} \qquad \text{for } E_{kk} \neq E_{jj}. \tag{13}$$

For $k = j$ and for degenerate eigenvalues the elements of f are taken equal to zero. Let us next consider the variation of the population-bond-order matrix, which, in the orthogonal case of eqn 1, is just:

$$P \equiv p + b = \sum_j C_{.j} \, n_j \, C_{.j}^\dagger, \tag{14}$$

where p and b are the diagonal (population) and the off-diagonal (bond-order) parts of P, respectively, and n_j is the occupation number of the j-th MO. From eqn 7 we find:

$$\Delta P = \sum_j C f_{.j} \, n_j \, C_{.j}^\dagger + \sum_j C_{.j} \, n_j \, f_{.j}^\dagger \, C^\dagger + \sum_j C_{.j} \, \Delta n_j \, C_{.j}^\dagger. \tag{15}$$

3.Derivation of electronegativity

Let us now specialize the above equations for the special case when only the population n_r of the r-th MO changes, and the reference scheme is a simple ω-technique
[10] applied to an extended-Hückel method, which is a highly simplified form of the
BMV procedure.

We start from a Hamiltonian $\mathbf{H}^\circ$ whose off-diagonal elements are assumed to form
a constant matrix γ and the diagonal elements depend on the net charges q_μ of the
individual AO's according to the expression

$$H^\circ_{\mu\mu} = \alpha_\mu + \omega q_\mu; \tag{16}$$

where

$$q_\mu = Z_\mu - p_{\mu\mu}, \tag{17}$$

α is a standard atomic parameter matrix, $\mathbf{Z}$ is the diagonal matrix of the AO occupations, and ω is a suitable constant. Finally, $p_{\mu\mu}$ is a diagonal element of the
population matrix associated to the given AO's. We adopt here the Löwdin popula-
tion analysis, i.e. assume that $\mathbf{P}$ (and therefore $\mathbf{p}$) is defined by eqn 14 in terms of
the coefficients of the Löwdin AO's associated to $\mathbf{H}^\circ$. If $\mathbf{S}$ is the AO overlap matrix,
then $\mathbf{H}$ of eqn 5 is given by

$$\mathbf{H} = \mathbf{\Lambda} \mathbf{H}^\circ \mathbf{\Lambda}, \qquad \text{with } \mathbf{\Lambda} = \mathbf{S}^{-1/2}, \tag{18}$$

and therefore, in virtue of eqns 16 and 17,

$$\mathbf{H} = \mathbf{\Lambda}\left(\alpha + \omega\,\mathbf{Z} + \gamma\right)\mathbf{\Lambda} - \omega\mathbf{\Lambda}\,\mathbf{p}\mathbf{\Lambda}. \tag{19}$$

This gives

$$\Delta\mathbf{H} = -\omega\sum_\varrho \mathbf{\Theta}^{(\varrho)}\,\Delta p_{\varrho\varrho}, \tag{20}$$

where

$$\Theta^{(\varrho)}_{\mu\nu} = \Lambda_{\mu\varrho}\Lambda_{\varrho\nu}. \tag{21}$$

Now, considering Δn_r as the only independent variation, and remembering that $\mathbf{f}$ is
an antisymmetric matrix, one gets from eqn 15

$$\Delta p_{\mu\mu} = \sum_{j,k} C_{\mu k}\,C_{\mu j}\,f_{kj}\,(n_j - n_k) + P^{(r)}_{\mu\mu}\,\Delta n_r, \tag{22}$$

where, for the sake of simplicity, the eigenvector coefficients have been assumed to
be real (as they are in molecular problems) and

$$\mathbf{P}^{(r)} = \mathbf{C}_{\cdot r}\,\mathbf{C}_{\cdot r}. \tag{23}$$

Equation 22 depends on $\Delta\mathbf{H}$ in virtue of eqn 13, and therefore does not define $\Delta p_{\mu\mu}$
completely. However, insertion of eqn 20 into eqn 13 transforms eqns 22 into a linear
system that can be solved for $\Delta p_{\mu\mu}$. We write eqn 13 for our special case in the form

$$f_{kj} = \sum_\varrho U_{kj,\varrho}\,\Delta p_{\varrho\varrho}, \tag{24}$$

with

$$U_{kj,\varrho} = -\omega \frac{T_{\varrho k} T_{\varrho j}}{E_{jj} - E_{kk}}, \tag{25}$$

for $E_{jj} \neq E_{kk}$, zero otherwise.

where $\mathbf{T} = \mathbf{\Lambda}\,\mathbf{C}$ is the eigenvector matrix of $\mathbf{H}^o$ in the original non-orthogonal basis. With this notation, eqn 22 becomes

$$\Delta p_{\mu\mu} = \sum_{\varrho} V_{\mu\varrho}\,\Delta p_{\varrho\varrho} + P_{\mu\mu}^{(r)}\,\Delta n_r, \tag{26}$$

where

$$V_{\mu\varrho} \equiv W_{\mu\mu}^{(\varrho)} \tag{27}$$

is a matrix formed by the diagonal elements of the matrices:

$$W_{\mu\nu}^{(\varrho)} = \sum_{j,k} C_{\mu k}\,U_{kj,\varrho}\,C_{\nu j}\,(n_j - n_k). \tag{28}$$

Equations 26 form a linear system which can be solved without any difficulty. Let us first of all divide eqn 26 by Δn_r and pass to the limit, so as to work directly in terms of partial derivatives. Let us then define the matrix:

$$\mathbf{X} = (1 - \mathbf{V})^{-1} - 1. \tag{29}$$

With this notation, we can write the very simple expression:

$$\frac{\partial p_{\mu\mu}}{\partial n_r} = \sum_{\alpha} X_{\mu\alpha}\,P_{\alpha\alpha}^{(r)} + P_{\mu\mu}^{(r)}. \tag{30}$$

For the off-diagonal part of $\mathbf{P}$ we have:

$$\frac{\partial b_{\mu\nu}}{\partial n_r} = \sum_{\varrho} W_{\mu\nu}^{(\varrho)} \frac{\partial p_{\varrho\varrho}}{\partial n_r} + P_{\mu\nu}^{(r)} =$$
$$= \sum_{\varrho,\alpha} W_{\mu\nu}^{(\varrho)}\,(\mathbf{X} + 1)_{\varrho\alpha}\,P_{\alpha\alpha}^{(r)} + P_{\mu\nu}^{(r)}, \tag{31}$$

with the matrices $\mathbf{W}$ and $\mathbf{X}$ defined by eqns 28 and 29, respectively. If we next define a set of matrices $\boldsymbol{\Xi}^{(\rho)}$ with elements

$$\Xi_{\mu\mu}^{(\rho)} = X_{\mu\rho} \qquad \Xi_{\mu\nu}^{(\rho)} = \sum_{\alpha} W_{\mu\nu}^{(\alpha)}\,(\mathbf{X} + 1)_{\alpha\rho}, \tag{32}$$

we can finally write

$$\frac{\partial \mathbf{P}}{\partial n_r} = \sum_{\rho} \boldsymbol{\Xi}^{(\rho)}\,P_{\rho\rho}^{(r)} + \mathbf{P}^{(r)}. \tag{33}$$

We are now ready for computing the electron chemical potential within the ω scheme. Since ours is a Hückel-like scheme, the total energy E_{tot} is the sum of the orbital energies multiplied by the pertinent occupations, and therefore

$$E_{tot} = Tr\, \mathbf{H}\mathbf{P}. \tag{34}$$

where Tr stands for the trace. Deriving the above expression with respect to n_r we obtain:

$$\frac{\partial E_{tot}}{\partial n_r} = Tr\, \frac{\partial \mathbf{H}}{\partial n_r}\mathbf{P} + Tr\, \mathbf{H}\frac{\partial \mathbf{P}}{\partial n_r}. \tag{35}$$

Substituting eqn 20, eqn 30 and eqn 31 into eqn 35 we find

$$Tr\, \frac{\partial \mathbf{H}}{\partial n_r}\mathbf{P} = -\omega\, Tr\, \sum_{\varrho}\Theta^{(\varrho)}\frac{\partial p_{\varrho\varrho}}{\partial n_r}\mathbf{P} = -\omega\sum_{\varrho\alpha}(\mathbf{X}+1)_{\varrho\alpha}\, P^{(r)}_{\alpha\alpha}\, Tr\, \Theta^{(\varrho)}\, \mathbf{P} \tag{36}$$

and

$$Tr\, \mathbf{H}\frac{\partial \mathbf{P}}{\partial n_r} = \sum_{\varrho}P^{(r)}_{\varrho\varrho}\, Tr\, \Xi^{(\varrho)}\, \mathbf{H} + Tr\, \mathbf{H}\mathbf{P}^{(r)}. \tag{37}$$

If we now apply the well known property $Tr\, \mathbf{H}\mathbf{P}^{(r)} = \epsilon_r$ (the latter being the energy of the r-th MO) and take into account that $Tr\, \Theta^{(\alpha)}\, \mathbf{P} = R_{\alpha\alpha}$, $\mathbf{R}$ being the density matrix over the non-orthogonal MO's $\mathbf{T}_{.j}$, (cf. eqn 26) obtained from $\mathbf{H}^{\circ}$, we find for eqn 35:

$$\frac{\partial E_{tot}}{\partial n_r} = \epsilon_r + \sum_{\alpha}P^{(r)}_{\alpha\alpha}[Tr\, \Xi^{(\alpha)}\, \mathbf{H} - \omega\sum_{\rho}(\mathbf{X}+1)_{\rho\alpha}\, R_{\rho\rho}]. \tag{38}$$

This is our final equation. A simplified form is found if the matrices $\mathbf{W}$ defined in eqn 28 are neglected (so that $\mathbf{X}$ and the matrices Ξ are ignored). This is possible, for example, in the case of large energy differences between MO's whose occupations are different. Then

$$\frac{\partial E_{tot}}{\partial n_r} = \epsilon_r - \omega\sum_{\alpha}P^{(r)}_{\alpha\alpha}\, R_{\alpha\alpha}. \tag{39}$$

4.Discussion

We have presented above the derivation of eqns 38 and 39 in great detail because it includes expressions of general utility, in particular the variation of the eigenvectors (eqns 7 and 24) of an MO problem after Löwdin orthogonalization and the resulting variation of the population matrix $\mathbf{P}$. The generalization to a Hamiltonian more complicated than that of eqn 19 is possible by following step by step the above derivation.

The physical meaning of our final equation is best seen on eqn 39. The term containing ω is essentially the self-energy correction introduced by Mulliken in his analysis of electronegativities to account for the average repulsion of electrons occupying the same orbital. In order to get an idea of the orders of magnitude, let us apply eqn 39 to a model computation of FeCO, made to compare the CIPSI results of Berthier et al. [11] with those of a simple orbital scheme. Consider one of the two π systems of FeCO, treated under the assumption of full localization (and therefore strict $\sigma-\pi$ separation) in an iterative MO-LCAO scheme using as an AO basis maximum localization hybrids [12], Hoffman's atomic parameters [13], and Cusachs' expression [14] for the off-diagonal elements of the Hamiltonian $\mathbf{H}^\circ$. Since this is just an illustration of the numerical aspects of the equations given above, we need not justify further the scheme used. The special features which make the present example especially suited for our purpose include the fact that in the case of the π system the MO occupation numbers must be given the values 2, 1.5, 0 (in the order of increasing orbital energies) in order to ensure the equivalence between the two degenerate π systems of our linear molecule. This feature is especially important since $W_{\mu\nu}^{(\varrho)}$ of eqn 28 is the sum of terms containing as factors the differences between MO occupation numbers $n_j - n_k$. This fact implies that only MO's with different occupation numbers play a role in the terms by which eqn 38 differs from the simpler form 39.

Table 1: Source data and $\partial \mathbf{P}/\partial n_r$ for the average π system of FeCO

a. Overlap matrix for one π system of FeCO

$$
\begin{array}{ccc}
1.0000 & 0.2783 & 0 \\
0.2783 & 1.0000 & 0.2414 \\
0 & 0.2414 & 1.0000
\end{array}
$$

b. Ham. matrix (eV) with σ correction at convergence ($\omega = -2.256$ eV)

$$
\begin{array}{ccc}
-9.6177 & -2.2081 & 0 \\
-2.2081 & -10.9490 & -2.7811 \\
0 & -2.7811 & -12.6638
\end{array}
$$

c. Eigenvalues (italics, eV) and Löwdin charge bond-order matrix

$$
\begin{array}{ccc}
\mathit{-12.7349} & \mathit{-11.4723} & \mathit{-9.4836} \\
0.7671 & 0.5442 & -0.0530 \\
0.5442 & -0.2747 & 0.0418 \\
-0.0530 & 0.0418 & -0.4924
\end{array}
$$

d. Derivative of $\mathbf{P}$ with respect to n_2 and diagonal elements of $\mathbf{P}^{(2)}$ (italics)

$$
\begin{array}{ccc}
0.2087 & -0.4060 & -0.0689 \\
-0.4060 & 0.7583 & 0.1859 \\
-0.0689 & 0.1859 & -0.0281 \\
\mathit{0.1502} & \mathit{0.8355} & \mathit{0.0142}
\end{array}
$$

The source matrices (for the sequence Fe C O) are presented in Table 1 together with the resulting derivative of the Löwdin population matrix $\mathbf{P}$. The electronegativities derived from the complete expression 38 and from the approximate expression 39 are -7.1690 eV and -7.3582 eV, respectively, thus suggesting that even in the unfavourable case here considered eqn 39 is a reasonable approximation of the exact expression. Two remarks may be added here. First, as was shown in a preceding paper [3], a correction must be added to the expression of the electron chemical potential whenever the given molecule is in the presence of another molecule or of a solid surface. Second, although we have referred to the ω scheme and to Löwdin's population analysis, no implication is made that the above analysis depends on either assumption. As has been mentioned, it has been designed for general all-valence SCF schemes. Also the introduction of Mulliken's population analysis is straightforward, since in that case

$$\mathbf{P}^{(Mu)} = (\mathbf{R}\,\mathbf{S} + \mathbf{S}\,\mathbf{R})/2 = (\mathbf{\Lambda}^{-1}\mathbf{P}^{(L\ddot{o})}\mathbf{\Lambda} + transpose)/2, \qquad (40)$$

and the whole derivation above can be applied to the Mulliken population without any difficulty.

Acknowledgement. The Author thanks the Italian National Research Council (CNR) and the Italian Ministry of Universities (MURST) for support.

References

1. J. Hinze, M. A. Whitehead, and H. H. Jaffé, *J. Am. Chem. Soc.*, **85**, 184 (1963) .

2. G. Del Re, *J. Chem. Soc. Far. Trans.*, **77**, 2067-2076 (1981).

3. R. S. Mulliken, *J. Chem. Phys.*, **2**, 782 (1934).

4. G. Berthier, Ph. Millié, and A. Veillard, *J. Chimie Phys.* , **62**, 8 (1965).

5. G. De Brouckère, *Theor. Chim. Acta*, **19**, 310 (1970).

6. G. Del Re, P. Otto, J. Ladik, *Isr. J. Chem.*, **19**, 265 (1970).

7. R. S. Mulliken, *J. Am. Chem. Soc.*, **72**, 600 (1950); **74**, 811 (1952).

8. G. Del Re, P. Otto, J. Ladik, *Int. J. Quant. Chem.*, **37**, 497 (1990).

9. R. T. Sanderson, *Science*, **114**, 670 (1951).

10. A. Streitwieser jr., Molecular Orbital Theory for Organic Chemists, Wiley & Sons, New York, 1961.

11. G. Berthier, A. Daoudi, M. Suard, *J. Mol. Struct. (Theochem)*, **179**, 407 (1988).

12. G. Del Re, *J. Mol. Struct. (Theochem)*, **169**, 487 (1988).

13. J.-Y. Saillard, R. Hoffmann, *J. Am. Chem. Soc.*, **106**, 2006 (1984), **74**, 811 (1952).

14. L. C. Cusachs, *J. Chem. Phys.*, **43**, S157 (1965).

Quasicrystals and Momentum Space

J.L. CALAIS
Quantum Chemistry Group, University of Uppsala
Box 518, S - 75120 - Uppsala, Sweden

1. Introduction

In November 1984 the world of crystallography was thoroughly shaken by the news that "forbidden" peaks characteristic of icosahedral symmetry had been recorded in electron diffraction diagrams of an Al-Mn-alloy [1]. According to "classical" crystallography long range order is compatible with rotations through multiples of $2\pi/6$ or $2\pi/4$, but not with rotations through $2\pi/5$ [2]. The point group of a space group must be one of the 32 crystallographic point groups and the icosahedral group is certainly not one of them. Scientific results of that nature are among the most interesting ones, since they open up qualitatively new perspectives.

A crystal is an extended system with (in principle) perfect long range order, which is invariant under all operations of a certain space group. At the other extreme we have disordered systems with a "completely" random arrangement of its constituent atoms. Intermediate cases with more or less short range order have been known for a long time [3]. What was unexpected in the paper by Shechtman et al. [1] was the combination of long range order and a non crystallographic point group. Already in 1902 the French mathematician Esclangon [4] pointed out, however, that arrangements which are aperiodic but non random are possible. And even though the paper by Shechtman et al. [1] must be regarded as the one which opened up this new field of crystallography, it seems that some Japanese results 20 years earlier [5] should also be interpreted as providing experimental evidence for the existence of quasi-periodic structures.

During the nearly ten years which have passed since the appearance of the "Shechtman paper" a large amount of both experimental and theoretical research has been carried out on quasiperiodic structures. For more material about quasicrystals we refer to a paper in *La Recherche* by the French collaborator in the Shechtman team [6], to a thesis by Dulea [7], and to a survey paper with a large number of references [8].

Last year a magnificent paper by Mermin appeared in the Reviews of Modern Physics [9], as the (so far) crowning contribution to a series of papers describing nothing less than a reformulation of crystallography [10 - 18]. Emphasising reciprocal space concepts Mermin and his collaborators have been able to treat both "classical" crystals and quasicrystals with the same method. As is often the case with truly original work this first of all throws new light on the theory of the "ordinary" space groups, which leads to a deeper understanding of notions and relationships believed to be well known. Then it provides a straightforward

127

Y. Ellinger and M. Defranceschi (eds.), Strategies and Applications in Quantum Chemistry, 127–138.
© *1996 Kluwer Academic Publishers.*

classification of both crystals and quasicrystals, as well as incommensurately modulated crystals and quasicrystals. The procedure offers a simple explanation of how seemingly contradictory concepts can in fact be combined in a perfectly consistent manner. This major achievement definitely deserves to become better known by all physicists and chemists who work with extended systems. One of the aims of the present paper is to contribute towards that goal.

Over the last few years there has been an increasing interest in using momentum space concepts for both molecules and polymers and also to perform explicit calculations directly in momentum space rather than making the detour over position space [19, 20]. Conceptually it is very valuable to work so to speak in parallel in position and momentum space, since corresponding concepts often help to "clarify each other". In the present paper we want to confront - in a preliminary way - the procedures proposed by Mermin and collaborators with certain momentum space notions. We expect first of all to get a better understanding of these procedures. And more specifically we want to use Mermin's results for investigating the symmetry properties of momentum wave functions for quasiperiodic systems. In this connection it is important to distinguish the closely related by still different notions of "reciprocal space" and "momentum space". In a certain sense these terms denote the same object. Momentum space and position space refer to different representations of wave functions related by Fourier transforms for any types of systems. The term "reciprocal space", on the other hand is normally used only in connection with solids. Until relatively recently this concept has been used only about crystals: solids with long range order which are invariant under one of the 230 space groups that can exist in three dimensions.

Mermin's "generalised crystallography" works primarily with reciprocal space notions centered around the density and its Fourier transform. Behind the density there is however a wave function which can be represented in position or momentum space. The wave functions needed for quasicrystals of different kinds have symmetry properties - so far to a large extent unknown. Mermin's reformulation of crystallography makes it attractive to attempt to characterise the symmetry of wave functions for such systems primarily in momentum space.

In the next setion we review some key concepts in Mermin's approach. After that we summarise in section III some aspects of the theory of (ordinary) crystals, which would seem to lead on to corresponding results for quasicrystals. A very preliminary sketch of a study of the symmetry properties of momentum space wave functions for quasicrystals is then presented in section IV.

2. Indistinguishability and Identity

As stressed by Mermin and collaborators [9 - 18] it is far too restrictive to define the structural indistinguishability of two mesoscopically homogeneous materials with reference to identical densities. Instead of the densities themselves one should study the properties of the *correlation functions*,

$$c_n(r_1, r_2, r_3, \dots r_n) = \frac{1}{V} \int dr \, \rho(r_1 - r)\rho(r_2 - r)\rho(r_3 - r)\dots\rho(r_n - r) \,. \quad (II.1)$$

Here V is the volume of the Born-von Kármán region, i.e. that part of position space which is repeated as a result of the fundamental periodic boundary conditions. The integration in (II.1) is carried out over that region, which we denote by BK.

For n = 1 we have for example,

$$c_1(r_1) = \frac{1}{V} \int dr\, \rho(r_1 - r) = \frac{N}{V}, \tag{II.2}$$

i.e. the average density of the system.

Two densities $\rho(r)$ and $\rho'(r)$ are said to be *indistinguishable* if all their correlation functions c_n and c_n' are identical. As shown by Mermin and collaborators their Fourier transforms then have some very interesting properties. We can always expand a density $\rho(r)$ in plane waves,

$$\rho(r) = \sum_k^{all} e^{ik\cdot r}\, \rho_k\,; \tag{II.3a}$$

$$\rho_k = \frac{1}{V} \int_{BK} dv\, \rho(r)\, e^{-ik\cdot r}. \tag{II.3b}$$

The wave vectors k can be expressed in terms of any basis vectors we choose. At the moment there is neither a direct nor a reciprocal lattice. Using (II.3a) in (II.1) we see that the Fourier components of two indistinguishable densities can differ only by a phase factor:

$$\rho'_k = e^{2\pi i\chi(k)}\, \rho_k\,. \tag{II.4}$$

The *gauge function* $\chi(k)$ is linear in its argument:

$$\chi(k_1) + \chi(k_2) = \chi(k_1 + k_2)\,. \tag{II.5}$$

A related concept is that of *phase function* $\phi_g(k)$ which relates the Fourier components of a density $\rho(r)$ and those of a transformed density obtained by letting a point group operation g work on r:

$$\rho(gr) = \sum_k^{all} e^{ik\cdot(gr)}\, \rho_k = \sum_{k'}^{all} e^{ik'\cdot r}\, \rho_{(gk')}\,. \tag{II.6}$$

If g is an element of the point group of the material meaning that $\rho(r)$ and $\rho(gr)$ are indistinguishable for all elements g in that group, corresponding Fourier components can differ only by a phase factor:

$$\rho_{(gk)} = e^{2\pi i\phi_g(k)}\, \rho_k\,. \tag{II.7}$$

A "generalised" space group is specified by a point group and the associated phase functions $\phi_g(k)$. The ordinary space groups constitute special cases of these generalised space groups.

Since $(gh)k = g(hk)$ we get with (II.7)

$$\rho_{((gh)k)} = e^{2\pi i\phi_{gh}(k)}\, \rho_k =$$

$$= \rho_{(g(hk))} = e^{2\pi i\phi_g(hk)}\, \rho_{(hk)} = e^{2\pi i\phi_g(hk)}\, e^{2\pi i\phi_h(k)}\, \rho_k\,, \tag{II.8}$$

which implies

$$\phi_{gh}(k) = \phi_g(hk) + \phi_h(k)\,. \tag{II.9}$$

From this *group compatibility condition* Mermin and his collaborators have derived both all the "ordinary" crystallographic and the quasicrystallographic space groups.

If $gk = k$, (II.7) implies that either the Fourier component ρ_k vanishes or the phase function $\phi_g(k)$ is an integer or zero. Another way of expressing that important result is to say, that given a phase function $\phi_g(k)$, those wave vectors k, for which that function is *not* equal to an integer or zero, determine a set of vanishing Fourier components ρ_k. The number of vanishing terms in the Fourier expansion (II.3a) of the density is a kind of measure of the degree of symmetry in the system.

3. Momentum space characteristics of crystals

The traditional characterisation of an electron density in a crystal amounts to a statement that the density is invariant under all operations of the space group of the crystal. The standard notation for such an operation is $\{R|m\}$, where R stands for the point group part (rotations, reflections, inversion and combinations of these) and the direct lattice vector m denotes the translational part. When such an operation works on a vector r we get

$$\{R|m\}r = Rr + m . \tag{III.1}$$

The details of the operation Rr can be further specified by the 3x3 matrix which represents the operation R in a suitably chosen coordinate system [2], in which also the vector r is expressed. For the operation on a *function of r* we need the inverse of the space group operation,

$$\{R|m\}^{-1} = \{R^{-1}|- R^{-1}m\} . \tag{III.2}$$

We thus have for an arbitrary function $f(r)$,

$$\{R|m\}f(r) = f(\{R|m\}^{-1}r) = f(R^{-1}r - R^{-1}m) . \tag{III.3}$$

A crystal characterised by a space group G has an electron density $\rho(r)$ which is invariant under all elements $\{R|m\}$ of G:

$$\{R|m\}\rho(r) = \rho(\{R|m\}^{-1}r) = \rho(r) . \tag{III.4}$$

The electron density is the diagonal element of the number density matrix $N(r,r')$, i.e the first order reduced density matrix after integration over the spin coordinates, :

$$\rho(r) = N(r,r) = \int d\zeta \, \gamma(r\zeta|r\zeta) . \tag{III.5}$$

A transformation of the number density matrix N under a space group operation means that both variables are transformed:

$$N(\{R|m\}^{-1}r, \{R|m\}^{-1}r') = N(R^{-1}r - R^{-1}m, R^{-1}r' - R^{-1}m) . \tag{III.6}$$

The following relation and its inverse hold between the elements of the number density matrices in momentum and position space [21]:

$$\underline{N}(p,p') = \frac{1}{8\pi^3} \int dr dr' \, N(r,r') \, e^{-i(p\cdot r - p'\cdot r')} . \tag{III.7}$$

The momentum space counterpart of (III.6) can therefore be written,

$$\frac{1}{8\pi^3} \int dr dr' \, N(\{R|m\}^{-1}r, \{R|m\}^{-1}r') \, e^{-i(p\cdot r - p'\cdot r')} =$$

$$= \underline{N}(R^{-1}p, R^{-1}p') \, e^{-i(p - p')\cdot m} . \tag{III.8}$$

Thus the point group part of the operation works on the momentum coordinates and the translation part gives rise to a phase factor. We notice that this phase factor reduces to 1 in the diagonal elements, or in general when the difference between the the two arguments of $\underline{N}(p,p')$ is a reciprocal lattice vector.

If the elements of the number density matrix in position space are invariant under all operations of the space group, i.e. if

$$N(\{R|m\}^{-1}r,\{R|m\}^{-1}r') = N(r,r') \text{ for } \{R|m\} \in \mathbf{G}, \tag{III.9}$$

we get with (III.8), that their momentum space counterparts satisfy

$$\underline{N}(R^{-1}p,R^{-1}p') \, e^{-i(p-p')\cdot m} = \underline{N}(p,p') . \tag{III.10}$$

The momentum distribution, i.e. the diagonal element of (III.10) then satisfies

$$\varrho(p) = \underline{N}(p,p) = \underline{N}(R^{-1}p,R^{-1}p) . \tag{III.11}$$

The reciprocal form factor [22] is the Fourier transform of the momentum distribution,

$$B(r) = \int dp \, \varrho(p) \, e^{ip\cdot r} . \tag{III.12}$$

Using (III.11) we see that the reciprocal form factor of a crystal which is invariant under a space group, satisfies the relations,

$$B(R^{-1}r) = B(r) , \tag{III.13}$$

for all point group elements R of the space group.

We notice that neither the momentum distribution nor the reciprocal form factor seems to carry any information about the translational part of the space group. The non diagonal elements of the number density matrix in momentum space, on the other hand, transform under the elements of the space group in a way which brings in the translational parts explicitly.

The number density matrix for a crystal with translation symmetry can be written in terms of its natural orbitals [23, 24], as

$$N(r,r') = \sum_{\mu}^{all} \sum_{i=1}^{2} \sum_{k}^{BZ} \phi_{\mu i}(k;r) n_\mu(k) \phi_{\mu i}^*(k;r') . \tag{III.14}$$

This is the most general expression obtained from a set of natural spin orbitals written in spinor form as

$$\psi_\mu(k;x) = \alpha(\zeta)\phi_{\mu 1}(k;r) + \beta(\zeta)\phi_{\mu 2}(k;r) =$$

$$= [\alpha(\zeta), \beta(\zeta)] \left[\phi_{\mu 1}(k;r), \phi_{\mu 2}(k;r) \right] = [\alpha(\zeta), \beta(\zeta)] \, \phi_\mu(k;r) . \tag{III.15}$$

The orbitals $\phi_{\mu i}(k;r)$ are Bloch functions labeled by a wave vector k in the first Brillouin zone (BZ), a band index μ, and a subscript i indicating the spinor component. The combination of k and μ can be thought of as a label of an irreducible representation of the space group of the crystal. The quantity $n_\mu(k)$ is the occupation function which measures the degree of occupation at wave vector k in band μ.

The momentum space counterpart of the Bloch orbital $\phi_{\mu i}(\mathbf{k};\mathbf{r})$,

$$\underline{\Phi}_{\mu i}(\mathbf{k};\mathbf{p}) = \frac{1}{\sqrt{8\pi^3}} \int dv \ \phi_{\mu i}(\mathbf{k};\mathbf{r}) \ e^{-i\mathbf{p}\cdot\mathbf{r}} , \qquad (III.16)$$

vanishes unless $\mathbf{k}$-$\mathbf{p}$ is a reciprocal lattice vector $\mathbf{K}$ [25]. In other words this function of the momentum variable $\mathbf{p}$ labeled by the wave vector $\mathbf{k}$, vanishes except when $\mathbf{p} = \mathbf{k}$, and at equivalent points $\mathbf{p} = \mathbf{k}+\mathbf{K}$ in the other Brillouin zones.

We expand the density (III.5) in a Fourier series,

$$\rho(\mathbf{r}) = \sum_{\mathbf{K}}^{all} e^{i\mathbf{K}\cdot\mathbf{r}} \rho_{\mathbf{K}} ; \qquad (III.17a)$$

$$\rho_{\mathbf{K}} = \frac{1}{V_{0a}} \int_{cell} dv \ \rho(\mathbf{r}) \ e^{-i\mathbf{K}\cdot\mathbf{r}} = \frac{1}{V} \int_{BK} dv \ \rho(\mathbf{r}) \ e^{-i\mathbf{K}\cdot\mathbf{r}} . \qquad (III.17b)$$

Here BK stands for "Born-von Kármán" and denotes the basic region of periodicity associated with the periodic boundary conditions. That "large period" must be carefully distinguished from the "small period" associated with the crystal lattice. BK contains N cells of volume V_{0a} and thus has the volume $V = NV_{0a}$. Wave functions have the "large period", but quantities like the density and the crystal potential have the "small period". We first notice the following connection between the Fourier component (III.17b) and the density matrix in momentum space, obtained from the inverse of (III.7):

$$\rho_{\mathbf{K}} = \frac{1}{V} \int d\mathbf{p} \ \underline{N}(\mathbf{p},\mathbf{p} - \mathbf{K}) = \frac{8\pi^3}{V^2} \sum_{\mathbf{P}}^{all} \underline{N}(\mathbf{p},\mathbf{p}-\mathbf{K}) . \qquad (III.18)$$

Combining the inverses of (III.14) and (III.16) we get the natural expansion for a general element of the number density matrix in momentum space:

$$\underline{N}(\mathbf{p},\mathbf{p}') = \sum_{\mu}^{all} \sum_{i=1}^{2} \sum_{\mathbf{k}}^{BZ} \underline{\Phi}_{\mu i}(\mathbf{k};\mathbf{p}) n_{\mu}(\mathbf{k}) \underline{\Phi}_{*,\mu i}(\mathbf{k};\mathbf{p}') =$$

$$= \sum_{\mathbf{k}}^{BZ} \underline{N}_{\mathbf{k}}(\mathbf{p},\mathbf{p}') . \qquad (III.19a)$$

Here the component of the number density matrix associated with the wave vector $\mathbf{k}$ is thus

$$\underline{N}_{\mathbf{k}}(\mathbf{p},\mathbf{p}') = \sum_{\mu}^{all} \sum_{i=1}^{2} \underline{\Phi}_{\mu i}(\mathbf{k};\mathbf{p}) n_{\mu}(\mathbf{k}) \underline{\Phi}_{*,\mu i}(\mathbf{k};\mathbf{p}') . \qquad (III.19b)$$

Substituting (III.19) in (III.18) and using the special properties of (III.14) we can then write the Fourier component of the density as

$$
\rho_K = \frac{8\pi^3}{V^2} \sum_{\mu}^{all} \sum_{i=1}^{2} \sum_{k}^{BZ} \sum_{K'}^{all} \phi_{\mu i}(k;k+K')n_{\mu}(k)\phi_{+,\mu i}(k;k+K'-K)
$$

(III.20)

A more condensed expression is obtained using (III.14) and (III.17b):

$$
\rho_K = \frac{1}{V} \sum_{\mu}^{all} \sum_{i=1}^{2} \sum_{k}^{BZ} n_{\mu}(k) \int_{BK} dv \, |\phi_{\mu i}(k;r)|^2 \, e^{-iK\cdot r} =
$$

$$
= \sum_{\mu}^{all} \sum_{i=1}^{2} \sum_{k}^{BZ} n_{\mu}(k) \, u_{\mu i}(k;K) .
$$

(III.21)

Here $u_{\mu i}(k;K)$ is the Fourier component of the square of the absolute value of the Bloch orbital $\phi_{\mu i}(k;r)$, which can be written as a product of a plane wave and a function $u_{\mu i}(k;r)$ having the periodicity of the lattice:

$$
|\phi_{\mu i}(k;r)|^2 = |u_{\mu i}(k;r)|^2 =
$$

$$
= \sum_{K}^{all} e^{iK\cdot r} \, u_{\mu i}(k;K) .
$$

(III.22)

Using (III.16) we can also write this Fourier component in terms of the momentum space orbitals as

$$
u_{\mu i}(k;K) = \frac{1}{V} \sum_{p}^{all} \phi_{\mu i}(k;p)\phi_{\mu i}^{*}(k;p-K) =
$$

$$
= \frac{1}{V} \sum_{K'}^{all} \phi_{\mu i}(k;k+K')\phi_{\mu i}^{*}(k;k+K'-K) .
$$

(III.23)

If the density is invariant under the space group operation $\{R|m\}$ we have with (III.4) and (III.17b),

$$
\rho_K = e^{-iK\cdot m} \rho_{(R^{-1}K)} = \rho_{(R^{-1}K)} .
$$

(III.24)

It is important to distinguish between symmetry properties of wave functions on one hand and those of density matrices and densities on the other. The symmetry properties of wave functions are derived from those of the Hamiltonian. The "normal" situation is that the Hamiltonian commutes with a set of symmetry operations which form a group. The eigenfunctions of that Hamiltonian must then transform according to the irreducible representations of the group. *Approximate* wave functions with the same symmetry properties can be constructed, and they make it possible to simplify the calculations.

In the case of a perfect crystal the Hamiltonian commutes with the elements of a certain space group and the wave functions therefore transform under the space group operations according to the irreducible representations of the space group. Primarily this means that the wave functions are Bloch functions labeled by a wave vector **k** in the first Brillouin zone. Under pure translations they transform as follows

$$\{1|m\}\ \phi_{\mu i}(\mathbf{k};\mathbf{r})\ =\ \phi_{\mu i}(\mathbf{k};\mathbf{r})\ e^{-i\mathbf{k}\cdot\mathbf{m}}\ . \tag{III.25}$$

This implies that a density built up from such Bloch functions [cf (III.5) and (III.14)] is invariant under all such translations [the "little" period]:

$$\{1|m\}\ \rho(\mathbf{r})\ =\ \rho(\mathbf{r}-\mathbf{m})\ =\ \rho(\mathbf{r})\ . \tag{III.26}$$

Corresponding relations for arbitrary space group elements $\{R|m\}$ show that if the orbitals $\phi_{\mu i}(\mathbf{k};\mathbf{r})$ which make up the density transform asthe irreducible representations of the space group, the density is invariant under all the operations of that group.

It is also of interest to study the "inverse" problem. If something is known about the symmetry properties of the density or the (first order) density matrix, what can be said about the symmetry properties of the corresponding wave functions? In a one electron problem the effective Hamiltonian is constructed either from the density [in density functional theories] or from the full first order density matrix [in Hartree-Fock type theories]. If the density or density matrix is invariant under all the operations of a space group, the effective one electron Hamiltonian commutes with all those elements. Consequently the eigenfunctions of the Hamiltonian transform under these operations according to the irreducible representations of the space group. We have a scheme which is selfconsistent with respect to symmetry.

The symmetry properties of the density show up experimentally as properties of its Fourier components $\rho_{\mathbf{k}}$. If those components vanish except when the wave vector **k** equals one of the lattice vectors **K** of a certain reciprocal lattice, the general plane wave expansion of the density,

$$\rho(\mathbf{r})\ =\ \sum_{\mathbf{k}}^{\text{all}} e^{i\mathbf{k}\cdot\mathbf{r}}\ \rho_{\mathbf{k}}\ ; \tag{III.27}$$

reduces to (III.17a). Since $\mathbf{K}\cdot\mathbf{m} = 2\pi$ times an integer, we then have

$$\{1|m\}\ \rho(\mathbf{r})\ =\ \sum_{\mathbf{K}}^{\text{all}} e^{i\mathbf{K}\cdot(\mathbf{r}-\mathbf{m})}\ \rho_{\mathbf{K}}\ =$$

$$=\ \sum_{\mathbf{K}}^{\text{all}} e^{i\mathbf{K}\cdot\mathbf{r}}\ \rho_{\mathbf{K}}\ =\ \rho(\mathbf{r})\ . \tag{III.28}$$

If (III.24) holds we get the corresponding result for arbitrary space group elements $\{R|m\}$.

The symmetry properties of the momentum space wave functions can be obtained either from their position space counterparts or more directly from the counterpart of the Hamiltonian in momentum space.

4. Momentum space characteristics of quasicrystals

One of the main points in the papers by Mermin and his collaborators [9 - 18] is the insistence on the primacy of reciprocal space. The properties of the Fourier transform of the density rather than the density itself determine those properties which are of importance for "generalized" crystallography. As pointed out by Mermin that point view was stressed in a paper by Bienenstock and Ewald already in 1962 [26].

Irrespective of the type of extended system we are interested in we impose periodic boundary conditions in position space - "the large period": BK. Such conditions imply a discretisation of momentum and reciprocal space [27] which means that integrations are replaced by summations:

$$\int d\mathbf{p}\, f(\mathbf{p}) \;\rightarrow\; \frac{8\pi^3}{V} \sum_{\mathbf{p}} f(\mathbf{p}) \; . \qquad\qquad (IV.1)$$

The discrete momenta can be written as

$$\mathbf{p} \;=\; \frac{2\pi}{G} (\mathbf{b}_1 v_1 + \mathbf{b}_2 v_2 + \mathbf{b}_3 v_3) \; , \qquad\qquad (IV.2)$$

where the v_i are positive or negative integers or zero, and the very large even integer G characterizes the BK region ($G^3 = N$). The reciprocal basis vectors $\mathbf{b}_j$ do not require any actual physical lattice, but can be seen as just providing a suitable framework. We have used (IV.1) several times in the previous section, but there we had lattices both in direct and in reciprocal space, and then this procedure may have seemed more natural. In the present section there is definitely no lattice in direct space and the "lattice" in reciprocal space may be of a different nature from the ordinary ones. Because of the periodic boundary conditions, (IV.1) should still be used, however.

The Fourier expansion of the density in an extended system which does not have any particular symmetry is

$$\rho(\mathbf{r}) \;=\; \sum_{\mathbf{k}}^{all} e^{i\mathbf{k}\cdot\mathbf{r}}\, \rho_{\mathbf{k}} \; . \qquad\qquad (IV.3)$$

This sum over *all reciprocal space vectors* of the form (IV.2) should be carefully distinguished from the expansion (III.4) of the density of a periodic crystal. If the density has the "little period", the expansion (IV.3) reduces to a sum over all reciprocal *lattice vectors*. The general case (IV.3) and the periodic case (III.4) actually represent two extreme cases. The presence of "more and more symmetry" in the density can be gauged by the disappearance of more and more Fourier components $\rho_{\mathbf{k}}$ in (IV.3). If some of the Fourier components in (IV.3) vanish, but not necessarily all which do not correspond to a set of reciprocal lattice vectors, we have a Fourier expansion of a density with another type of long range order than the one known from traditional crystals. There are *quasicrystals, incommensurately modulated crystals or incommensurately modulated quasicrystals* [9].

Using the inverse of (III.7) we can write the density of an arbitrary extended system as

$$\rho(\mathbf{r}) = \frac{8\pi^3}{V^2} \sum_{\mathbf{k}}^{all} e^{i\mathbf{k}\cdot\mathbf{r}} \sum_{\mathbf{p}}^{all} \underline{N}(\mathbf{p},\mathbf{p}-\mathbf{k}) , \qquad (IV.4)$$

which means that the Fourier component of the density can be written

$$\rho_{\mathbf{k}} = \frac{8\pi^3}{V^2} \sum_{\mathbf{p}}^{all} \underline{N}(\mathbf{p},\mathbf{p}-\mathbf{k}) . \qquad (IV.5)$$

This should be compared to (III.18) where the role of $\mathbf{k}$ in (IV.5) is played by a reciprocal *lattice vector* $\mathbf{K}$.

Mermin's conceptual starting point is a set of vectors $\mathbf{k}$ in reciprocal space which correspond to sharp Bragg peaks in the experimental diffraction pattern. The non vanishing Fourier components are then to be found for wave vectors which can be characterized as the set of all integral linear combinations of a certain finite set of D basis vectors $\mathbf{b}^{(i)}$, i = 1, 2, ...D. In an ordinary crystal D = 3 and the point group must be one of the 32 crystallographic point groups. If we have a non crystallographic point group the *rank* D of the lattice can be larger than three. Such a system is called a quasicrystal. A system with a crystallographic point group and a lattice with a rank D higher than three is called an incommensurately modulated crystal.

An important and interesting question is obviously whether for quasicrystals and incommensurately modulated crystals there is anything corresponding to the Bloch functions for crystals. Momentum space may be a better hunting ground in that connection than ordinary space, where we have no lattice. Not only is there no lattice, one cannot even specify the location of each atom yet [8].

A Bloch function for a crystal has two characteristics. It is labeled by a wave vector $\mathbf{k}$ in the first Brillouin zone, and it can be written as a product of a plane wave with that particular wave vector and a function with the "little" period of the direct lattice. Its counterpart in momentum space vanishes except when the argument $\mathbf{p}$ equals $\mathbf{k}$ plus a reciprocal lattice vector. For quasicrystals and incommensurately modulated crystals the reciprocal lattice is in a certain sense replaced by the D-dimensional lattice L spanned by the vectors $\mathbf{b}^{(i)}$. It is conceivable that what corresponds to Bloch functions in momentum space will be non vanishing only when the momentum $\mathbf{p}$ equals $\mathbf{k}$ plus a vector of the lattice L.

The problem is to "translate" the fact that certain terms are absent in the expansion (IV.3) to symmetry properties of the density in the sense of transformation properties under certain operations. We have a density with non vanishing Fourier components only for such wave vectors $\mathbf{k}$ which belong to the lattice L:

$$\rho(\mathbf{r}) = \sum_{\mathbf{k}}^{L} e^{i\mathbf{k}\cdot\mathbf{r}} \rho_{\mathbf{k}} . \qquad (IV.6)$$

Mermin [9, 18] has given a recipe for the construction of a set of Fourier components for a density characterised by a certain space group. The space group is then specified by a point group G, a lattice of wave vectors in the sense discussed above, and a set of phase functions $\phi_g(\mathbf{k})$, one for each element of the point group.

The Fourier components of the density are then obtained from the expression

$$\rho_{\mathbf{k}} = \sum_{h}^{G} f(h\mathbf{k})\, e^{-2\pi i \phi_h(\mathbf{k})} . \tag{IV.7}$$

Here f is a function on the lattice satisfying $f^*(\mathbf{k}) = f(-\mathbf{k})$ and such that $f(\mathbf{k})$ is the Fourier transform of a function with no symmetries whatever. That last condition is imposed in order to avoid that the density obtained from (IV.7) gets any symmetries which are not associated with the point group G, and also to prevent $\rho_{\mathbf{k}}$ from vanishing on a set of wave vectors so large that the lattice is thinned out to a sublattice for which the space group would have a different character. The components (IV.7) transform under the elements of the point group according to the fundamental rule (II.7).

An effective one electron Schrödinger equation with a local potential $V(\mathbf{r})$ in position space, (atomic units),

$$[\frac{\mathbf{p}^2}{2} + V(\mathbf{r})]\, \psi(\mathbf{r}) = E\, \psi(\mathbf{r}) , \tag{IV.8}$$

corresponds in momentum space to the following equation [19],

$$\frac{\mathbf{p}^2}{2}\underline{\psi}(\mathbf{p}) + \int d\mathbf{q}\, \underline{V}(\mathbf{p}\text{-}\mathbf{q})\, \underline{\psi}(\mathbf{q}) = E\, \underline{\psi}(\mathbf{p}) . \tag{IV.9}$$

Wave functions in position and momentum spacce are related as in (III.16), and the Fourier component of the potential is

$$\underline{V}(\mathbf{q}) = \frac{1}{8\pi^3} \int d\mathbf{v}\, V(\mathbf{r})\, e^{-i\mathbf{q}\cdot\mathbf{r}} . \tag{IV.10}$$

In density functional theories the potential is determined by the density, and consequently its Fourier components are related to those of the density. One can therefore connect the symmetry properties of the momentum functions, in other words the transformation properties of $\underline{\psi}(\mathbf{p})$ under the operations of the point group, with those of the Fourier components of the density, (II.7).

What has been sketched here is obviously just the bare framework of a general investigation of the symmetry properties of momentum space functions in quasicrystals. With all the information available in the papers by Mermin and collaborators it should however be a very tempting enterprise to go ahead along the lines sketched and learn about the details of the symmetry properties of those wave functions - both in momentum and in posiition space - which will be needed in quasiperiodic extended systems.

References

1.   D.Shechtman, I. Blech, D. Gratias and J.W. Cahn, *Phys. Rev. Letters,* **53**, 1951 (1984).
2.   J. F. Cornwell, Group Theory in Physics, Vol. 1, Academic Press, London (1989).
3.   See e.g. Electrons in Disordered Metals and at Metallic Surfaces, P. Phariseau, B.L Györffy and L. Scheire Eds., NATO Advanced Study Institute Series, Series B: Physics, Volume 42, Plenum Press New York and London (1979).
4.   M.E. Esclangon, *C.R. Acad. Sci. (Paris)* **135**, 891 (1902).
5.a   S. Tanisaki, *J. Phys. Soc. Japan* , **16**, 579 (1961).
  b   Y. Yamada, S. Shibuya and S. Hoshino, *J. Phys. Soc. Japan* , **18**, 1594 (1963).

6.   D. Gratias, *La Recherche*, **17**, 788 (1986).
7.   M. A. Dulea, Physical Properties of One-Dimensional Deterministic Aperiodic
     Systems, Linköping Studies in Science and Technology, No. 269, Linköping
     (1992).
8.   A.I. Goldman and M. Widom, *Annu. Rev. Phys. Chem.* **42**, 685 (1991).
9.   N.D. Mermin, *Rev. Mod. Phys.* **64**, 3 (1992).
10. D.S. Rokhsar, N.D. Mermin, and D.C. Wright, *Phys. Rev.* **B35**, 5487 (1987).
11. N.D. Mermin, D.S. Rokhsar, and D.C. Wright, *Phys. Rev. Lett.* **58**, 2099 (1987).
12. D.A. Rabson, T.L. Ho and, N.D. Mermin, *Acta Cryst.*, **A44**, 678 (1988).
13. D.S. Rokhsar, D.C. Wright, and N.D. Mermin, *Phys. Rev.* **B37**, 8145 (1988).
14. N.D. Mermin, D.A. Rabson, D.S. Rokhsar, and D.C. Wright, *Phys. Rev.* **B41**,
     10498 (1990).
15. N.D. Mermin, in Quasicrystals: The State of the Art, P.J. Steinhardt and D.P.
     DiVincenzo Eds. World Scientific, Singapore (1991).
16. N.D. Mermin, in Proceedings of the International Workshop on Modulated
     Crystals, Bilbao, Spain , World Scientific, Singapore 1991.
17. D.A. Rabson, N.D. Mermin, D.S. Rokhsar, and D.C. Wright, *Rev. Mod. Phys.*
     63, 699 (1991).
18. N.D. Mermin, *Rev. Mod. Phys.* **64**, 1163 (1992).
19. G. Berthier, M. Defranceschi and J. Delhalle, in Numerical Determination of the
     Electronic Structure of Atoms, Diatomic and Polyatomic Molecules,
     M. Defranceschi and J. Delhalle Eds. NATO Advanced Study Institute Series, C:
     Mathematical and Physical Sciences, Volume 271, Kluwer Academic Publishers,
     Dordrecht (1989).
20. J.-L. Calais, M. Defranceschi, J.G. Fripiat and J. Delhalle, *J. Phys.:Condens.
     Matter*, **4**, 5675 (1992).
21. See e.g. P. Kaijser and V.H. Smith, Jr, *Adv. Quantum Chem.* **10**, 37   (1977).
22. M. Bräuchler, S. Lunell, I. Olovsson and W. Weyrich, *Int. J. Quantum   Chem.*
     35, 895 (1989) and references therein.
23. P.- O. Löwdin, *Phys. Rev.* **97**, 1474 (1955).
24. J.- L. Calais and J. Delhalle, *Phys. Scripta* **38**, 746 (1988).
25. J.- L. Calais, *Coll. Czechoslovak Chem. Commun.* **53**, 1890 (1988).
26. A. Bienenstock and P.P. Ewald, *Acta Crystallogr.* **15**, 1253 (1962).
27. J.-L. Calais and W. Weyrich, to be published.

Quantum Chemistry Computations in Momentum Space

M. DEFRANCESCHI [1], J. DELHALLE [2], L. DE WINDT [1], P. FISCHER [1, 3],
J.G. FRIPIAT [2]

[1] *Commissariat à l'Energie Atomique, CE-Saclay, DSM/DRECAM/SRSIM,
F-91191 Gif-sur-Yvette Cedex, France*
[2] *Facultés Universitaires Notre-Dame de la Paix, Laboratoire de Chimie Théorique
Appliquée, Rue de Bruxelles, 61, B-5000 Namur, Belgium*
[3] *Université de Paris-Dauphine, Ceremade, Place Maréchal de Lattre de Tassigny,
F-75016 Paris, France*

1. Introduction

In quantum mechanics, the state of a physical system is described by a vector ψ of an Hilbert space, represented by a linear superposition of eigenvectors of Hermitian operators which result from a particular choice of a maximal set of commuting observables [1,2]. The various representations obtained in this way are connected by a generalized Fourier transformation. The so-called Schrödinger method, normally used for systems of electrons and nuclei, starts in an Hilbert space by taking the components of particle coordinates as a maximal set ; consequently, the state function ψ of the system is written in the coordinate representation, and this leads to the familiar Schrödinger equation for determining the possible energies of atoms and molecules as eigenvalues of the total Hamiltonian operator in position space. The Schrödinger equation can be expressed in other representations as well ; e.g. by referring to the various particles in terms of momenta $\mathbf{p}_i$ instead of position vectors $\mathbf{r}_i$. The state function ϕ in momentum space representation becomes the ordinary Fourier transform of the state function in position space, with appropriate $\hbar$ factor :

$$\phi(\mathbf{p}) = \left[\phi(\mathbf{r})\right]^T(\mathbf{p}) = \left(\frac{1}{2\pi\hbar}\right)^{3/2} \int e^{-i\mathbf{p}\cdot\mathbf{r}/\hbar} \phi(\mathbf{r}) d\mathbf{r} \tag{1}$$

Taking the Fourier transform of the ordinary Schrödinger equation yields, in atomic units,

139

Y. Ellinger and M. Defranceschi (eds.), Strategies and Applications in Quantum Chemistry, 139–158.
© 1996 Kluwer Academic Publishers.

$$\left(\frac{\mathbf{p}^2}{2} - E\right)\phi(\mathbf{p}) = -\int d\mathbf{q}\phi(\mathbf{p} - \mathbf{q})\left[\left(\frac{1}{2\pi}\right)^{3/2}\int e^{-i\mathbf{q}\cdot\mathbf{r}}V(\mathbf{r})d\mathbf{r}\right] \tag{2}$$

where E is the total energy and $V(\mathbf{r})$ represents the electron-nucleus attraction potential and the electron-electron repulsion potential.

Except for a few situations related to scattering problems where observables typically involve momenta, physical quantities are defined in position space ($\mathbf{r}$-representation) even where the momentum space representation ($\mathbf{p}$-representation) would be more natural. For instance, experiments such as Compton profiles and (e,2e) measurements [3,4] are compared with theoretical momentum space distribution obtained by Fourier transformation of wavefunctions [5] computed in the position space. The lack of wave functions directly evaluated in momentum space is no doubt due to the development of techniques using the Schrödinger equation in the $\mathbf{r}$-representation for a large variety of situations. At least two other factors contribute to dissuade the physicists and chemists from considering momentum space as an interesting direction for solving their problems. First, interpretation and visualization can be more difficult in momentum space and, second, the Schrödinger equation, and approximations to it, e.g. the Hartree-Fock (HF) equation, are expressed as integral equations in the $\mathbf{p}$-representation instead of differential equations in the $\mathbf{r}$-representation. In spite of these barriers, momentum space offers advantages which should not be ignored. For instance, it provides an interesting alternative way for solving electronic structure problems of atoms and molecules, traditionally addressed in position space [6,7]. This aspect is central to this work.

As far as in the thirties the possibility of calculating wave functions in momentum space has been recognized ; in 1932, Hylleraas [8] treated the problem of a one-electron atom, the solutions of which for discrete and continuous spectra are well known [9]. In 1949, McWeeny and Coulson [10,11] tried to generalize this approach to many-electron systems involving electron repulsion terms. Starting with fixed trial functions, they applied the iterative method developed by Svartholm [12] for the case of nuclear systems to solve variationally the integral momentum space wave equation of helium atom and hydrogen molecule H_2^+ and H_2. Owing to convergence difficulties found in the simplest systems, they concluded that direct calculations of electronic wave functions in momentum space were hopeless ; and so the subject disappeared from Quantum Chemistry literature for nearly 30 years. The situation changed in 1981, when two crystallographers, Navaza and

Tsoucaris, decided to treat by Fourier transformation, not the Schrödinger equation itself, but one of its most popular approximate forms for electron systems, namely the Hartree-Fock equations. The form of these equations was known before, in connection with electron-scattering problems [13], but their advantage for Quantum Chemistry calculations was not yet recognized.

The work by Navaza and Tsoucaris on the H_2 molecule [7] proved the feasibility of direct numerical molecular orbitals computations, i.e. without atomic basis functions contrary to what happens in **r**-space where it is difficult to obtain accurate Hartree-Fock solutions for atoms, molecules and solids due to the need of representing the solutions in terms of a finite basis of known functions, e.g. the linear combination of atomic orbitals (LCAO) approximation. For chemists interested in polyatomic molecules, the momentum method is quite attractive because it is not limited to systems whose geometry determines the coordinates to be used for integrating the position space equations, as for example polar coordinates for AH_n molecules [14] because they have approximate spherical symmetry and/or spheroidal coordinates for diatomic molecules, see e.g. Ref. [15]. During the last years, we have contributed to demonstrate that direct momentum space calculations are in principle feasible for any molecule by studying hydrogen systems of increasing complexity : the H_2 ground state at the SCF and MC-SCF level [16], an H_3 open-shell system [17] and a chain of H atoms including an infinite number of electrons and nuclei [18,19]. More complex systems have also been studied : atoms up to neon [20-28], cations [22,23, 28-31], anions [22,23,27,28], symmetric molecules [16, 17,32-36] as well as asymmetric molecules such as HeH^+ [37] or HF [38].

The advantages of the momentum approach are not only limited to the opportunity for direct numerical calculations for chemical systems, but it also offers the prospect of selecting better bases of atomic functions on which rely almost all first principle quantum mechanical calculations.

2. MOMENTUM SPACE EQUATIONS FOR A CLOSED-SHELL SYSTEM

The Fourier transformation method enables us to immediately write the momentum space equations as soon as the SCF theory used to describe the system under consideration allows us to build one or several effective Fock Hamiltonians for the orbitals to be determined. This includes a rather large variety of situations :

♦ Closed-shell systems as defined in the standard Hartree-Fock theory [39-40].

♦ Unrestricted monodeterminantal treatments using different orbitals for different spins for open-shell systems (free radicals, triplet states, etc.) [41,42].

♦ Roothaan open-shell treatments involving a closed-shell subsystem and outer unpaired electrons interacting through two-index integrals of Coulomb and exchange type only [43].

♦ MC-SCF treatments written in terms of coupled Fock equations [44]. The simplest examples are the two-configuration SCF theory [45] used in s^2p^n and p^{n+2} atomic mixing [46], or bonding-antibonding molecular problems [47], and more generally the Clementi-Veillard electron-pair MC-SCF theory [48].

♦ SCF treatments for infinite chains having translational symmetry [49,50].

In the recent past, we have investigated and published examples illustrating the different cases. For instance in Ref. [17] a Roothaan open-shell system, H_3, has been detailed, in Refs. [18, 19] a SCF treatment for infinite chains and finally in Ref [16] a MC-SCF treatment were proposed.

In this contribution our purpose is to review the principles and the results of the momentum space approach for quantum chemistry calculations of molecules and polymers. To avoid unnecessary complications, but without loss of generality, we shall consider in details the case of closed-shell systems.

2.1. RESTRICTED HARTREE-FOCK EQUATIONS

Since both position and momentum formulations contain exactly the same information, it is convenient to start from the familiar position space expression and express it in momentum space. In the case of a closed-shell system of $N=2n_0$ electrons in the field of M nuclear charges (Z_A) located at fixed positions R_A (Born-Oppenheimer approximation), the n_0 doubly occupied orbitals φ_n of the Hartree-Fock model in the position space are obtained from the second-order differential equation of the form $(F-\varepsilon_n)\varphi_n = 0$, if we assume -as usual - that the off-diagonal Lagrange multipliers $\varepsilon_{nn'}$ ensuring the orthogonality of the φ_n's have been eliminated by an appropriate unitary transformation inside the closed set. The F operator giving the φ orbitals iteratively is a one-electron Hamiltonian including a kinetic term and an effective potential in which the electron-nucleus attraction is balanced by the Coulomb-exchange potential approximating the real electron-electron interaction. In atomic units, we have :

$$F(\mathbf{r})\varphi_i(\mathbf{r}) = \left[-\frac{1}{2}\sum_i \nabla^2(\mathbf{r}) - \sum_{A=1}^{M} \frac{Z_A}{|\mathbf{r} - \mathbf{R}_A|} + \sum_{j=1}^{N/2} 2\int d\mathbf{r}' \frac{\varphi_j^*(\mathbf{r}')\varphi_j(\mathbf{r}')}{|\mathbf{r} - \mathbf{r}'|} \right] \varphi_i(\mathbf{r})$$

$$- \sum_{j=1}^{N/2} \int d\mathbf{r}' \frac{\varphi_j^*(\mathbf{r}')\varphi_i(\mathbf{r}')}{|\mathbf{r} - \mathbf{r}'|} \varphi_j(\mathbf{r}) = \varepsilon_i \varphi_i(\mathbf{r})$$

(3)

and by applying the Fourier transform to Eq. 3, we get the momentum space RHF equation. The linearity of the Fourier transformation allows a separate treatment of each of the terms occurring in the Hartree Fock equation.

Kinetic energy term.

$$\left[-\frac{\nabla^2(\mathbf{r})}{2}\varphi(\mathbf{r}) \right]^T (\mathbf{p}) = (2\pi)^{-3/2} \int d\mathbf{r} e^{-i\mathbf{p}\cdot\mathbf{r}} \left[-\frac{\nabla^2(\mathbf{r})}{2}\varphi(\mathbf{r}) \right]$$

(4)

The integral in Eq. 4 is readily evaluated if $\varphi(\mathbf{r})$ is replaced by its inverse Fourier transform. After rearrangement of the terms, one finds that the integral over $\mathbf{r}$ yields the delta function $\delta(\mathbf{p}-\mathbf{q})$. Carrying out the remaining integral yields the final expression.

$$\left[-\frac{\nabla^2(\mathbf{r})}{2}\varphi(\mathbf{r}) \right]^T (\mathbf{p}) = (2\pi)^{-3/2} \int d\mathbf{r} e^{-i\mathbf{p}\cdot\mathbf{r}} \left[-\frac{\nabla^2(\mathbf{r})}{2}(2\pi)^{-3/2} \int d\mathbf{q} e^{i\mathbf{q}\cdot\mathbf{r}}\phi(\mathbf{q}) \right]$$

$$= (2\pi)^{-3} \int d\mathbf{r} e^{-i\mathbf{p}\cdot\mathbf{r}} \int \frac{\mathbf{q}^2}{2} d\mathbf{q} e^{i\mathbf{q}\cdot\mathbf{r}}\phi(\mathbf{q}) = (2\pi)^{-3} \int \frac{\mathbf{q}^2}{2} d\mathbf{q}\phi(\mathbf{q})(2\pi)\delta(\mathbf{p}-\mathbf{q}) = \frac{\mathbf{p}^2}{2}\phi(\mathbf{p})$$

(5)

By convention $\mathbf{p}^2$ is a shorthand for the dot product $\mathbf{p}\cdot\mathbf{p} = (p_x^2 + p_y^2 + p_z^2)$; both $|\mathbf{p}|$ and p will be used to denote the length of vector $\mathbf{p}$.

Nuclear attraction, electron-electron repulsion, and exchange terms.

$$\left[-\sum_{A=1}^{M} \frac{Z_A}{|\mathbf{r} - \mathbf{R}_A|}\varphi(\mathbf{r}) \right]^T (\mathbf{p}) = -\sum_{A=1}^{M} Z_A(2\pi)^{-3/2} \int d\mathbf{r} \frac{e^{-i\mathbf{p}\cdot\mathbf{r}}}{|\mathbf{r} - \mathbf{R}_A|}\varphi(\mathbf{r})$$

(6)

and

$$\left[\int dr' \frac{\varphi_i^*(\mathbf{r}')\varphi_j(\mathbf{r}')}{|\mathbf{r}'-\mathbf{r}|}\varphi_k(\mathbf{r})\right]^T(\mathbf{p}) = (2\pi)^{-3/2}\int d\mathbf{r}\, e^{-i\mathbf{p}\cdot\mathbf{r}}\left[\int d\mathbf{r}' \frac{\varphi_i^*(\mathbf{r}')\varphi_j(\mathbf{r}')}{|\mathbf{r}'-\mathbf{r}|}\right]\varphi_k(\mathbf{r}) \qquad (7)$$

The above integrals are most conveniently reduced if $|\mathbf{r}|^{-1}$ (resp. $|\mathbf{r}-\mathbf{r}'|^{-1}$) is substituted by the inverse Fourier transform of $[\,|\mathbf{r}|^{-1}]^T(\mathbf{p})$ (resp. $[\,|\mathbf{r}-\mathbf{r}'|^{-1}]^T(\mathbf{p})$). The steps for the final expression of the nuclear term and the electron-electron repulsion term in $\mathbf{p}$-representation are summarized below :

$$\left[-\sum_{A=1}^{M}\frac{Z_A}{|\mathbf{r}-\mathbf{R}_A|}\varphi(\mathbf{r})\right]^T(\mathbf{p}) = -\sum_{A=1}^{M}Z_A(2\pi)^{-3/2}\int d\mathbf{r}\, e^{-i\mathbf{p}\cdot\mathbf{r}}\left(\frac{1}{2\pi^2}\right)\int\frac{d\mathbf{q}}{q^2}\varphi(\mathbf{r})e^{i\mathbf{q}\cdot(\mathbf{r}-\mathbf{R}_A)}$$

$$= -\sum_{A=1}^{M}\frac{Z_A}{2\pi^2}\int\frac{d\mathbf{q}}{q^2}e^{-i\mathbf{q}\cdot\mathbf{R}_A}\left[(2\pi)^{-3/2}\int d\mathbf{r}\, e^{-i(\mathbf{p}-\mathbf{q})\cdot\mathbf{r}}\varphi(\mathbf{r})\right] = -\frac{1}{2\pi^2}\int\frac{d\mathbf{q}}{q^2}F(\mathbf{q})\phi(\mathbf{p}-\mathbf{q})$$

$$(8)$$

where $F(\mathbf{q}) = \sum_{A}^{M}Z_A e^{-i\mathbf{q}\cdot\mathbf{R}_A}$

$$\left[\int dr' \frac{\varphi_i^*(\mathbf{r}')\varphi_j(\mathbf{r}')}{|\mathbf{r}'-\mathbf{r}|}\varphi_k(\mathbf{r})\right]^T(\mathbf{p}) = (2\pi)^{-3/2}\int d\mathbf{r}\, e^{-i\mathbf{p}\cdot\mathbf{r}}\varphi_k(\mathbf{r})\int d\mathbf{r}'\,\varphi_i^*(\mathbf{r}')\varphi_j(\mathbf{r}')$$

$$\cdot\left(\frac{1}{2\pi^2}\right)\int\frac{d\mathbf{q}}{q^2}e^{i\mathbf{q}\cdot(\mathbf{r}-\mathbf{r}')} = \frac{1}{2\pi^2}\left[\int\frac{d\mathbf{q}}{q^2}\int d\mathbf{r}'\,\varphi_i^*(\mathbf{r}')\varphi_j(\mathbf{r}')e^{-i\mathbf{q}\cdot\mathbf{r}'}\right]\phi_k(\mathbf{p}-\mathbf{q})$$

$$(9)$$

Using the convolution theorem the content of the square brackets in Eq. 9 is rewritten as :

$$\int d\mathbf{r}'\, e^{-i\mathbf{q}\cdot\mathbf{r}'}\varphi_i^*(\mathbf{r}')\varphi_j(\mathbf{r}') = (2\pi)^{-3/2}\int d\mathbf{r}'\,\varphi_i^*(\mathbf{r}')\int d\mathbf{p}\,\phi_j(\mathbf{p})e^{i(\mathbf{p}-\mathbf{q})\cdot\mathbf{r}'}$$

$$= \int d\mathbf{p}\,\phi_j(\mathbf{p})\left[(2\pi)^{-3/2}\int d\mathbf{r}'\,\varphi_i(\mathbf{r}')e^{-i(\mathbf{p}-\mathbf{q})\cdot\mathbf{r}'}\right]^* = \int d\mathbf{p}\,\phi_i^*(\mathbf{p}-\mathbf{q})\phi_j(\mathbf{p}) \qquad (10)$$

Defining a quantity $W_{ij}(\mathbf{q})$,

$$W_{ij}(\mathbf{q}) = \int d\mathbf{q}\,\phi_i^*(\mathbf{p})\phi_j(\mathbf{p}-\mathbf{q}) \qquad (11)$$

and introducing it in Eq. 9 leads to the expression for the two-electron term :

$$\left[\int d\mathbf{r}' \frac{\varphi_i^*(\mathbf{r}')\varphi_j(\mathbf{r}')}{|\mathbf{r}-\mathbf{r}'|}\varphi_k(\mathbf{r})\right]^T(\mathbf{p}) = \frac{1}{2\pi^2}\int \frac{d\mathbf{q}}{\mathbf{q}^2} W_{ji}^*(\mathbf{q})\phi_k(\mathbf{p}-\mathbf{q}) \tag{12}$$

With the above results, it is possible to write the expanded momentum space form of the Hartree-Fock equations :

$$\frac{\mathbf{p}^2}{2}\phi_i(\mathbf{p}) - \frac{1}{2\pi^2}\int \frac{d\mathbf{q}}{\mathbf{q}^2} F(\mathbf{q})\phi_i(\mathbf{p}-\mathbf{q}) + \frac{1}{2\pi^2}\sum_{j=1}^{N/2} 2\int \frac{d\mathbf{q}}{\mathbf{q}^2} W_{jj}^*(\mathbf{q})\phi_i(\mathbf{p}-\mathbf{q})$$

$$-\frac{1}{2\pi^2}\sum_{j=1}^{N/2}\int \frac{d\mathbf{q}}{\mathbf{q}^2} W_{ji}^*(\mathbf{q})\phi_j(\mathbf{p}-\mathbf{q}) = \varepsilon_i\phi_i(\mathbf{p}) \tag{13}$$

The equations to be fulfilled by momentum space orbitals contain convolution integrals which give rise to momentum orbitals $\phi(\mathbf{p}-\mathbf{q})$ shifted in momentum space. The so-called form factor F and the interaction terms W_{ij} defined in terms of current momentum coordinates are the momentum space counterparts of the core potentials and Coulomb and/or exchange operators in position space. The nuclear field potential transfers a momentum to electron i, while the interelectronic interaction produces a momentum transfer between each pair of electrons in turn. Nevertheless, the total momentum of the whole molecule remains invariant thanks to the contribution of the nuclear momenta [7].

2.2 ORBITAL AND TOTAL ENERGIES

The calculation of ε in momentum space is analogous to that in position space. Starting with the **r**-representation, and expressing the quantity $F(\mathbf{r})\varphi_i(\mathbf{r})$ as the inverse Fourier transform of $[F(\mathbf{r})\,\varphi_i(\mathbf{r})]^T(\mathbf{p})$, one easily finds that :

$$\varepsilon_i = \int d\mathbf{p}\varphi_i^*(\mathbf{p})[F(\mathbf{p})\varphi_i(\mathbf{p})] \tag{14}$$

The one-electron energy ε_i has the same expression in the **p**-representation as in the position space where the different contributions can be expressed as follows :

Kinetic energy term. Its expression is straightforward to write :

$$\int d\mathbf{p}\phi_i^*(\mathbf{p})\frac{\mathbf{p}^2}{2}\phi_i(\mathbf{p}) \tag{15}$$

Nuclear attraction, electron-electron repulsion and exchange terms. Using the Eqs. 6 and 7, these contributions are respectively written in terms of the quantity $W_{ij}(q)$ previously defined in Eq. 11 :

$$\frac{1}{2\pi^2}\int d\mathbf{p}\phi_i^*(\mathbf{p})\int\frac{d\mathbf{q}}{\mathbf{q}^2}\left[-\sum_{A=1}^{M}Z_A e^{-i\mathbf{p}.\mathbf{R}_A}\right]\phi_i(\mathbf{p}-\mathbf{q}) = \frac{1}{2\pi^2}\int\frac{d\mathbf{q}}{\mathbf{q}^2}F(\mathbf{q})W_{ii}(\mathbf{q}) \qquad (16)$$

$$\int d\mathbf{p}\phi_i^*(\mathbf{p})\left[\frac{1}{2\pi^2}\int\frac{d\mathbf{q}}{\mathbf{q}^2}W_{jj}^*(\mathbf{q})\phi_i(\mathbf{p}-\mathbf{q})\right] = \frac{1}{2\pi^2}\int\frac{d\mathbf{q}}{\mathbf{q}^2}W_{jj}^*(\mathbf{q})W_{ii}(\mathbf{q}) \qquad (17)$$

and

$$\int d\mathbf{p}\phi_i^*(\mathbf{p})\left[\frac{1}{2\pi^2}\int\frac{d\mathbf{q}}{\mathbf{q}^2}W_{ij}^*(\mathbf{q})\phi_j(\mathbf{p}-\mathbf{q})\right] = \frac{1}{2\pi^2}\int\frac{d\mathbf{q}}{\mathbf{q}^2}W_{ij}^*(\mathbf{q})W_{ij}(\mathbf{q}) \qquad (18)$$

The detailed expression for ε is then :

$$\varepsilon_i = \int d\mathbf{p}\phi_i^*(\mathbf{p})\frac{\mathbf{p}^2}{2}\phi_i(\mathbf{p}) - \frac{1}{2\pi^2}\int\frac{d\mathbf{q}}{\mathbf{q}^2}\left[F(\mathbf{q}) - 2\sum_{j=1}^{N/2}W_{jj}^*(\mathbf{q})\right]W_{ii}(\mathbf{q})$$

$$-\frac{1}{2\pi^2}\sum_{j=1}^{N/2}\int\frac{d\mathbf{q}}{\mathbf{q}^2}W_{ij}^*(\mathbf{q})W_{ij}(\mathbf{q}) \qquad (19)$$

The momentum space equivalent of the total energy E, is :

$$E = 2\sum_{i=1}^{N/2}\left\{\int d\mathbf{p}\phi_i^*(\mathbf{p})\frac{\mathbf{p}^2}{2}\phi_i(\mathbf{p}) - \frac{1}{2\pi^2}\int\frac{d\mathbf{q}}{\mathbf{q}^2}\left[F(\mathbf{q}) - \sum_{j=1}^{N/2}W_{jj}^*(\mathbf{q})\right]W_{ii}(\mathbf{q})\right\}$$

$$-\frac{1}{4\pi^2}\sum_{i=1}^{N/2}\sum_{j=1}^{N/2}\int\frac{d\mathbf{q}}{\mathbf{q}^2}W_{ij}^*(\mathbf{q})W_{ij}(\mathbf{q}) + \frac{1}{2}\sum_{A=1}^{M}\sum_{B=1}^{M}\frac{Z_A Z_B}{|\mathbf{R}_A - \mathbf{R}_B|} \qquad (20)$$

3. Principles for numerical resolutions

Because of the terms $|\mathbf{r}-\mathbf{R}_A|^{-1}$ and $|\mathbf{r}-\mathbf{r'}|^{-1}$ explicit solutions to Eq. 3 cannot be obtained in position space. In such cases approximate solutions are usually expressed as truncated linear combinations of basis functions (LCAO expressions). In spite of its successes, the LCAO approximation experiences various difficulties (truncation limits, nature of the basis functions, etc.) hard to estimate and which are not entirely controllable [51].

Furthermore their incidence are very dependent upon the nature of the properties [52,53]. Due to computer limitations, basis sets cannot be extended indefinitely and direct numerical evaluations seem the ultimate solution for molecules [54]. In position space this is a viable alternative for diatomic molecules [55,56], but it cannot be extended easily to polyatomic systems. Formulated in momentum space, the HF equations have not explicit solutions and the difficulties to express them in terms of basis functions are analogous to those encountered in the **r**-space. However the momentum space HF equations give way to numerical approaches in which Coulombic interactions become tractable even for polyatomic molecules [7] ; among other advantages, these equations, Eqs. 13 and 20, do not require coordinate systems adapted to the geometry of the molecules to remove Coulombic singularities. In both equations the only singular contribution comes from the $\mathbf{q}^{-2}$ factor.

3.1. VARIATION-ITERATION PROCEDURE

In both position and momentum spaces, iterative procedures are necessary to solve the HF equations. Starting from a trial orbital $\phi_i^{(0)}(\mathbf{p})$, an approximate orbital, $\phi_i^{(k+1)}(\mathbf{p})$, is obtained after k+1 iterations from Eq. 13 rewritten as :

$$\phi_i^{(k+1)}(\mathbf{p}) = \frac{1}{\left(\dfrac{p^2}{2} - \varepsilon_i^{(k)}\right)} \left[\frac{1}{2\pi^2} \int \frac{d\mathbf{q}}{\mathbf{q}^2} F(\mathbf{q})\phi_i^{(k)}(\mathbf{p}-\mathbf{q}) \right.$$

$$\left. + \frac{1}{2\pi^2} \sum_{j=1}^{N/2} 2\int \frac{d\mathbf{q}}{\mathbf{q}^2} W_{jj}^{*(k)}(\mathbf{q})\phi_i^{(k)}(\mathbf{p}-\mathbf{q}) - \frac{1}{2\pi^2} \sum_{j=1}^{N/2} \int \frac{d\mathbf{q}}{\mathbf{q}^2} W_{ji}^{*(k)}(\mathbf{q})\phi_j^{(k)}(\mathbf{p}-\mathbf{q}) \right]$$

$$(21)$$

The procedure is repeated until convergence is reached. Since we are interested in bound states where $\varepsilon_i < 0$, no problem of divergence or cusps conditions is raised. But the method can be adapted to more general situations by introducing a translation of the energy origin in Eq. 21.

Numerical and computational problems associated with the implementation of the approach for routine use fall in two main categories : (a) numerical integration and (b) enforcement of the orthogonality and renormalization of the numerical orbitals during the iteration steps. Many different integration schemes have been considered in the past, some of which will be detailed in the section 3.2. As concerns orthonormalization, at each step the new iterates $\phi_i^{(k+1)}(\mathbf{p})$'s, even if initially orthonormal, need to be renormalized and orthogonalized to form true canonical HF orbitals. Great care must be exercised in selecting orthogonalization procedures, for instance the so-called Löwdin's symmetric orthogonalisation procedure [57], often used in Quantum Chemistry, mixes all the orbitals simultaneously, tends to contaminate all the iterates, and impairs the convergence of the iterative steps. Schmidt orthogonalization does better (since it allows to choose the sequence of orthogonalization) but looses track of the symmetry of these orbitals. Finally, the canonical orthogonalization performs a maximum in mixing of all states. We have shown [31] for the ground state of Be and B$^+$ that the Gram-Schmidt procedure turns out to be more appropriate in most cases, but with very good trial functions, Löwdin's symmetric procedure yields equivalent results. In all cases reported in Table 1, Gram-Schmidt orthonormalization has been used.

3.2.   NUMERICAL TECHNIQUES

Different integration schemes have been considered. To cancel the $\mathbf{q}^{-2}$ singularity factor in Eq. 13 by the integration volume element, Navaza and Tsoucaris have proposed the use of spherical polar coordinates. However, because of the convolution integrals, interpolation schemes are needed in these coordinates since arguments $(\mathbf{p}\text{-}\mathbf{q})$ do not necessarily belong to the grid points. The computation time increases as the square, N^2, of the number of points of the integration grid, and for large systems, this time becomes prohibitive. Another point of view has been to focus on these convolution integrals and treat them via a more economical fast Fourier transform procedure. In this case, the computation time increases only as $N\log_2 N$, but at the expense of an approximate treatment of the $\mathbf{q}^{-2}$ singular factor [58,59]. Variants [60,61] based on the Fock transformation have also been proposed to deal with the infinite limits of integration resorting to a one-to-one correspondence ($p = \tan\alpha$) between intervals $(0, \infty)$ and $(0, \pi/2)$. At the present time, none of the approaches has been satisfactory enough to bring the fully numerical momentum quantum chemistry calculations beyond a stage of prematurity. Furthermore, computational tests [25] on helium atom have shown the importance of accuracy and convergence of the integrals. It seems that straightforward numerical calculations are not readily applicable and our work is now directed toward mixed numerical and analytical procedures.

3.3. SEMI-ANALYTICAL TECHNIQUE

The method presented here allows, starting with trial gaussian functions, a partial analytical treatment which we have used to improve the LCAO-GTO orbitals (trial functions) essentially obtained from all ab initio quantum chemistry programs. As in $\mathbf{r}$-representation, trial functions $\phi_i^{(k)}(\mathbf{p})$ (Eq. 21) are conveniently expressed as linear combinations of ω functions $\chi_t(\mathbf{p})$ themselves written as linear combinations of G_t gaussian functions (LCAO-GTO approximation) $g_{ta}(\mathbf{p})$,

$$\phi_i^{(0)}(\mathbf{p}) = \sum_{t=1}^{\omega} C_{ti}\chi_t(\mathbf{p}) \tag{22}$$

and

$$\chi_t(\mathbf{p}) = \sum_{a=1}^{G_t} d_{at}g_{ta}(\mathbf{p}) \tag{23}$$

where $g_{ta}(\mathbf{p})$ is a normalized gaussian function expressed in momentum space. As $\{\chi_t\}t \in N^*$ belong to the Sobolev space $H^1(R^3)$, direct Fourier transformation leads to a set $\{\chi_t\}t \in N^*$ that fulfills the criterion about the convergence of the energy and wave function (the completeness of the orbital bases $\{\chi_t\}t \in N^*$ only in $L^2(R^3)$ is not sufficient to guarantee the convergence of the energy and wave function in the norm of $L^2(R^3)$; to ensure this convergence the set $\{\chi_t\}t \in N^*$ must be complete in $H^1(R^3)$). The expression for the first iterate $\phi_i^{(1)}(\mathbf{p})$ based on trial functions $\phi_i^{(0)}(\mathbf{p})$ expressed as LCAO-GTO expansions is thus :

$$\phi_i^{(1)}(\mathbf{p}) = \frac{1}{\left(\dfrac{\mathbf{p}^2}{2} - \varepsilon_i^{(k)}\right)} \Bigg[\frac{1}{2\pi^2} \int \frac{d\mathbf{q}}{\mathbf{q}^2} \Bigg[F(\mathbf{q}) - 2\sum_{j=1}^{N/2} W_{jj}^{(0)}(\mathbf{q}) \Bigg] \Bigg\{ \sum_{t=1}^{\omega} C_{ti} \sum_{a=1}^{G_t} d_{at}g_{ta}(\mathbf{p}-\mathbf{q}) \Bigg\}$$

$$-\frac{1}{2\pi^2} \sum_{j=1}^{N/2} \int \frac{d\mathbf{q}}{\mathbf{q}^2} W_{ji}^{*(0)}(\mathbf{q}) \Bigg\{ \sum_{t=1}^{\omega} C_{tj} \sum_{a=1}^{G_t} d_{at}g_{ta}(\mathbf{p}-\mathbf{q}) \Bigg\} \Bigg]$$

$$\tag{24}$$

The various quantities entering Eq. 24 are deduced when the trial orbitals $\phi_i^{(0)}(\mathbf{p})$ are expressed as linear combinations of Gaussian functions, they are expressible in terms of known transcendental functions. In the case of s-functions, two basic integrals, respectively denoted by I_1 and I_2, have to be solved to obtain $\phi_i^{(1)}(\mathbf{p})$:

$$I_1(\alpha,\mathbf{R}_A,\beta,\mathbf{R}_B,\mathbf{q}) = \int d\mathbf{p} e^{-\alpha p^2} e^{i\mathbf{p}.\mathbf{R}_A} e^{-\beta(\mathbf{p}-\mathbf{q})^2} e^{i(\mathbf{p}-\mathbf{q}).\mathbf{R}_B} \tag{25}$$

$$I_2(\alpha,\mathbf{R}_A,\beta,\mathbf{R}_B,\mathbf{q}) = \int \frac{d\mathbf{p}}{p^2} e^{-\alpha p^2} e^{i\mathbf{p}.\mathbf{R}_A} e^{-\beta(\mathbf{p}-\mathbf{q})^2} e^{i(\mathbf{p}-\mathbf{q}).\mathbf{R}_B} \tag{26}$$

The details of the calculations can be found in Ref.35, the final expressions are :

$$I_1(\alpha,\mathbf{R}_A,\beta,\mathbf{R}_B,\mathbf{q}) = \left(\frac{\pi}{(\alpha+\beta)}\right)^{3/2}.\exp\left(-\frac{\alpha\beta}{\alpha+\beta}q^2\right).\exp\left(-\frac{(\mathbf{R}_A+\mathbf{R}_B)^2}{4(\alpha+\beta)}\right)$$

$$.\exp\left(i\frac{\mathbf{q}.(\beta\mathbf{R}_A-\alpha\mathbf{R}_B)}{\alpha+\beta}\right) \tag{27}$$

$$I_2(\alpha,\mathbf{R}_A,\beta,\mathbf{R}_B,\mathbf{q}) = \frac{2\pi^{3/2}}{(\alpha+\beta)^{1/2}}.\exp\left(-\frac{\alpha\beta}{\alpha+\beta}q^2\right).\exp\left(-\frac{(\mathbf{R}_A+\mathbf{R}_B)^2}{4(\alpha+\beta)}\right)$$

$$.\exp\left(i\frac{\mathbf{q}.(\beta\mathbf{R}_A-\alpha\mathbf{R}_B)}{\alpha+\beta}\right).\frac{\mathrm{Daw}(z)}{z} \tag{28}$$

with $z^2 = \dfrac{\left(\beta q + i\dfrac{\mathbf{R}_A+\mathbf{R}_B}{2}\right)^2}{\alpha+\beta}$

and $\mathrm{Daw}(z) = \exp(-z^2)\int_0^\infty e^{x^2} dx$ is the so-called Dawson function. The individual terms appearing in Eq. 24 are :

Nuclear attraction term :

$$\int \frac{d\mathbf{q}}{q^2} \exp(i\mathbf{q}.\mathbf{R}_A) N_{ar} \exp\left(-\alpha_{ar}(\mathbf{p}-\mathbf{q})^2\right) \exp(i(\mathbf{p}-\mathbf{q}).\mathbf{R}_r) = N_{ar} I_2(0,\mathbf{R}_A,\alpha_{ar},-\mathbf{R}_r,\mathbf{p}) \tag{29}$$

Fourier transform of orbital products :

$$W_{ij}^{(0)}(\mathbf{q}) = \sum_{r=1}^{\omega} \sum_{s=1}^{\omega} C_{ri}^* C_{sj} \sum_{b=1}^{G_r} \sum_{c=1}^{G_s} d_{br} d_{cs} N_{br} N_{cs} I_1(\alpha_{br}, \mathbf{R}_r, \alpha_{cs}, -\mathbf{R}_s, \mathbf{q}) \tag{30}$$

Electronic repulsion term :

$$\int \frac{d\mathbf{q}}{q^2} W_{jj}^{*(0)}(\mathbf{q}) \exp\left(-\alpha_{at}(\mathbf{p}-\mathbf{q})^2\right) = N_{at} \sum_{r=1}^{\omega} \sum_{s=1}^{\omega} C_{rj}^* C_{sj} \sum_{b=1}^{G_r} \sum_{c=1}^{G_s} d_{br} d_{cs} N_{br} N_{cs}$$

$$\cdot \left[\frac{\pi}{\alpha_{br} + \alpha_{cs}}\right]^{3/2} \exp\left(-\frac{(\mathbf{R}_r - \mathbf{R}_s)^2}{4(\alpha_{br} + \alpha_{cs})}\right) I_2\left(\frac{\alpha_{br}\alpha_{cs}}{\alpha_{br} + \alpha_{cs}}, \frac{\alpha_{cs}\mathbf{R}_r + \alpha_{br}\mathbf{R}_s}{\alpha_{br} + \alpha_{cs}}, \alpha_{at}, -\mathbf{R}_s, \mathbf{p}\right) \tag{31}$$

Exchange term :

$$\int \frac{d\mathbf{q}}{q^2} W_{ij}^{*(0)}(\mathbf{q}) \exp\left(-\alpha_{at}(\mathbf{p}-\mathbf{q})^2\right) = N_{at} \sum_{r=1}^{\omega} \sum_{s=1}^{\omega} C_{ri}^* C_{sj} \sum_{b=1}^{G_r} \sum_{c=1}^{G_s} d_{br} d_{cs} N_{br} N_{cs}$$

$$\cdot \left[\frac{\pi}{\alpha_{br} + \alpha_{cs}}\right]^{3/2} \exp\left(-\frac{(\mathbf{R}_r - \mathbf{R}_s)^2}{4(\alpha_{br} + \alpha_{cs})}\right) I_2\left(\frac{\alpha_{br}\alpha_{cs}}{\alpha_{br} + \alpha_{cs}}, \frac{\alpha_{cs}\mathbf{R}_r + \alpha_{br}\mathbf{R}_s}{\alpha_{br} + \alpha_{cs}}, \alpha_{at}, -\mathbf{R}_s, \mathbf{p}\right) \tag{32}$$

So the first iteration transforms the trial wave functions expressed as linear combinations of gaussian functions into an expression which involves Dawson functions [62,63]. We have not been able to find a tabular entry to perform explicitly the normalization of the first iterate, accordingly this is carried out numerically by the Gauss-Legendre method [64].

One of the drawbacks of the first iteration, however, is that computation of energy quantities, e.g. orbital and total energies, requires to evaluate the integrals occurring in Eq. 3 on the basis of the $\phi_i^{(1)}(\mathbf{p})$. Unfortunately, the transcendental functions in terms of which the $\phi_i^{(1)}(\mathbf{p})$ are expressed at the end of the first iteration do not lead to closed form expressions for these integrals and a numerical procedure is therefore needed. This constitutes a barrier to carry out further iterations to improve the orbitals by approaching the HF limit. A compromise has been proposed between a fully numerical scheme and the simple first iteration approach based on the fact that at the end of each iteration the $\phi_i^{(k)}(\mathbf{p})$'s entail the main qualitative characteristics of the exact solution and most importantly the right asymptotic decay. The idea is thus to fit the iterated analytical functions $\phi_i^{(k)}(\mathbf{p})$ obtained at the kth step on a finite set of gaussian functions and then use these fitted functions as a new set of trial functions $\phi g_i^{(k)}(\mathbf{p})$. The advantage is twofold. First, with exponents and linear coefficients specific for each orbital, energies and functions are quickly improved. Second, the problematic convolution products and integrals are efficiently computed in terms of the gaussian functions obtained to represent the $\phi g_i^{(k)}(\mathbf{p})$'s. The analytical functions $\phi_i^{(1)}(\mathbf{p})$ are represented as linear combinations of gaussian functions, $\phi g_i^{(1)}(\mathbf{p})$. This fit is carried out using a modified version of the Gausfit package [65] developed by Stewart [66] for gaussian fits of Slater functions. The resulting functions are analytically orthonormalized.

For atoms, the radial part of $\phi g_i(p)$ is expressed as a linear combination of spherical gaussians, which, in the case of 2p orbitals writes as :

$$g_i^g(p) = \sum_{r=1}^{N} d_{ri}\, p\exp\left(-\alpha_{ri}p^2\right) \tag{33}$$

Given a radial function $\phi(p)$ to fit, one minimizes the variance,

$$\int_0^{\infty} dp\, p^2\, \omega(p)\left[\phi_i(p) - \phi_i^g(p)\right]^2 \tag{34}$$

where $\omega(p)$ is a function which weights the contributions to the integral according their expected importance [28]. From several tests on Be and Ne we have found that the following weight functions are quite efficient :

$$\omega_{1s}(p) = 1 + p\,, \quad \text{and} \quad \omega_{2s\ and\ 2p}(p) = 1 + p^2\,. \tag{35}$$

Gaussian functions do not have the right asymptotic decay due to too low amplitudes in regions of large p values, therefore representations in terms of gaussians are of much slower convergence than Slater functions. Since contributions from high momenta are essential to the energy, a second degree polynomial, Eq. 35, is used to enforce them in the valence orbitals.

A set of nine gaussians allows a satisfactory fit with low variance, Eq. 34, values : about 10^{-7} for the 1s and 2p orbitals and 10^{-6} for the 2s orbital. The valence orbitals having node(s) are slightly more difficult to fit. Under these conditions, the iterative scheme converges to results close to HF limit, but obviously it cannot approach it completely because the fit is based on a limited number of gaussian functions.

4. Results and advantages

In Tab. 1 are given the various results obtained in our group ; the precision of the method used for the resolution as well as the main interest of these results are summarized. When results have been obtained numerically (section 3.1) the method is denoted *num-SCF* or *num-MCSCF* according to the level of theory used ; when an analytical treatment (section 3.2) has been performed, the denotation is *analyt-gauss* if the trial functions were expressed as linear combinations of gaussian functions or *analyt-Slater* if the trial functions were expressed as linear combinations of Slater functions. Finally when a semi-analytical treatment (section 3.3) has been done the method is called *analyt-gauss**. Results fall into three categories : the first one corresponds to pure numerical results on He, H_2, H_3, Li_2 and -$(H)_n$- which have demonstrated the feasibility of numerical calculations. They have also provided momentum wavefunctions for physical quantities such as Compton profiles [17], (e,2e) cross-sections [26]. In the second category we have investigated the possibilities of using a variation-iteration procedure defined in momentum space to improve the one-electron states for various chemical systems expressed as linear combinations of gaussian functions. Significant improvements in energy quantities and properties sensitive to the shape of the wave function (Compton profile, momentum distribution, etc.) were indeed noted. In particular, the first iteration transforms the trial wave function expressed as linear combinations of gaussian functions in an expression which involves Dawson functions. An asymptotic analysis carried on the first iterate discloses a behavior quite close to the exact one. In the third category, the semi-numerical approach is used to provide physical quantities. Similarly to the position space approach it is based on the variation principle which guides the changes of the wavefunction : the closer the energy E to E_0, the nearer the trial wave function ϕ the ground state ϕ_0. In LCAO-SCF-MO schemes however, the function obtained by minimizing the total energy does not necessarily give a good description of properties such as multipole moments, while in momentum space due to the capacity of the method to improve the quality of a wavefunction significant improvements have been obtained e.g. for the dipole moment of the hydrogen fluoride [38].

CHEMICAL SYSTEMS	RESOLUTION SCHEME	PROPERTY OF INTEREST	REFERENCES
H	analyt-gauss	first evidence of the right asymptotic decay	20-22
H⁻	analyt-gauss	two-electron atomic system, importance of the orthogonalisation	22,23
He	analyt-gauss	two-electron atomic system, importance of the orthogonalisation	22,23
	analyt-slater	accuracy of numerical calculations	24
	num-SCF		25
		(e,2e) cross section	26
Li⁺	analyt-gauss	two-electron atomic system, importance of the orthogonalisation	22,23
Li⁻	analyt-gauss *	four-electron atomic system, importance of the orthogonalisation, variation-iteration on an anionic system	27,28
Be	analyt-gauss	four-electron atomic system, convergence of the variation-iteration method, importance of the orthogonalisation	22,29
			30
			31
	analyt-gauss*	convergence of the variation-iteration method	28
B⁺	analyt-gauss	four-electron atomic system, convergence of the variation-iteration method, importance of the orthogonalisation	29
			30
			31
	analyt-gauss *	convergence of the variation-iteration method	28

CHEMICAL SYSTEMS	RESOLUTION SCHEME	PROPERTY OF INTEREST	REFERENCES
F^-	analyt-gauss *	variation-iteration on a ten electron anionic system	27,28
Ne	analyt-gauss *	variation-iteration on a ten electron anionic system	28
Na^+	analyt-gauss *	variation-iteration on a ten electron anionic system	28
H_2	num-SCF	test of the method	16, 32-34
	analyt-gauss	first molecular analytical treatment	35,37
	num-MCSCF	Compton profile	16,32,33
Li_2	num-SCF		33,36
HeH^+	analyt-gauss	first asymptotic molecular treatment	37
HF	analyt-gauss *	dipole moment	38
H_3	num-SCF	open-shell system	17,32-34,36
H_n	eqs	infinite system	18,19,33

Tab.1 : Chemical systems (atoms, molecules or chain) treated in momentum space : methods and properties.

analyt-gauss : analytical one-step iteration of a SCF procedure starting with trial functions expressed as linear combinations of gaussian functions

analyt-gauss * : semi-analytical treatment

analyt-slater : analytical one-step iteration of a SCF procedure starting with trial functions expressed as linear combinations of Slater functions

num-SCF : fully numerical iterative SCF procedure

num-MCSCF : fully numerical iterative MC-SCF procedure

eqs : momentum space equations

References

1.   P.A.M. Dirac, The principles of Quantum Mechanics, 4th ed, The Clarendon Press, Oxford, 1958, pp. 89-97

2.   P. Roman, Advanced Quantum Theory, Addison Wesley, Reading, 1965, p. 11

3.   J. H. Moore, J. A. Tossel, M. A. Coplan, *Acc. Chem. Res.*, **15**, 192, (1982)

4.   C. E. Brion, *Int. J. Quantum Chem.*, **29**, 1397, (1986)

5.   E. Weigold, Momentum Space Wave Functions, American Institute of Physics, vol. 86, Adelaide, 1982

6.   B. K. Novosadov, *J. Mol. Struct.*, **54**, 269, (1979)

7.   J. Navaza, G. Tsoucaris, *Phys. Rev.*, **A24**, 683, (1981)

8.   E.A. Hylleraas, *Z. Physik*, **74**, 216, (1932)

9.   H.A. Bethe, E.E. Salpeter, Quantum Mechanics of One and Two-electrons Atoms, Springer-Verlag, Berlin, 1957, pp. 36-47

10. R. McWeeny, C.A. Coulson, *Proc. Phys. Soc.*, **162**, 509, (1949)

11. R. McWeeny, *Proc. Phys. Soc.*, **162**, 519, (1949)

12. N.V. Svartholm, The Binding Energies of the Lightest Atomic Nuclei, Thesis, Lund, 1945

13. E.N. Lassettre, *J. Chem. Phys.*, **58**, 1991, (1973)

14. C. Carter, *Proc. Roy. Soc.*, **A235**, 321, (1956)

15. E.A. McCullough Jr., *J. Chem. Phys.*, **62**, 3991, (1975) and L. Laaksonen, P. Pyykkö, D. Sundholm, *Chem. Phys. Lett.*, **96**, 1, (1983)

16. M. Defranceschi, M. Suard, G. Berthier, *C. R. Acad. Sci. Paris*, **299**, 9, (1984)

17. M. Defranceschi, M. Suard, G. Berthier, *C. R. Acad. Sci. Paris*, **296**, 1301, (1983) and *Int. J. Quantum Chem*, **25**, 863, (1984)

18. M. Defranceschi, J. Delhalle, *C.R. Acad. Sc. Paris*, **301**, 1405, (1985)

19. M. Defranceschi, J. Delhalle, *Phys. Rev.* **B34**, 5862, (1986)

20. J. Delhalle, J.G. Fripiat and M. Defranceschi, *Annales Soc. Scient. Bruxelles* , **101**, 9, (1987)

21. J.G. Fripiat, M. Defranceschi, J. Delhalle in Numerical Determination of the Electronic Structure of Atoms, Diatomic and Polyatomic Molecules, M.Defranceschi, J. Delhalle (eds), NATO-ASI Series C vol. 271, Kluwer Academic Publishers, Dordrecht, 1989, pp. 245-250

22. M. Defranceschi, L. De Windt, J.G. Fripiat, J. Delhalle, *J. Mol. Struct.(Theochem)*, **258**, 179, (1992)

23. J. Delhalle, J.G. Fripiat, M. Defranceschi, *Bull. Soc. Chim. Belg.*, **99**, 135, (1990)

24. M. Defranceschi, J. Delhalle, *Eur. J. Phys.*, **11**, 172, (1990)

25. J. Delhalle and M. Defranceschi, *Int. J. Quantum Chem.*, **S 21**, 425, (1987)

26. M. Defranceschi, A Lahmam-Bennani, *J. Elect. Spectr. Rel. Phenom.*, **48**, 1,(1989)

27. L. De Windt, M. Defranceschi, J. Delhalle, *Theoret. Chim. Acta*, **86**, 487, (1993)

28. L. De Windt, P. Fischer, M. Defranceschi, J. Delhalle, J.G. Fripiat, *J. Comp. Phys.*, **111**, 266, (1994)

29. L. De Windt, J.G. Fripiat, J. Delhalle, M. Defranceschi, *J. Mol. Struct.(Theochem)*, **254**, 145, (1992)

30. L. De Windt, M. Defranceschi, J. Delhalle, *Int. J. Quantum Chem.*, **45**, 609, (1993)

31. L. De Windt, M. Defranceschi, J.G. Fripiat, J. Delhalle, *Annales Soc. Scient. Bruxelles*, **105**, 89, (1991)

32. G. Berthier, M. Defranceschi, J. Navaza, M. Suard, G. Tsoucaris, *J. Mol. Struct. (Theochem)*, **120**, 243, (1985)

33. G. Berthier, M. Defranceschi, J. Delhalle, in Numerical Determination of the Electronic Structure of Atoms, Diatomic and Polyatomic Molecules, M. Defranceschi and J.Delhalle (eds) NATO-ASI Series C vol 271, Kluwer Academic Publishers, Dordrecht, 1989, pp. 209-238

34. G. Berthier, M. Defranceschi, J. Delhalle in Self-Consistent Field : Theory and Applications, R. Carbo, M. Klobukowski Eds, Elsevier, Amsterdam 1990, "Studies in Physical and Theoretical Chemistry ", n 70, p. 387

35. P. Fischer, M. Defranceschi, J. Delhalle, *Numerische Mathematik*, **63**, 67, (1992)

36. M. Defranceschi, M. Suard, G. Berthier, *Folia Chimica Theoretica Latina*, **XVII,** 65, (1990)

37. P. Fischer, L. De Windt, M. Defranceschi, J. Delhalle, *J. Chem. Phys.,* **99**, 7888, (1994)

38. L. De Windt, M. Defranceschi, J. Delhalle, to be published

39. C.C.J. Roothaan, *Rev. Mod. Phys.*, **23**, 69, (1951)

40. G.G. Hall, *Proc. Roy. Soc.*, **A205**, 541, (1951)

41. G. Berthier, *C. R. Acad. Sci. Paris*, **238**, 91, (1954) ; *J. Chim. Phys.*, **51**, 363, (1954) and **52**, 141, (1955)

42. J.A. Pople, R.K. Nesbet, *J. Chem. Phys.*, **22**, 571, (1954)

43. C.C.J. Roothaan, *Rev. Mod. Phys.*, **32**, 179, (1960)

44. J. Hinze, *J. Chem. Phys.*, **59**, 6424, (1973)

45. A. Veillard, *Theoret. Chim. Acta*, **4**, 22, (1967)

46. D.R. Hartree, W. Hartree, B. Swirles, *Phil. Trans. Roy. Soc.*, **A238**, 229, (1939)

47. G. Das, A.C. Wahl, *J. Chem. Phys.*, **44**, 87, (1970)

48. A. Veillard, E. Clementi, *Theoret. Chim. Acta*, **7**, 133, (1967)

49. G. Del Re, J. Ladik, G. Biczo, *Phys. Rev.*, **155**, 977, (1967)

50. J.M. André, L. Gouverneur, G. Leroy, *Int. J. Quantum Chem.*, **1**, 427, (1967)

51. G. Fonte, *Theoret. Chim. Acta* , **59**, 533, (1981)

52. E.R. Davidson, D. Feller, *Chem. Rev.*, **86**, 681, (1986)

53. B. Klahn, J.D. Morgan, *J. Chem. Phys.*, **81**, 410, (1984)

54. C.L. Davis, H.J. Monkhorst, *Chem. Phys. Lett*, **111**, 526, (1984)

55. L. Laaksonen, P. Pyykkö, D. Sundholm, *Comput. Phys. Rep.*, **4**, 313, (1986)

56. E.A. McCullough, Comput. Phys. Rep., **4**, 265, (1986)

57. P.O. Löwdin, *J. Chem. Phys.*, **18**, 365, (1950)

58. S.A. Alexander, H.J. Monkhorst, *Int. J. Quantum Chem.*, **32**, 361, (1987)

59. S.A. Alexander, R.L. Coldwell, H.J. Monkhorst, *J. Comput. Phys.*, **76**, 263, (1988).

60. W. Rodriguez, Y. Ishikawa, *Chem. Phys. Lett.*, **146**, 515, (1988)

61. Y. Ishikawa, I.L. Aveponte-Allevanet, S.A. Alexander, *Int. J. Quantum Chem*, **S23**, 209, (1989).

62. N.N. Lebedev, Special Functions and their Applications, Dover, New York, 1972

63. J. Spanier, K.B. Oldham, An Atlas of Functions, Hemisphere, Washington, 1987

64. M. Defranceschi, M. Sarrazin, *Comput. Phys. Com*, **52**, 409, (1989)

65. Gausfit, M. Herman, R.E. Stanton, *Quantum Chemistry Program Exchange*, **11**, 237, (1973)

66. R.F. Stewart, *J. Chem. Phys.*, **50**, 2485, (1969)

Core-Valence Separation in the Study of Atomic Clusters

O. SALVETTI
Dipartimento di Chimica e Chimica Industriale. Università di Pisa
Via Risorgimento 35, 56126 Pisa, Italy

The study of clusters containing an increasing number of atoms provides an interesting theoretical way of understanding the properties of solid matter.

In particular it allows us to consider in a simple way possible irregularities of structure, the existence of non stoichiometric compounds, and the possibility of replacing one atom by another.

A study of the variation of properties with cluster size is also of great importance, especially in view of experimentally observed variations, which may amount to almost a change of phase, in clusters ranging from 10 to 50 atoms [1].

The main difficulty in the theoretical study of clusters of heavy atoms is that the number of electrons is large and grows rapidly with cluster size. Consequently, ab initio "brute force" calculations soon meet insuperable computational problems. To simplify the approach, conserving atomic concept as far as possible, it is useful to exploit the classical separation of the electrons into "core" and "valence" electrons and to treat explicitly only the wavefunction of the latter. A convenient way of doing so, without introducing empirical parameters, is provided by the use of generalyzed product function, in which the total electronic wave function is built up as antisymmetrized product of many group functions [2-6].

This scheme is very appealing, since it allows us to reduce drastically the numbers of electrons to be considered, thus making possible essentially "ab initio" calculation, even for large systems,.

If a cluster is built from various separated atoms A, B, ... with N_A, N_B, ... "core" electrons, descibed by the functions $\Phi_A(1, ... N_A)$, $\Phi_B(1, ... N_B)$, ..., the generalized product for the total number of electrons will be given by the following expression:

$$\Psi(1,2,...N) = M\hat{A}\left[\Phi_A(1,...N_A)\Phi_B(N_A+1,...N_A+N_B)...\Phi_V(N_\gamma+1,...N_\gamma+N_V)\right] \quad (1)$$

where N_γ is the total number of "core" electrons, N_V and Φ_V are the total number and the wave function of the "valence" electrons, $\hat{A}$ is the operator that antisymmetrizes the product, and M is a normalization factor.

The strong orthogonality requirement among the wave functions of different groups, is satisfied for the "core" groups, because they are localized in different spatial sites, but it must be imposed between Φ_V and each "core" function. It is well known that this last condition is equivalent to assuming that the function Φ_V is built up using spin-orbitals drawn from a set orthogonal to all orbitals of the "core" functions.

Y. Ellinger and M. Defranceschi (eds.), Strategies and Applications in Quantum Chemistry, 159–164.
© 1996 *Kluwer Academic Publishers.*

Let us first consider a single atom, A. We can study this atom with high accuracy and prepare some atomic quantities useful in subsequent calculations. Also for this atom we suppose that the total wave function is given by a product

$$\Psi_A = M\hat{A}\left[\Phi_A(1,...N_A)\Phi_{VA}(N_A+1,...N_A+N_{VA})\right] \tag{2}$$

where the meaning of the symbols is obvious.

We suppose that the core functions Φ_A is built up from orbitals σ_1, σ_2, ... which satisfy the following relation

$$\left\langle\sigma_i|\sigma_j\right\rangle_{SA} \cong \left\langle\sigma_i|\sigma_j\right\rangle = \delta_{ij} \tag{3}$$

where $<\ >_{SA}$ means that the integration is extended to a sphere SA around A. This sphere SA is supposed much smaller than the Van der Waals sphere of the atom A.

The Φ_{VA} function is then built up from the orbitals σ_1, σ_2, ... χ_1, χ_2... , where the χ_i functions extend well beyond SA.

Since we are dealing with a monocentric problem, the function Φ_{VA} can be easily studied as accurately as required. In order to determine the valence function Φ_{VA} we generally use an open shell H.F. method and so obtain a set of orthonormal orbitals φ_{VAi}, which satisfy the conditions of strong orthogonality to "core" functions. Each of these orbitals will be of the following form

$$\varphi_{VAi} = \sum_j a_{ij}\sigma_j + \sum_s b_{is}\chi_s \tag{4}$$

and if φ_{CAj}, is one of the orbitals appearing in Φ_A the following relation holds:

$$\left\langle\varphi_{VAi}|\varphi_{CAj}\right\rangle_{SA} \cong \left\langle\varphi_{VAi}|\varphi_{CAj}\right\rangle = 0 \tag{5}$$

The "core" orbital φ_{CAj} is built up from function σ_1, σ_2, ...
The following atomic quantities can now be calculated

$$E_{CA} = \left\langle\Phi_A\left|\sum_i\left(-\frac{1}{2}\nabla_i^2 - \frac{z_A}{r_{iA}}\right) + \frac{1}{2}\sum_{i,j}'\frac{1}{r_{ij}}\right|\Phi_A\right\rangle \tag{6}$$

$$q_{Anml} = \left\langle\varphi_{VAn}\left|r_A^l\right|\varphi_{VAm}\right\rangle_{SA}$$

$$q'_{Anml} = \left\langle\chi_{An}\left|r_A^l\right|\chi_{Am}\right\rangle_{SA} \qquad (l=0,1,2,...) \tag{7}$$

$$\rho_{As} = 2\sum_j\left\langle\varphi_{CAj}\left|r_A^s\right|\varphi_{CAj}\right\rangle \tag{8}$$

$$\varepsilon_{Anm} = \left\langle\varphi_{VAn}\left|-\frac{1}{2}\nabla^2 - \frac{z_A}{r_A} + \sum_j\left(2J_{CAj} - K_{CAj}\right)\right|\varphi_{VAm}\right\rangle_{SA}$$

$$\varepsilon'_{Anm} = \left\langle\chi_{An}\left|-\frac{1}{2}\nabla^2 + \frac{-z_A + 2n_{CA}}{r_A}\right|\chi_{Am}\right\rangle_{SA} \qquad \left(n_{CA} = \frac{N_A}{2}\right) \tag{9}$$

$$\alpha_{Aijkl} = \left\langle f_i f_j \middle| f_k f_l \right\rangle_{SA}$$

$$\beta_{Aijkl} = \left\langle \chi_i^0 \chi_j^0 \middle| \chi_k^0 \chi_l^0 \right\rangle_{SA} \tag{10}$$

$$\gamma_{Akl} = \left\langle \chi_k^0 \middle| \frac{1}{r_A} \middle| \chi_l^0 \right\rangle_{SA} \tag{11}$$

$$\rho_{Aij} = \left\langle f_i \middle| f_j \right\rangle_{SA} \tag{12}$$

$$\rho_{Aij}^0 = \left\langle \chi_i^0 \middle| \chi_j^0 \right\rangle_{SA} \tag{13}$$

To build up Φ_v in the cluster function (1) we use the functions $\varphi_{VA1}, \varphi_{VA2} \cdots \varphi_{VB1}, \varphi_{VB2}, \cdots$, all of which satisfy the strong orthogonality condition in the sense of to (2), but do not satisfy the strong orthogonality needed for (1) We therefore consider the linear combination

$$f_i = \sum_R \sum_j c_{rij} \varphi_{VRj} \tag{14}$$

and require that this function be a linear combination of the functions φ_{VPj} in each sphere SP. This condition can be only approximatively satisfied and it is useful to have a measure of the goodness of the approximation. To obtain such a criterion we consider the quantity

$$T_i = \sum_R \left\langle f_i - \sum_j u_{rij} \varphi_{VRj} \middle| f_i - \sum_j u_{rij} \varphi_{VRj} \right\rangle_{SR} \tag{15}$$

This quantity measures the error in the orthogonality of f_i to all the group functions. Since the u_{rij} are arbitrary coefficients, we can put $u_{rij} = c_{rij}$ and so obtain

$$T_i = \sum_P \sum_l c_{pil} \sum_Q \sum_m c_{qim} \sum_{R \neq, P, Q} \left\langle \varphi_{VPl} \middle| \varphi_{VQm} \right\rangle_{SR} \tag{16}$$

As noted in previous papers [7-11], by considering the matrix $\mathbb{T}$

$$T_{pl,qm} = \sum_{R \neq, P, Q} \left\langle \varphi_{VPl} \middle| \varphi_{VQm} \right\rangle_{SR} \quad (P, Q = A, B \ldots) \tag{17}$$

we can find the minimum of (15) by diagonalizing the $\mathbb{T}$ matrix. The eigenvectors, ordered according to the corresponding increasing eigenvalues, give functions less and less orthogonal to the "core" orbitals. The associated eigenvalues give us a measure of the goodness of the functions obtained. One must keep only functions corresponding to eigenvalues smaller than some chosen threshold.
In this way we obtain n functions $f_1, f_2 \ldots f_n$

$$f_i = \sum_R \sum_j c_{rij} \varphi_{VRj} = \sum_R \left[\sum_j a_{rij} \sigma_{Rj} + \sum_s b_{ris} \chi_{Rs} \right] = \sigma_i^0 + \chi_i^0 \tag{18}$$

with the property

$$f_i(r) = \sum_s c_{ris} \varphi_{VRs} \quad (r \in SR) \tag{19}$$

From (1) and (6), (8) we then obtain

$$E = E_c + E_v \tag{20}$$

$$E_c = \sum_R E_{cR} + \frac{1}{2} \sum_{P,Q}{}' \sum_{nm} \beta_{nm} \frac{P_{pn} P_{qm}}{R_{pq}^{n+m+1}} \tag{21}$$

The $\beta_{l_p l_q}$ in (21), are the coefficients of the two center expansion of the potential energy of interaction of two non overlapping charge distributions. [12].

To obtain the E_V contribution only from calculations over χ type functions and from atomic data one needs a more detailed analysis of the equations. Let us consider the following form of Fock operator

$$F = \left\{ -\frac{1}{2}\nabla^2 + \sum_R \left[-\frac{z_R}{r_R} + \sum_i \left(2J_{CRi} - K_{CRi} \right) \right] \right\} + \sum_j \left(2J_{Vj} - K_{Vj} \right) = \hat{h} + G_v \tag{22}$$

To obtain matrix elements of $\hat{h}$ we have

$$\left\langle f_i \middle| \hat{h} \middle| f_j \right\rangle = \sum_R \left\langle f_i \middle| \hat{h} \middle| f_j \right\rangle_{SR} + \left\langle f_i \middle| \hat{h} \middle| f_j \right\rangle_{Vext} \tag{23}$$

But from (18-19) and (7-13) one obtains

$$\left\langle f_i \middle| \hat{h} \middle| f_j \right\rangle_{Vext} = \left\langle \chi_i^0 \middle| \hat{h} \middle| \chi_j^0 \right\rangle_{Vext}$$

$$\left\langle \chi_i^0 \middle| \hat{h} \middle| \chi_j^0 \right\rangle_{Vext} = \left\langle \chi_i^0 \middle| \hat{h} \middle| \chi_j^0 \right\rangle - \sum_p \left\langle \chi_i^0 \middle| \hat{h} \middle| \chi_j^0 \right\rangle_{SP}$$

$$\left\langle f_i \middle| \hat{h} \middle| f_j \right\rangle = \sum_p \left\langle f_i \middle| \hat{h} \middle| f_j \right\rangle_{SP} + \left\langle \chi_i^0 \middle| \hat{h} \middle| \chi_j^0 \right\rangle - \sum_p \left\langle \chi_i^0 \middle| \hat{h} \middle| \chi_j^0 \right\rangle_{SP}$$

$$= \left\langle \chi_i^0 \middle| \hat{h} \middle| \chi_j^0 \right\rangle + \sum_p \left[\left\langle f_i \middle| \hat{h} \middle| f_j \right\rangle_{SP} - \left\langle \chi_i^0 \middle| \hat{h} \middle| \chi_j^0 \right\rangle_{SP} \right]$$

$$= \sum_{PQ} \sum_{nm} b_{pin} b_{qjm} \left\langle \chi_{Pn} \middle| -\frac{1}{2}\nabla^2 + \sum_R \frac{-z_R + 2n_{CR}}{r_R} \middle| \chi_{Qm} \right\rangle \tag{24}$$

$$+ \sum_P \sum_{nm} \left\{ c_{pin} c_{qjm} \left[\varepsilon_{pnm} + \sum_{R \ast P} \sum_{ls} \beta_{ls} \frac{q_{pnml} P_{rs}}{r_{PR}^{l+s+1}} \right] - b_{pin} b_{qjm} \left[\varepsilon'_{pnm} + \sum_{R \ast P} \sum_{ls} \beta_{ls} \frac{q'_{pnml} P_{rs}}{r_{PR}^{l+s+1}} \right] \right\}$$

In order to derive Gv matrix elements, we consider the bielectronic integral $\left\langle f_i f_j \middle| f_k f_l \right\rangle$ which can be written

$$\langle f_i f_j | f_k f_l \rangle = \sum_{R,P} {}_{SR}\langle f_i f_j | f_k f_l \rangle_{SP} + \sum_R \left\{ {}_{SR}\langle f_i f_j | f_k f_l \rangle_{Vext} \right.$$

$$+ {}_{Vext}\langle f_i f_j | f_k f_l \rangle_{SR} \left. + {}_{Vext}\langle f_i f_j | f_k f_l \rangle_{Vext} \right\}$$

since

$$_{Vext}\langle f_i f_j | = {}_{Vext}\langle \chi_i^0 \chi_j^0 |$$

$$| f_k f_l \rangle_{Vext} = | \chi_k^0 \chi_l^0 \rangle_{Vext}$$

one has

$$_{Vext}\langle \chi_i^0 \chi_j^0 | \chi_k^0 \chi_l^0 \rangle_{Vext} = \langle \chi_i^0 \chi_j^0 | \chi_k^0 \chi_l^0 \rangle - \sum_{R,P} {}_{SR}\langle \chi_i^0 \chi_j^0 | \chi_k^0 \chi_l^0 \rangle_{SP}$$

$$- \sum_R \left({}_{SR}\langle \chi_i^0 \chi_j^0 | \chi_k^0 \chi_l^0 \rangle_{Vext} + {}_{Vext}\langle \chi_i^0 \chi_j^0 | \chi_k^0 \chi_l^0 \rangle_{SR} \right)$$

$$\langle f_i f_j | f_k f_l \rangle = \langle \chi_i^0 \chi_j^0 | \chi_k^0 \chi_l^0 \rangle + \sum_{RP} \left({}_{SR}\langle f_i f_j | f_k f_l \rangle_{SP} - {}_{SR}\langle \chi_i^0 \chi_j^0 | \chi_k^0 \chi_l^0 \rangle_{SP} \right)$$

$$+ \sum_R \left({}_{SR}\langle f_i f_j | \chi_k^0 \chi_l^0 \rangle_{Vext} + {}_{Vext}\langle \chi_i^0 \chi_j^0 | f_k f_l \rangle_{SR} - {}_{SR}\langle \chi_i^0 \chi_j^0 | \chi_k^0 \chi_l^0 \rangle_{Vext} - {}_{Vext}\langle \chi_i^0 \chi_j^0 | \chi_k^0 \chi_l^0 \rangle_{SR} \right)$$

$$\cong \langle \chi_i^0 \chi_j^0 | \chi_k^0 \chi_l^0 \rangle + \sum_R ({}_{SR}\langle f_i f_j | f_k f_l \rangle_{SR} - {}_{SR}\langle \chi_i^0 \chi_j^0 | \chi_k^0 \chi_l^0 \rangle_{SR})$$

$$+ \sum_{R \neq P} \sum_{nm} \beta_{nm} \frac{q_{rijn} q_{pklm} - q'_{rijn} q'_{pklm}}{r_{RP}^{n+m+1}} + \sum_R \left(\left[\langle \chi_k^0 | 1 / r_R | \chi_i^0 \rangle - \langle \chi_k^0 | 1 / r_R | \chi_i^0 \rangle_{SR} \right] \right.$$

$$\times \left[\rho_{Rij} - \rho_{Rij}^0 \right] + \left[\langle \chi_i^0 | 1 / r_R | \chi_j^0 \rangle - \langle \chi_i^0 | 1 / r_R | \chi_j^0 \rangle_{SR} \right] \left[\rho_{Rij} - \rho_{Rij}^0 \right] \right)$$

$$= \sum_{PQRT} \sum_{i' j' k' l'} b_{Pii'} b_{Qjj'} b_{Rkk'} b_{Tll'} \langle \chi_{Pi'} \chi_{Qj'} | \chi_{Rk'} \chi_{Tl'} \rangle + \sum_R \left(\alpha_{Rijkl} - \beta_{Rijkl} \right)$$

$$+ \sum_{R \neq P} \sum_{nm} \beta_{nm} \frac{q_{rijn} q_{pklm} - q'_{rijn} q'_{pklm}}{r_{RP}^{n+m+1}}$$

$$+ \sum_R \left\{ \left[\langle \chi_k^0 | 1 / r_R | \chi_i^0 \rangle - \gamma_{Rkl} \right] \left(\rho_{Rij} - \rho_{Rij}^0 \right) + \left[\langle \chi_i^0 | 1 / r_R | \chi_j^0 \rangle - \gamma_{Rij} \right] \left(\rho_{Rkl} - \rho_{Rkl}^0 \right) \right\} \quad (25)$$

From (24) and (25) it follows that all the matrix elements $\langle f_i | F | f_j \rangle$ can be obtained by calculation over the set χ of "valence" functions, with the addition of terms relating to single atoms.

Numerical applications to particular clusters are wery encouraging [7-11].

References

1.   U.Even, N.Ben-Horin and J.Jortner, *Phys.Rev.Lett.* **62**, 140, (1989).
2.   R.McWeeny, *Proc.R.Soc.London Ser.* **A 253**, 242, (1959).
3.   R.McWeeny and B.T.Sutcliffe, *Proc.R.Soc.London Ser.* **A 273**, 103, (1963).
4.   Y.Ohrn and R.McWeeny, *Arch. Phys.* **31**, 461, (1966).
5.   P.D.Drace and R.McWeeny, *Proc.R.Soc.London Ser.* **A 317**, 435, (1970).
6.   M.Klessinger and R.McWeeny, *J.Chem.Phys.* **42**, 3343, (1965).
7.   R.Colle, A.Fortunelli and O.Salvetti, *J.Chem.Phys.* **80**, 2654, (1984).
8.   R.Colle, A.Fortunelli and O.Salvetti, *Mol.Phys.* **57**, 1305, (1986).
9.   A.Fortunelli and O.Salvetti, *Mol.Phys.* **75**, 1191, (1992).
10.   A.Fortunelli, O.Salvetti and G.Villani, *Surface Sci.* **244**, 355, (1991).
11.   A.Fortunelli, A.Desalvo, O.Salvetti and E.Albertazzi, <u>Cluster Models for Surface and Bulk Phenomena</u>, Plenum Press,(1992).
12.   J.O.Hirschfelder, C.F.Curtiss and R.B.Bird, <u>Molecular Theory of Gases and Liquids</u>, Wiley (1965).

Core-Hole States and the Koopmans Theorem

C. AMOVILLI and R. McWEENY

Dipartimento di Chimica e Chimica Industriale, Via Risorgimento 35, 56100 Pisa, Italy

1. Introduction

The theorems of Brillouin [1,2] and Koopmans [3], in both their original and generalized forms, have provided a recurring theme in the work of Gaston Berthier who always showed a profound appreciation of their significance and importance (see, for example, [4,5]). Both theorems have been of immense value in the calculation and interpretation of a wide range of molecular properties. But both are 'first-order' theorems, based originally on the Hartree-Fock model, and refer to the first-order effect of perturbations that are considered 'small'. When the perturbations become large the theorems lose their value, except as a basis for rough approximations, bu. the violations themselves are also of considerable practical importance. In particular, as every quantum chemist knows, the ionization energy for removal of an electron from orbital ϕ_k is related to the Hartree-Fock orbital energy ϵ_k, according to the Koopmans theorem, by

$$I_k = -\epsilon_k \tag{1}$$

where ϵ_k is calculated using the 'zero-order' ϕ_k for the unperturbed (*neutral*) system. The perturbation of the *orbitals* in passing from the neutral to the ionized system is irrelevant to the first-order result. To calculate second- and higher-order corrections to equation (1), however, it is necessary to allow the orbitals of the ionized system to 'relax' in order to describe the perturbation of the Hartree-Fock field caused by the change in occupation number of orbital ϕ_k. Such relaxation effects are often rather small and the Koopmans result (1) can give a fairly satisfactory interpretation of the ionization processes observed in valence-electron photoelectron spectroscopy (PES); but for 'deep' ionizations, as observed in ESCA experiments (see, for example, Siegbahn et al [6]) where electrons are knocked out of atomic inner shells, the relaxation effects can be very large. The electron distribution tends to 'collapse' towards the 'core hole' – roughly equivalent to an increased nuclear charge – and the use of (1) commonly yields ionization energies in error by 20–30 eV.

This note is concerned with the alternative procedure in which (1) is replaced by

$$I_k = E_k^+ - E, \tag{2}$$

where E (the electronic energy of a neutral molecule) and E_k^+ (that for the molecule in a 'core-hole' state) are both calculated independently. It must be remarked at

165

Y. Ellinger and M. Defranceschi (eds.), Strategies and Applications in Quantum Chemistry, 165–173.

the outset that since the energy of a 'core-hole' state normally lies high in the continuum, relative to the lowest energy state in which the hole has been filled by an Auger transition from a valence orbital, there are severe problems in calculating the energy E_k^+ by conventional bound-state methods: indeed the corresponding 'state' is not a true bound state at all, being at best metastable and subject to spontaneous decay, with filling of the hole and ejection of a second (Auger) electron. A completely satisfactory calculation would thus require the inclusion in the basis of continuum functions, to admit the possibile presence of a scattered electron, and would employ propagator methods which are well adapted to the description of such processes (see, for example, Agren [7]). Nevertheless, bound-state methods have been widely and successfully used in the interpretation of PES, ESCA and Auger spectra. In particular, the formulation of SCF methods for systems containing incompletely occupied shells (McWeeny [8]) was applied by Firsht and McWeeny [9] to free atoms and ions, with inner-shell holes, yielding results of much higher accuracy than those based on the Koopmans theorem. The present paper reports applications of similar methods to some small molecules.

2. Formulation

For inner-shell ionizations, where the energy change may be several hundred eV, it is sufficient to use ensemble averaging (Slater [10]; McWeeny [8]) over the various states of a configuration - which differ relatively little in energy. The corresponding formulation of many-shell SCF theory is fully described elsewhere (McWeeny [11]) and will be summarized only briefly. We use $\{\phi_k^K\}$ to denote the m_K orbitals of Shell K, containing n_K electrons, and express the orbitals of all shells in terms of a common set of m basis functions $\{\chi_i\}$: thus, collecting the functions in row matrices,

$$\phi = \chi \mathbf{T}, \qquad (3)$$

where $\mathbf{T}$ is an $m \times n$ matrix, n being the total number of orbitals employed. It is also convenient to partition the row matrix ϕ into subsets $\phi_\mathbf{K}$, and the rectangular matrix $\mathbf{T}$ into corresponding $m \times n_K$ blocks $\mathbf{T}_K$. The set of occupation numbers $(n_A, n_B, ...n_K ...)$ then defines the electron configuration, while the average energy (for all states with the same partitioning of electrons among shells) is given by

$$E_{\mathrm{av}} = \sum_K \nu_K \mathrm{tr} \mathbf{R}_K (\mathbf{h} + \tfrac{1}{2} \mathbf{G}_K^{\mathrm{av}}). \qquad (4)$$

Here $\nu_K = n_K/2m_K$ is the *fractional* occupation number of the $2m_K$ *spin*-orbitals of Shell K and $\mathbf{G}_K^{\mathrm{av}}$ is a suitably averaged electron interaction matrix (cf. the usual Roothaan 'G matrix') and depends on the density matrices $\mathbf{R}_K(= \mathbf{T}_K \mathbf{T}_K^\dagger)$ of all shells: in fact

$$\mathbf{G}_K = \mathbf{G}(\nu_K' \mathbf{R}_K) + \sum_{L(\neq K)} \mathbf{G}(\nu_L \mathbf{R}_L), \qquad (5)$$

where the modified occupation number (removing the self-interaction when $n_K = 1$) is $\nu_K' = 2(n_K - 1)/(2m_K - 1)$. The matrix $\mathbf{G}(2\mathbf{R})$ coincides with the usual $\mathbf{G}$ matrix for a closed-shell system, while $\mathbf{h}$ in (4) is the usual 1-electron Hamiltonian matrix. All matrices are defined with respect to the basis functions in χ.

The energy expression (4) applies when the orbitals are orthonormal and in seeking a stationary value it is thus necessary to introduce constraints to maintain orthonormality during a variation. When this is done, the orbitals that give a stationary

point turn out to be eigenfunctions of a certain 'effective' Hamiltonian (which embodies the constraints); and this leads to an iterative procedure parallel to that of
the usual closed-shell SCF theory. For molecules, these 'canonical' orbitals are normally the delocalized MOs which extend over the whole molecular framework; but
E_{av} is invariant against unitary mixing of the orbitals within each shell (which leaves
the matrices $\mathbf{R}_K$ unchanged) and this freedom may be exploited in the usual way
to obtain alternative orbitals with a high degree of localization in different regions
of the molecule (e.g. inner shells, bonds, lone pairs). Clearly, in discussing phenomena related to physically well-defined regions, we shall be more concerned with
the localized orbitals than the canonical MOs. The question that then arises is that
of what localization criterion to adopt: the one to be used in working with atomic
inner shells is simply that the inner-shell orbital be constructed from basis functions
located on the atom in question. All other orbitals are easily orthogonalized against
inner shells (e.g. by the Schmidt method) and among themselves (e.g. by the Löwdin
$\mathbf{S}^{-\frac{1}{2}}$ transformation).

Instead of using repeated solution of a suitable eigenvalue equation to optimize the
orbitals, as in conventional forms of SCF theory, we have found it more convenient
to optimize by a gradient method based on direct evaluation of the energy functional
(4), orthonormalization being restored after every parameter variation[1]. Although
many iterations are required, the energy evaluation is extremely rapid, the process is
very stable, and any constraints on the parameters (e.g. due to spatial symmetry or
choice of some type of localization) are very easily imposed.It is also a simple matter
to optimize with respect to non-linear parameters such as orbital exponents.

3. Some results

We have considered K-shell ionizations from the atoms of carbon, oxygen, and nitrogen in a series of small molecules, typically using basis sets of 'double-zeta' quality
(as tabulated by, for example, Dunning [12]), with the addition of polarization functions for the smallest systems. The total energies for the neutral molecules and some
of their core-hole positive ions are collected in Table 1. Energies for the molecular
ground states, as calculated by standard (RHF) SCF methods, are also shown for
comparison. It is evident that the localization constraint for the inner shells has a
negligible effect on the energies.

The energies in the last column of Table 1 show the effect of modifying the basis
set, after the ionization, to allow for the increased central field to which the valence
electrons are then exposed. Some of the early work on the interpretation of ESCA
and Auger spectroscopy employed an 'equivalent-core' approximation (Shirley [13])
in which, with a minimal basis set, valence orbitals were given exponents appropriate
to an effective atomic number $Z + 1$ instead of Z: the inner shell, with only one
1s electron, was thus 'modelled' by an 'equivalent core' with two 1s electrons but
one extra unit of positive charge on the nucleus. This simple model has been found
equally effective in the case of a DZ basis: in describing the valence electrons of
an atom with a core hole it is sufficient to use contracted gaussians with tabulated
exponents and contraction coefficients for the atom of atomic number $Z + 1$ instead
of Z. For the 1s orbital, on the other hand, appeal to 'screening constant' rules
(Slater [14]; Clementi and Raimondi [15]) suggests that the Gaussian exponents for

---

[1]The valence set is orthogonalized against the core set, so as not to 'contaminate' the core
orbitals, while symmetric ($\mathbf{S}^{-\frac{1}{2}}$) orthonormalization is employed within each set.

the neutral atom should be multiplied by a scale factor

$$\eta = \left(\frac{Z^* + 0.31}{Z^*}\right)^2 \tag{6}$$

following removal of an electron, Z^* being the recommended effective nuclear charge for the atom in question. This value (close to the actual nuclear charge) proves to be perfectly satisfactory. In fact, the procedure just described leads to energy values which are not appreciably affected by further parameter optimization. The results in the Table confirm the need to re-optimize the basis following ionization, the resultant drop in energy being quite significant. The agreement with experiment at this level is now probably as good as can be expected, bearing in mind the extreme simplicity of the theoretical model on which the calculations are based.

Tab.1 - Energies (hartree) of ground and hole states of some small molecules.

Molecule	Basis set[1]	E_{RHF}	E_{av}	E_{av}^+	$E_{av}^{+[2]}$
O_2	DZ+P	-149.59970	-149.56585	-129.54438	-129.61425
N_2	DZ+P	-108.95849	-108.95750	-93.80383	-93.88673
NO	DZ+P	-129.26776	-129.26205	-114.06148	-114.13374
				-109.25956	-109.33144
CO	DZ+P	-112.75847	-112.75559	-101.76264	-101.84031
				-92.77571	-92.84943
H_2O	DZ	-76.00984	-76.00879	-56.13243	-56.19138
NH_3	DZ	-56.17629	-56.17527	-41.22097	-41.26973
CH_4	DZ	-40.18528	-40.18493	-29.44749	-29.48871
HCHO	DZ	-113.82741	-113.82321	-102.93218	-102.97407
				-93.97469	-94.03449
CH_3OH	DZ	-115.01394	-114.99129	-104.19377	-104.21490
				-95.16137	-95.21276

(1) DZ refers to the Dunning [12] contractions of the Huzinaga [17] Gaussian
    sets. The polarization function in all cases is a d Gaussian with exponent 0.8.
(2) Modified basis set (see text).

The results for homonuclear molecules are of particular interest in so far as they exhibit "symmetry breaking". For N_2, for example, removal of an electron from the $1\sigma_g$ MO of a ground-state SCF calculation, with re-optimization of all orbitals subject to symmetry constraints, leads to an energy value of -93.49612 hartree for the positive ion in which the core hole is symmetrically 'shared' between the two atoms. But when the symmetry constraint is relaxed the energy falls to -93.88673 hartree, corresponding to localization of the hole on one centre alone: this is the result expected on physical grounds, given a sufficiently short time scale for the process of electron removal – the valence-electron distribution responding immediately to the enhanced attraction towards the core with the hole. Of course, as already remarked, the resultant metastable state will decay rapidly: the symmetry constrained wavefunction describes the stationary state, in a long-time limit, of a fictitious model system in which the hole appears on either centre with equal probability. The spectroscopic

observations are in fact consistent with the short-time situation, before relaxation of
the electron distribution has taken place.

The inner-shell ionization energies are collected in Table 2 and compared with the
Koopmans estimates (which are seen to be seriously in error) and the best available
experimental values. Whilst the Koopmans approximation is clearly incapable of
giving good ionization energies and must therefore be used with caution in predicting
the 'chemical shifts' in going from one molecule to another, the ionization energies
based on (2) are rather satisfactory.

Tab.2 - Binding energies (eV) of core electrons.

Molecule	Orbital	KT[1]	IP[2]	IP[2,3]	exp[4,5]
O_2	O1s	564.6	544.8	542.9	543.1(q)
					544.2(d)
NO	O1s	563.1	544.3	542.3	543.3(t)
					544.0(s)
CO	O1s	562.3	543.7	541.7	542.1
H_2O	O1s	559.2	540.9	539.3	539.7
HCHO	O1s	559.9	540.1	538.5	539.4
CH_3OH	O1s	558.7	539.6	538.2	538.9
N_2	N1s	427.0	412.3	410.1	409.9
NO	N1s	427.9	413.6	411.6	410.3(t)
					411.8(s)
NH_3	N1s	422.8	406.9	405.6	405.6
CO	C1s	309.1	299.1	297.0	295.9
HCHO	C1s	308.7	296.4	295.2	294.5
CH_3OH	C1s	306.7	293.8	293.2	292.3
CH_4	C1s	305.1	292.2	291.0	290.7

(1) Koopmans' theorem.
(2) Ionization potentials obtained from the energies of Tab.1.
(3) Modified basis set (see text).
(4) In parentheses the multiplicity of the final states (s-singlet,
     d-doublet, t-triplet, q-quadruplet).
(5) Siegbahn et al [6], Carroll and Thomas [18].

In view of possible applications of the method to much larger molecules, where the
use of extended basis sets may be impracticable, it is worth asking whether good
results might also be obtained with only a minimal basis: in this case, with the
reduced flexibility of the basis, it would clearly be desirable to optimize the orbital
exponents. To investigate this possibility, a study has been made of the inner-shell
ionization energies of the carbon atom in the series of fluoro-substituted methanes
$CH_x F_y$ $(x + y = 4)$, using the MIDI-4 basis of Tatewaki and Huzinaga [16]. The
approximation (6) was used for the 1s orbitals, but the scale factors for the valence
orbitals had to be optimized (for both the neutral and ionized systems) since the
tabulated MIDI-4 values refer to the isolated atoms. Table 3 shows the resultant
scale factors for the carbon 2s and 2p orbitals, which are most strongly affected by

the core ionization. The absolute energy values are of course somewhat inferior to those obtained with a DZ basis (cf. the results for CH_4 in Table 1) but the ionization energies, shown in Table 4, are still in very satisfactory agreement with experiment. Even the chemical shifts for fluoro-substitution are very close to those observed: what is more surprising is that the *shifts* predicted by the Koopmans theorem are also quite satisfactory, even though the ionization energies are in error by 10-15 eV. Whether the Koopmans theorem will retain its apparent predictive value in situations where the chemical shift is much smaller remains an open question.

Tab.3 - Calculated energies (hartree) of ground and carbon hole states for fluorinated methanes and optimized scale factors.

Molecule	η_{2s}	η_{2p}	E_{RHF}	E_{av}	η_{2s}	η_{2p}	E_{av}^+
CF_4	1.344	1.543	-434.88864	-434.87627	1.461	1.905	-423.74009
CHF_3	1.300	1.500	-336.18893	-336.17818	1.420	1.843	-325.16163
CH_2F_2	1.270	1.427	-237.48610	-237.47700	1.345	1.772	-226.57747
CH_3F	1.231	1.335	-138.79063	-138.78305	1.309	1.629	-127.99446
CH_4	1.181	1.185	-40.11401	-40.10897	1.200	1.585	-29.41466

Tab.4 - Binding energies (eV) of carbon core electrons in fluorinated methanes.

Molecule	KT	c.s.[1]	IP	c.s.	exp[2]	c.s.
CF_4	316.6	12.1	303.0	11.9	301.7	11.0
CHF_3	313.3	8.8	299.8	8.7	299.0	8.3
CH_2F_2	310.0	5.5	296.6	5.5	296.3	5.6
CH_3F	306.7	2.2	293.6	2.5	293.5	2.8
CH_4	304.5	0.0	291.1	0.0	290.7	0.0

(1) Chemical shift.
(2) Adams and Clark [19].

4. Conclusions and further applications

It is evident that the method of calculation used in this work provides an extremely simple approach to the interpretation of ESCA results for ionization from an atomic K shell, in spite of the fact that the state of the ion plus the ejected electron lies high in the energy continuum of the neutral molecule. More sophisticated methods of dealing with such states are of course available (see Agren et al [7]) but, whilst capable of giving excellent results for valence electron ionizations (including also intensities and vibrational fine structure), encounter considerable difficulty in treating core-hole states, where relaxation effects are very severe. The simple model used here, on the other hand, is particularly well adapted to the study of these 'deep' ionizations and gives an immediate and transparent interpretation of the relaxation effects in terms of scaling (contraction) of the valence orbitals. It is also possible to extend the approach in various ways; for example, for open-shell molecules, the states of the configuration with a core hole (which arise from differences of spin coupling between the core and valence electrons) can readily be studied by using the optimized orbitals for the configurational average energy to set up the secular equations that will lead to the individual states.

Another application is to the study of the 'Auger states' in which a further electron ionization of attachment may occur, leaving the system with holes in more than one shell. Such states were considered in some detail by Firsht and McWeeny [9] for free atoms: here we have made a preliminary application to the nitrogen molecule. The initial aim is simply to identify and assign the principal peaks and satellites in the Auger spectrum of gaseous N_2.

Tab.5 - Calculated average energies of different configurations of N_2.

Configuration	E_{av}(hartree)	charge	type	$E_{av} - E_{av}^{GS}$(eV)	exp[1]
G.S.	-108.958	0	-	0.0	
$N1s^* \, 2\pi^1$	-94.235	0	Ke	400.6	400.8
$1\pi^*_u$	-108.392	1	L	15.4	16.8
$3\sigma^*_g$	-108.374	1	L	15.9	15.5
$2\sigma^*_u$	-108.216	1	L	20.2	18.6
$2\sigma^*_g$	-107.568	1	L	37.8	37.3
$N1s^*$	-93.887	1	K	410.1	409.9
$N1s^* \, 1\pi^* \, 2\pi^1$	-93.583	1	KLe	418.4	
$3\sigma^*_g \, 1\pi^*_u$	-107.374	2	LL	43.1	
$1\pi^*_u \, 1\pi'^*_u$	-107.372	2	LL	43.2	
$1\pi^{**}_u$	-107.306	2	LL	45.0	
$3\sigma^{**}_g$	-107.286	2	LL	45.5	
$3\sigma^*_g \, 2\sigma^*_u$	-107.248	2	LL	46.5	
$2\sigma^*_u \, 1\pi^*_u$	-107.226	2	LL	47.1	
$3\sigma^{**}_u$	-106.976	2	LL	53.9	
$3\sigma^*_g \, 2\sigma^*_g$	-106.533	2	LL	66.0	
$2\sigma^*_g \, 1\pi^*_u$	-106.513	2	LL	66.5	
$2\sigma^*_u \, 2\sigma^*_g$	-106.404	2	LL	69.5	
$2\sigma^{**}_g$	-105.496	2	LL	94.2	
$N1s^* \, 1\pi^*$	-92.796	2	KL	439.8	
$1\pi^{**}_u \, 1\pi'^*_u \, 1\pi^1_g$	-106.931	2	LLLe	55.2	
$1\pi^{**}_u \, 1\pi'^*_u$	-105.788	3	LLL	86.3	

(1) Nakamura et al [20], Siegbahn et al [6].

The calculations were performed using a double-zeta basis set with addition of a polarization function and lead to the results reported in Table 5. The notation used for each state is of typical hole-particle form, an asterisc being added to an orbital (or shell) containing a hole, a number (1) to one into which an electron is promoted. In the same Table we show also the frequently used 'letter' symbolism in which 'K' indicates an inner-shell hole, 'L' a hole in the valence shell, and 'e' represents an excited electron. The more commonly observed ionization processes in the Auger spectra of N_2 are of the type K–LL (a 'normal' process, 'core-hole state' ↔ 'double-hole state');

Figure 1: Experimental gas phase N_2 Auger spectrum (Siegbahn et al [6]) and estimated transition energies for normal processes (vertical lines) and for the other most important processes (dashed vertical lines).

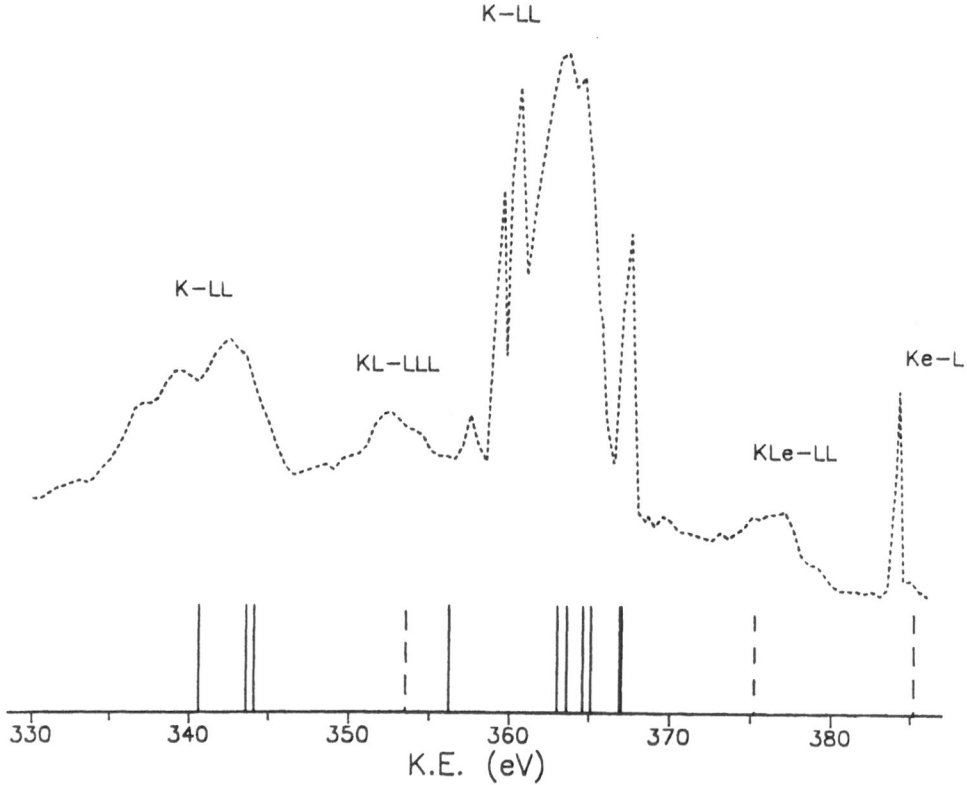

KL–LLL ('monopole ionizations'); and KLe–LL or Ke–L ('high-energy satellites'). From Table 5 it is possible to estimate the energies of transition for various pairs of states, even though the 'state' energies are stricly speaking configurational averages. Figure 1 shows (vertical lines) the estimated values of these transition energies, superimposed on the experimental spectrum. It is noteworthy that, even if the vertical lines are not actually coincident with the peak positions, the *assignment* of the peaks for various processes is substantially in accord with that made by Moddeman et al [21] on the basis of the experimental data.

References

1. L. Brillouin, *Actualités Sci. et Ind.*, No. 71, (1933).

2. L. Brillouin, *Actualités Sci. et Ind.*, No. 159, (1934).

3. T. A. Koopmans, *Physica* **1**, 104, (1933).

4. G. Berthier, in <u>Current Aspects of Quantum Chemistry</u>, R. Carbó (Ed.), Elsevier, Amsterdam, 1982, pp. 145-156.

5. G. Berthier, in Quantum Chemistry - Basic Aspects, Actual Trends, R. Carbó (Ed.), Elsevier, Amsterdam, 1989, pp. 91-102.

6. K. Siegbahn, C. Nordling, G. Johansson, J. Hedman, P. F. Heden, K. Hamrin, U. Gelius, T. Bergmark, L. O. Werme, R. Manne and Y. Baer, ESCA Applied to Free Molecules, North-Holland, Amsterdam, 1969.

7. H. Agren, A. Cesar and C. M. Liegener, *Adv. Quant. Chem.* **23**, 1, (1992).

8. R. McWeeny, *Mol. Phys.* **28**, 1273, (1974).

9. D. Firsht and R. McWeeny, *Mol. Phys.* **32**, 1637, (1976).

10. J. C. Slater, Quantum Theory of Atomic Structure, Vol. I, McGraw-Hill, New York, 1960.

11. R. McWeeny, Methods of Molecular Quantum Mechanics, 2nd ed., Academic, London, 1989.

12. T. H. Dunning, *J. Chem. Phys.* **53**, 2823, (1970).

13. D. A. Shirley, *Chem. Phys. Lett.* **16**, 220, (1972).

14. J. C. Slater, *Phys. Rev.* **36**, 57, (1930).

15. E. Clementi and D. L. Raimondi, *J. Chem. Phys.* **38**, 2686, (1963).

16. H. Tatewaki and S. Huzinaga, *J. Comput. Chem.* **1**, 205, (1980).

17. S. Huzinaga, *J. Chem. Phys.* **42**, 1293, (1965).

18. T. X. Carroll and T. D. Thomas, *J. Electron Spectrosc. Relat. Phenom.* **10**, 215, (1977).

19. D. B. Adams and D. T. Clark, *Theor. Chim. Acta* **31**, 171, (1973).

20. M. Nakamura, M. Sasanuma, S. Sato, M. Watanabe, H. Yamashita, Y. Iguchi, A. Ejiri, S. Nakai, S. Yamaguchi, T. Sagawa, Y. Nakai and T. Oshio, *Phys. Rev.* **178**, 80, (1969).

21. W. E. Moddeman, T. A. Carlson, M. O. Krause, B. P. Pullen, W. E. Bull and G. K. Schweitzer, *J. Chem. Phys.* **55**, 2317, (1971).

An Application of the Half-Projected Hartree-Fock Model to the Direct Determination of the Lowest Singlet and Triplet Excited States of Molecular Systems

Y. G. SMEYERS, P. FERNANDEZ-SERRA[1] and M. B. RUIZ
Instituto de Estructura de la Materia, C.S.I.C., c/Serrano, 123, E-28006-Madrid, Spain

1.Introduction

Among the many ways to go beyond the usual Restricted Hartree-Fock model in order to introduce some electronic correlation effects into the ground state of an electronic system, the Half-Projected Hartree-Fock scheme, (HPHF) proposed by Smeyers [1,2], has the merit of preserving a conceptual simplicity together with a relatively straigthforward determination. The wave-function is written as a DODS Slater determinant projected on the spin space with S quantum number even or odd. As a result, it takes the form of two DODS Slater determinants, in which all the spin functions are interchanged. The spinorbitals have complete flexibility, and should be determined from applying the variational principle to the projected determinant.

The difficulty of determining the Half-Projected Hartree-Fock function has somewhat hampered its utilization [3-10]. Some calculations, however, exist in literature. At present time, because of the increasing computing facilities, as well as the introduction of more powerful convergence techniques, the HPHF model is expected to play a more important role, especially in the field of medium size molecules, in which the use of more sophisticated procedure are not yet possible [9-10].

In addition, since the HPHF wavefunction exhibits a two-determinantal form, this model can be used to describe singlet excited states or triplet excited states in which the projection of the spin momentum $M_S=0$. The HPHF approximation appears thus as a simple method for the direct determination of excited states (with $M_S=0$) such as the usual Unrestricted Hartree Fock model does for determining triplet excited states with $M_S=\pm 1$.

In the present paper, we propose the use of the HPHF approximation for the direct calculation of excited states, in which $M_S=0$, just as Berthier [11], and Pople and Nesbet [12] did for the determination of states in which $M_S \neq 0$. We give some examples of such calculations, either when the excited state wavefunction is orthogonal or not by symmetry to that of the ground state.

---

[1]Permanent Address: Departamento de Ingeniería de Circuitos y Sistemas, E.U.I.T. de Telecomunicación, Universidad Politécnica de Madrid, E-28031-Madrid, Spain

Y. Ellinger and M. Defranceschi (eds.), Strategies and Applications in Quantum Chemistry, 175–188.
© 1996 *Kluwer Academic Publishers.*

2. Theory

2.1. GENERAL APPROACH

The HPHF wavefunction for an $2n$ electron system, in a ground state of S quantum number, even or odd, is written as a linear combination of only two DODS Slater determinants, built up with spinorbitals which minimize the total energy [1-2]:

$$\psi^{HPHF} = 1/2\{|a_1\bar{b}_1a_2\bar{b}_2.....a_n\bar{b}_n| + (-1)^{S+n}|\bar{a}_1b_1\bar{a}_2b_2.....\bar{a}_nb_n|\} \tag{1}$$

where a_i and $\bar{b}_i$ are two spinorbitals of opposite spin belonging to a same electron pair, so that $a_i \approx b_i$.

This linear combination is obtained by projection of one of the determinants on the spin eigenstates with S even or odd:

$$\hat{A}(S)D_{00} = 1/2[1 + (-1)^{S+n}\hat{P}_{\alpha\beta}]D_{00} = 1/2[D_{00} + (-1)^{S+n}D_{nn}] \tag{2}$$

where $\hat{P}_{\alpha\beta}$ is a permutation operator which interchanges all the α and β spin functions in the D_{00} initial determinant.

Since the HPHF wavefunction for singlet states does not contain any triplet contamination, this model was seen to produce relatively good results for singlet ground states, very close to those of the fully projected one [1-10].

The Brillouin's theorem has been shown to hold in the case of the HPHF function [2]. As a result, any variations of the orbitals which minimizes the HPHF total energy, can be expressed as:

$$\frac{\partial E}{\partial \varepsilon_{it}} =< \Psi^{HPHF}|\hat{H} - E|\Psi_{it}^{HPHF} >= 0. \tag{3}$$

where Ψ_{it}^{HPHF} is the HPHF function in which an a_i occupied orbital has been replaced for an a_t virtual one.

Introducing the HPHF wave-function expression (1) in (3), and taking into account the idempotency of operator $\hat{A}(s)$, the following equation may be obtained:

$$\frac{\partial E}{\partial \varepsilon_{it}} =< D_{00}|\hat{H}|(D_{00}^{it} + (-1)^{S+n}D_{nn}^{it}) > - < D_{00}|(D_{00}^{it} + (-1)^{S+n}D_{nn}^{it}) > E \tag{4}$$

where D^{it} is a Slater determinant in which one i occupied orbital has been substituted by a t virtual one.

In order to solve equation (4), the following matrix elements between Slater Determinants have to be considered:

$$< D_{00}|\hat{H}|D_{00}^{it} >, \qquad < D_{00}|D_{00}^{it} > \tag{5}$$

$$< D_{00}|\hat{H}|D_{nn}^{it} >, \qquad < D_{00}|D_{nn}^{it} > \tag{6}$$

Since D_{00} and D_{00}^{it} are constructed with the same set of orthonormal spinorbitals, the two first matrix elements can easily rewritten, according to the Slater's rules [13], as:

$$< D_{00}|\hat{H}|D_{00}^{it} >=< a_i|\hat{F}^a|a_t >$$
$$< D_{00}|D_{00}^{it} >= 0.$$

$$(7)$$

where D_{00}^{it} is a Slater determinant in which an a_i orbital is replaced by an a_t one. In this expression, the $\hat{F}_a$ operator is the usual Fock operator of the Unrestricted Hartree-Fock method [14]:

$$< a_m|\hat{F}^a|a_l >=< a_m|\hat{h}|a_l > +\sum_{i}^{n}[< a_m a_l|a_i a_i > + < a_m a_l|b_i b_t > - $$
$$< a_m a_i|a_i a_l >]$$

$$(8)$$

In this equation, the Mulliken notation for the repulsion integrals is used, that is:

$$< a_m a_l|a_i a_j >=< a_{(1)}^m a_{(2)}^i|\frac{1}{r_{12}}|a_{(1)}^l a_{(2)}^j >$$

$$(9)$$

and $\hat{h}$ stands for the well known monoelectronic operator:

$$\hat{h}_{(1)} = -\frac{1}{2}\Delta_1 +\sum_{A}\frac{Z_A}{r_{1A}}$$

$$(10)$$

A similar operator as (8) can be written when a b_i orbital is substituted by a b_t virtual one:

$$< b_m|\hat{F}^b|b_l >=< b_m|\hat{h}|b_l > +\sum_{n}^{i}[< b_m b_l|b_i b_i > + < b_m b_l|a_i a_i > - $$
$$< b_m b_i|b_i b_l >]$$

$$(11)$$

The calculation of the cross matrix elements (6) is somewhat more difficult, because the Slater Determinants involved in them are constructed with two sets of non-orthonormal spinorbitals. This calculation, however, may be greatly simplified, if the two sets are assumed to be *corresponding*, that is, if they fulfill the following condition [14]:

$$< a_i|b_j >= s_i\delta_{ij}$$

$$(12)$$

As well known, this condition is not a restriction whenever the wavefunction is invariant under an unitary transformation [2].

Taking into account (12), the matrix elements (6), with their phase factor in (4), may be written in the following way:

$$(-1)^{S+n} < D_{00}|\hat{H}|D_{nn}^{it} >=$$

$$= \frac{S}{s_i} < b_i|\hat{F}^{ba}|a_t > -\frac{1}{s_i} < b_i|a_t > E_2 - \frac{S}{s_i}\sum_j^n \frac{1}{s_j} < b_j|a_t >< b_i|\hat{F}^{ba}|a_j >(13)$$

and

$$(-1)^{S+n} < D_{00}|D_{nn}^{it} >= \frac{S}{s_i} < b_i|a_t > \tag{14}$$

In these equations, **S** is the overlap between the two determinants built up with *corresponding* orbitals (12)(multiplied by a phase factor according the multiplicity required):

$$\mathbf{S} = (-1)^{S+n} < D_{00}|D_{nn} >= (-1)^S \prod_i^n s_i^2 \tag{15}$$

and

$$E_2 = (-1)^{S+n} < D_{00}|\hat{H}|D_{nn} >= \mathbf{S}\{\sum_i^n \frac{2}{s_i}[< a_i|\hat{h}|b_i > +$$

$$+ \sum_i^n \sum_j^n \frac{1}{s_i s_j}[2 < a_i b_i|a_j b_j > - < a_i b_j|a_j a_i >]\} \tag{16}$$

$\hat{F}^{ba}$ is defined as a cross operator analogous to the Fock operator (8):

$$< b_m|\hat{F}^{ba}|a_l >=< b_m|\hat{h}|a_l > + \sum_i^n \frac{1}{s_i}[< b_m a_l|b_i a_i > + < b_m a_l|a_i b_i >$$

$$- < b_m a_i|b_i a_l >] \tag{17}$$

A similar operator can be written for the case in which a b_i orbital is replaced by a virtual one, b_t:

$$< a_l|\hat{F}^{ab}|b_m >=< a_l|\hat{h}|b_m > + \sum_i^n \frac{1}{s_i}[< a_l b_m|a_i b_i > + < a_l b_m|b_i a_i >$$

$$- < a_l b_i|a_i b_m >] \tag{18}$$

Introducing now expressions (13) and (14) in (4) , the following equation is obtained:

$$\frac{\partial E}{\partial \varepsilon_{it}} =< a_i|\hat{F}^a|a_t > +\frac{1}{s_i} < b_i|a_t > (E_2 - SE) + \frac{S}{s_i} < b_i|\hat{F}^{ba}|a_t > -$$

$$-\frac{S}{s_i}\sum_j^n \frac{1}{s_j} < b_j|a_t >< b_i|\hat{F}^{ba}|a_j > \tag{19}$$

which is the expression of the Generalized Brillouin Theorem for the HPHF function written as a function of the orbitals. A similar equation can be deduced when a b_i orbital is replaced by a b_t one. The next step will now be to write equations (19) as pseudo-eigenvalue equations to be solved in an iterative way just as in the Unrestricted Hartree-Fock method.

For this purpose, let us define the following density projection operators:

$$\hat{R}^{ab} = \sum_i^n \frac{1}{s_i} |a_i ><b_i| \qquad \hat{R}^{ba} = \sum_i^n \frac{1}{s_i} |b_t ><a_i|$$

(20)

and let us introduce them in (19). After some straightforward operations, we obtain:

$$\frac{\partial E}{\partial \varepsilon_{it}} =< a_i|\hat{F}^a|a_t > + < a_i|\hat{R}^{ab}|a_t > (E_2 - \mathbf{S}E) +$$

$$+\mathbf{S} < a_i|\hat{R}^{ab}\hat{F}^{ba}[1 - \hat{R}^{ab}]|a_t >= 0$$

(21)

From this equation the following HPHF Fock operator for determining the a_i orbitals can be extracted:

$$\hat{H}^a = \hat{F}^a + \hat{R}^{ab}(E_2 - \mathbf{S}E) + \mathbf{S}\hat{R}^{ab}\hat{F}^{ba}[1 - \hat{R}^{ab}]$$

(22)

A similar $\hat{H}^b$ operator for determining the b_i orbitals can be obtained in the same way.

Let us remark that operator (21) is not symmetric. But, it can be symmetrized easily just by adding the adjoint of the asymmetric part:

$$\hat{R}^{ba}(E_2 - \mathbf{S}E) + \mathbf{S}[1 - \hat{R}^{ba}]\hat{F}^{ab}\hat{R}^{ba}$$

(23)

In addition, since the action of $\hat{R}^{ba}$ operator on a virtual orbital is zero, it is seen that this adjoint will not affect the results. So that the complete $\hat{H}^a$ operator may be written as:

$$\hat{H}^a = \hat{F}^a + \hat{R}^{ab}(E_2 - \mathbf{S}E) + \mathbf{S}\hat{R}^{ab}\hat{F}^{ba}[1 - \hat{R}^{ab}] + \hat{R}^{ba}(E_2 - \mathbf{S}E) + \mathbf{S}[1 - \hat{R}^{ba}]\hat{F}^{ab}\hat{R}^{ba}$$

(24)

2.2. APPLICATION TO EXCITED STATES

The HPHF wavefunction for an excited state ($b_k \rightarrow b_u$) is constructed by substituting in the HPHF ground state wavefunction (1) an b_k occupied spinorbital by an b_u virtual one. In order to avoid the possible collapsing of to so-constructed excited wavefunction onto the ground state one during the variational process, it is convenient that the excited function should be orthogonal to the former. In some cases, this orthogonality requirement is automatically achieved, when both wavefunctions exhibit different multiplicities or different spatial symmetries. In the second case, the promoved and excited spinorbitals, b_k (or a_k) and b_u possess also different symmetries. When both wavefunctions exhibit the same multiplicity and the same spatial symmetry, it is convenient that the excited function should be orthogonal to the fundamental one [15]. One way to achieve partially this requirement is orthogonalized the excited orbital b_u to its companion a_k, at each step of the iterative procedure. Remember that a_k and b_k possess the same symmetry.

In any cases, the orthogonality requirement applied to the orbitals:

$$< a_k|b_u >= 0 \tag{25}$$

implies some modifications in the formulae of the previous paragraph in order to avoid some singularities [7]. In particular, new cross Fock operators have to be redefined:

$$< b_m|\hat{F}^{ba}|a_l >=< b_m|\hat{h}|a_l > + \sum_{i\neq k}^{n}\frac{1}{s_i}[< b_m a_l|b_i a_i > + < b_m a_l|a_i b_i >$$
$$- < b_m a_i|b_i a_l >] \tag{26}$$

and

$$< a_l|\hat{F}^{ab}|b_m >=< a_l|\hat{h}|b_m > + \sum_{i\neq k}^{n}\frac{1}{s_i}[< a_l b_m|a_i b_i > + < a_l b_m|b_i a_i >$$
$$- < a_l b_i|a_i b_m >] \tag{27}$$

in which the sumations are restricted to the nonorthogonal orbitals.

In addition, partial cross Fock operators are also to be defined for evaluating the matrix elements in which the orthogonal orbitals are involved:

$$< b_m|\hat{f}^{ba}|a_l >= [2 < b_m a_l|b_u a_k > - < b_m a_k|b_u a_l >] \tag{28}$$

and

$$< a_l|\hat{f}^{ab}|b_m >= [2 < a_l b_m|a_k b_u > - < a_l b_u|a_k b_m >] \tag{29}$$

For the same reason, new density projection operators are redefined:

$$\hat{R}^{ab} = \sum_{i\neq k}^{n}\frac{1}{s_i}|a_i >< b_i| \qquad \hat{R}^{ba} = \sum_{i\neq k}^{n}\frac{1}{s_i}|b_i >< a_i| \tag{30}$$

as well as limited projection operators to the k or u orbital space:

$$\hat{r}^{ab} = |a_k >< b_u| \qquad \hat{r}^{ba} = |b_u >< a_k| \tag{31}$$

Finally, a restricted overlap between the two determinants limited to the nonorthogonal orbitals is defined:

$$S' = (-1)^S \prod_{i \neq k}^n s_i^2 \tag{32}$$

In order to deduce the new pseudo-eigenvalue equations (22), we have to distinguish the two possibilities:

When $a_i \neq a_k$, the Brillouin Theorem equation (20) is reduced to:

$$< a_i|\hat{F}^a|a_t > + \frac{< b_i|a_t >}{s_i} E_2 - \frac{S'}{s_i} < b_u|a_t >< b_i|\hat{f}^{ba}|a_k > \tag{33}$$

where the cross energy term, E_2', between the two Slater determinants, takes the form of a simple repulsion integral:

$$E_2' = S' < a_k b_u|a_k b_u > \tag{34}$$

In contrast, when $a_i = a_k$, the following expression is found:

$$< a_k|\hat{F}^a|a_t > + S'\{< b_u|\hat{f}^{ba}|a_t > + < b_u|a_t >< a_k|\hat{F}^{ab}|b_u > -$$
$$- \sum_{j \neq k}^n \frac{< b_j|a_t >}{s_j} < b_u|\hat{f}^{ba}|a_j >\} = 0. \tag{35}$$

From equations (33) and (35), a general HPHF Fock operator for determining the a_i orbitals of excited states can be extracted after some straightforward transformations:

$$\hat{H}^a = \hat{F}^a + \hat{R}^{ab} E_2' + S'\{\hat{r}^{ab}\hat{f}^{ba} + \hat{r}^{ab}\hat{F}^{ba}\hat{r}^{ab} - \hat{r}^{ab}\hat{f}^{ba}\hat{R}^{ab} - \hat{R}^{ab}\hat{f}^{ba}\hat{r}^{ab}\} \tag{36}$$

Since equation (36) is not symmetric, it is symmetrized by addition of the ad joint of the asymmetric part. We obtain the new expression:

$$\hat{H}^a = \hat{F}^a + (\hat{R}^{ab} + \hat{R}^{ba})E_2' + S'\{\hat{r}^{ab}\hat{f}^{ba} + \hat{r}^{ab}\hat{F}^{ba}\hat{r}^{ab} - \hat{r}^{ab}\hat{f}^{ba}\hat{R}^{ab} -$$
$$- \hat{R}^{ab}\hat{f}^{ba}\hat{r}^{ab} + \hat{f}^{ab}\hat{r}^{ba} + \hat{r}^{ba}\hat{F}^{ab}\hat{r}^{ba} - \hat{R}^{ba}\hat{f}^{ab}\hat{r}^{ba} - \hat{r}^{ba}\hat{f}^{ab}\hat{R}^{ba}\} \tag{37}$$

A similar equation can be deduced for the b_i orbitals.

3.Calculation

In order to determine the HPHF wave-functions, the HPHF Fock operators (24) and (37) for the ground and excited states, respectively, have to be expressed in matrix form, in which the orbitals are developed on a basis function set. So, we have for the ground state:

$$\mathbf{H}^{\mathbf{a}} = \mathbf{F}^{\mathbf{a}} + \mathbf{s}(\mathbf{Q}^{\mathbf{ab}} + \mathbf{Q}^{\mathbf{ba}})\,\mathbf{s}\,(E_2 - SE) +$$
$$+ S\{\mathbf{s}\mathbf{Q}^{\mathbf{ab}}\mathbf{F}^{\mathbf{ba}}[\mathbf{I} - \mathbf{Q}^{\mathbf{ab}}\mathbf{s}] + [\mathbf{I} - \mathbf{s}\mathbf{Q}^{\mathbf{ba}}]\mathbf{F}^{\mathbf{ab}}\mathbf{Q}^{\mathbf{ba}}\mathbf{s}\} \tag{38}$$

where $\mathbf{H}^a, \mathbf{F}^a, \mathbf{F}^{ba}, \mathbf{Q}^{ab}, ..$ are the matrix representations of the corresponding operators. In particular, $\mathbf{Q}^{ab}$ is defined as $\mathbf{C}^a \Lambda \mathbf{C}^{b+}$, where Λ is now the inverse diagonal matrix of the overlap between the occupied *corresponding* orbitals, a_i and b_i, and $\mathbf{C}^a$ and $\mathbf{C}^b$ the coefficient matrices of two set of occupied orbitals. Finally, s is the overlap matrix between the basis functions.

For the excited states, with an orthogonal orbital pair, we have:

$$\mathbf{H}^a = \mathbf{F}^a + s(\mathbf{Q}^{ab} + \mathbf{Q}^{ba})s\,E_2' + S'\{sq^{ab}f^{ba} + sq^{ab}\mathbf{F}^{ba}q^{ab}s - sq^{ab}f^{ba}\mathbf{Q}^{ab}s -$$
$$-s\mathbf{Q}^{ab}f^{ba}q^{ab}s + f^{ab}q^{ba}s + sq^{ba}\mathbf{F}^{ab}q^{ba}s - s\mathbf{Q}^{ba}f^{ab}q^{ba}s - sq^{ba}f^{ab}\mathbf{Q}^{ba}s\} \quad (39)$$

where $\mathbf{Q}^{ab}$ is defined as before except for the diagonal element, s_k^{-1} of the Λ, corresponding to the orthogonal orbitals which is missing, and where $q^{ab} = c_k^a c_u^{b+}$. c_k^a and c_u^b are the column vectors of the coefficients of the a_k and b_u orbitals, respectively.

3.1. THE METHYLENE MOLECULE

As well known, the methylene lowest state is a $^3B_1$ triplet, with electronic configuration $(1a_1)^2 (2a_1)^2 (1b_2)^2 (3a_1) (1b_1)$, which lies somewhat below the fundamental singlet state, $^1A_1$. In addition, the companion singlet state, $^1B_1$, is also known. To

Table 1: Optimized energy values for the $^3B_1$ and $^1B_1$ states of the methylene as function of the HCH bending angle.

$^3B_1$ State			$^1B_1$ State		
θ	r	Energy	θ	r	Energy
180	1.056	-38.905859	180	1.056	-38.850284
170	1.057	-38.907761	170	1.057	-38.851125
160	1.061	-38.912535	160	1.060	-38.853195
150	1.065	-38.918169	150	1.063	-38.855367
140	1.069	-38.922733	140	1.068	-38.856294
130	1.073	-38.924777	140.54	1.068	-38.856296
128.88	1.073	-38.924800	130	1.073	-38.854753
120	1.076	-38.923256	120	1.078	-38.849739
110	1.080	-38.917369	110	1.084	-38.840415
100	1.085	-38.906442	100	1.091	-38.826023
90	1.090	-38.889860	90	1.000	-38.805808
80	1.096	-38.867050	80	1.109	-38.778970
70	1.104	-38.837472	70	1.120	-38.744633

θ in degrees, r in Å and energy in Hartrees.

study the potential energy functions of both B_1 states, as a function of the bending

Table 2: Theoretical geometrical values fully optimized of methylene versus experimental data.

		Calculations[a]		Experimental [16,17]		$\langle \hat{S}^2 \rangle$
HPHF	$^1B_1$	$\alpha = 140.54$	$r = 1.068$	$\alpha = 139.3$	$r = 1.070$	0.0126
HPHF	$^3B_1$	$\alpha = 128.88$	$r = 1.073$	$\alpha = 134.04$	$r = 1.0766$	2.0000
barrier	$^1B_1$	1319.5		1617		
barrier	$^3B_1$	4157.1		1931 ± 30		

[a] Huzinaga-Dunning basis set + pol. α in degrees, r in Å and the barrier height in cm^{-1}.

angle, was for many years an attractive problem for many scientists, in order to analyze the spectrum structure of the $^1A_1 \rightarrow^1 B_1$ transition which is allowed [16,17]. Next, and as an example of HPHF calculations, the potential energy curves for the bending in the $^3B_1$ and $^1B_1$ states of methylene were determined into the HPHF approximation, using the Huzinaga-Dunning valence basis set with polarization orbitals (d on the C atom, and p on the H atoms with 0.8 and 1.1 as exponent respectively), and full geometry optimization. For this purpose, an $3a_1$ orbital was substituted by an $1b_1$ one, into the fundamental configuration. The energy variations are given in Table 1 for both states. The geometrical parameters encountered in this way are in very good agreement with the experimental data. These are gathered in Table 2.

3.2. THE METHANAL MOLECULE

Methanal (formol) presents an additional degree of freedom: the out-of-plane wagging of the oxygen atom. In its singlet ground, X^1A_1, this molecule is planar. But, in its triplet and singlet lowest exited a^3A'' and $A^1A''(n, \pi^*)$, this molecule is pyramidal with the possibility of inversion, because its π electronic system is destabilized by a π^* antibonding orbital. In order to test the performance of the HPHF model, to determine the potential energy curves for the inversion seems to be illustrative.
HPHF calculations with full geometry optimization were performed for these lowest excited states using the Huzinaga-Dunning valence basis set with polarization orbitals. The potential energy curves are given in Figure 1. As expected, a double potential energy well is obtained for both excited states of methanal. In Table 3, the geometrical parameters, as well as the inversion barrier, obtained with different basis sets and different approaches are given for the first singlet state, S_1 (n, π^*). It is seen that the values for the barrier height are very basis dependent. Anyway, the HPHF approach gives the best values into the same basis set, except for the SCF-CI calculation which involves 17,000 configurations [18].
In Table 4, similar results are gathered for the first triplet excited state, $T_1(n, \pi^*)$. Here, the UHF model ($M_S= 1$) is seen to furnish slightly better results. The HPHF model, however, yields a better value for $< \hat{S}^2 >$.

Figure 1: Potential energy curves for inversion in the first triplet and singlet excited states of methanal.

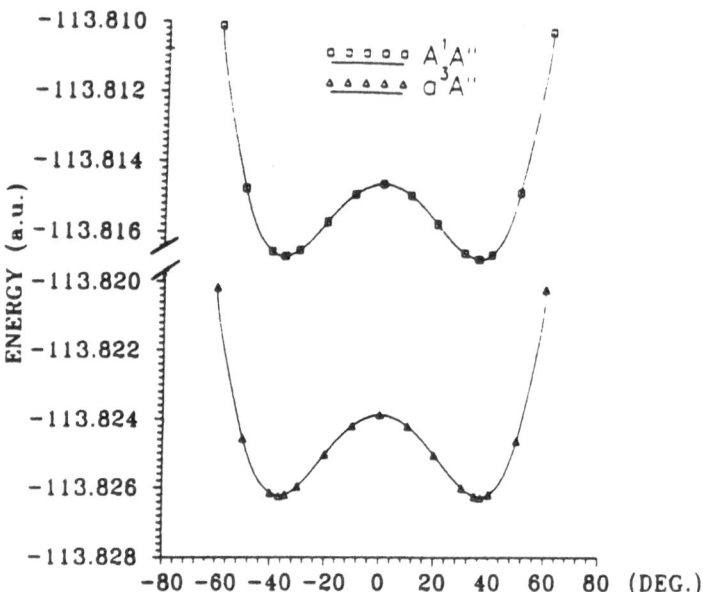

Table 3: Geometry obtained for the methanal $S_1(n, \pi*)$ excited state. Distances in Å, angles in degrees, energies in Hartrees, and barrier height in cm^{-1}.

	r(CO)	α(wag.)	α(HCH)	r(CH)	Energy	V(cm^{-1})
RHF[a]	1.360	39.97	118.6	1.074	-113.807423	788.1
UHF[b]	1.345	36.59	119.14	1.079	-113.794325	526.0
UHF[c]	1.349	36.12	119.34	1.008	-113.821417	487.0
HPHF[a]	1.391	28.93	121.980	1.072	-113.628891	183.6
HPHF[b]	1.353	36.00	120.287	1.079	-113.789977	496.0
HPHF[c]	1.356	35.60	119.552	1.081	-113.816664	463.6
SCF-CI[c]					-114.052539	393.37
Exp.	1.3252	33.6	118.0	1.0947	–	356.2

bases [a] 4-31G [b] 6-31G** [c] Huzinaga-Dunning + pol.

Table 4: Methanal $T_1(n, \pi^*)$ excited state. Comparison of HPHF and UHF calculations and experimental data.

	r(CO)	r(CH)	α(HCH)	α(wag.)	Energy	Barrier	$\langle \hat{S}^2 \rangle$
UHF[a]	1.340	1.080	118.923	38.45	-113.825744	648.6	2.0150
HPHF[a]	1.339	1.081	119.022	36.8	-113.826209	525.02	2.0001
Exp.[b]	1.3069	1.0962	118.0	37.9		775.6	

Distances in Å, angles in degrees, Energy in Hartrees and cm^{-1}.
[a]Calculations obtained with Huzinaga-Dunning + pol. basis. [b][19].

3.3. TRANS-BIACETYL MOLECULE

In order to test the HPHF model with a larger system, we have considered the trans-biacetyl molecule, which contains six atoms of the second row and six hydrogen atoms. The energy values of the two extremal configurations for the double rotation of the methyl groups were determined on the potential energy surface, with full geometry optimization except for the molecular frame which is constrained to be planar. In the same calculation, the barrier height was determined.

Table 5: Energy values obtained for the ground and first singlet and triplet excited states of trans-biacetyl in two extremal conformations of the methyl groups, as well as torsional barriers.

States	Model	(Basis)	Energy (a.u.) $(0°, 0°)$	Energy (a.u.) $(60°, 60°)$	barrier to rotation (cm^{-1})
$S_0(^1A_g)$	RHF	(4-31G)	-304.232787	-304.230217	564.05
	RHF	(6-31G**)	-304.699302	-304.696224	675.44
	HPHF	(4-31G)	-304.278350	-304.276056	503.5
$S_1(^1A_u)$	HPHF	(4-31G)	-304.104061	-304.107051	656.3
$T_1(^3A_u)$	UHF	(4-31G)	-304.126692	-304.128994	505.08
	UHF	(6-31G**)	-304.584153	-304.586074	421.55
	HPHF	(4-31G)	-304.142880	-304.143456	126.3

The RHF and UHF values are taken from [20].

In Table 5, we give the HPHF energy values obtained for the singlet ground state, and the first triplet and singlet excited states, as well as the RHF and UHF values obtained elsewhere [20] with different basis sets. It is seen that whereas 0°,0°, conformation is the preferred in the singlet ground state, the 60°, 60° one is the most stable in the excited states in accordance with the large band progressions observed in the electronic spectrum, as well as the band assignments [20]. This change of conformational preference can be also interpreted on the basis of the destabilization of the π electronic system.

Here, once more the barrier height values are seen to be very basis dependent. The value encountered for the singlet excited state, however, is found to be in relatively good agreement with the experimental value: 550 cm^{-1} [20].

3.4. FORMIC ACID

Finally, the HPHF approach is applied to the formic acid molecule, in its first singlet excited state, which is not orthogonal any more to the singlet ground state in a random conformation.

The calculations were performed into two basis sets, with full geometry optimization except for the torsional angles α and θ. Two non planar conformations were considered, which correspond to minima on the potential energy surface into the GVB approximation [21]. In these conformations, the molecule adopts a pyramidal conformation, as in methanal. In addition, the hydroxilic group is rotated up or down the OCO plane.

In Table 6, the formation energy values for these two preferred conformations are given, together with the corresponding values for the planar conformations *syn* and *anti*. It is seen that one of the minima is only slightly more stable than the other when calucated with the larger basis, but much more stable than the planar conformations in accordance with the GVB calculations [21].

Table 6: Energy values (in a.u.) of $^1A(n, \pi^*)$ excited state of formic acid for the two minima and two planar conformations.

θ(tor.)	α(wag.)	E(HPHF/4-31G)	E(HPHF/6-31G**)
-47.91	41.32	-188.371380	-188.643616
63.66	45.76	-188.367798	-188.643132
0.0	0.0	-188.361620	-188.631021
180.0	0.0	-188.361081	-188.632521

The geometrical parameters found for these four conformations are gathered in Table 7. It is seen that the carbon atom of the preferred conformations exhibits an sp^3 hybridization because of the destabilization of the π electronic system by the π^* antibonding orbital, whereas the carbon atom of the planar conformations shows mostly an sp^2 one.

Table 7: Geometrical parameters for four conformations of the $^1A(n, \pi*)$ singlet excited state of formic acid.

r(C=0)	r(CO)	r(OH)	r(CH)	α(OCO)	α(COH)	α(OCH)	θ(tor.)	α(wag.)
1.417	1.385	0.957	1.073	109.29	115.20	120.93	-47.91	41.36
1.397	1.385	0.957	1.077	109.32	115.20	120.88	63.66	45.76
1.408	1.373	0.961	1.064	123.97	114.71	116.91	0.0	0.0
1.392	1.369	0.948	1.067	111.86	114.90	120.98	180.0	0.0

Distances in Å and angles in degrees.

4.Discussions and conclusions

In the present paper, the Half-Projected model is applied to the direct determination of the lowest singlet and triplet excited states in which $M_S = 0$, just as the usual UHF method is employed for states in which $M_S = 1$ [11,12]. As examples, the method is successfully applied to the calculation of some molecular properties of methylene, methanal, dimethylglyoxal and formic acid, in these excited states.

In the case of methylene, methanal or dimethylglyoxal, in which the excited wave-function is always orthogonal by symmetry to that of the singlet ground state, the procedure converges well without any complication. In this case, the HPHF method could be regarded as an extension of the usual UHF procedure for the states in which $M_S = 0$. This extension appears to be especially interesting for the direct determination of the lowest singlet excited states of medium size molecules, for which no simple and efficient method exists.

In the case of formic acid, in a random conformation, the excited wave-function possesses the same symmetry as that of the ground state. In the present paper, we propose to orthogonalize the excited orbital to its companion in order to avoid the variational collapsing of the excited state into the fundamental one. The procedure converges more slowly, but procedures for accelerating convergence may be used. Notice that the excited wave-function is not necessarily orthogonal to the fundamental one in this way of proceeding. But, both functions have not to be orthogonal because the Hamiltonian operators (24) and (37) are not the same. They may expected, however, to be nearly orthogonal. It may be added here that a complete orthogonalization will yield probably worst results [15].

It may be concluded thus that the Half-Projected Hartree-Fock model proposed more than two decades ago for introducing some correlation effects in the ground state wave-function [1,2], could be employed advantageously for the direct determination of the lowest triplet and singlet excited states, in which $M_S = 0$. This procedure could be especially suitable for the singlet excited states of medium size molecules for which no other efficient procedure exists.

References

1. Y.G. Smeyers, An. Fis. (Madrid), **67**, 12 (1971).

2. Y.G. Smeyers and L. Doreste, *Int. J. Quant. Chem.* , **7**, 687 (1973).

3. Y.G. Smeyers and G. Delgado-Barrio, *Int. J. Quant. Chem.* , **8**, 733 (1974).

4. P. A. Cox and M.H. Wood, *Theoret. Chim. Acta* , **41**, 269 (1976).

5. Y.G. Smeyers and A.M. Bruceña, *Int. J. Quant. Chem.* , **14**, 641 (1978).

6. B.H. Lengsfield III, D.H. Phillips and J.C. Schug, *J. Chem. Phys.* , **74**, 5174 (1981).

7. S. Olivella and J. Salvador, *Int. J. Quant. Chem.* , **37**, 713 (1990).

8. M.B. Ruiz, P. Fernández-Serra and Y.G. Smeyers, *Fol. Chim. Theor. Lat.*, **19**, 85 (1991).

9. S. Olivella and J. Salvador, *J. Comp. Chem.*, **12**, 792 (1991).

10. R.G.A. Bone and P. Pulay, *Int. J. Quant. Chem.* , **45**, 133 (1992).

11. G. Berthier, *J. Chim. Phys.* , **51**, 363 (1954).

12. J.A. Pople and R.K. Nesbet, *J. Chem. Phys.* , **22**, 571 (1954).

13. J.C. Slater, in Quantum Theory of Atomic Structure, McGraw-Hill, New York, 1960.

14. A.T. Amos and G.G. Hall, *Proc. Roy. Soc.* , **A263**, 483 (1961).

15. R. Colle, A. Fortunelli and O. Salvetti, *Theoret. Chim. Acta* , **71**, 467 (1987).

16. G. Duxbury and Ch. Jungen, *Mol. Phys.* ,**6**, 981 (1988).

17. P.R. Bunker and T.J. Sears, *J. Chem. Phys.* , **83**, 4866 (1985).

18. M. Baba, U. Nagashima and I. Hanazaki, *Chem. Phys.* , **93**, 425 (1985).

19. V.T. Jones and J.B. Coon, *J. Mol. Spectrosc.* , **31**, 137 (1969).

20. M.L. Senent, D.Moule, Y.G. Smeyers, A. Toro-Labbé and F.J. Peñalver, *J. Mol. Spectrosc.*, in press.

21. Fr. Ioannoni, Fluorescence Spectra and Ab Initio Study of HCOOH, Master Degree Thesis, Brock University, St-Catharines, 1989.

FSGO Hartree-Fock Instabilities of Hydrogen in External Electric Fields

J.M. ANDRE, G. HARDY, D. H. MOSLEY and L. PIELA
Facultés Universitaires Notre-Dame de la Paix, Laboratoire de Chimie Théorique Appliquée, 61 rue de Bruxelles, B-5000 Namur, Belgium and University of Warsaw, Quantum Chemistry Laboratory, Pasteura 1, 02-093 Warsaw, Poland

1. Introduction

In the early sixties, it was shown by Roothaan [1] and Löwdin [2] that the symmetry adapted solution of the Hartree-Fock equations (i.e. belonging to an irreducible representation of the symmetry group of the Hamiltonian) corresponds to a specific extreme value of the total energy. A basic fact is to know whether this value is associated with the global minimum or a local minimum, maximum or even a saddle point of the energy. Thus, in principle, there may be some symmetry breaking solutions whose energy is lower than that of a symmetry adapted solution.

The Hartree-Fock description of the hydrogen molecule requires two spinorbitals, which are used to build the single-determinant two-electron wave function. In the *Restricted Hartree-Fock* method (RHF) these two spinorbitals are created from the same spatial function (orbital) but differ only by its multiplication by the α or β spin basis functions.

It is common knowledge that, in the case of the hydrogen molecule studied in a minimal basis set, the correlation error can be explained by the existence of ionic species in the hydrogen dissociation products:

$$H_2 \rightarrow \{ H \cdot + \frac{1}{2} (H^- + H^+) \}$$

This is an artefact due to the non-zero probability of the restricted wave-function of finding two electrons of opposite spins at the same spatial position.

FSGO's (Floating Spherical Gaussian Orbital) were introduced by Frost [3] in the mid 1960s. With FSGO's one abandons the idea of atomic orbitals centred on nuclear positions to arrive at an even more compact basis set than a minimal one. FSGO's correspond to s-type Gaussians that are not fixed at the atomic centers but are able to "float" in space so as to optimally represent each localized pair of electrons. Because only one function is needed for each electron pair, the basis set used is often referred to as being "subminimal".

Y. Ellinger and M. Defranceschi (eds.), Strategies and Applications in Quantum Chemistry, 189–202.
© *1996 Kluwer Academic Publishers.*

With respect to correlation, the behaviour of the hydrogen molecule studied in a subminimal FSGO basis set is still more striking than the one observed in a minimal basis set. By symmetry arguments, the single FSGO which describes the electron pair of the hydrogen molecule is centred at the middle of the H-H bond. As the internuclear distance increases and ultimately when the molecule dissociates, such a description would lead to a physical nonsense. Indeed, at the dissociation limit, this would correspond to two protons ($2H^+$) and an isolated pair of electrons ($2e^-$).

Thus, we understand that, in the FSGO model, for some critical distance, the single Gaussian will jump from its symmetric position at the middle of the H-H bond to a dissymetric one represented below. Thus, the FSGO dissociation scheme corresponds to one electron pair on one of the proton (H^-) and no electron on the second proton (H^+):

$$H_2 \rightarrow H^- + H^+$$

This behaviour is a nice example of the symmetry dilemma in the conventional Hartree-Fock scheme and is intimately connected with the question of Hartree-Fock instability.

In this paper, we analyze the instabilities which appear in the Hartree-Fock method. Our analysis is made in the framework of basis sets. In the hydrogen molecule, the single floating Gaussian orbital (FSGO) desribing the electron pair has its optimal position in the middle of the hydrogen molecule only for small internuclear distances. For large enough distances its optimum position is close to one of the nuclei and a broken-symmetry solution is thus preferred. Application of an external electric field along the molecular axis induces some additional instabilities in the lowest-energy solution with respect to the electric field value. Here, both types of instabilities are investigated analytically as well as numerically.

This paper is, thus, a double tribute to Professor Berthier. On one side, G. Berthier has provided excellent analysis of quantum mechanical instabilities [4], while additionally being at the origin of the interest of the Namur group for studies of (hyper)polarizabilities in organic molecules and chains.

2. Subminimal basis set Hartree-Fock-type calculations of the hydrogen molecule

In a FSGO basis set, the Gaussian orbital is defined by:

$$\phi_i(r) = (\frac{2\alpha_i}{\pi})^{3/4} \exp(-\alpha_i(r-R_i)^2)$$

where both, the Gaussian exponent and the Gaussian center are optimized in a FSGO calculation. Depending on the complexity of the calculation, we could have to compute the following integrals:

one-electron integrals:

$$<\phi_i|h^N|\phi_i> = \int\phi_i(r)\{-\frac{1}{2}\nabla^2(r) - \frac{1}{|r_a|} - \frac{1}{|r_b|}\}\phi_i(r)\ dr$$

$$<\phi_i|h^N|\phi_j> = \int\phi_i(r)\{-\frac{1}{2}\nabla^2(r) - \frac{1}{|r_a|} - \frac{1}{|r_b|}\}\phi_j(r)\ dr$$

$$<\phi_i|\phi_j> = \int\phi_i(r)\ \phi_j(r)\ dr$$

two-electron integrals:

$$(\phi_i\phi_j|\phi_k\phi_l) = \iint \phi_i(r_1)\,\phi_j(r_1)\,\frac{1}{|r_{12}|}\,\phi_k(r_2)\,\phi_l(r_2)\,dr_1\,dr_2$$

In the RHF method, the FSGO describing the electron pairs is doubly occupied and the wave-function has the form:

$$\psi(r_1,\omega_1,r_2,\omega_2) = \frac{1}{\sqrt{2}}\begin{vmatrix} \phi(r_1)\alpha(\omega_1) & \phi(r_1)\beta(\omega_1) \\ \phi(r_2)\alpha(\omega_2) & \phi(r_2)\beta(\omega_2) \end{vmatrix}$$

$$= \phi(r_1)\,\phi(r_2)\,\frac{1}{\sqrt{2}}\{\alpha(\omega_1)\,\beta(\omega_2) - \beta(\omega_1)\,\alpha(\omega_2)\}$$

Its associated RHF energy is, at a given internuclear distance (R):

$$E(RHF\text{-}FSGO) = 2 <\phi|h^N|\phi> + (\phi\phi|\phi\phi) + \frac{1}{|R|}$$

Table 1. shows the total energies obtained using the RHF method for: 1. LCAO minimal basis set STO-1G for the sake of comparison with FSGO, 2. FSGO in its symmetric and broken symmetry solutions and, 3. LCAO minimal basis set STO-3G in order to allow a safer comparison with the quality of the subminimal basis used in the FSGO technique. The dissociation curves are given in Figure 1.

Table 1. Total energy as a function of the internuclear distance R, in STO-1G and 3G, in FSGO (symmetric and broken symmetry solutions). (data in a.u.)

R	TOTAL ENERGY			
	STO-1G	FSGO symmetric	FSGO broken symmetry	STO-3G
1.0	- 0.8746	- 0.8745	(- 0.8745)	- 1.0660
1.5	- 0.9771	- 0.9558	(- 0.9558)	- 1.1117
2.0	- 0.9333	- 0.9190	(- 0.9190)	- 1.0492
2.5	- 0.8612	- 0.8570	(- 0.8570)	- 0.9658
3.0	- 0.7859	- 0.7934	(- 0.7934)	- 0.8853
3.5	- 0.7125	- 0.7346	(- 0.7346)	- 0.8163
4.0	- 0.6480	- 0.6820	(- 0.6820)	- 0.7611
4.5	- 0.5988	- 0.6353	(- 0.6353)	- 0.7185
5.0	- 0.5657	- 0.5941	(- 0.5941)	- 0.6864
5.5	- 0.5451	- 0.5576	(- 0.5576)	- 0.6626
6.0	- 0.5323	- 0.5250	- 0.5327	- 0.6451
6.5	- 0.5239	- 0.4960	- 0.5162	- 0.6322
7.0	- 0.5177	- 0.4698	- 0.5030	- 0.6228
7.5	- 0.5127	- 0.4462	- 0.4920	- 0.6156
8.0	- 0.5085	- 0.4249	- 0.4827	- 0.6100

The "equilibrium" FSGO-hydrogen molecule is found for the parameters:

$R = 1.474$ a.u.   (could be compared to the experiment: 1.401 a.u.)
$R_G = 0.0$ a.u.   (center of the molecule)

$\alpha = 0.31862$ a.u.

$E = - 0.95593$ a.u.

One observes that the energy minimum at the calculated equilibrium internuclear distance always corresponds to the symmetric solutions ($R_G=0$). For the values given between parentheses in the above table, the broken symmetry solution does not exist; the single Gaussian remains centered at the middle of the H-H bond. However, for interatomic distances greater than 5.6 a.u., the broken symmetry solutions (Gaussians centered near $R_G \approx 2$ a.u.) give the absolute minimum while the symmetric solution has a higher energy; this is a further example that, for approximate wave functions, the basic symmetry properties do not follow automatically from the variation principle and consequently do not have necessarily the full symmetry of the nuclear framework.

In the HF scheme, the first origin of the correlation between electrons of antiparallel spins comes from the restriction that they are forced to occupy the same orbital (RHF scheme) and thus some of the same location in space. A simple way of taking into account the basic effects of the electronic correlation is to release the constraint of double occupation (UHF scheme = Unrestricted HF) and so use Different Orbitals for Different Spins (DODS scheme which is the European way of calling UHF). In this methodology, electrons with antiparallel spins are not found to doubly occupy the same orbital so that, in principle, they are not forced to coexist in the same spatial region as is the case in usual RHF wave functions.

A UHF wave function over different orbitals ϕ_α and ϕ_β is then:

$$\psi(r_1,\omega_1,r_2,\omega_2) = \frac{1}{\sqrt{2}} \begin{vmatrix} \phi_\alpha(r_1)\alpha(\omega_1) & \phi_\beta(r_1)\beta(\omega_1) \\ \phi_\alpha(r_2)\alpha(\omega_2) & \phi_\beta(r_2)\beta(\omega_2) \end{vmatrix}$$

$$= \frac{1}{\sqrt{2}} \{\phi_\alpha(r_1)\, \phi_\beta(r_2)\alpha(\omega_1)\, \beta(\omega_2) - \phi_\beta(r_1)\, \phi_\alpha(r_2)\beta(\omega_1)\, \alpha(\omega_2)\}$$

The wave function obtained corresponds to the Unrestricted Hartree-Fock scheme and becomes equivalent to the RHF case if the orbitals ϕ_α and ϕ_β are the same. In this UHF form, the UHF wave function obeys the Pauli principle but is not an eigenfunction of the total spin operator and is thus a mixture of different spin multiplicities. In the present two-electron case, an alternative form of the wave function which has the same total energy, which is a pure singlet state, but which is no longer antisymmetric as required by the Pauli principle, is:

$$\psi(r_1,\omega_1,r_2,\omega_2) = \phi_\alpha(r_1)\, \phi_\beta(r_2)\, \frac{1}{\sqrt{2}} \{\alpha(\omega_1)\, \beta(\omega_2) - \beta(\omega_1)\, \alpha(\omega_2)\}$$

In both cases, the energy formula (E(UHF-FSGO)) is the same:

$$E(UHF\text{-}FSGO) = <\phi_\alpha|h^N|\phi_\alpha> + <\phi_\beta|h^N|\phi_\beta> + (\phi_\alpha\phi_\alpha|\phi_\beta\phi_\beta) + \frac{1}{|R|}$$

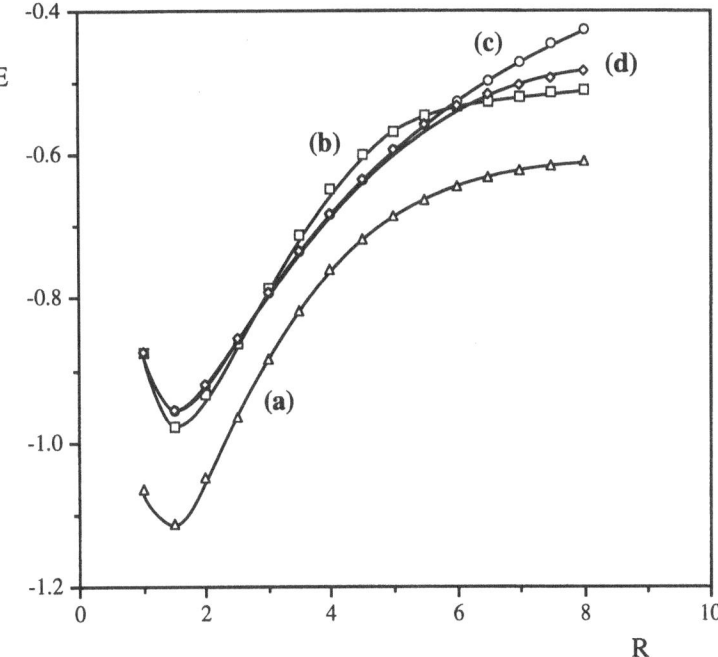

Figure 1: Dissociation curve for the hydrogen molecule (energy and distances in a.u.)
  (a) STO-3G
  (b) STO-1G
  (c) FSGO, symmetric solution
  (d) FSGO, broken symmetry solution

Table 2 gives the values of E(RHF-FSGO) and E(UHF-FSGO) for internuclear distances from 1.0 a.u. to 7 a.u. (step 0.5 a.u.) and also the value for R = ∞. We observe that we have the "correct" dissociation behavior for the UHF case ($H_2 \rightarrow 2 H\cdot$).

Since in those forms of the UHF wave functions, one drops a constraint (either the need of a pure spin state in the first case or the Pauli antisymmetry rule in the second case), it is expected that the resulting wave function will give a lower energy than in the RHF case and thus introduce a part of the correlation energy. As shown in the table above, there is no splitting of the ϕ_α and ϕ_β orbitals for small interatomic distances (<2.5 a.u.). The single Gaussian describing the different spinorbitals remains located at the same central position. For larger distances, however, a progressive splitting of the ϕ_α and ϕ_β orbitals exists, with the orbitals tending to localize near each hydrogen atom leading to correct dissociation into two hydrogen atoms.

Further improvements to the previous UHF schemes can be obtained by using the Projected (PHF) and Extended (EHF) Hartree-Fock schemes. Löwdin has shown that if one carries out a component analysis of the non-pure UHF wave function, there is at least one component which restores the symmetry and which has a lower energy than the UHF wave function.

The form of the PHF wave function is, in this case:

$$\psi(r_1,\omega_1,r_2,\omega_2) = \frac{1}{2} \begin{vmatrix} \phi_\alpha(r_1)\alpha(\omega_1) & \phi_\beta(r_1)\beta(\omega_1) \\ \phi_\alpha(r_2)\alpha(\omega_2) & \phi_\beta(r_2)\beta(\omega_2) \end{vmatrix} - \begin{vmatrix} \phi_\alpha(r_1)\beta(\omega_1) & \phi_\beta(r_1)\alpha(\omega_1) \\ \phi_\alpha(r_2)\beta(\omega_2) & \phi_\beta(r_2)\alpha(\omega_2) \end{vmatrix}$$

$$= \frac{1}{\sqrt{2}} \{\phi_\alpha(r_1)\phi_\beta(r_2) + \phi_\beta(r_1)\phi_\alpha(r_2)\} \frac{1}{\sqrt{2}} \{\alpha(\omega_1)\beta(\omega_2) - \beta(\omega_1)\alpha(\omega_2)\}$$

Note that the PHF wave function is no longer a single determinant and is a sum of two terms. This PHF function both satisfies the Pauli principle and is a pure singlet state. The energy formula E(PHF-FSGO) is easily derived:

$$E(PHF\text{-}FSGO) = \frac{1}{|R|} +$$

$$\frac{<\phi_\alpha|h^N|\phi_\alpha> + <\phi_\beta|h^N|\phi_\beta> + 2<\phi_\alpha|h^N|\phi_\beta><\phi_\alpha|\phi_\beta> + (\phi_\alpha\phi_\alpha|\phi_\beta\phi_\beta) + (\phi_\alpha\phi_\beta|\phi_\alpha\phi_\beta)}{1 + <\phi_\alpha|\phi_\beta>^2}$$

In the PHF method, the variational procedure is applied to the UHF wave function and subsequently the projection is performed on the UHF ϕ_α and ϕ_β orbitals. If no splitting is obtained during the UHF step ($\phi_\alpha = \phi_\beta$), the RHF, UHF and PHF wave functions are equivalent and have the same total energy.

Table 2. Total energy as a function of the internuclear distance R, in FSGO (RHF, UHF, PHF, and EHF) solutions). (All data in a.u.)

R	TOTAL ENERGY			
	FSGO RHF	FSGO UHF	FSGO PHF	FSGO EHF
1.0	- 0.8745	(- 0.8745)	(- 0.8745)	- 0.8965
1.5	- 0.9558	(- 0.9558)	(- 0.9558)	- 0.9922
2.0	- 0.9190	(- 0.9190)	(- 0.9190)	- 0.9719
2.5	- 0.8570	- 0.8755	- 0.9255	- 0.9301
3.0	- 0.7934	- 0.8574	- 0.8926	- 0.8933
3.5	- 0.7346	- 0.8513	- 0.8682	- 0.8695
4.0	- 0.6819	- 0.8495	- 0.8568	- 0.8573
4.5	- 0.6353	- 0.8490	- 0.8518	- 0.8518
5.0	- 0.5941	- 0.8489	- 0.8497	- 0.8497
5.5	- 0.5576	- 0.8488	- 0.8490	- 0.8490
6.0	- 0.5250	- 0.8488	- 0.8488	- 0.8488
6.5	- 0.4960	- 0.8488	- 0.8488	- 0.8488
7.0	- 0.4698	- 0.8488	- 0.8488	- 0.8488
∞	- 0.3547	- 0.8488	- 0.8488	- 0.8488

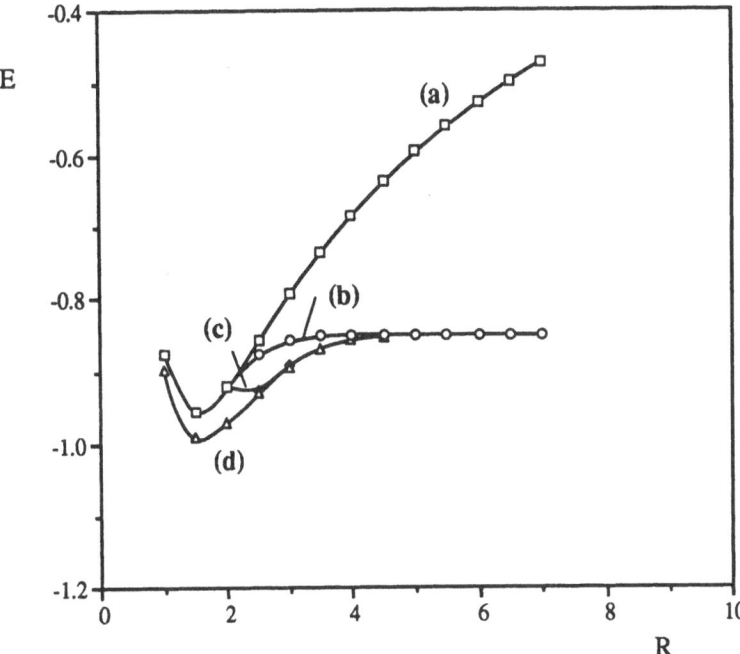

Figure 2: Dissociation curve for the hydrogen molecule (energy and distances in a.u.)
   (a) FSGO-RHF
   (b) FSGO-UHF
   (c) FSGO-PHF
   (d) FSGO-EHF

In the Extended Hartree-Fock (EHF) technique, the minimization is performed on the form of the PHF wave function. This type of wave function should produce for each interatomic distance a further lowering of the energy with respect to the RHF, UHF, and PHF total energies. The values of E(PHF-FSGO), and E(EHF-FSGO) for internuclear distances from 1.0 a.u. to 7 a.u. (step 0.5 a.u.) are also given in Table 2. As in the UHF case, we have the "correct" dissociation behavior (H$_2$ → 2H·).

Indeed, the extrapolated values converge to the expected values; we have shown that in the RHF symmetric case, the limiting situation is H$^-$ + H$^+$, i.e., an asymptotic energy which should correspond to the H$^-$ described whose two electrons are described by a single Gaussian:

$$E_{H^-}(\alpha) = 2\left(\frac{3}{2}\alpha\right) - 2\sqrt{\frac{8\alpha}{\pi}} - \sqrt{\frac{\alpha}{\pi}}$$

With the optimisation condition:

$$\frac{dE_{H^-}(\alpha)}{d\alpha} = 0$$

we obtain:

$$\alpha_{opt} = \frac{\left(\sqrt{8}-1\right)^2}{9\pi} = 0.118239 \text{ a.u.}$$

$$E_{H^-}^{opt}(\alpha) = -0.35471 \text{ a.u.}$$

In all the other schemes (UHF, PHF, EHF), the dissociation limit is the correct one corresponding to two neutral hydrogen atoms (2H·); each FSGO-hydrogen atom energy is thus obtained by the simple variational procedure:

$$E_{H\cdot}(\alpha) = \frac{3}{2}\alpha - \sqrt{\frac{8\alpha}{\pi}}$$

$$\alpha_{opt} = \frac{8}{9\pi} = 0.28294 \text{ a.u.} \qquad E_{H\cdot}^{opt}(\alpha) = \frac{3}{2}\left(\frac{8}{9\pi}\right) - \sqrt{\frac{8}{\pi}\left(\frac{8}{9\pi}\right)} = -0.42441 \text{ a.u.}$$

dissociation limit for two hydrogens: $2 \, (-0.42441) = -0.84882$ a.u.

As a final comment, it is interesting to note that this FSGO study of the hydrogen molecule offers a new and simple illustration of the behavior of sophisticated Hartree-Fock schemes like UHF, PHF and EHF. Furthermore, it provides a very efficient numerical example of instabilities in the standard Hartree-Fock method. It is important to see that the UHF, PHF and EHF schemes all correct the wrong RHF behavior and lead to the correct dissociation limit. However, the UHF and PHF schemes only correct the wave function for large enough interatomic distances and the effect of projection in the PHF scheme even results in a spurious minimum. The EHF scheme is thus the only one which shows a lowering of the energy with respect to RHF for all interatomic distances.

3. Subminimal basis set Hartree-Fock-type calculations of the hydrogen molecule in an external electric field.

When applying an external electrical field to the FSGO model of the hydrogen molecule, one expects that the floating gaussian will be moved in accordance with the polarity of the field, i.e., displaced towards the positive pole. Thus, near equilibrium internuclear distances, a minimum should be obtained close to the middle of the molecule. On the other hand, continuing to move the floating gaussian towards the positive pole, a barrier should appear close to the hydrogen atom the gaussian is floating towards. After having passed that barrier, the "energy catastrophe" of the unbound perturbing potential should produce an infinitely negatively stable position. This is the type of behaviour which is listed in Table 3 and illustrated in Figure 3. For the equilibrium internuclear distance ($R = 1.474$ a.u.) and the optimal exponent ($\alpha = 0.318655$ a.u.), we compute the energy as a function of the orbital position (z_G) for various strength of the external electrical field ($F = 0.0, 0.001, 0.01, 0.1, 0.25$ and 0.5 a.u.). The energy formulae can be obtained from the Hamiltonian for the hydrogen molecule in the electric field $F = (0,0,F)$ (z being the axis of the molecule):

$$H = H_0 - \mu_z F$$

where H_0 is the Hamiltonian of the isolated molecule and μ_z is the z component of the dipole moment operator (in a.u.)

$$\mu_z = + \sum_\alpha Z_\alpha z_\alpha - \sum_i z_i$$

with the α summation over nuclei and the i summation over electrons. Z_α stands for the nuclear charges, while z denotes the z-coordinate.

In the RHF case with the doubly occupied orbital ϕ, one obtains the mean value of the energy:

$$E(RHF) = 2 <\phi|h^N|\phi> + (\phi\phi|\phi\phi) + \frac{1}{|R|} - \sum_\alpha Z_\alpha z_\alpha F + <\phi|z|\phi> F$$

$$= 2 <\phi|-\frac{1}{2}\Delta|\phi> + 2 <\phi|-\frac{1}{r_a} - \frac{1}{r_b}|\phi> + (\phi\phi|\phi\phi) + \frac{1}{|R|} - \sum_\alpha Z_\alpha z_\alpha F + <\phi|z|\phi> F$$

Thus, we observe that when applying a finite electric field to a molecule, in addition to the instability with respect to the internuclear distance R, one obtains an instability connected to a change of the electric field strentgh F, as intuitively explained previously. In a more general way, the possible instabilities can be rationalized as follows: at some fixed values of the internuclear distance R and the FSGO exponent α the energy may be viewed as a function, $E(z_G)$, of the FSGO orbital position only. In view of the cylindrical symmetry of the problem, the position is determined by the z coordinate of the FSGO center, z_G . The formula for the function $E(z_G)$ is:

$$E(z_G) = \beta + 2 <\phi|-\frac{1}{r_a} - \frac{1}{r_b}|\phi> + 2 z_G F$$

where β is a constant equal to:

$$\beta = 3\alpha + 2 \sqrt{\frac{\alpha}{\pi}} + \frac{1}{|R|} - \sum_\alpha Z_\alpha z_\alpha F$$

which contains the mean value of the kinetic energy, the electron-electron and the nuclear repulsions, as well as the nuclear dipole moment interaction with the electric field. As one can see, if the field is positive, the energy goes to $-\infty$ when $z_G = -\infty$.

When expressing the nuclear attraction integrals in the FSGO basis, one has explicitly:

$$E(z_G) = \beta - 2^{5/2} \sqrt{\frac{\alpha}{\pi}} \{F_0[2\alpha(\frac{R}{2} + z_G)^2] + F_0[2\alpha(z_G - \frac{R}{2})^2]\} + 2 z_G F$$

where Fo is the standard error function:

$$F_0 (u) = \int_0^1 \exp(-ux^2) dx$$

The optimal positions of the FSGO orbital correspond to the minima of the function $E(z_G)$. They may be found by imposing the necessary condition:

$$\frac{dE}{dz_G} = 0$$

The derivative of the error function is given by:

$$\frac{dF_0}{du} = -F_1(u)$$

Table 3. Total energy as a function of the position of the floating Gaussian (z_G) for six strengths of the external electric field ($F = 0.0, 0.001, 0.01, 0.1, 0.25,$ and 0.5 a.u.). (All data in a.u., equilibrium distance $= 1.474$ a.u., orbital exponent $= 0.318655$ a.u., the center of the molecule corresponds to $z_G = 0.0$)

z_G	TOTAL ENERGY					
	$F = 0.0$	$F = 0.001$	$F = 0.01$	$F = 0.1$	$F = 0.25$	$F = 0.5$
7.0	1.6935	1.7075	1.8335	3.0935	5.1935	8.6935
6.0	1.5944	1.6064	1.7144	2.7944	4.5944	7.5944
5.0	1.4535	1.4635	1.5535	2.4535	3.9535	6.4535
4.0	1.2363	1.2443	1.3163	2.0363	3.2363	5.2363
3.0	0.8617	0.8677	0.9217	1.4617	2.3617	3.8617
2.0	0.2022	0.2062	0.2422	0.6022	1.2022	2.2022
1.5	-0.2139	-0.2109	-0.1839	0.0860	0.5361	1.2860
1.0	-0.5981	-0.5961	-0.5781	-0.3981	-0.0981	0.4018
0.5	-0.8629	-0.8619	-0.8529	-0.7629	-0.6129	-0.3629
0.0	-0.9559	-0.9559	-0.9559	-0.9559	-0.9559	-0.9559
-0.5	-0.8629	-0.8639	-0.8729	-0.9629	-1.1129	-1.3629
-1.0	-0.5981	-0.6001	-0.6181	-0.7981	-1.0981	-1.5981
-1.5	-0.2139	-0.2169	-0.2439	-0.5139	-0.9639	-1.7139
-2.0	0.2022	0.1982	0.1622	-0.1977	-0.7977	-1.7977
-3.0	0.8617	0.8557	0.8017	0.2617	-0.6382	-2.1382
-4.0	1.2363	1.2283	1.1563	0.4363	-0.7637	-2.7636
-5.0	1.4535	1.4435	1.3535	0.4535	-1.0465	-3.5464
-6.0	1.5944	1.5824	1.4744	0.3944	-1.4056	-4.4055
-7.0	1.6935	1.6795	1.5535	0.2935	-1.8065	-5.3064

where:

$$F_1(u) = \int_0^1 x^2 \exp(-ux^2)\, dx$$

By performing the derivative of the error function, one finds easily that the optimal position z_G of the FSGO necessarily satisfies the following equation:

$$-\frac{F}{8\,\alpha}\left(\frac{\pi}{2\alpha}\right)^{1/2} = S(z_G)$$

where:

$$S(z_G) = \{F_1[2\alpha(\tfrac{R}{2} + z_G)^2](\tfrac{R}{2} + z_G) + F_1[2\alpha(z_G - \tfrac{R}{2})^2](z_G - \tfrac{R}{2})\}$$

It is convenient to define a further function, $S'(z_G)$, for the discussion which follows, in which

$$S'(z_G) = \{F_1[2\alpha(\tfrac{R}{2} + z_G)^2](\tfrac{R}{2} + z_G) + F_1[2\alpha(z_G - \tfrac{R}{2})^2](z_G - \tfrac{R}{2})\} + \frac{F}{8\,\alpha}(\frac{\pi}{2\alpha})^{1/2}$$

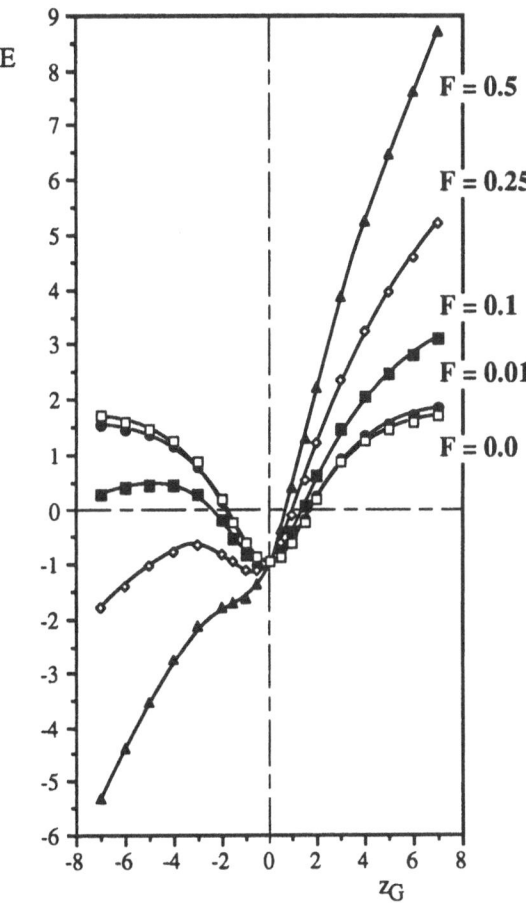

Figure 3. Total energy as a function of the position of the floating Gaussian (z_G) for five strengths of the external electric field (F = 0.0, 0.01, 0.1, 0.25, and 0.5 a.u.). For the sake of clarity of the figure, the curve for F = 0.001 has been omitted. (All data in a.u., equilibrium distance = 1.474 a.u., orbital exponent = 0.318655 a.u., the center of the molecule corresponds to z_G = 0.0)

At first sight, we are forced to solve this equation numerically, but its overall form allows a qualitative insight into the number of solutions and their approximate values. For example, one easily see that S represents a sum of two identical quasi-atomic (one-dimensional) functions each centered on the corresponding hydrogen nucleus. The functions are quite similar to $2p_z$ Gaussian functions, but they differ by their one-dimensionality and by a different radial dependence. Indeed, instead of the usual exponential behaviour, one has the $F_1(2\alpha z^2)$ function that is positive, even with respect to z, and, as it seen from the previous equation, has its maximum at $z = 0$ while it vanishes for large z.

This information is sufficient to analyze the qualitative behaviour of $E(z_G)$. Indeed, two limiting cases may be considered.

For one limiting case (small values of $R^2\alpha$, i.e., close to equilibrium distances), the S' function can be easily evaluated. By putting $R = 0$, it looks like a single one-dimensional p_z-type orbital centered in the symmetry center of the molecule. As one can see for $F = 0$, there is only one solution, $z_G = 0$ of the S'-equation and it corresponds obviously to a single minimum of the energy (symmetric solution) as seen in Figure 3. For very large values of $F > 0$, the S'-equation cannot be longer satisfied and does not attain negative values. This corresponds to the energy changing monotonically qualitatively. The energy minimum at $F = 0$ is clearly unstable when F increases. As one sees from Figure 4 for moderate values of F, one should observe two z_G values satisfying the equation. One of them corresponds to a minimum and the other to a maximum of the energy. One example of a single minimum is given in Figure 3 for the cases, $F = 0.01, 0.1$, and 0.25 a.u. Thus, there is a certain critical electric field value for which the energy curve changes qualitatively from the one having a single minimum to that with no minimum at all. From Figure 3, one easily see the existence of one and no minima in the curve $E = f(z_G)$ according to the strength of the field. For large values of F, there is no root in the S-function and no minimum is found for the $E(z_G)$. This is the case for $F = 0.5$ in Figure 3. Since the coordinate origin has been placed at the center of the molecule, the contribution of the nuclei to the dipole moment is 0 and the total dipole moment is equal to $-2z_G$ (in a.u.). At large values of z_G, the energy of the molecule mainly comes from the interaction of the dipole moment with the electric field and therefore has the asymptotic form $+2z_GF$. From Figure 3, it is also seen that the energy changes linearly with z_G for large z_G. For small z_G values, one has a strong influence of the nuclei (through the first term of the $E = f(z_G)$ equation). Also, it is easily proved that the asymptotic behaviour of the energy comes from its linear dependence of z_G for large values of this variable and that its asymptotic slope is 2F.

The other limiting case corresponds to the limit of large values of $R\alpha^2$. In order to remain concise, this has not been illustrated. For $F = 0$ one finds three values of z_G for which $S' = S = 0$. The first one is $z_G = 0$ and the remaining two are close to $z_G = R/2$ and $z_G = -R/2$ (the two positions of the nuclei). They correspond to the broken symmetry behaviour analyzed in section 2. For very large values of $F>0$, no value of z_G satisfies the equation, and the S' function does not attain negative values. This means that the energy $E(z_G)$ has no extremum.

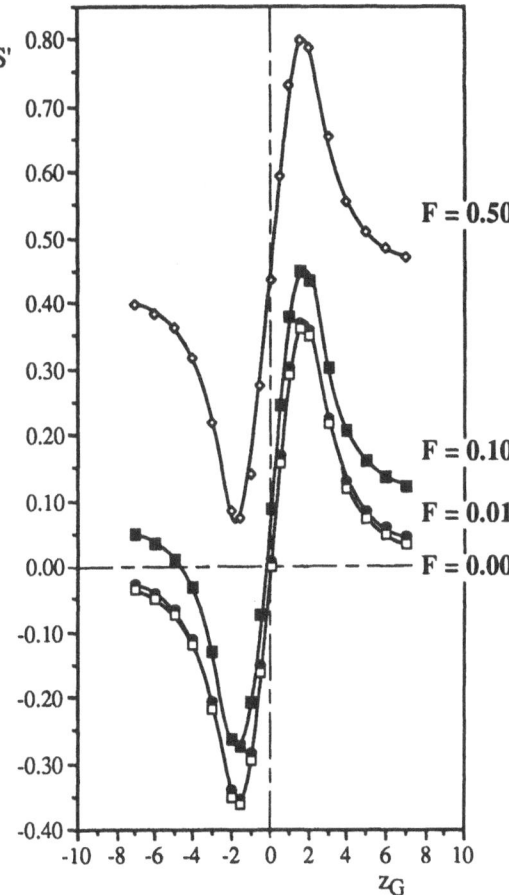

Figure 4: S'-function as a function of the position of the floating Gaussian (z_G) for four strengths of the external electric field (F = 0.0, 0.01, 0.1, and 0.5 a.u.). For sake of clarity of the figure, the curve for F = 0.001 has been omitted since it superposes almost exactly with that corresponding to F = 0.0. (All data in a.u., equilibrium distance = 1.474 a.u., orbital exponent = 0.318655 a.u., the middle of the molecule corresponds to z_G = 0.0)

For intermediate values of F, either four or two z_G values satisfy the equation. This corresponds to one or two energy minima, respectively, both of which are unstable when F increases. When broken symmetry solutions are observed in the absence of external electric field (F = 0.0), one might expect a single minimum appearing either at nucleus a or b, by turning on the electric field. In fact, for F>0, the minimum eventually appears only at the nucleus a, because the S' values for its left hand side (at the nucleus a) negative branch are slightly smaller than for its right hand side (at the nucleus b) positive branch. Both instabilities (with respect to F) should however be extremely close one to another. A more precise numerical analysis confirms also the possibility of two minima at low electric field and at a sufficiently large distance to allow for broken symmetry solutions.

Acknowledgements

The authors would like to acknowledge the support of the convention n° 9.4593.92 from the Belgian National Science Foundation (FNRS = Fonds National de la Recherche Scientific and of its associated Foundation (FRFC = Fonds de la Recherche Fondamentale et Collective). One of us (L.P.) acknowledges also the support by the BST-439/23/93 project of the Department of Chemistry, University of Warsaw, Poland. D.H.M. thanks the Belgian Ministry of Science (SPPS) for financial support.

References

1.   C.C.J. Roothaan, *Rev. Mod. Phys.* **32**, 179 (1960).
2.   P.-O. Löwdin, *J. Appl. Phys. Suppl.* **33**, 251 (1962).
3.   A.A. Frost, *J. Chem. Phys.* **47**, 3707 (1967).
4.   G. Berthier, *Folia Chimica Acta*, **15**, 67 (1987)

Electronic Charge Density of Quantum Systems in the Presence of an Electric Field: a Search for Alternative Approaches

G.P. ARRIGHINI and C. GUIDOTTI
Dipartimento di Chimica e Chimica Industriale. Università di Pisa
Via Risorgimento 35, 56126 Pisa, Italy

1. Introduction

Many fundamental properties of atoms and molecules could come within our reach through the "simple" knowledge of the electronic distribution density $n(\vec{r})$. Among these properties we limit ourselves to quote as particularly significant the various multipole moments of the distribution itself, electric potential and field generated by it in the surrounding space. The list of properties that we could master grows longer if we were in a position to establish how the electronic distribution density is polarized under the action of external electric and magnetic fields, inasmuch as one might evaluate also various kinds of (generally nonlinear) response parameters of matter, a piece of information that is nowadays of utmost importance for a vast series of research programs endowed with prominent technological significance (for instance, oriented toward the very ambitious goal of "designing" molecularly-thought materials in such strategic fields as photonics, optoelectronics, etc. [1-3]). Although the electronic density $n(\vec{r})$ is physically defined in a space of low dimensionality D ($D = 1; 2; 3$, according to the proper modeling adopted for the system under investigation), the canonical approach to the computation of such fundamental quantity involves the preliminary obtainment of the electronic wavefunction, a solution to the Schrödinger equation depending on the totality of the DN_e space coordinates associated with the N_e present electrons. Without going into the slightest detail, we simply restrict our comments on this point to re-emphasize what is well known even to quantum chemistry students, the fact that the usually accepted descriptions of the quantum behaviour of many-electron systems correspond to approximate solutions to the Schrödinger equation, most frequently built up in terms of one-electron wavefunctions, i.e. orbitals. Hartree-Fock (HF) orbitals constitute almost invariably the output of ab initio molecular calculations carried out by today's computer program packages up to $N_e \cong O(10^2)$ and lead obviously to an orbital picture of the electronic distribution as a sum of contributions from each of the occupied orbitals. Overcoming the HF accuracy so as to take into account effects beyond the mean-field approximation (electron correlation) is permitted at the cost of handling much smaller molecules, while for systems containing a very large number of electrons even the HF level of description becomes untenable and one has to turn to empirical or semiempirical models.

Y. Ellinger and M. Defranceschi (eds.), Strategies and Applications in Quantum Chemistry, 203–218.
© *1996 Kluwer Academic Publishers.*

The concerns we have expressed are bound to get even more acute if the problem under study demands that we are able to adequately describe distortion effects induced in the electron distribution by external fields. The evaluation of linear (and, still more, non linear) response functions [1] by perturbation theory then forces one to take care also of the nonoccupied portion of the complete orbital spectrum, which is entrusted with the role of representing the polarization caused by the external fields in the unperturbed electron distribution [4].

A still more outstanding role in quantum many-particle systems is assigned to the electron density $n(\vec{r})$ by the Hohenberg and Kohn theorem [5], a not obvious statement affirming the existence of a rigorous theoretical framework where one is allowed to obtain ground state properties of the system in terms of the ground state density alone. Unfortunately, although the electron kinetic and exchange-correlation energy contributions are shown to be universal functionals of the density $n(\vec{r})$, the theorem does not offer any practical guide to their actual construction. In view of the extremely attracting perspective of treating many-electron systems at an accuracy level beyond the HF one, without making recourse to wavefunction approaches, it is quite understandable that many efforts have been addressed to the development of density functional theories (DFT's) [6-8]. There exists possibly general agreement that the most satisfactory DFT approach presently implemented, suggested by Kohn and Sham [9], actually fails the original program, because it involves a return to an orbital picture (Kohn-Sham orbitals) as a rescue from the difficulties posed by our insufficient knowledge of the basic universal functionals inherent of the procedure, particularly the kinetic energy one. As a consequence, troubles met with large molecules, that we presumed to be able to leave outdoors thanks to the novel approach, again enter home from the windows, thus challenging to a substantial extent applications concerning most of the chemically and technologically interesting problems.

The present (very preliminary) investigation follows a research line closer to the true spirit of the DFT's, moving in the same direction as some recent papers where the attention is focused on the development of a formalism able to lead to the electron density $n(\vec{r})$ without invoking wavefunctions, orbitals in particular [10-15]. It is right to recall that the seminal ideas of this approach are anything but new, their origin dating back to the atomic statistical model put forward more than sixty years ago by Thomas and Fermi. Without pretending to review the concerned literature during such a long period of time (but a very complete bibliography is collected in ref. [7]), we limit ourselves to point out as particularly relevant to the present work some additional papers [16-26] where the manifest intent of revitalizing an old subject proceeds through the development of a general formalism that contemplates the Thomas-Fermi theory as a low-order level of approximation.

By the present paper we intend to start to explore the possibility of generating explicit, approximate ways for calculating the electron density $n(\vec{r})$ of a quantum system subjected to an external homogeneous and static electric field, without invoking, in the construction, orbitals as basic ingredients. Although the electronic distribution of the system is at the outset assumed to be describable in terms of (unspecified) occupied orbitals, we immediately shirk the orbital approach in favor of an integral representation of the electronic density $n(\vec{r})$ involving the knowledge of the quantum mechanical propagator (QMP) [27-30]. A drastic ansatz for the latter quantity based on the known QMP of a particle moving in a linear potential field is the key-step of the whole procedure, by which we attain, without any further approximations, an explicit final result for $n(\vec{r})$ expressed in terms of Airy function and its first derivative. The only applications of the result thus deduced will be restricted to the case of a model of independent particles moving in a quadratic potential while simultaneously acted by a static electric field: the predicted electric dipole moment induced by the field in the system is shown to be the exact value, despite the fact that the electronic charge density resulting from the approach is only an approximation to the correct one.

2. An approximate approach to the electronic density

The system under study is assumed to consist of $2N_e$ electrons, possibly in the presence of a nuclear framework. An orbital picture of the quantum behaviour of the system is then introduced on accepting the validity of an independent-particle model where each electron moves in the field of an effective potential $v(\vec{r})$, which afterwards is left substantially unspecified. We emphasize, however, that the choice of $v(\vec{r})$ is an essential step of any modeling. Besides semiempirical forms, effective potentials $v[n(\vec{r})]$ functionally dependent on the electron numeral density $n(\vec{r})$ are intuitively bound to play a prominent role in applications.

The one-electron Hamiltonian operator $\hat{h} = \hat{h}_o + \hat{v}$, with $\hat{h}_o$ kinetic energy operator, generates a complete spectrum of orbitals $|\phi_j\rangle$ according to the Schrödinger equation

$$\left[\hat{h}_o + \hat{v}\right]|\phi_j\rangle = \varepsilon_j|\phi_j\rangle \tag{2.1}$$

The ground state of the system corresponds to $2N_e$ electrons occupying the N_e lowest-energy ε_j levels, so that the electron numeral density $n(\vec{r})$ is

$$n(\vec{r}) = 2\sum_{j=1}^{N_e}\left|\phi_j(\vec{r})\right|^2 \tag{2.2}$$

To attain an expression of $n(\vec{r})$ which does not make explicit reference to the occupied orbitals, we rewrite eq. (2.2) in the form

$$n(\vec{r}) = 2\langle\vec{r}|\left(\sum_{j=1}^{N_e}|\phi_j\rangle\langle\phi_j|\right)|\vec{r}\rangle = 2\langle\vec{r}|\left(\sum_{j=1}^{\infty}\Theta(\varepsilon_F - \varepsilon_j)|\phi_j\rangle\langle\phi_j|\right)|\vec{r}\rangle \tag{2.3}$$

where we have introduced the Heaviside function $\Theta(x)$ and the Fermi level energy ε_F. If we make use of the following standard representation of $\Theta(x)$,

$$\Theta(x) = -\frac{i}{2\pi}\int_{-\infty}^{+\infty}dt\,\frac{e^{ixt}}{t - io^+} \tag{2.4}$$

from eq. (2.1) and the completeness of the spectrum of orbitals supported by $\hat{h}_o + \hat{v}$ the electron density $n(\vec{r})$, eq. (2.3), can be expressed as

$$n(\vec{r}) = -2\frac{i}{2\pi}\int_{-\infty}^{+\infty}dt\,\frac{e^{i\varepsilon_F t}}{t - io^+}\langle\vec{r}|e^{-it\left[\hat{h}_o + \hat{v}\right]}|\vec{r}\rangle \tag{2.5}$$

(atomic units with $e = m_e = \hbar = 1$ are used throughout this paper). Eq. (2.5) is a well known result [12,14-17,20,24,25], that makes evident the key-role played by the diagonal matrix element of the QMP $K(\vec{r},t;\vec{r}') \equiv \langle \vec{r}|e^{-it[\hat{h}_o+\hat{v}]}|\vec{r}'\rangle$ in determining the electronic density. Considering that the QMP knowledge allows one, in principle, to solve the problem of the time-evolution of any arbitrary initial quantum state, the obtainment of $K(\vec{r},t;\vec{r}')$ is to be regarded in general as a true piece of skill. It is a fact that we have at disposal only very few exact QMP expressions in analytical closed form [29], despite tremendous advances in quantum dynamics, particularly during the past fifteen years [31]. Much progress in the QMP evaluation has been realized following mainly the idea that the propagator for an arbitrary time t can be rigorously expressed in terms of short-time propagators, for which simple approximations are available [31]. The latter procedure has actually been developed in some of the papers quoted [10-15,25], which should therefore be regarded as more rigorous contributions to the problem of representing the electronic density according to eq. (2.5), even though it is right to say that the implementation of the formalism to explicit calculations has not kept the pace with theory.

Our more rudimentary approach is basically founded on an ansatz choice for the quantity $K(\vec{r},t;\vec{r}) \equiv \langle \vec{r}|e^{-it[\hat{h}_o+\hat{v}]}|\vec{r}\rangle$, of higher quality with respect to the short-time approximation $K(\vec{r},t;\vec{r}) \cong e^{-itv(\vec{r})}\langle \vec{r}|e^{-it\hat{h}_o}|\vec{r}\rangle$ which neglects all quantum effects arising from the noncommutativity of the operators $\hat{h}_o$ and $\hat{v}$. In order to appreciate the nature of the approximation, let us consider the case where the energy potential $v(\vec{r}) = v_o + \vec{A}\cdot\vec{r}$, with v_o and $\vec{A}$ constant quantities. Although the QMP for a particle subjected to a constant force is one of the few cases explicitly known [32], for our convenience we adopt the following exact alternative representation of the propagator for a particle moving in a linear potential [see Appendix A, eq. (A.8)]

$$K(\vec{r},t;\vec{r}) = (2\pi)^{-3}\exp\left\{-i\left[t\left(v_o + \vec{A}\cdot\vec{r}\right)+\frac{t^3}{6}A^2\right]\right\}\int d\vec{p}\,e^{-i\frac{t}{2}p^2}e^{\frac{i}{2}t^2\vec{A}\cdot\vec{p}}$$

$$= (2\pi)^{-3}\exp\left\{-i\left[tv(\vec{r})+\frac{t^3}{6}(\nabla v)^2\right]\right\}\int d\vec{p}\,e^{-i\frac{t}{2}p^2}e^{i\frac{t^2}{2}\nabla v\cdot\vec{p}} \tag{2.6}$$

The ansatz for the diagonal matrix element of the QMP appearing in eq. (2.5) corresponds to assume the validity of eq. (2.6) also for potentials $v(\vec{r})$ other than the linear one. Taking, moreover, into account that a homogeneous and static electric field $\vec{E}$ is associated with a potential energy $\vec{E}\cdot\vec{r}$, the propagator ansatz for the system subjected simultaneously to the action of an electric field generalizes in a straightforward way from eq. (2.6) to yield

$$K\left(\vec{r},t;\vec{r}\big|\vec{E}\right) = (2\pi)^{-3}\exp\left\{-it\left[v(\vec{r})+\vec{E}\cdot\vec{r}\right]-i\frac{t^3}{6}\left[\nabla v + \vec{E}\right]^2\right\}\int d\vec{p}\,e^{-i\frac{t}{2}p^2}e^{i\frac{t^2}{2}[\nabla v+\vec{E}]\cdot\vec{p}} \tag{2.7}$$

Eq. (2.7) is the starting point of the procedure we are going to develop. The neglect of the exponentials involving t^2 and t^3 leads to the same result obtainable according to the Trotter formula [30]; as easily verified, such short-time approximation is the basis for recovering from eq. (2.5) the same result predicted by the Thomas-Fermi theory.

Thanks to eq. (2.7), the electronic density expression given by eq. (2.5) can be cast into the (obviously approximate) form

$$n(\vec{r};\vec{E}) = 2\frac{i}{(2\pi)^4}\int d\vec{p}J(\vec{p};\vec{r},\vec{E}) \tag{2.8}$$

with

$$J(\vec{p};\vec{r},\vec{E}) \equiv \int_{-\infty}^{+\infty}\frac{dt}{t+io^+}\exp\left\{-it\left[\varepsilon_F - \frac{p^2}{2} - v(\vec{r}) - \vec{E}\cdot\vec{r}\right] + i\frac{t^2}{2}(\nabla v + \vec{E})\cdot\vec{p} + i\frac{t^3}{6}(\nabla v + \vec{E})^2\right\} \tag{2.9}$$

A series of manipulations [see Appendix B, eq. (B.4)] allows the function $J(\vec{p};\vec{r},\vec{E})$ to be expressed as follows

$$J(\vec{p};\vec{r},\vec{E}) = -2\pi i\left(\frac{2}{F^2}\right)^{\frac{1}{3}}\exp\left\{i\left[\frac{(\vec{F}\cdot\vec{p})^3}{3F^4} - \left(\frac{p^2}{2} + v(\vec{r}) + \vec{E}\cdot\vec{r} - \varepsilon_F\right)\frac{\vec{F}\cdot\vec{p}}{F^2}\right]\right\}\cdot$$

$$\int_0^\infty d\omega\exp\left[-i\omega\frac{\vec{F}\cdot\vec{p}}{F^2}\right]Ai\left[\left(\frac{2}{F^2}\right)^{\frac{1}{3}}\left(\omega - \varepsilon_F + \frac{p^2}{2} + v(\vec{r}) + \vec{E}\cdot\vec{r} - \frac{(\vec{F}\cdot\vec{p})^2}{2F^2}\right)\right] \tag{2.10}$$

where we have set

$$\vec{F} = \nabla v + \vec{E} \tag{2.11}$$

and $Ai[x]$ denotes the Airy function of argument x [33].

The replacement of the result for $J(\vec{p};\vec{r},\vec{E})$, eq. (2.10), into eq. (2.8) yields for the electronic density $n(\vec{r};\vec{E})$,

$$n(\vec{r};\vec{E}) = 2\left(\frac{2}{F^2}\right)^{\frac{1}{3}}\frac{1}{(2\pi)^3}\int_0^\infty d\omega\Omega(\omega;\vec{r},\vec{E}) \tag{2.12}$$

where

$$\Omega(\omega;\vec{r},\vec{E}) = \int d\vec{p}\exp\left\{i\left[\frac{(\vec{F}\cdot\vec{p})^3}{3F^4} - \left(\frac{p^2}{2} + v(\vec{r}) + \vec{E}\cdot\vec{r} - \varepsilon_F\right)\frac{\vec{F}\cdot\vec{p}}{F^2} - \omega\frac{\vec{F}\cdot\vec{p}}{F^2}\right]\right\}\cdot$$

$$\cdot Ai\left[\left(\frac{2}{F^2}\right)^{\frac{1}{3}}\left(\omega - \varepsilon_F + \frac{p^2}{2} + v(\vec{r}) + \vec{E}\cdot\vec{r} - \frac{(\vec{F}\cdot\vec{p})^2}{2F^2}\right)\right] \tag{2.13}$$

The expression for $\Omega(\omega;\vec{r},\vec{E})$ can be elaborated rather simply,[see Appendix C, eq. (C.4)]

$$\Omega(\omega;\vec{r},\vec{E}) = (2\pi)^2\left|\vec{F}\right|\int_a^\infty du\ [Ai(u)]^2 \tag{2.14}$$

$$a = \left(\frac{2}{F^2}\right)^{\frac{1}{3}}\left[\omega + v(\vec{r}) + \vec{E}\cdot\vec{r} - \varepsilon_F\right]$$

Indefinite integrals involving products of Airy functions and/or their derivatives can be evaluated without many difficulties [34]. Thus,

$$\Omega(\omega;\vec{r},\vec{E}) = -(2\pi)^2\left|\vec{F}\right|\left\{a\ [Ai(a)]^2 - [Ai'(a)]^2\right\}$$

and successively

$$n(\vec{r};\vec{E}) = -2\,|\vec{F}|\left(\frac{2}{F^2}\right)^{\frac{1}{3}}\frac{1}{2\pi}\int_0^\infty d\omega\,\left\{a[Ai(a)]^2 - [Ai'(a)]^2\right\}$$

$$= -2\,\frac{|\vec{F}|}{2\pi}\int_b^\infty da\,\left\{a[Ai(a)]^2 - [Ai'(a)]^2\right\} \tag{2.15}$$

$$b = \left(\frac{2}{F^2}\right)^{\frac{1}{3}}\left[v(\vec{r}) + \vec{E}\cdot\vec{r} - \varepsilon_F\right]$$

After carrying out the integrations involved in eq. (2.15) [34], we finally obtain the result

$$n(\vec{r};\vec{E}) = \frac{2}{3\pi}\,|\vec{F}|\,\left\{b^2[Ai(b)]^2 - b[Ai'(b)]^2 - \frac{1}{2}Ai(b)Ai'(b)\right\} \tag{2.16}$$

which allows one to calculate the numerical electronic density in terms of both the potential $v(\vec{r})$ characterizing the one-electron model assumed and the electric field $\vec{E}$ polarizing the electron distribution itself. It should be evident from the derivation that the effect of the field $\vec{E}$ has not been taken into account according to a perturbative treatment; eq. (2.16) is an approximate result for the electron density that includes at infinite order the polarization distortion caused by the external field.

Eq. (2.16) is not an entirely new result. After this work had been concluded and we were looking around in search of bibliographical material, we came upon a paper by Englert and Schwinger [24] dealing with the introduction of quantum corrections to the Thomas-Fermi statistical atom. These authors attain the same result expressed by eq. (2.16) (for the case $\vec{E} = 0$) by resorting to a somewhat more general assumption about the adopted QMP $K(\vec{r},t;\vec{r}')$ as compared to our choice.

For a quantum system with a single degree of freedom (dimensionality $D=1$), a procedure parallel to that sketched above leads to the following result

$$n(x;E) = 2(2|F|)^{\frac{1}{3}}\left\{[Ai'(b)]^2 - b[Ai(b)]^2\right\} \tag{2.17}$$

where

$$F = \frac{\partial v(x)}{\partial x} + E$$

$$\tag{2.18}$$

$$b = \left(\frac{2}{F^2}\right)^{\frac{1}{3}}\left[v(x) + Ex - \varepsilon_F\right]$$

The Fermi level energy ε_F appearing in eq. (2.16) [or eq. (2.17)] through the argument b of the Airy function and its derivative is fixed by the normalization requirement

$$\int d\vec{r}\,n(\vec{r};\vec{E}) = 2N_. \tag{2.19}$$

[or the analogous one-dimensional stemming from eq. (2.17)]. Obviously ε_F depends on the external field $\vec{E}$ amplitude.

Unperturbed electron densities descend naturally from the above formalism by letting $\vec{E}$ vanish.

3. An elementary application of the formalism

As a very simple application of the approach presented in sect. 2, we confine our attention to a model system consisting of $2N_e$ independent charged particles ("electrons"), moving in a one-dimensional harmonic effective potential $v(x) = \frac{1}{2}\omega_o^2 x^2$ while simultaneously acted by a static, homogeneous electric field E. An exact treatment of this standard problem is sketchily reviewed in Appendix D for reasons of completeness.

The approximate numeral density $n(x;E)$ is that obtained from eqs. (2.17), (2.18) with

$$F = \omega_o^2 x + E$$

$$b = \left[\frac{2}{\left(\omega_o^2 x + E\right)^2} \right]^{\frac{1}{3}} \left(\frac{1}{2}\omega_o^2 x^2 + Ex - \varepsilon_F \right) \tag{3.1}$$

Typical properties of the charge distribution are summarized by its various electric multipole moments. The electric dipole moment μ induced in the system by the external field is obviously

$$\mu(E) = -\int_{-\infty}^{\infty} dx \ x \ n(x;E) \tag{3.2}$$

For further progress, it is convenient to change integration variable from x to F, eq. (3.1), so that

$$\mu(E) = -\frac{1}{\omega_o^4} \int_{-\infty}^{\infty} dF (F - E) n\big[b(F;E)\big] \tag{3.3}$$

with

$$b(F;E) = \left(\frac{2}{F^2} \right)^{\frac{1}{3}} \left[\frac{F^2}{\omega_o^2} - \left(2\varepsilon_F + \frac{E^2}{\omega_o^2} \right) \right] \tag{3.4}$$

After noting that $b(-F;E) = b(F;E)$, we are simply left with

$$\mu(E) = \frac{E}{\omega_o^4} \int_{-\infty}^{\infty} dF \ n\big[b(F;E)\big] \tag{3.5}$$

Use of the normalization condition finally leads to

$$\mu(E) = \frac{E}{\omega_o^2}(2N_e) \tag{3.6}$$

a result coincident with the exact prediction [see Appendix D, eq. (D.6)].
The static electric dipole polarizability α of the model system investigated is therefore

$$\alpha = \left(\frac{\partial \mu}{\partial E} \right)_{E=0} = \frac{2N_e}{\omega_o^2} \tag{3.7}$$

while higher-order polarizabilities (hyperpolarizabilities) vanish rigorously.
In view of the result just found, it is interesting to contrast exact and approximate behaviour of the density $n(x;E)$ [eqs. (D.5) and (2.17), respectively]. Some insight into the nature of the approximations contained in our treatment is gained through the inspection of Table 1, which collects Fermi-level energy values calculated for several electron occupation numbers and two different electric field amplitudes. The entries have

been obtained by an easily feasible trial-and-trial procedure until satisfying the normalization requirement [eq. (2.19)] evaluated by numerical quadrature. The field-free values $\varepsilon_F(E = 0)$ are seen to be spaced nearly uniformly by $\hbar\omega_o$, according to the

Table 1. Fermi-level energy ε_F predicted for the harmonic-well model $(\omega_o = 1.\ a.u.)$ for different electron numbers $(2N_e)$ and different electric field amplitudes $E\ (a.u.)$.

$2N_e$	$E=0.$	$E=1.$
2	0.9985	0.4985
4	2.0000	1.5000
6	3.0000	2.5000
8	4.0000	3.5000
40	20.0000	19.5013
10^2	50.0022	49.5022
$5 \cdot 10^2$	250.005	249.505
$1 \cdot 10^3$	500.007	499.507

well known behaviour of the quantum harmonic oscillator spectrum. The lowest energy value (corresponding to $2N_e = 2$) and consequently all the others, however, are shifted upward about $\frac{1}{2}\hbar\omega_o$, thus suggesting the picture of a ladder spectrum of discrete energy values dephased with respect to the exact one. Some progressive deterioration seems to creep slowly into such harmonic picture as the electron number increases more and more. For a given number of electrons, moreover, the dependence of the Fermi-level energy shift $\varepsilon_F - \varepsilon_F(E = 0)$ on the electric field E is in perfect agreement with the exact prediction, $\varepsilon_F - \varepsilon_F(E = 0) = -\dfrac{E^2}{2\omega_o^2}$ [see eq. (D.4)].

Figs. 1-4 allow one to gain some further feeling about the quality of the approximation upon which our derivation of eq. (2.17) for $n(x;E)$ has been founded. Figs. 1 and 2 represent the electron density as a function of the coordinate x, in the absence of external electric field, for $2N_e=8$ and $2N_e=40$ respectively. Excellent overall agreement between exact and approximate profiles of $n(x;E = 0)$ is immediately recognized. In particular, there is a perfect reproduction of the electron distribution in the outer region, while for the central core the approximation leads to a "simulation" of the exact behaviour, able to represent only in some average manner the typical spatial oscillations of the quantum density. Such "simulation" becomes seemingly more adequate as the electron number increases from $2N_e=8$ to $2N_e=40$; the two cases actually display a similar behaviour, the drawing in Fig. 2 being unable to put in evidence details because of the too coarse-grain scale employed. Fig. 3 refers to the same situation illustrated in Fig. 2, the only change corresponding to the presence of an external electric field $E=1$. a.u. It should be manifest that the effect of switching on the external field is simply to translate uniformly the whole electron distribution toward more negative x values, the shift amounting to $-E / \omega_o^2$, in accordance with the exact prediction [eq. (D.5)].

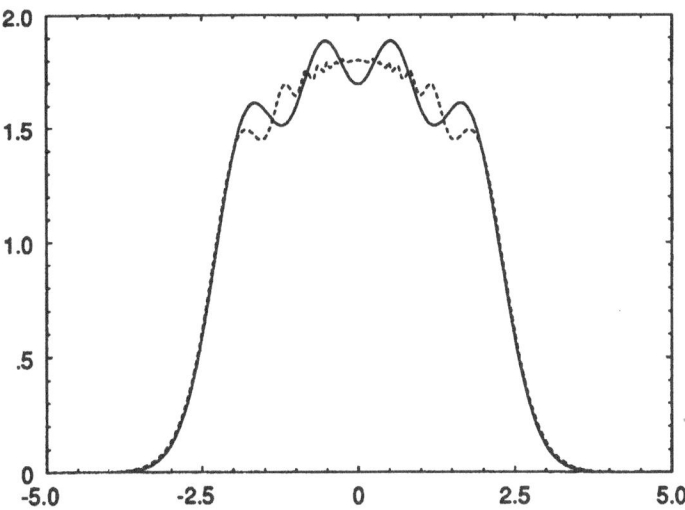

Fig. 1- Behaviour of the density $n(x; E = 0)$ as a function of x for a harmonic potential well with $\omega_O = 1$. a.u. Electron number $2N_e = 8$ (solid line: exact result; dashed line: approximate result).

The behaviour of the approximate density $n(x; E = 0)$ at both large and small x values can be understood considering the analytical properties of the function $f_1(b) \equiv [Ai'(b)]^2 - b[Ai(b)]^2$, eq. (2.17) [24,33]. As $x \to \pm\infty$, in fact, $b \to \infty$ and $f_1(b) \to 0$ exponentially, since asymptotically $f_1(b) \approx b^{-1/2} \exp[-(4/3)b^{3/2}]$ as $b \gg 1$. On the other hand, from the asymptotic behaviour $f_1(b) \approx (-b)^{1/2} / \pi$ valid for $-b \gg 1$ [24], as $x \to 0$, $b \to -\infty$ according to $-(2/x^2)^{1/3} \varepsilon_F$, so that we deduce $f_1 \to 2^{1/6} \varepsilon_F^{1/2} / (\pi x^{1/3})$, a divergent result. The density at $x=0$, however, is finite, its value from eq. (2.17) being $n(0; E = 0) \approx [8\varepsilon_F(E = 0)]^{1/2} / \pi$. Very rapid small oscillations which characterize both $Ai(z)$ and $Ai'(z)$ at large negative values of their arguments become concentrated in the region of small x values around $x=0$. Such unphysical oscillations, which arise from the approximate nature of the QMP utilized in our approach, do not result evident in the figures because of scale reasons, but can be partially perceived. Exact and approximate physical oscillations exhibited by $n(x;E)$ are compared on a magnified scale in Fig. 4 for the case $2N_e=40$ and $E=1$. a.u.

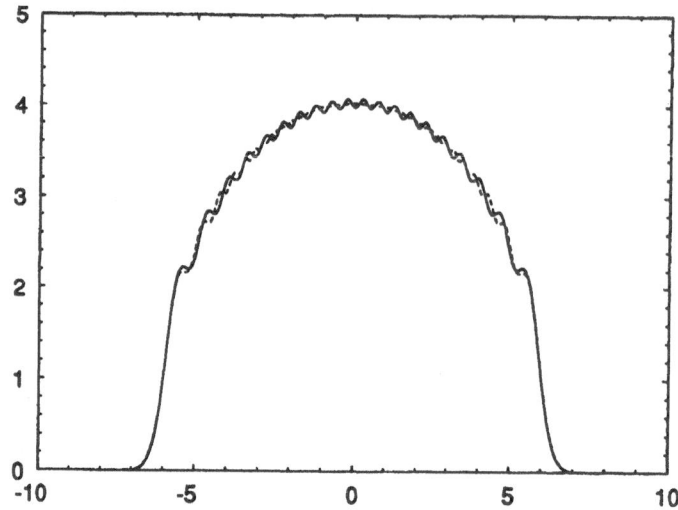

Fig. 2 - The same as in Fig. 1, the only change being $2N_e=40$.

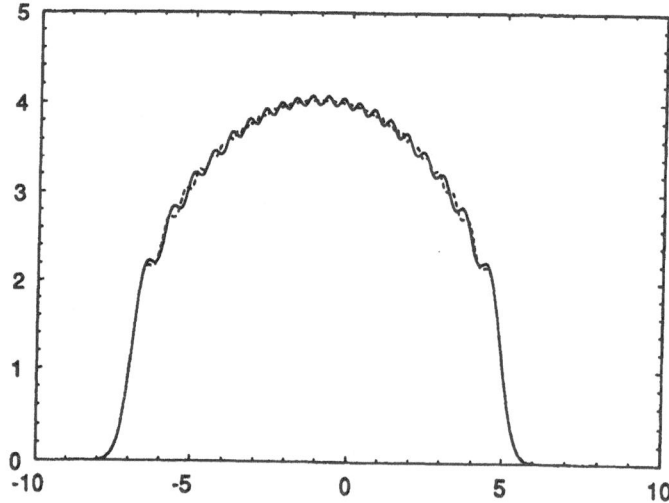

Fig. 3 - Behaviour of the density $n(x;E)$ as a function of x for the same potentiial as in Fig. 1. $E=1$. a.u., $2N_e=40$.

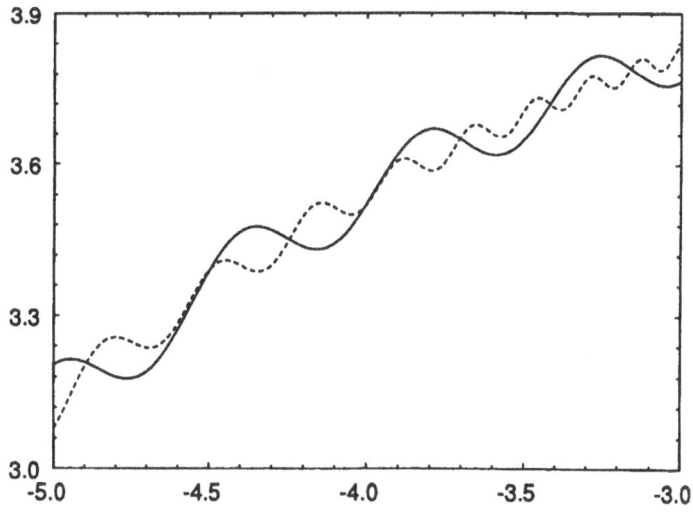

Fig. 4 - Oscillatory behaviour of the density $n(x;E)$ represented in Fig. 3
(magnified view of the region -5.0<x<-3.0 a.u.).

Appendix A

The time-evolution operator $\hat{U}(t)$ for a single electron moving in a 3D-linear potential
can be expressed in the form

$$\hat{U}(t) = e^{-it\left[\hat{h}_o + \hat{v}\right]} = e^{-it\left[\frac{\hat{p}^2}{2} + v_o + \left(A_x \hat{x} + A_y \hat{y} + A_z \hat{z}\right)\right]} = \hat{U}_x(t)\hat{U}_y(t)\hat{U}_z(t)e^{-iv_o t} \tag{A.1}$$

where

$$\hat{U}_j(t) = e^{-it\left[\frac{1}{2}\hat{p}_j^2 + A_j \hat{r}_j\right]} \qquad (j \equiv x, y, z) \tag{A.2}$$

A useful manipulation of the operator $\hat{U}_j(t)$ can be carried out by setting

$$\hat{U}_j(t) = e^{-it\left[\frac{1}{2}\hat{p}_j^2 + A_j \hat{r}_j\right]} = e^{-itA_j \hat{r}_j} e^{-i\frac{t}{2}\hat{p}_j^2} \hat{G}_j(t) \tag{A.3}$$

$\hat{G}_j(t)$ being an unknown operator, with $\hat{G}_j(0) = \hat{1}$. A differential equation for $\hat{G}_j(t)$ is
easily obtained by deriving with respect to t both sides of eq. (A.3):

$$\frac{\partial \hat{G}_j(t)}{\partial t} = \frac{i}{2}\hat{p}_j^2 \hat{G}_j(t) - \frac{i}{2}e^{i\frac{t}{2}\hat{p}_j^2}\left(e^{itA_j \hat{r}_j}\hat{p}_j^2 e^{-itA_j \hat{r}_j}\right)e^{-i\frac{t}{2}\hat{p}_j^2}\hat{G}_j(t) \tag{A.4}$$

Noting that $\left(e^{itA_j \hat{r}_j}\hat{p}_j^2 e^{-itA_j \hat{r}_j}\right) = \left(e^{itA_j \hat{r}_j}\hat{p}_j e^{-itA_j \hat{r}_j}\right)^2 = \left(\hat{p}_j - tA_j\right)^2$, it follows

$$\frac{\partial \hat{G}_j(t)}{\partial t} = i\left(tA_j \hat{p}_j - \frac{1}{2}t^2 A_j^2\right)\hat{G}_j(t) \tag{A.5}$$

and therefore

$$\hat{G}_j(t) = \exp\left[\frac{i}{2}t^2 A_j \hat{p}_j - \frac{i}{6}t^3 A_j^2\right] \tag{A.6}$$

Taking into account that operators associated with different labels j commute between each other, form eqs. (A.1), (A.3), (A.6) we get

$$\hat{U}(t) = \exp\left[-it\left(v_o + A_x\hat{x} + A_y\hat{y} + A_z\hat{z}\right)\right] e^{-i\frac{t}{2}\hat{p}^2} e^{\frac{i}{2}t^2\left(A_x\hat{p}_x + A_y\hat{p}_y + A_z\hat{p}_z\right)} e^{-\frac{i}{6}t^3 A^2} \tag{A.7}$$

The diagonal matrix element $K(\vec{r},t;\vec{r})$ of the QMP can therefore be written in the form

$$K(\vec{r},t;\vec{r}) \equiv \langle \vec{r}|\hat{U}(t)|\vec{r}\rangle = e^{-it\left(v_o + \vec{A}\cdot\vec{r}\right)} e^{-\frac{i}{6}t^3 A^2} \langle \vec{r}|e^{-i\frac{t}{2}\hat{p}^2} e^{\frac{i}{2}t^2\left(A_x\hat{p}_x + A_y\hat{p}_y + A_z\hat{p}_z\right)}|\vec{r}\rangle$$

$$= (2\pi)^{-3} e^{-it\left(v_o + \vec{A}\cdot\vec{r}\right)} e^{-\frac{i}{6}t^3 A^2} \int d\vec{p}\, e^{-i\frac{t}{2}p^2} e^{\frac{i}{2}t^2 \vec{A}\cdot\vec{p}} \tag{A.8}$$

The evaluation of the simple integral contained in eq. (A.8) would lead, of course, to the known expression for the QMP of the system under study [32].

Appendix B

The function $J(\vec{p};\vec{r},\vec{E})$ of eq. (2.9) is conveniently expressed as follows

$$J(\vec{p};\vec{r},\vec{E}) = \int_{-\infty}^{\infty} dt \exp\left\{i\left[\frac{t^2}{2}(\nabla v + \vec{E})\cdot\vec{p} + \frac{t^3}{6}(\nabla v + \vec{E})^2\right]\right\} \cdot I(t)$$

with

$$I(t) = \int_{-\infty}^{\infty} dt' \frac{\delta(t-t')}{t' + io^+} \exp\left\{-it'\left[\varepsilon_F - \frac{p^2}{2} - \left(v(\vec{r}) + \vec{E}\cdot\vec{r}\right)\right]\right\}$$

The Fourier representation of the Dirac delta function leads then to the result

$$I(t) = (2\pi)^{-1} \int_{-\infty}^{\infty} d\omega' e^{i\omega' t} \int_{-\infty}^{\infty} dt' \frac{\exp\left\{-it'\left[\varepsilon_F - \frac{p^2}{2} + \omega' - \left(v(\vec{r}) + \vec{E}\cdot\vec{r}\right)\right]\right\}}{t' + io^+}$$

$$= -i \int_{-\infty}^{\infty} d\omega' e^{i\omega' t} \Theta\left[\varepsilon_F - \frac{p^2}{2} + \omega' - \left(v(\vec{r}) + \vec{E}\cdot\vec{r}\right)\right]$$

and therefore,

$$J(\vec{p};\vec{r},\vec{E}) = -i \int_{-\infty}^{\infty} d\omega' \Theta\left[\varepsilon_F - \frac{p^2}{2} + \omega' - \left(v(\vec{r}) + \vec{E}\cdot\vec{r}\right)\right] \cdot$$

$$\cdot \int_{-\infty}^{\infty} dt \exp\left\{i\left[\omega' t + \frac{t^2}{2}(\nabla v + \vec{E})\cdot\vec{p} + \frac{t^3}{6}(\nabla v + \vec{E})^2\right]\right\} \tag{B.1}$$

The cubic form in t appearing in the exponential of eq. (B.1) is expressible as

$$Y(t) \equiv \omega' t + \frac{t^2}{2}(\nabla v + \vec{E})\cdot\vec{p} + \frac{t^3}{6}(\nabla v + \vec{E})^2$$

$$= \frac{1}{6}\left(tF^{\frac{2}{3}} + \frac{\vec{F}\cdot\vec{p}}{F^{\frac{4}{3}}} \right)^3 + \left[\omega' - \frac{\left(\vec{F}\cdot\vec{p}\right)^2}{2F^2} \right] t - \frac{\left(\vec{F}\cdot\vec{p}\right)^3}{6F^4}$$

where we have set $\vec{F} = \nabla v + \vec{E}$.
Thus the integration in the time variable involved in eq. (B.1) yields

$$\int_{-\infty}^{\infty} dt e^{iY(t)} = 2\pi \left(\frac{2}{F^2}\right)^{\frac{1}{3}} \exp\left\{ -i\left[\frac{\left(\vec{F}\cdot\vec{p}\right)^3}{6F^4} + \left(\omega' - \frac{\left(\vec{F}\cdot\vec{p}\right)^2}{2F^2} \right)\frac{\vec{F}\cdot\vec{p}}{F^2} \right] \right\} \cdot$$

$$\cdot Ai\left[\left(\frac{2}{F^2}\right)^{\frac{1}{3}} \left(\omega' - \frac{\left(\vec{F}\cdot\vec{p}\right)^2}{2F^2} \right)\frac{\vec{F}\cdot\vec{p}}{F^2} \right]$$

(B.2)

where we have utilized the definition of the Airy function [33]

$$Ai[x] = \frac{1}{2\pi} \int_{-\infty}^{\infty} dy e^{-ixy} e^{-iy^3/3}$$

(B.3)

From the latter result, eq. (B.1) is finally cast into the form

$$J\left(\vec{p};\vec{r},\vec{E}\right) = -2\pi i \left(\frac{2}{F^2}\right)^{\frac{1}{3}} \exp\left\{ i\left[\frac{\left(\vec{F}\cdot\vec{p}\right)^3}{3F^4} - \left(\frac{p^2}{2} + v(\vec{r}) + \vec{E}\cdot\vec{r} - \varepsilon_F \right)\frac{\vec{F}\cdot\vec{p}}{F^2} \right] \right\} \cdot$$

$$\cdot \int_0^{\infty} d\omega \exp\left[-i\omega\frac{\vec{F}\cdot\vec{p}}{F^2} \right] Ai\left[\left(\frac{2}{F^2}\right)^{\frac{1}{3}} \left(\omega - \varepsilon_F + \frac{p^2}{2} + v(\vec{r}) + \vec{E}\cdot\vec{r} - \frac{\left(\vec{F}\cdot\vec{p}\right)^2}{2F^2} \right) \right]$$

(B.4)

Appendix C

Eq. (2.14) in the text can be derived in a straightforward way after choosing the z-axis along the direction of the vector $\vec{F}$, so that $\vec{F}\cdot\vec{p} = |\vec{F}|p_z$. Then $\Omega\left(\omega;\vec{r},\vec{E}\right)$, eq. (2.13), can be written down in the form

$$\Omega\left(\omega;\vec{r},\vec{E}\right) = \int_{-\infty}^{\infty} dp_x \int_{-\infty}^{\infty} dp_y Ai\left[\left(\frac{2}{F^2}\right)^{\frac{1}{3}} \left(\omega - \varepsilon_F + v(\vec{r}) + \vec{E}\cdot\vec{r} + \frac{p_x^2 + p_y^2}{2} \right) \right] \cdot$$

$$\cdot \int_{-\infty}^{\infty} dp_z \exp\left\{ -i\left[\frac{p_z^3}{6|\vec{F}|} + \frac{p_z}{|\vec{F}|}\left(v(\vec{r}) + \vec{E}\cdot\vec{r} - \varepsilon_F + \omega + \frac{p_x^2 + p_y^2}{2} \right) \right] \right\}$$

(C.1)

By a simple variable change the integration in p_z is expressible in terms of the Airy function of proper argument [see eq. (B.3)], so that

$$\Omega\left(\omega;\vec{r},\vec{E}\right) = 2\pi\left(2|\vec{F}|\right)^{\frac{1}{3}} \int_{-\infty}^{\infty} dp_x \int_{-\infty}^{\infty} dp_y \left(Ai\left[\left(\frac{2}{F^2}\right)^{\frac{1}{3}} \left(\omega + v(\vec{r}) + \vec{E}\cdot\vec{r} - \varepsilon_F + \frac{p_x^2 + p_y^2}{2} \right) \right] \right)^2$$

(C.2)

Transforming to plane polar coordinate we get

$$\Omega\left(\omega;\vec{r},\vec{E}\right) = (2\pi)^2 \left|\vec{F}\right| \int_a^\infty du \left[Ai(u)\right]^2 \tag{C.3}$$

where

$$a = \left(\frac{2}{F^2}\right)^{\frac{1}{3}}\left[\omega + v(\vec{r}) + \vec{E}\cdot\vec{r} - \varepsilon_F\right] \tag{C.4}$$

Appendix D

For a one-dimensional model system of $2N_e$ independent particles moving in a harmonic effective potential $v(x) = \frac{1}{2}\omega_0^2 x^2$ and simultaneously subjected to an electric field E, the numeral distribution density $n(x,E)$ is given by

$$n(x;E) = 2\sum_{s=0}^{N_e-1}\left|\phi_s(x;E)\right|^2 \tag{D.1}$$

where $\phi_s(x;E)$ is an orbital fully "dressed" by the external field E, a solution to the Schrödinger equation

$$\left[-\frac{1}{2}\frac{d^2}{dx^2} + \frac{1}{2}\omega_0^2 x^2 + Ex\right]\phi_s(x;E) = \varepsilon_s\phi_s(x;E) \tag{D.2}$$

After recognizing that $\frac{1}{2}\omega_0^2 x^2 + Ex = \frac{1}{2}\omega_0^2\left(x + \frac{E}{\omega_0^2}\right)^2 - \frac{E^2}{2\omega_0^2}$, a variable change

$x \rightarrow y = x + \dfrac{E}{\omega_0^2}$ leads to the following Schrödinger equation for the orbitals required,

$$\left[-\frac{1}{2}\frac{d^2}{dy^2} + \frac{1}{2}\omega_0^2 y^2\right]\phi_s'(y) = \left[\varepsilon_s + \frac{E^2}{2\omega_0^2}\right]\phi_s'(y) \tag{D.3}$$

i.e. the same equation one should solve for orbitals in the absence of external field. Therefore,

$$\varepsilon_s = \left(s + \frac{1}{2}\right)\omega_0 - \frac{E^2}{2\omega_0^2}$$

$$\phi_s'(y) = N_s e^{-\omega_0 y^2/2} H_s\left(\sqrt{\omega_0}\,y\right) \tag{D.4}$$

$$s = 0, 1, 2, \ldots$$

The exact numeral density of the model system is consequently

$$n(x;E) = 2e^{-\omega_0\left(x+\frac{E}{\omega_0^2}\right)^2}\sum_{s=0}^{N_e-1}N_s^2 H_s^2\left[\sqrt{\omega_0}\left(x + \frac{E}{\omega_0^2}\right)\right] \tag{D.5}$$

The electric dipole moment induced in the system by the electric field is therefore

$$\mu(E) = -\int_{-\infty}^{\infty} dx\ x\ n(x;E) = \frac{E}{\omega_0^2}(2N_e) \tag{D.6}$$

Acknowledgments

Financial support by the Italian National Research Council (CNR) and by the Ministry of University and Scientific and Technological Research (MURST) is acknowledged. This work was developed under the CNR "Progetto Finalizzato Materiali Speciali per Tecnologie Avanzate" and MURST "40%" funds.

References

1. P.N. Prasad and D.J.Williams, Introduction to Nonlinear Optical Effects in Molecules & Polymers, J. Wiley & Sons, New York (1991).
2. P.N. Prasad and B.A. Reinhardt, *Chem. Mater.* **2**, 660 (1990).
3. A. D'Andrea, A. Lapiccirella, G. Marletta and S. Viticoli (Eds.), Materials for Photonic Devices, World Scientific, Singapore (1991).
4. C.E.Dykstra, Ab Initio Calculation of the Structures and Properties of Molecules, Elsevier, Amsterdam (1988).
5. P.C. Hohenberg and W. Kohn, *Phys. Rev.* **136 B**, 864 (1964).
6. R.G. Parr and W. Yang, Density-Functional Theory of Atoms and Molecules, Oxford U.P., New York (1989).
7. E.S. Kryachko and E.V. Ludeña, Energy Density Functional Theory of Many-Electron Systems, Kluwer Academic Publ., Dordrecht (1990).
8. R.M. Dreizler and E.K.U. Gross, Density Functional Theory, Springer Verlag, Berlin (1990).
9. W. Kohn and L.J. Sham, *Phys. Rev.* **140 A**, 1133 (1965).
10. R.A. Harris and L.R. Pratt, *J. Chem. Phys.* **82**, 856 (1985).
11. G.G. Hoffman, L.R. Pratt and R.A. Harris, *Chem. Phys. Letters* **148**, 313 (1988).
12. L.R. Pratt, G.G. Hoffman and R.A. Harris, *J. Chem. Phys.* **88**, 1818 (1988).
13. L.R. Pratt, G.G. Hoffman and R.A. Harris, *J. Chem. Phys.* **92**, 6687 (1990).
14. W. Yang, *Phys. Rev. A* **38**, 5494 (1988); *ibid. A* **38**, 5504 (1988); *ibid. A* **38**, 5512 (1988).
15. W. Yang, *Adv. Quantum Chem.* **21**, 293(1990).
16. S. Golden, *Rev. Mod. Phys.* **32**, 322 (1960).
17. G.S. Handler, *J. Chem. Phys.* **43**, 252 (1965).
18. G.S. Handler and P.S.C. Wang, *J. Chem. Phys.* **56**, 1546 (1972).
19. G.S. Handler, *J. Chem. Phys.* **58**, 1 (1973).
20. J.C. Light and J.M. Yuan, *J. Chem. Phys.* **58**, 660 (1973).
21. G. Starkschall and J.C. Light, *J. Chem. Phys.* **61**, 3417 (1974).
22. J. Schwinger, *Phys. Rev. A* **24**, 2353 (1981).
23. R.A. Harris and J.A. Cina, *J. Chem. Phys.* **79**, 1381 (1983).
24. B.G. Englert and J. Schwinger, *Phys. Rev. A* **29**, 2339 (1984).
25. R.A. Harris and L.R. Pratt, *J. Chem. Phys.* **82**, 5084 (1985).
26. L. Spruch, *Rev. Mod. Phys.* **63**, 151 (1991).
27. R.P. Feynman and A.R. Hibbs, Quantum Mechanics and Path Integrals, McGraw-Hill, New York (1965).
28. S.M. Binder, in International Review of Science, Physical Chemistry Series Two, Vol. 1 Theoretical Chemistry, A.D. Buckingham and C.A. Coulson Ed., Butterworth, London , p. 1 (1975).
29. D.C. Khandekar and S.V. Lawande, *Phys. Reports* **137**, 115 (1986).
30. L.S. Schulman, Techniques and Applications of Path Integration, J. Wiley & Sons, New York (1981).

31.   K.C. Kulander, <u>Time-Dependent Methods for Quantum Dynamics (A Thematic Issue of Computer Physics Communications)</u>, *Comp. Phys. Commun.* **63**, 1(1991).
32.   C.A. Moyer, *J. Phys. C* **6**, 1461(1973).
33.   M. Abramowitz and I.A. Stegun, <u>Handbook of Mathematical Functions</u>, Dover Publications, Inc., New York (1965).
34.   J.R. Albright, *J. Phys. A* **10**, 485 (1977).

How Much Correlation Can We Expect to Account for in Density Functional Calculations ? Case Studies of Electrostatic Properties of Small Molecules

J. WEBER, P. JABER, P. GULBINAT and P.-Y. MORGANTINI
Université de Genève, Département of Chimie Physique, 30 quai Ernest-Ansermet, 1211 Genève 4, Switzerland

1. Introduction

It is well known that the traditional ab initio techniques of quantum chemistry are able to incorporate many-electron effects through expansions of the many-particle wavefunction, which leads in principle to systematic procedures to take correlation effects into account. However, the computational challenge issued by these post-Hartree-Fock calculations is generally a formidable task, as for both variational configuration interaction (CI) and size-consistent many-body perturbation theory (MBPT) techniques, the amount of computations required to reach chemical accuracy is enormous. In addition, these methods, and in particular those of multiconfiguration self-consistent field (MCSCF) and CI type, are sophisticated and in virtually no case they can be used as black boxes. Indeed, the problem is that, unless the system investigated is small enough so as to allow for a full CI treatment, truncated CI expansions have to be used and, according to the qualified statement of Berthier et al. [1], "the choice of an appropriate molecular orbital (MO) basis set in then a considerable concern".

On the other hand, it is indispensable for most molecular properties to account for correlation effects so as to achieve quantitative, or even sometimes qualitative, predictions as the neglect of instantaneous repulsions introduces an error which may be significant [2]. Fortunately, substantial efforts have been made in the last twenty years in order to develop correlated quantum chemical methods and there is ample choice among them today for a given problem. For example, most of the popular semiempirical models offer the possibility to include some CI using, e.g., the Pople-Pariser-Parr formalism, as implemented in the AMPAC series of programs [3]. As far as they are concerned, the techniques based on density functional theory (DFT) are able to incorporate some treatment of correlation through the energy functional used to solve the Kohn-Sham equations [4,5]. However, the degree of correlation introduced in these methods depends on the form of the so-called exchange-correlation potential and it is difficult to estimate how much correlation is present in the results unless performing comparative ab initio calculations. As to the latter ones, they have the advantage to allow in principle for a progressively increasing treatment of correlation through enlarging the N-electron basis set in CI calculations or through the introduction of higher orders of perturbation in MBPT studies. It is thus possible to rather accurately quantify (generally in percent) how much correlation is introduced in, e.g., a CI study, by comparing the calculated correlation energy with the difference between Hartree-Fock and "experimental" (when available), or full CI with a very large one-electron basis

219

Y. Ellinger and M. Defranceschi (eds.), Strategies and Applications in Quantum Chemistry, 219–228.
© 1996 *Kluwer Academic Publishers.*

set (when possible), total energies [6]. It is of course impossible to implement such a procedure using the DFT methodology, as there is no equivalent to the Hartree-Fock energy. In other words, one extracts from such a calculation a total energy which contains "some" correlation contribution, without the possibility to separate it from the SCF energy. To elucidate this point, one has therefore to compare molecular properties calculated using the DFT formalism with their values predicted by ab initio computations performed at various levels of approximation.

It has been recently pointed out that DFT models are in general adequate to describe to a good extent, through the standard exchange-correlation potentials generally used, the correlated movement of electrons at short interelectronic distance, i.e. the so-called dynamic correlation [5]. This is even possible in the simple formalism of the local density approximation (LDA) [7], using a potential such as that of Vosko, Wilk and Nusair (VWN) which is now commonly employed for many DFT applications [8]. It has indeed been shown that such calculations incorporate dynamic correlation effects at least to the same extent as second-order Moller-Plesset (MP2) MBPT [9], which represents now a standard for post-Hartree-Fock ab initio calculations. However, there is a second category of correlation, known as static or long-range correlation, which accounts for near-degeneracy effects in the wavefunction [10]. Whereas it can be accounted for in ab initio calculations through the MCSCF procedure, it is more difficult to describe in DFT as it requires the use of involved exchange-correlation potentials, with the risk of a double counting of correlation corrections [11]. Alternatively, long-range correlation can be introduced in the DF formalism by combining CI or MCSCF with DF through a scaling of the electron density by a factor depending on Hartree-Fock and CI (or MCSCF) two-electron density matrices calculated in the same one-electron basis set [12].

In the present work, we shall investigate the problem of the amount of correlation accounted for in the DF formalism by comparing the molecular electrostatic potentials (MEPs) and dipole moments of CO and N_2O calculated by DF and ab initio methods. It is indeed well known that the calculated dipole moment of these compounds is critically dependent on the level of theory implemented and, in particular, that introduction of correlation is essential for an accurate prediction [13,14]. As the MEP property reflects reliably the partial charges distribution on the atoms of the molecule, it is expected that the MEP will exhibit a similar dependence and that its gross features correlate with the changes in the value of dipole moment when switching from one level of theory to the other. Such a behavior has indeed been reported recently by Luque et al. [15], but their study is limited to the ab initio method and we found it worthwhile to extend it to the DF formalism. Finally, the proton affinity and the site of protonation of N_2O, as calculated by both DF and ab initio methods, will be reported.

2. Computational Details

For the DFT calculations, the linear combination of Gaussian - type orbitals - density functional (LCGTO-DF) method and its corresponding deMon program package [16] have been used. In all calculations, the VWN exchange-correlation potential was employed [8] and all the core and valence electrons were explicitly taken into account. To enable a meaningful comparison with the ab initio results, the same one-electron basis set has been used in all the calculations, i.e. 6-31+G(2d,2p) [17] which has been recently found adequate for calculating the dipole moment of CO and N_2O [14]. The auxiliary basis sets required by the LCGTO-DF model to fit the electron density and the exchange-correlation

potential have been chosen as C(5,3;5,3), O(4,4;4,4), N(4,3;4,3) and H(3,1;3,1).
The ab initio calculations have been performed using the Gaussian 90 program package
[18]. The 6-31+G(2d,2p) has been employed throughout at both self-consistent field (SCF)
and MP2 levels of theory. In all the calculations, the proton affinities (PAs) have been
obtained as the difference between total energies of optimized unprotonated and protonated
species and no zero-point nor thermodynamic contributions have been introduced. This
means that we are more interested in comparing the PAs deduced from the theoretical
models at various levels of theory than in performing accurate comparisons with the
experimental values.

3. Results and Discussion

Table 1 presents the results obtained for the bond distance and dipole moment of carbon
monoxide.

Table 1: Theoretical and experimental results obtained for CO

	E_{tot}/h	$d_{CO}/Å$	μ^a/D
SCF	-112.75007	1.106	-0.270
MP2	-113.07599	1.140	0.284
LCGTO-DF	-112.41630	1.134	0.202
Experimental	...	1.128[b]	0.122[c]

[a] Calculated at the experimental geometry: a positive dipole corresponds to C$^-$-O$^+$
[b] Ref.[19]
[c] Ref.[20]

As expected, it is seen that the SCF bond distance is somewhat too short and that the MP2
calculation leads to a significant bond lengthening, which is in agreement with the well-
known trend that introduction of correlation effects substantially increases calculated bond
lengths [21]. As expected, the LCGTO-DF result is in very good agreement with
experiment, even though the so-called nonlocal corrections have not been introduced [16],
the predicted bond distance being intermediate between the experimental value and the MP2
prediction. Actually, the LCGTO-DF value for d_{CO} lies much closer to the MP2 result than
to the SCF one, which is a first indication for some correlation being taken into account.
This conclusion is strengthened by examination of the dipole moment of CO (Table 1). It is
indeed seen that, whereas the ab initio method at the SCF level predicts, as is well known
[13], the wrong sign, both the MP2 and LCGTO-DF values lead to a C$^-$ - O$^+$ dipole
moment in agreement with experiment. In addition, the DFT result is substantially smaller
than the MP2 one, which makes it closer to the 0.122 D experimental value. As the dipole
moment of CO is very sensitive to the amount of correlation introduced in the
wavefunction, this suggests that our LCGTO-DF calculation incorporates a correlation
contribution roughly equivalent to the MP2 level.

Let us turn to the MEP of CO which we have calculated using the same methods as those
employed for the dipole moment (Fig. 1). We expect that these MEPs will reflect the same

trend as the dipole moment, inasmuch as a multicenter multipole expansion of the molecular electron charge density can be used to generate a MEP which is a good approximation to that obtained from the wavefunction [22]. It has been shown indeed that the contributions arising from the monopolar and dipolar terms are generally preponderant [23], which suggests that the MEP and molecular dipole moment calculated from the same wavefunction should exhibit the same trend within a series of compounds or when evaluated at various levels of theory.

It is seen in Fig. 1a that the MEP calculated at the ab initio SCF level exhibits two roughly equivalent minima on both carbon and oxygen ends, lying at -14.0 and -11.4 kcal/mol, respectively, along the CO bond axis. From the point of view of electrostatics, both atoms behave therefore similarly at the SCF level towards an incoming proton. This picture is drastically modified when examining the MP2 result (Fig. 1b). In this case, the minimum on carbon (-18.0 kcal/mol) is significantly lower than that on oxygen (-3.8 kcal/mol), which indicates that the electron density calculated at the MP2 level is substantially different from that resulting from the SCF calculation. In particular, this dissymmetry in the MEP minima on both atoms suggests that part of the electron density is shifted towards the carbon atom, which simultaneously allows to rationalize the change of sign of the calculated dipole moment when going from SCF to MP2 and the C^- - O^+ polarity evaluated in this latter case. It is therefore of interest to examine the LCGTO-DF MEP (Fig. 1c) in order to confirm the shift of electron density induced by introduction of correlation. It is seen that indeed the LCGTO-DF result is very similar to the MP2 map, with MEP minima lying at -18.8 kcal/mol on the C-end at -6.7 kcal/mol on the O-end. However, the difference between C and O MEP minima is about 2 kcal/mol larger in MP2 than in LCGTO-DF, which suggests that the MP2 electron density of CO leads to a slightly more polar molecule than in the LCGTO-DF case and hence to a larger C^- - O^+ dipole moment (see Table 1). In spite of these small differences, there is no doubt when examining Fig. 1 that the LCGTO-DF electron density calculated for CO incorporates correlation to an extent comparable to MP2, which confirms the results of previous investigations [4,9,16].

Let us turn to the results obtained for nitrous oxide N_2O. From a theoretical point of view, this is an interesting molecule as properties such as its dipole moment and protonation site have been found very difficult to calculate accurately [14,24]. N_2O is a linear species whose predicted bond lengths and dipole moment are presented in Table 2.

Table 2: Theoretical and experimental results obtained for N_2O

	E_{tot}/h	$d_{NN}/Å$	$d_{NO}/Å$	μ^a/D
SCF	-183.69919	1.085	1.172	-0.621
MP2	-184.29011	1.160	1.185	0.063
LCGTO-DF	-183.29794	1.139	1.180	0.110
Experimental	...	1.128[b]	1.184[b]	-0.161[c]

[a] Calculated at the experimental geometry: a positive dipole corresponds to $^+NNO^-$
[b] Ref.[25]
[c] Ref.[26]

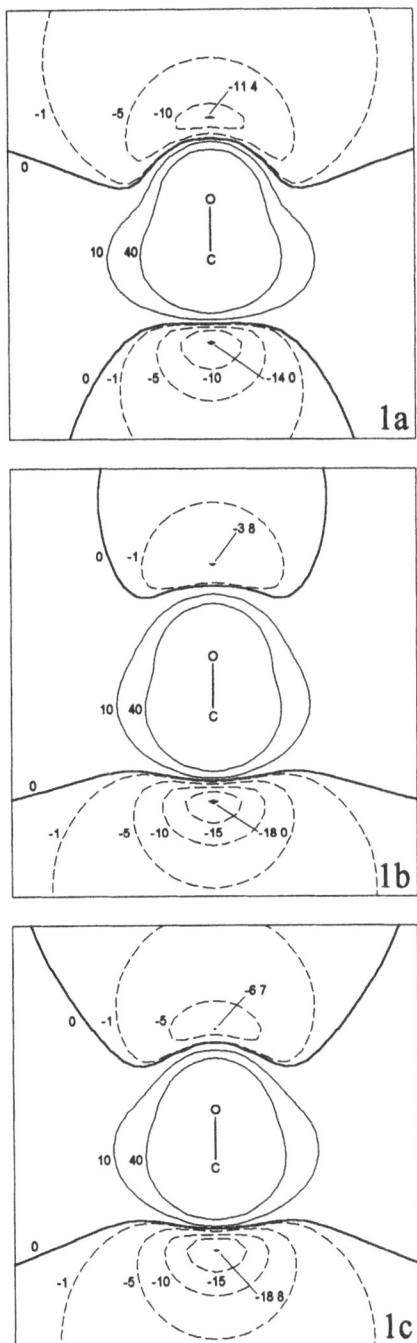

Fig. 1. MEP contour maps calculated for the CO molecule using the SCF (1a), MP2 (1b) and LCGTO-DF (1c) methods. The contour levels and MEP minima are given in kcal/mol.

It is seen in this Table that, similarly to CO, the bond lengths of N_2O increase significantly when going from SCF to MP2. It is also noteworthy that the introduction of correlation at the ab initio level leads to a much better agreement with experiment, particularly for the NO bond distance. Again, the LCGTO-DF results lie much closer to the MP2 than to the SCF ones, the LCGTO-DF bond distances being actually on an average closer to the experimental values than both SCF and MP2 results. Examination of this part of Table 2 suggests therefore again that a roughly equivalent amount of electron correlation is present in both MP2 and LCGTO-DF calculations. This conclusion is strengthened by the results obtained for the dipole moment : it is seen indeed in Table 2 that, whereas the ab initio SCF result leads to a $^+NNO^-$ dipole moment, in agreement with the experimental sign, though the absolute value is much too large, both MP2 and LCGTO-DF predictions exhibit the wrong sign but significantly smaller absolute values. Actually, the dipole moment of N_2O has been shown to be even more sensitive than that of CO to the level of theory and calculation parameters employed [14]. In particular, Moller-Plesset calculations performed at various orders of perturbation exhibit oscillations with a change of sign of l at each order from the SCF case to MP4 [14], which clearly indicates a very slow and difficult convergence of this property. Actually, infinite-order methods such as quadratic CI are necessary for a correct prediction of the dipole moment of N_2O. Coming back to Table 2, the fact that both MP2 and LCGTO-DF results for the dipole moment of N_2O exhibit the same sign is a clear indication of a similar amount of correlation present in both methods, though one might argue that the +0.150 D value obtained by Frisch and del Bene at the MP4 level using the same one-electron basis set [14] is actually closer to the LCGTO-DF prediction than the MP2 result. However, further comparative calculations would be clearly needed to better substantiate this point. Actually, these and previous [9] results allow us to conclude that correlation contributions equivalent to *at least* the MP2 level are most probably included in DFT calculations such as those performed here.

The MEPs of N_2O, as calculated at the SCF, MP2 and DFT levels of theory, are presented in Fig. 2. As expected from the results obtained for the dipole moment of this compound, considerable differences are observed between the SCF and MP2 MEP maps. Indeed, whereas all the MEPs exhibit a single minimum on the terminal nitrogen lying on the molecular axis and a crown region of out-of-axis minima on oxygen, the energy values of these minima are quite different. In the SCF case, the absolute minimum is found on oxygen (-16.2 kcal/mol) and the minimum on nitrogen is only a local one lying at -6.6 kcal/mol, which is in accordance with the $^+NNO^-$ dipole moment calculated at this level of theory. However, the energies of the minima in both MP2 and LCGTO-DF cases are in sharp contrast with this picture, as the absolute minimum is found on nitrogen, the minimum on oxygen being only a local one lying at a much higher energy. In other words, both MP2 and LCGTO-DF results are characteristic of a situation where a significant amount of electron density has been shifted from the oxygen to the terminal nitrogen atom with respect to the SCF description, which is in line with the $^-NNO^+$ dipole moment predicted by these models. Again, the resemblance between MP2 and LCGTO-DF MEPs is striking as both maps exhibit practically the same energy minima on oxygen and terminal nitrogen atoms. On the other hand, it immediately seen when comparing Figs. 2a and 2c that SCF and LCGTO-DF MEPs are quite dissimilar, which underlines the substantial amount of correlation introduced into LCGTO-DF wavefunctions.

Finally, we turn to the problem of protonation of N_2O. In principle, this molecule has two possible sites of protonation : the oxygen and the nitrogen ends. While experimental results are not really able to discriminate between these two sites [27], Amano has provided rotational constants for the protonated species [28] which were used by Rice et al. [29] in their theoretical study to support protonation on oxygen. This result has been recently confirmed by Ekern et al. [30], who report, however, that this prediction is very sensitive to the level of electron correlation introduced in the calculations : at MP3, MP4SDQ and QCISD(T) levels, O-protonation is preferred, whereas in the MP2 and MP4SDTQ cases, N-protonation is predicted. Actually, the fact that the protonated N_2O species is a difficult case for quantum chemistry is not unexpected on the basis of the sensitivity of the calculated dipole moment and MEP of this compound upon the level of electron correlation introduced.

Table 3: Calculated proton affinities for the N_2O molecule[a] [kcal/mol]

	N-end	O-end
SCF	128.8	149.1
MP2	135.5	132.9
LCGTO-DF	136.8	129.6

[a] Calculated at the corresponding optimized geometry;
   experimental value: 142.0 kcll/mol[27]

Table 3 presents the proton affinities calculated (without zero-point and thermodynamic contributions) for both N- and O-ends of N_2O. It is seen that indeed the SCF result leads unambiguously to O-protonation, which is consistent with both $^+NNO^-$ dipole moment and MEP minima calculated at this level of theory. However, both MP2 and LCGTO-DF results predict erroneously N-protonation and there is probably no doubt that introduction of zero-point and thermodynamic contributions would not modify this conclusion [30]. Our calculations confirm therefore the theoretical results previously obtained for protonated N_2O [29,30] and concluding that correlation effects have to be introduced at a higher level of theory than MP2 and LCGTO-DF so as to perform a reliable prediction of the protonation site. Again it is noteworthy that MP2 and LCGTO-DF protonation energies lie very close one another. Undoubtedly, the LCGTO-DF model incorporates correlation effects to an extent similar to MP2 and not MP3, as in the latter case O-protonation would be favored.

As a conclusion, the present investigation has shown that it is possible to estimate the amount of correlation accounted for in density functional theory by performing comparative calculations of selected properties of small molecules. Among the properties studied, the MEP presents the advantage being a local property, which leads to visual comparisons between maps calculated in molecular planes. It is thus possible to rapidly evaluate the similarities between SCF, post-Hartree-Fock and DFT calculations and to deduce general conclusions as to the main characteristics of the corresponding wavefunctions. In particular, it is expected that such a procedure would be useful in comparing various exchange-correlation potentials commonly employed in DFT calculations, and further investigations in this direction are in progress.

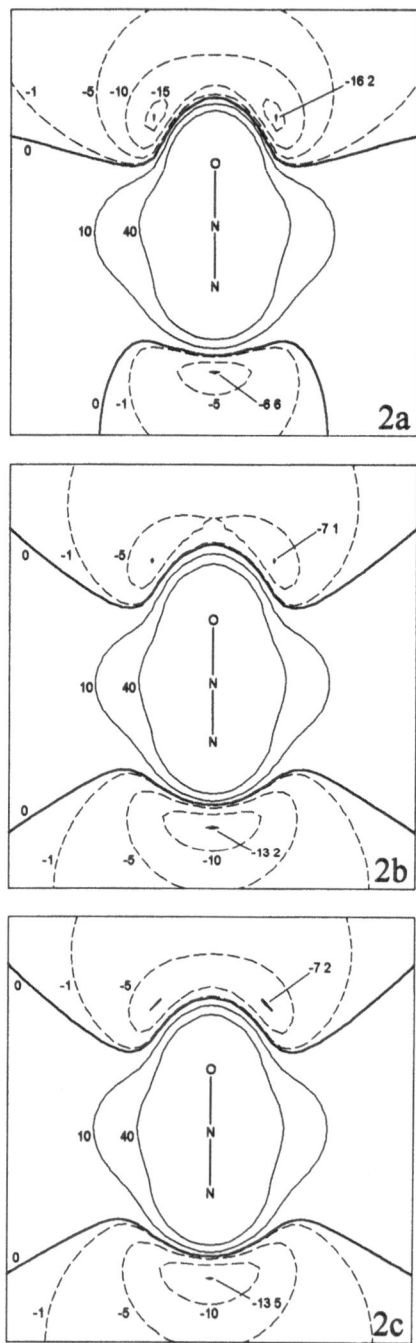

Fig. 2. MEP contour maps calculated for the N_2O molecule using the SCF (2a), MP2 (2b) and LCGTO-DF (2c) methods. The contour levels and MEP minima are given in kcal/mol.

Acknowledgements

The authors are grateful to Professor D. Salahub for providing a copy of the deMon program. This work is part of Project 20-36131.92 of the Swiss National Science Foundation.

References

1. G. Berthier, A. Daoudi and J.P. Flament, *J. Mol. Struct. Theochem* **166**, 81 (1988).
2. C.E Dykstra, J.D. Augspurger, B. Kirtman and D.J. Malik, in *Reviews in Computational Chemistry*, vol. I, K.B. Lipkowitz and D.B. Boyd, Eds., VCH, New York (1990).
3. M.J.S. Dewar, in *Modern Techniques in Computational Chemistry : MOTECC-91*, E. Clementi, Ed., Escom, Leiden (1990).
4. M. Cook and M. Karplus, *J. Phys. Chem.* **91**, 31 (1987).
5. T. Ziegler, *Chem. Rev.* **91**, 651 (1991).
6. A.C. Hurley, Electron Correlation in Small Molecules, Academic Press, London, (1976).
7. W. Kohn and L.J. Sham, *Phys. Rev.* **140**, A1133 (1965).
8. S.J. Vosko, L. Wilk and M. Nusair, *Can. J. Phys.* **58**, 1200 (1980).
9. J. Andzelm and E. Wimmer, *J. Chem. Phys.* **96**, 1280 (1992).
10. B.O. Roos, M. Szulkin and M. Jaszunski, *Theor. Chim. Acta* **71**, 375 (1987).
11. A. Savin, *Int. J. Quantum Chem. Symp.* **22**, 59 (1988).
12. R. Colle and O. Salvetti, *Theor. Chim. Acta* **53**, 55 (1979).
13. G.E. Scuseria, M.D. Miller, F. Jensen and J. Geertsen, *J. Chem. Phys.* **94**, 666 (1991).
14. M. J. Frisch and J.E. Del Bene, *Int. J. Quantum Chem. Symp.* **23**, 363 (1989).
15. F.J. Luque, M. Orozco, F. Illas and J. Rubio, *J. Am. Chem. Soc.* **113**, 5203 (1991).
16. D.R. Salahub, R. Fournier, P. Mlynarski, I. Papai, A. St-Amant and J. Ushio, in *Density Functional Methods in Chemistry*, J.K. Labanowski and J.W. Andzelm, Eds., Springer, New York (1991).
17. M.J. Frisch, J.A. Pople and J.S. Binkley, *J. Chem. Phys.* **80**, 3265 (1984).
18. M.J. Frisch, M. Head-Gordon, G.W. Trucks, J.B. Foresman, H.B. Schlegel, K. Raghavachari, M. Robb, J.S. Binkley, C. Gonzales, D.F. DeFrees, D.J. Fox, R.A. Whiteside, R. Seeger, C.F. Melius, J. Baker, R.L. Martin, L.R. Kahn, J.J.P. Stewart, S. Topiol and J.A. Pople, Gaussian 90, Gaussian Inc., Pittsburgh (1990).
19. K.P. Huber and G. Herzberg, Constants of Diatomic Molecules, Van Nostrand, New York (1979).
20. J. S. Muenter, *J. Mol. Spectrosc.* **55**, 490 (1975)
21. B.H. Besler, G.E. Scuseria, A.C. Scheiner and H.F. Schaefer, *J. Chem. Phys.* **89**, 360 (1988).
22. J.R. Rabinowitz, K. Namboodiri and H. Weinstein, *Int. J. Quantum Chem.* **29**, 1697 (1986).
23. J. Langlet, P. Claverie, F. Caron and J.C. Boeuve, *Int. J. Quantum Chem.* **20**, 299 (1981).
24. W.D. Allen, Y. Yamaguchi, A.G. Csaszar, D.A. Clabo, R.B. Remington and H.F. Schaefer, *Chem. Phys.* **145**, 427 (1990).

25. CRC Handbook of Chemistry and Physics, 55th ed., R.C. Weast, Ed. Chemical Rubber, Cleveland (1974).
26. H. Jalink, D.H. Parker and S. Stolte, *J. Mol. Spectry* **121**, 236 (1987).
27. T.B. McMahon and K. Kebarle, *J. Chem. Phys.* **83**, 3919 (1985).
28. T. Amano, *Chem. Phys. Lett.* **127**, 101 (1986).
29. J.E. Rice, T.J. Lee and H.F. Schaefer, *Chem. Phys. Lett.* **130**, 333 (1986).
30. S.P. Ekern, A.J. Illies and M.L. McKee, *Mol. Phys.* **78**, 263 (1993)

Applications of Nested Summation Symbols to Quantum Chemistry: Formalism and Programming Techniques

R. CARBÓ and E. BESALÚ
Institute of Computational Chemistry, University of Girona
Albereda 5, 17071 Girona, Spain

1. Introduction

Our research on various Quantum Chemistry areas has been directed in a great extend to the construction of general useful algorithms based on, as elementary as possible, mathematical concepts [1,2]. We tried along this past period to obtain computational procedures with sufficiently interesting features leading to a three fold purpose. First, the results must be pedagogically adequate. Second, the algorithmic structure must be susceptible of easy implementation to high level programming languages. Third, the development must benefit the computational side of Chemistry as Physics and be solidly grounded of Applied Mathematics principles.

In this sense, our intention was, such that the final working schemes can serve to connect mathematical general formulae writing and computationally valid general program structures. Thus, programming techniques can also be assisted by means of this process, as well as Artificial Intelligence [1c] algorithms may use partially the results of our outline in order to increase the performances of formulae generation and translation programs.

With all this conditioning principles in mind, the present work tries to describe in a first place the definition and properties of two fundamental symbols: Logical Kronecker Deltas (LKD's) and Nested Sums. The authors hope these symbol forms turn to be as useful to the scientific community as they had been in the development of their quest of a valid computational scheme based on PC machinery, whose main features had been already explained by one of us, see for example reference [3].

Consequently, here are studied under the formulation of the Nested Summation Symbols (NSS's) symbolism some Quantum Chemical problems and topics.

2. Definition and Properties of the NSS

2.1 LOGICAL KRONECKER DELTA SYMBOL DEFINITION

Let us define a generalization of the Kronecker delta symbol and call it a *Logical Kronecker Delta* (LKD). This symbol is written as $\delta(L)$ and corresponds to a function that can return two possible values: 1 if the logical argument L is true or 0

Y. Ellinger and M. Defranceschi (eds.), Strategies and Applications in Quantum Chemistry, 229–248.
© 1996 *Kluwer Academic Publishers.*

otherwise. The Kronecker delta symbol is a particular case of a LKD, where in the logical expression L there is involved an equivalence symbol.

2.2 DEFINITION OF THE NSS

The NSS concept corresponds to an operator attached to an arbitrary number of nested sums. In other words, a NSS represents a set of summation symbols where the number of them can be variable. In a general notation one can write a NSS as $\Sigma_n(j=i,f,s,L)$ where the meaning of this convention corresponds to perform all the sums involved in the generation of all the possible values of the index vector j under the fulfillment of the set of logical expressions collected in the components of the vector L. The elements of the vector j have the following limits:

$$\{i_k \leq j_k \leq f_k, \ if \ s_k \geq 0\} \ or \ \{i_k \geq j_k \geq f_k, \ if \ s_k \leq 0\} \ ; \ \forall k=1,n \ , \tag{1}$$

where the j_k indices can be incremented or decremented respectively in steps of length s_k. The index n is the *dimension of the NSS*, that is: the number of summation symbols embedded in the operator, and thus the dimension of the involved vectors j, i, f and s. The set of all the vectors appearing as arguments of the NSS can be named *parameters of the NSS*.

The logical vector L is of the type $\{\delta(L_i)\}$. The delta symbol corresponds to a LKD. In this manner, the indices of the vector L are 0's or 1's. So, the convention of a NSS stands for the generation of all the possible forms of the index vector j that are attached to the logical vector $L=1=(1,1,...,1)$.

A NSS has a computational implementation we have called a GNDL [1,4]. The Fortran code of the algorithm implementing a GNDL can be found described in **Program 1** below. The GNDL algorithm constitutes the link between the mathematical notation of the NSS and the computer codification of this operator.

2.3 SIMPLIFIED NESTED SUM NOTATION

Despite the general form adopted here to write a NSS, sometimes it is superfluous to explicit all the involved parameters. When this circumstance does occur, some parts of the general form can be dropped in an arbitrary manner. The most important cases are:

a) If the logical vector L is not specified it will mean that all the possible forms of the vector j must be generated with no restriction. In all the remaining text this NSS form will be used.

b) When the step vector increment is irrelevant only the initial and final vector index parameters need to be explicitly written. In this case, the $\Sigma_n(j=i,f)$ notation can be employed. Frequently, the step vector s is a n-dimensional vector such as $s=1$.

c) The same can be said of the final parameter vector **f**, which may be a product by a scalar of vector **1**. As an example the notation: $\Sigma_n(j=1,m1)$, displays the symbol which is constructed by n nested sums, whose indices take the same values within each sum in the interval $\{1,m\}$.

d) When initial, final and increment values are implicit in the nested sum, a simplified symbol such as $\Sigma_n(j)$ may be also used.

e) When the vector dimension n is obvious, then the n subscript can be omitted from the sum, as in: $\Sigma(j=i,f)$, for instance.

2.4 MATHEMATICAL NSS PROPERTIES

Following from NSS's definition, some properties of these operators arise and have to be considered. Here are listed some of them:

a) NSS's can be recognized as *linear operators* with respect to any general expression placed at the right side of the symbol.

b) A product of two NSS's of dimensions n and m is another NSS of dimension n+m, or:

$$\Sigma_n(j=i,f,s,L)\ \Sigma_m(j'=i'f',s',L') = \Sigma_{n+m}(j\oplus j'=i\oplus i'\ f\oplus f',s\oplus s',L\oplus L'),\qquad (2)$$

where the new index vectors are constructed using the direct sum of the original vectors appearing in the product.

c) The symbol $\Sigma_0(j=i,f,s,L)$ can be made by convention equivalent to the unit operator.

d) The classical summation symbol $\overset{f}{\underset{j=i}{\Sigma}}$ is a particular case of the NSS one, it can be written as: $\Sigma_1(j=i,f,1)$.

e) Einstein's convention, by which a set of nested sums are omitted from an expression, corresponds to obviate writing a NSS like $\Sigma_n(j=1,m1)$.

3. Computational implementation of a NSS: the GNDL algorithm

3.1 GENERAL CONSIDERATIONS

In standard high level language programming the dimension of the NSS: n, signals the number of nested do loops which are necessary to reproduce the structure in a computational environment. But the mathematical usefulness of this entity can be easily recognized when the particular characteristic of this symbolic unit is analyzed: the involved vector parameters could be chosen with *arbitrary and variable dimensions*. There are many scientific and mathematical formulae which will benefit of this property, when written in a paper or computationally implemented.

NSS symbolism constitutes a link between mathematical formalism and program implementation techniques, because successive generation of **j** index vector elements can be programmed in a general but simple way under any high level language. This can be achieved using a unique **do** or **for** loop statement construct, which is *general and independent of the dimension of the involved nested sums*. This kind of programming structure constitutes the GNDL algorithm.

NSS have *not a direct* translation to the usual high level languages. Present day compilers or standard language rules ignore such an interesting feature, see for example the practical final form of the standard Fortran 90 language [5]. Even high level language compilers have no capacity of processing more than a limited number of classical do loops in a nest, for example VAX Fortran and NDP Fortran compilers [6] have a limit of 20 nested do loops. Thus, the GNDL structure is a good candidate to circumvent these limitations in any compiler.

It looks simple to introduce GNDL in the family of repetitive sentences found in high level languages. So we feel that a claim in this direction to language and compiler builders can be made here. Some immediate computational benefits in order to construct *really* general programs may be obtained.

3.2 A SCHEMATIC GNDL PROGRAM EXAMPLE

In order to show in a practical manner the computational implementation of a NSS, **Program 1** represents a Fortran source code corresponding to the NSS structure. The NSS implementation using a GNDL algorithm generates all the possible forms of vector **j**. According to this, **Program 1** generates the indices of the n-dimensional NSS $\Sigma_n(j=i,f,s)$. The dimension n and the initial, final and step index values collected in the vectors **i**, **f**, and **s** have not been specified and the question mark symbol stands for their possible values. These values depend on the concrete application given to the algorithm. Here it is assumed that the step vector **s** has all its components positive definite.

```
          Parameter (n=?) ! Dimension of the NSS
          Integer j(n),i(n),f(n),s(n)
*  < Put initial values on index vectors >
          do k=1,n
             i(k)=?
             f(k)=?
             s(k)=?
             j(k)=i(k)
          end do
*  < Nested sum procedure >
          k=n
          do while (k.gt.0)
             if (j(k).gt.f(k)) then
                j(k)=i(k)
                k=k-1
             else
                call Application(n,j)
                k=n
             end if
             if (k.gt.0) j(k)=j(k)+s(k) ! Step
          end do
          END
```

Program 1: Example of a NSS implementation: $\Sigma_n(j=i,f,s)$. See text for its meaning.

There, *Application* is a called procedure where the n nested loops converge and where their leading indices can be arbitrarily used in the desired internal application. The j index values generation is sequential but the execution of *Application* can be performed into separate CPU's, each one controlling the process attached to one of the forms of the vector j. In this manner, the full computation can be parallelized if desired. In fact, this is a general algorithm, enabling to perform a parallel *Application* implementation if the nature of the problem asks for such a process and the available hardware allows to run it in this manner. A previous tentative description on GNDL, in a sequential programming framework, was initially made by Carbó and Bunge [4].

Various application examples have been constructed by the authors. Some Fortran source codes on combinatorial analysis, product of an arbitrary number of matrices and determinant evaluation in a parallel transputer environment [7] have been tested and encouraging results obtained.

4. Mathematical application examples

As an illustration of the possible use of the described symbols, there will be presented first a set of possible purely mathematical application examples of NSS.

4.1 GENERATION OF VARIATIONS AND COMBINATIONS

A NSS can be used to generate variations and combinations of m elements belonging to an arbitrary set of mathematical objects. It is only necessary numbering in a canonical order, from 1 to m, all the elements in the set. This will produce a completely formal development which can be occasionally used for immediate implementation on any high level language. Although this direct translation will obviously lack of programming refinement in the first bulk program scheme, it may be considered a not too bad starting point in order to obtain a given optimized code.

Then, one can easily describe the expressions that stands for the generation of some combinatorial entities. It is required the implementation of the following NSS: $\Sigma_n(j=1,m1,L)$. Depending on the definition of the logical vector L they are obtained different entities:

a) If L is obviated the NSS then represents the generation of all the m^n possible variations with repetition, which can be formed making groups of n elements out of the m element set.

b) When L is defined as $L(j)=\{\delta(j_p \neq j_q); \forall p \neq q=1,n\}$ then they are reproduced the $m!/(m-n)!$ variations without repetition, which can be formed making groups of n objects taken from the m element set inside the nested sum. It is required the condition $m \geq n$.

c) If L is defined as $L(j)=\{\delta(j_p < j_q); \forall p < q=1,n\}$, then the NSS creates the $m!/n!(m-n)!$ combinations related to a set of m elements, when they are taken in groups of n out of the m element set.

d) Combinations generation can be also performed by means of the implementation of the NSS obviating the logical vector **L** and defining the parameter vectors **i** and **f** as $\mathbf{i}=(1,j_1+1,j_2+1,...,j_{n-1}+1)$ and $\mathbf{f}=(m-n+1,m-n+2,m-n+3,...,m)$, respectively. This last choice implies to rewrite **Program 1** in some special manner, where also the initial indices are modified, while the GNDL is executed.

4.2 EXPLICIT EXPRESSION OF THE DETERMINANT OF AN ARBITRARY SQUARE MATRIX

Using the NSS, one can reformulate the expression which gives the determinant of an arbitrary (n×n) square matrix **A**, $Det|\mathbf{A}|$. A compact formula of $Det|\mathbf{A}|$ can be written in this way as:

$$Det|\mathbf{A}| = \Sigma_n(j=1,n1,L)\ S(j)\ \Pi(j,\mathbf{A})\ , \tag{3}$$

where the logical vector **L** is a function of the **j** vector indices and is defined as:

$$L(j) = \{\ \delta(j_p \neq j_q)\ ;\ \forall p \neq q = 1,n\ \} \tag{4}$$

and the S(j) factor is a sign, which can be expressed by:

$$S(j) = (-1)^{P(j)}\ , \tag{5}$$

being P(j) the parity of the order of the values of the index of the vector **j**. This parity value can be expressed as:

$$P(j) = \delta(\ \dot{2} \neq \sum_{p=1}^{n-1}\sum_{q=p+1}^{n}\delta(j_p > j_q)\ )\ . \tag{6}$$

Finally, the last term in equation (3) is a product of the elements of the matrix:

$$\Pi(j,\mathbf{A}) = \prod_{i=1}^{n} a_{i,j_i}\ . \tag{7}$$

Although this final determinant structure can easily lead to an immediate construction of sequential or parallel Fortran subroutines, there cannot be a claim such that this procedure will be better, from a computational point of view, than well established numerical ones, based on other grounds as Cholesky decomposition, see references [8] for more details. One can recall again the remarks already made at the beginning of section 3.1 above, and stress once more the *formal nature* of the programming immediate translation capabilities of NSS's.

However, the previous determinant development form can be used as a very general interpretative formula, which can compete pedagogically and practically with other widespread alternatives, for example these usually employed in Quantum Chemistry, see reference [9] as a guide. In section 5 this determinant form is used, for example, to deal with Slater determinants.

One can easily see that, despite all criticisms which can arise from the programming technical side, the nested sum formalism permits to solve in a very elegant manner the following problem: *Program in a chosen high level language a function procedure which can be used to compute the determinant of a general square matrix using the direct Laplace determinant definition* [10].

4.3 TAYLOR SERIES EXPANSION OF A n-VARIABLE FUNCTION

The complete formula for the Taylor series expansion attached to a n-variable function $f(x)$ in the neighbourhood of the point x_0, possess the following peculiar simple structure when using NSS's:

$$f(x) = \sum_{m=0}^{\infty} \frac{1}{m!} \sum_m (j=1,n1) \ \Pi^{(m)}(j, x-x_0) \ \partial^{(m)}(j)[f(x_0)] \ . \tag{8}$$

The $\Pi^{(m)}(j,z)$ terms are defined by means of the product:

$$\Pi^{(m)}(j,z) = \prod_{i=1}^{m} z_{j_i} \ ; \ m \neq 0 \quad \wedge \quad \Pi^{(0)}(j,z) = 1 \ . \tag{9}$$

Finally, $\partial^{(m)}(j)[f(x_0)]$ is a short symbol expressing the m-th order partial derivative operators, acting first over the function $f(x)$ and then, the resultant function, evaluated at the point x_0. The differential operators can be defined in the same manner as the terms present in equation (9), but using as second argument the nabla vector:

$$\partial^{(m)}(j) = \Pi^{(m)}(j,\nabla) \quad ; \ m \neq 0 \quad \wedge \quad \partial^{(0)}(j) = \hat{1} \ . \tag{10}$$

The expression (8) is very useful in the sense one can control the series truncation. This is so because the parameter m gives the order of the derivatives appearing in the expansion.

Although there are some general textbook approaches to equation (8), see reference [11] for example, we have not found the expression of the Taylor expansion in full as simple as it has been presented here. Moreover, many potential Taylor expansions are used in various physical and chemical applications; for instance in theoretical studies of molecular vibrational spectra [12] and other quantum chemical topics, see for example reference [13]. Then, the possibility to dispose of a compact and complete potential expression may appear useful.

5. Quantum chemical application examples

Several Quantum Chemical application examples of NSS's follow. Some of them had been chosen because they are related to the actual research in this field in our Laboratory.

We do not pretend to give here an exhaustive account of all the possible applications of NSS's into Quantum Chemistry. Some areas, which for sure can be studied from the nested summation point of view, like the Coupled Cluster Theory [14], are not included here.

In fact, our interest in the present formulation, the use of NSS's and LKD's, has been aroused when studying the integrals over Cartesian Exponential Type Orbitals [1a,b] and Generalized Perturbation Theory [1d,e]. The use of both symbols in this case has been extensively studied in the above references, so we will not repeat here the already published arguments. Instead we will show the interest of using nested sums in a wide set of Quantum Chemical areas, which in some way or another had been included in our research interests [1c].

5.1 SLATER DETERMINANTS

As it is shown in section 4.2, using NSS terminology, the general expression for any determinant can be obtained. In this manner, this formulation can be transferred into the Slater determinants [9], constructed by n spinorbitals associated to n electrons. Adopting the following structure and notation for unnormalized Slater determinants:

$$D(n) = D(1,2,...,n) = Det|\psi_1\psi_2...\psi_n| = \Sigma_n(j=1,n1,L) \ S(j) \ \Psi(j) \ , \qquad (11)$$

where the logical vector L, defined in equation (4), is needed in order to obtain all the variations without repetition of the values of the vector j indices. Here, a term constructed by means of spinorbital products is present:

$$\Psi(j) = \prod_{k=1}^{n} \psi_{j_k}(k) \ . \qquad (12)$$

A similar definition of the symbol (12) can be taken into account, just using the products of $\psi_k(j_k)$.

The Slater determinant expression of equations (11) and (12) will be taken as implicit in this paper from now on.

An operator, depending of an arbitrary number of electron coordinates, has an easily expressible set of matrix elements, using two Slater determinants $D(j)$ and $D(k)$.

The term $D(j)$ can be taken as a Slater determinant, formed by n functions chosen from a set of m available spinorbitals, and ordered following the actual internal values of the j index vectors. That is:

$$D(j) = Det|\psi_{j_1}\psi_{j_2}...\psi_{j_n}|, \tag{13}$$

where the usual abbreviated form for a Slater determinant has been used as in equation (11). Both determinants $D(j)$ and $D(k)$ can be considered built up in the same manner. The number of different spinorbitals appearing in both determinants, can produce a zero result for the matrix element, as it is well known for one and two electron operators, see reference [9]. Generalization to integrals over any number of electrons can be performed as follows.

Suppose a r-electron operator to be written as $\Omega(r)$, with the r-dimensional vector **r** representing the coordinates of the canonically ordered r ($\leq$n) electron set: $(r_1,r_2,...,r_r)$. The matrix element between two Slater determinants can be written as:

$$<D(j)|\Omega(r)|D(k)> = \Sigma_n(p=1,n1,L(p))\ \Sigma_n(q=1,n1,L(q))\ S(p)\ S(q)$$
$$\cdot <\Psi(j[p])|\Omega(r)|\Psi(k[q])> , \tag{14}$$

where the symbol $j[p]$ means that a permutation **p** has been performed over the parameter vector **j** subindices. Here must be noted that the expression (14) above can be written with a unique summation symbol, using the property outlined in equation (2). Then, the integral over the spinorbital products, appearing as the rightmost term of equation (14) can be now simplified. Because in the spinorbital products appearing in equation (11), the canonical ordering of the electrons is preserved by convention in equation (12), as discussed before, one can write the integral using only the first r spinorbitals of the successive products, which will be the ones connected with $\Omega(r)$, the r-electron operator:

$$<\Psi(j[p])|\Omega(r)|\Psi(k[q])> = <\prod_{i=1}^{r} \psi_{j[p_i]}(i)|\Omega(r)|\prod_{i=1}^{r} \psi_{k[q_i]}(i)>$$
$$\cdot \delta(j[p_j]=k[q_j] ;\ \forall j=r+1,n) . \tag{15}$$

The logical Kronecker delta, which appears when integration is performed over the coordinates of the remaining n-r electrons, can be easily substituted by the equivalent logical expression:

$$\delta(j[p_j]=k[q_j] ;\ \forall j=r+1,n) = \delta(\| j[p]-k[q] \|_1 = \sum_{i=1}^{r} |j[p_i]-k[q_i]|) , \tag{16}$$

where the Minkowski norm of the difference, between the permuted vectors $j[p]$ and $k[q]$, must be equal to the sum of the absolute values of the differences between the first r-th components of both vectors.

The right hand part of the last equality (16), may be substituted in equation (15) and the resulting formula transferred into the expression (14). The final result indicates fairly well one can have *at least r differences* between the spinorbitals involved in constructing both determinants in order that the integral becomes not automatically zero. This result encompass the well described zero-, one- and two-electron operator cases [9], generalizing in this way the rules governing the calculation of operator matrix elements between two Slater determinants. One can say that the general rule in order to prevent automatic integral nullity is: "*r-electron operators allow a maximal amount of r spinorbital differences*". This rule is connected to the Brillouin theorem [15].

The same expression can be used with the appropriate restrictions to obtain matrix elements over Slater determinants made from non-orthogonal one-electron functions. The logical Kronecker delta expression, appearing in equation (15) as defined in (16) must be substituted by a product of overlap integrals between the involved spinorbitals.

5.2 CI WAVEFUNCTIONS

Using the approach already described for combination generation, one can formulate in a short but completely general form the CI wavefunctions [16].

This kind of wavefunctions, in the complete CI framework, as Knowles and Handy [16e] have proved feasible, for a system of m spin-orbitals and n ($\leq$m) electrons can be written within the NSS formalism:

$$\Phi = \Sigma_n(j=1,m1,L) \; C(j) \; D(j) \; , \tag{17}$$

where the logical vector L is defined according to the combinations generation and the terms $D(j)$ are Slater determinants constructed as the one defined in equation (13). The $C(j)$ factors are the variational coefficients attached to each Slater determinant.

Also, an alternative formulation of equation (17) can be conceived if one wants to distinguish between ground state, monoexcitations, biexcitations, ... and so on. Such a possibility is symbolized in the following CI wavefunction expression for n electrons, constructed as to include Slater determinants up to the p-th (p$\leq$n) excited order. One can initially start from n occupied spinorbitals $\{\psi_j\}_{j=1,n}$ and m ($\geq$p) unoccupied ones $\{\varphi_k\}_{k=1,m}$. Then, the CI wavefunction is written in this case as the linear combination:

$$\Phi = \sum_{e=0}^{p} \Sigma_{n-e}(j=1,n1,L(j)) \; \Sigma_e(k=1,m1,L(k)) \; C(j \oplus k) \; D(j \oplus k) \; , \tag{18}$$

where the index e, appearing in the first classical sum, signals the excitation order. That is, for e=0 one has the ground state, for e=1 the monoexcitations are obtained and so on.

In equation (18) the $D(j \oplus k)$ terms are n-electron Slater determinants formed by the spin-orbitals numbered by means of the direct sum: $j \oplus k$ of the vector index parameters attached to the involved nested sums and to the occupied-unoccupied orbitals respectively. That is:

$$D(j \oplus k) = Det| \, \psi_{J_1} \psi_{J_2} \cdots \psi_{J_{n-r}} \phi_{k_1} \phi_{k_2} \cdots \phi_{k_r} \, | \, .$$

(19)

This two general CI function expressions, along with the results obtained in the section 5.1 above, permit to compute the expected value form of any quantum mechanical operator in a most complete general way.

5.3 DENSITY FUNCTIONS

Density functions can be obtained up to any order from the manipulation of the Slater determinant functions alone as defined in section 5.1 or from any of the linear combinations defined in section 5.2. Density functions of any order can be constructed by means of Löwdin or McWeeny descriptions [17], being the diagonal elements of the so called m-th order density matrix, as was named by Löwdin the whole set of possible density functions. For a system of n electrons the n-th order density function is constructed from the square modulus of any n-electron wavefunction attached to the n-electron system somehow.

5.3.1. *Density functions over Slater determinants*

Using a unnormalized n-electron Slater determinant D(j) as system wavefunction, constructed as discussed in section 5.1, then one can write the n-th order density function $\rho^{(n)}(j)$ as:

$$\rho^{(n)}(j) = |D(j)|^2 = \Sigma_n(p=1,n1,L(p)) \; \Sigma_n(q=1,n1,L(q)) \; S(p) \; S(q)$$

$$\cdot \; |\Psi(j[p])><\Phi(j[q])| \; .$$

(20)

A recurrent procedure can be defined in order to obtain the remaining lesser order density functions. The (n-1)-th order density function is obtained from the n-th order one, integrating over the coordinates of the n-th electron (or the first) the right hand side of equation (20). The result is:

$$\rho^{(n-1)}(j) = \int |D(j)|^2 dr_n$$

$$= \Sigma_n(p=1,n1,L(p)) \; \Sigma_n(q=1,n1,L(q)) \; S(p) \; S(q)$$

$$\cdot \; |\Psi(j[p'])><\Psi(j[q'])| \; \delta(j[p_n]=j[q_n])$$

$$= n\Sigma_{n-1}(p'=1,(n-1)1,L(p')) \; \Sigma_{n-1}(q'=1,(n-1)1,L(q')) \; S(p') \; S(q')$$

$$\cdot \; |\Psi(j[p'])><\Psi(j[q'])| \; ,$$

(21)

where the primed index vectors mean that the n-th element has been erased from the initial unprimed vector.

Thus, there is the possible relationship between both n-th and (n-1)-th order density functions:

$$\rho^{(n-1)}(j) = \int \rho^{(n)}(j) \ dr_n \ , \tag{22}$$

It is straightforward to deduce, in general, how to obtain the (n-m)-th term of the sequence:

$$\rho^{(n-m)}(j) = \int \rho^{(n-m+1)}(j) \ dr_{n-m+1} = \int \rho^{(n)}(j) \ dr_{n-m+1}...dr_n \ . \tag{23}$$

The zero-th order term being, finally, the norm of the Slater determinant, which by means of equation (23) becomes n!, a well known result.

Generalization of this one determinant function to linear combinations of Slater determinants, defined for example as these discussed in the previous section 5.2, is also straightforward. The interesting final result concerning m-th order density functions, constructed using Slater determinants as basis sets, appears when obtaining the general structure, which can be attached to these functions, once spinorbitals are described by means of the LCAO approach.

5.3.2.   *LCAO expression of density functions*

Taking into account equation (23), and supposing the Slater determinants normalized, one can write, calling the initial constant factor $v(n,m)=1/(n-m)!$:

$$\rho^{(n-m)}(j) = v(n,m) \ \Sigma_{n-m}(p=1,(n-m)1,L(p)) \ \Sigma_{n-m}(q=1,(n-m)1,L(q))$$

$$\cdot \ S(q) \ S(q) \ |\Psi(j[p])><\Psi(j[q])| \ , \tag{24}$$

and using the LCAO approach for the spinorbitals, written as:

$$\psi_k = \sum_{a=1}^{M} c_{ak} \ \chi_a \ , \tag{25}$$

where each spinorbital has been expressed as a linear combination of atomic spinorbitals from a given M-dimensional basis set $\chi=(\chi_1,\chi_2,...\chi_M)$. Then, a product of spinorbitals like (12) can be structured by means of the linear combination (25) as:

$$\Psi(j[p]) = \prod_{i=1}^{n-m} \sum_{a_i=1}^{M} c_{a_i, j[p_i]} \chi_{a_i} = \Sigma_{n-m} \ (a=1,M1) \ C(a,j[p]) \ X(a) \ , \tag{26}$$

where $C(a,j[p])$ and $X(a)$ are products of the coefficients and the basis functions respectively, appearing in the linear combinations (25) for every spinorbital. Now using (26) in the spinorbital product appearing in the rightmost side of (24), one obtains using a simplified NSS notation:

$$|\Psi(j[p])><\Psi(j[q])| = \Sigma(a) \ \Sigma(b) \ C(a,j[p]) \ C^*(b,j[q]) \ |X(a)><X(b)| \ . \tag{27}$$

Finally the density function of (n-m)-th order can be expressed in terms of the atomic spinorbitals as:

$$\rho^{(n-m)}(j) = \Sigma(a) \ \Sigma(b) \ P^{(n-m)}(a,b) \ |X(a)><X(b)| \ , \tag{28}$$

being the (n-m)-th order charge and bond order hypermatrices, $P^{(n-m)}(a,b)$, defined as:

$$P^{(n-m)}(a,b) = \Sigma_{n-m} \ (p=1,(n-m)1) \ \Sigma_{n-m}(q=1,(n-m)1)$$
$$\cdot \ \Omega^{(n-m)}(p,q) \ C(a,j[p]) \ C^*(b,j[q]) \ , \tag{29}$$

using the hypermatrix elements:

$$\Omega^{(n-m)}(p,q) = v(n,m) \ S(p) \ S(q) \ . \tag{30}$$

The equation (28) has the same structure as the well known LCAO form of the first order density function [9]. Thus, it can be concluded that density functions of any order exhibit the same formal structure. In this manner, it can be seen that NSS's lead to an interesting mnemotechnical rule.

5.4 PERTURBATION THEORY

In order to define the notation which we will use from now on, let us consider the application of the perturbation theory to a system which has a perturbed hamiltonian H composed by an unperturbed one, H^0, plus a perturbation operator λV, where $\lambda \approx 0$:

$$H = H^0 + \lambda V. \tag{31}$$

From here, the goal consists to find the eigenvalues and the eigenvectors of the perturbed system, which we denote as the sets $\{E_i\}$ and $\{ | i> \}$ respectively. That is, the target is focused into solving the eigenvalue problem:

$$H|i> = E_i|i>. \tag{32}$$

The eigenvalues and eigenvectors of the unperturbed hamiltonian are assumed to be known:

$$H^0|0;i> = E_i^{(0)}|0;i> \tag{33}$$

and the ket $|0;i>$ stands for the unperturbed eigenfunction of the i-th state and $E_i^{(0)}$ is the corresponding energy. Also it is assumed that this system has an energy spectrum with a simple structure.

The perturbed energies for the i-th state can be expressed as:

$$E_i = \sum_{n=0}^{\infty} \lambda^n E_i^{(n)} , \tag{34}$$

and the corresponding wavefunction is:

$$|i> = \sum_{n=0}^{\infty} \lambda^n |n;i> , \tag{35}$$

where the index n signals the correction order in expressions (34) and (35).

On the other hand, the n-th order energy correction can be written using the form:

$$E_i^{(n)} = <i;0|V|n-1;i> ; \quad n>0 , \tag{36}$$

provided that the orthogonality condition holds between the unperturbed state wavefunction and the corrections of any order:

$$<i;0|n;i> = \delta(n=0) , \tag{37}$$

where $\delta(n=0)$ stands for a LKD.

5.4.1.   *Brillouin-Wigner perturbation theory*

In the Brillouin-Wigner perturbation formalism, the following identity is used [18]:

$$H^0|n;i> + V|n-1;i> = E_i|n;i> + E_i^{(n)}|0;i> . \tag{38}$$

Combining equations (36) and (38) it can be easily found that the n-th order wavefunction correction is given by [18]:

$$|n;i> \; = \; \sum_{J_1}'\sum_{J_2}'...\sum_{J_n}' \frac{|0;j_1> \, U_{J_1J_2}U_{J_2J_3}...U_{J_n0}}{(E_i-E_{J_1}^{(0)})(E_i-E_{J_2}^{(0)})...(E_i-E_{J_n}^{(0)})} \quad ; \; n>0 \; , \tag{39}$$

being the vector $|0;i>$ defined in equation (33) and where the terms $U_{ij}=<i;0\,|\,V\,|\,0;j>$ constitute the representation of the perturbation operator V within the characteristic basis set of the unperturbed hamiltonian H^0. In equation (39) the primed summation symbols are attached to sums performed over all index values except the i-th.

The n-th order correction for the energy takes the form [18]:

$$E_i^{(n)} \; = \; \sum_{J_1}'\sum_{J_2}'...\sum_{J_{n-1}}' \frac{U_{iJ_1}U_{J_1J_2}...U_{J_{n-1}i}}{(E_i-E_{J_1}^{(0)})(E_i-E_{J_2}^{(0)})...(E_i-E_{J_{n-1}}^{(0)})} \quad ; \; n>1 \; , \tag{40}$$

being $E_i^{(0)}$ and $E_i^{(1)}=U_{ii}$ defined in equations (33) and (36) respectively.

Equations (39) and (40) can be rewritten using the NSS formalism. The corrections for the wavefunction take now the simple form:

$$|n;i> \; = \; \sum_n (j=1,\infty 1,L)R_i(j)|0;i> \quad ; \; n>0 \; , \tag{41}$$

and the corrections over the energies are expressed by equation (36).

In equation (41) the vectors $\mathbf{1}$ and $\mathbf{L}$ are n-dimensional and $\mathbf{L}$ components are LKD's of the type $\{\delta(j_k\neq i) \; ; \; \forall k=1,n\}$. The operator $R_i(j)$ is written as:

$$R_i(j) \; = \; \prod_{p=1}^{n} Z_{iJ_p} \; , \tag{42}$$

where $Z_{p,q}$ is a projector-like operator defined in turn as:

$$Z_{p,q} \; = \; |0;q><q;0|V(E_p-E_q^{(0)})^{-1} \; . \tag{43}$$

Thus, one can see NSS as a useful device which permits to write in a compact manner equations (39) and (40). Also it allows to easily obtain these formulae by means of the NSS straightforward implementation, the GNDL algorithm.

5.4.2.   General Rayleigh-Schrödinger perturbation theory

As it can be seen in equation (41), the NSS notation permits to write some equations in an elegant and compact manner. This is due to the fact that NSS opens a new door in order to obtain algebraic expressions. In this sense we propose that the use of NSS as an ideal framework to construct a *really general perturbation theory* scheme. Next discussion will try to prove this.

Let us write a perturbed hamiltonian by a set of k independent perturbation operators using the following expression involving a NSS:

$$H = \Sigma_k(p=0,K) \ \lambda(p) \ H(p) \ , \tag{44}$$

where the vectors **s** and **L** of the NSS are omitted, assuming that **s=1** and all the possible forms of vector **p** have to be generated. In equation (44) the first parameter vector value gives the unperturbed hamiltonian **H(0)**, thus the convention $\lambda(0)=1$ must hold, and any other vector index **p** structure generates a set of perturbation operators $\{H(p); p\neq 0\}$. The final parameter vector **K** contains the maximal order of the perturbation, which can be different for every operator. The symbol $\lambda(p)$ is an element of the scalar set of perturbation parameters. Both **H(p)** and $\lambda(p)$ can be considered products of perturbation operators and the attached parameters.

That is:

$$H(p) = \prod_{i=1}^{k} H_i^{(p_i)} \tag{45}$$

and

$$\lambda(p) = \prod_{i=1}^{k} \lambda_i^{p_i} \ . \tag{46}$$

The adequate technique here is to substitute the usual Rayleigh-Schrödinger scalar perturbation order by a *vector perturbation order* **n**.

The perturbed energies and wavefunctions for the i-th system state can be expressed in a similar way as in scalar perturbation theory:

$$E_i = \Sigma_k(n=0,\infty 1) \ \lambda(n) \ E_i(n) \tag{47}$$

and

$$|i> = \Sigma_k(n=0,\infty 1) \ \lambda(n) \ |n;i> \ , \tag{48}$$

being the expressions (47) and (48) the generalization of equations (34) and (35) respectively.

Substituting equations (44), (47) and (48) into the perturbed Schrödinger secular equation produces the **n**-th order equation:

$$\Sigma_k(p \oplus q, \delta(p+q=n)) \ H(p) \ |q;i> = \Sigma_k(p \oplus q, \delta(p+q=n)) \ E_i(p) \ |q;i> \ , \tag{49}$$

which when **n=0** yields the unperturbed Schrödinger equation.

Thus, the n-th order energy correction for the i-th system's state can be written as:

$$E_i(n) = \Sigma_k(p{=}0,n,\delta(p{\neq}0)) \, <i;0|H(p)|n{-}p;i> , \tag{50}$$

provided that the orthogonality condition:

$$<i;0|p;i> = \delta(p{=}0) \tag{51}$$

holds between the unperturbed state wavefunction and their perturbation corrections up to any order.

The wavefunction corrections can be obtained similarly through a resolvent operator technique which will be discussed below. The n-th wavefunction correction for the i-th state of the perturbed system can be written in the same manner as it is customary when developing some scalar perturbation theory scheme: by means of a linear combination of the unperturbed state wavefunctions, excluding the i-th unperturbed state. That is:

$$|n;i> = \Sigma'_j \, a_{ji}|0;i> . \tag{52}$$

Using expression (52) into equation (49), after some straightforward manipulation, one can obtain the equivalent rule in order to construct the n-th order wavefunction correction:

$$|n;i> = \Sigma_k(p,\delta(p{\neq}0)) \, R_i(p) \, |n{-}p;i> , \tag{53}$$

where a set of Resolvent Operators $\{R_i(p)\}$ for the i-th state are easily defined as follows:

$$R_i(p) = Z_i(0) \, (H(p){-}E_i(p)) , \tag{54}$$

with the weighted projector sum $Z_i(0)$ defined in turn as:

$$Z_i(0) = \Sigma'_j \, (E_i(0){-}E_j(0))^{-1} \, P_j(0) , \tag{55}$$

being $\{P_j(0)\}$ the set of projectors over the unperturbed states:

$$P_j(0) = |0;j><j;0| . \tag{56}$$

In this context equations (50) and (53) can be considered forming a completely general perturbation theory for *nondegenerate* systems, although a recent development permits to extend the formalism to degenerate states [1e].

6. Conclusions

A mathematical device, the NSS, which can be related to Artificial Intelligence techniques, has been defined and applied in order to solve or reformulate some quantum chemical problems. This symbol is related to computer formulae generation. It has been shown that by means of the use of NSS's many applications of such symbols can be found in mathematics as well as in Mathematical Chemistry in particular.

Apart of being able to simplify typographical structures, the NSS symbols constitute the basic elements of a completely general framework, allowing to write mathematical formulae, in such a manner that *immediate translation* to any high level programming language is feasible, producing a *complete general* code, which can be kept sequential or parallelized in a simple manner.

Pedagogical and in many cases mnemotechnical formula structures appear to be also deduced at a very generic level as a consequence of the use of this kind of devices.

The obtained mathematical patterns seems to be also fairly well adapted to Artificial Intelligence formula writing programming philosophy.

An assorted set of purely mathematical and Quantum Chemical application examples prove the generalization power and flexibility of this presently described symbolic framework.

When NSS's together with LKD's are adopted as working tools, both structures appear to trigger some sort of thinking machine, in such a way that once a given problem is solved, new study areas immediately appear to be a promising future application field in the focus of the imagination eye.

One can conclude that a *robust and powerful theoretical machinery* has been described, possessing general, far reaching imaginative possibilities.

Perhaps there are hidden in the symbolic limbo other possible similar tools, even better than these described here. We are confident in that this paper will stimulate the research interest in this direction.

7. Acknowledgments

This work is a contribution of the "Grup de Química Quàntica de l'Institut d'Estudis Catalans" and it has been financed by the "Comissió Interdepartamental per a la Recerca i Innovació Tecnològica" of the "Generalitat de Catalunya" through a grant: #QFN91-4206. E.Besalú benefits of a grant of the "Departament d'Ensenyament de la Generalitat de Catalunya".

8. References

1. a) R.Carbó and E.Besalú, *Adv. Quantum Chem.*, **24**, 115-237, (1992)
   b) R.Carbó and E.Besalú, *Can. J. Chem.*, **70**, 353-361, (1992)
   c) E.Besalú and R.Carbó, *Intern.J.Quantum Chem.*, (submitted)
   d) R.Carbó and E.Besalú, *J.Math.Chem.*, (in press)
   e) E.Besalú and R.Carbó, *J.Math.Chem.*, (submitted)

2. a) R.Carbó and J.M.Riera, <u>A General SCF Theory</u>, Springer Verlag. Berlin, 1978.
   b) R.Carbó and O.Gropen, *Adv. Quantum Chem.*, **12**, 159-187, (1980)
   c) R.Carbó,Ll.Domingo and J.J.Peris, *Adv. Quantum Chem.*, **15**, 215-265, (1982)
   d) R.Carbó,J.Miró,J.J.Novoa and Ll.Domingo, *Adv. Quantum Chem.*, **20**, 375-441, (1989)

3. R.Carbó and B.Calabuig, "A project for the development of a computational system, based on PC-compatible computers to be used in Quantum Chemistry teaching and research", pp. 73-90 in: R.Carbó (Editor), <u>Quantum Chemistry, Basic Aspects, Actual Trends. Studies in Physical and Theoretical Chemistry</u>, Vol. **62**, Elsevier, Amsterdam, 1989.

4. R.Carbó and C.Bunge, *PC Actual Magazine*, 124-126, September, (1989).

5. Fortran 90. X3J3 internal document s8.118. Submitted as text for ANSI x3.198-1991. May 1991.

6. a) VAX Fortran Document AA-D034D-TE. Page E28. Digital Equipment Corporation. Maynard. Mass (1984)
   b) NDP Fortran 386 v3.0. User's Manual. Microway Inc. Kingston (1990)

7. Microway. Quadputer2. Owner's Manual. Microway Inc. Kingston (1989)

8. a) J.H.Wilkinson and C.Reinsch, <u>Linear algebra</u>, Springer-Verlag, Berlin, 1971.
   b) R.Carbó and Ll.Domingo, <u>Algebra Matricial y Lineal</u>, McGraw-Hill, Madrid, 1987.

9. R.G.Parr, <u>The Quantum Theory of Molecular Electronic Structure</u>, A.Benjamin Inc., New York, 1963.

10. a) F.Ayres, <u>Theory and Problems of Matrices</u>, Schaum Pub., New York, 1962.
    b) P.A.Horn and Ch.A.Johnson, <u>Matrix Analysis</u>, Cambridge Univ. Press, Cambridge, 1985.

11. a) D.H.Menzel (Editor), <u>Fundamental Formulas of Physics</u>, Vol. **1**, Dover, New York, 1960.
    b) M.R.Spiegel, <u>Mathematical Handbook of Formulas and tables</u>, McGraw-Hill, New York, 1968.

12. S.Califano, <u>Vibrational States</u>, John Wiley & Sons, London, 1976.

13. D.M.Hirst, <u>A Computational Approach to Chemistry</u>, Blackwell Scientific Publications, Oxford, 1990.

14. K.Jankowski, "Electron Correlation in Atoms", pp 1-116 in: S.Wilson (Editor), Methods in Computational Chemistry. Electron Correlation in Atoms and Molecules, Vol. 1, Plenum Press, New York, 1987.

15. G.Berthier, "The three theorems of the Hartree-Fock theory and their extensions ", pp. 91-102 in the same reference as 3.

16. a) B.Roos, "The Configuration Interaction Method", pp. 251-297 in: G.H.F.Diercksen, B.T.Sutcliffe and A.Veillard (Editors), Computational Techniques in Quantum Chemistry and Molecular Physics, D. Reidel Pub. Dordrecht, 1975.
    b) I.Shavitt, "The Method of Configuration Interaction", pp. 189-275 in: H.F.Schaefer (Editor), Methods of Electronic Structure Theory, Vol 3. Plenum Press, New York, 1977.
    c) P.E.M.Siegbahn, "The externally contracted CI method", pp. 65-79 in: R.Carbó (Editor), Current Aspects of Quantum Chemistry 1981, Elsevier, Amsterdam, 1982.
    d) P.E.M.Siegbahn, "The direct CI method", pp. 189-207 in: G.H.F. Diercksen and S.Wilson (Editors), Metyhods in Computational Molecular Physics, D. Reidel Pub., Dordrecht, 1983.
    e) P.J.Knowles and N.C.Handy, *J.Chem.Phys.*, **91**, 2396-2398, (1989)
    f) A.Szabo and N.S.Ostlund, Modern Quantum Chemistry, McGraw-Hill, Inc., New York, 1989.

17. a) P.O.Löwdin, *Phys. Rev.* **97**, 1474-1489, (1955)
    b) R.McWeeny, *Proc. Roy. Soc.*, A **232**, 114-135, (1955)
    c) R.McWeeny, *Proc. Roy. Soc.*, A **253**, 242-259, (1956)

18. C.H.Wilcox (Editor), Perturbation theory and its applications in Quantum Mechanics, John Wiley & Sons, Inc. New York, 1966.

J'espère donner par là une preuve de ce qu'ont avancé des chimistes très distingués, qu'on n'est peut-être pas éloigné de l'époque à laquelle on pourra soumettre au calcul la plupart des phénomènes chimiques.

J. L. Gay-Lussac
Mémoires de Physique et de Chimie,
de la Société d'Arcueil, 31 Décembre 1808

Applications to Physical Phenomena

Some 180 years after ....

Vibrational Modulation Effects on EPR Spectra

V. BARONE, A. GRAND, C. MINICHINO and R. SUBRA
Dipartimento di Chimica, Universita Federico II,Via Mezzocannone 4, 80134 Napoli, Italy
SESAM, CEN Grenoble, BP 85X, F-38041 Grenoble, France

1. Introduction

Hyperfine coupling constants provide a direct experimental measure of the distribution of unpaired spin density in paramagnetic molecules and can serve as a critical benchmark for electronic wave functions [1,2]. Conversely, given an accurate theoretical model, one can obtain considerable information on the equilibrium structure of a free radical from the computed hyperfine coupling constants and from their dependence on temperature. In this scenario, proper account of vibrational modulation effects is not less important than the use of a high quality electronic wave function.

Semirigid molecules can be described in terms of normal modes by well known perturbative treatments [3]. This approach is, however, ill-adapted to treat large amplitude vibrations, in view of their strong curvilinear character and of poor convergency in the Taylor expansion of the potential [4]. These situations demand, especially in the case of lareg (i.e. containing more than four atoms) molecules, some separation between the active large amplitude motions (LAM) and the "spectator" small amplitude ones. On these grounds, the influence of vibrational effects on EPR parameters has been studied at the ab-initio level for a series of radicals [5-14], using different basis sets, correlation expansions, and treatments of vibrational averaging. In our opinion the key limitation of these approaches is their lack of generality. In fact, the use of global internal coordinates and of analytical kinetic energies leads to quite complicated formalisms specific to a reduced class of systems [12-15], unless oversimplified metrics are used [11,13,14]. We have recently proposed a general numerical procedure [16] to treat the nuclear motion taking into the proper account the variation of the reduced mass along any kind of curvilinear LAM. Here we apply this approach to the radicals CH_3 and CF_3, whose inversion motion is governed by quite different potential wells. In order to focus attention on general trends, avoiding specific technical details, we have used a standard polarised basis set (6-311G**) and treatment of correlation (MP2). The more so as for localized pseudo-π radicals, this level of theory appears completely adequate and readily applicable to large systems [17].

2. Methods

All the electronic calculations were performed with the GAUSSIAN/90 [18] and GAUSSIAN/92 [19] codes and the vibrational studies by the DiNa package [16]. Electronic wave functions were generated by the Unrestricted Hartree-Fock (UHF) formalism,

251

Y. Ellinger and M. Defranceschi (eds.), Strategies and Applications in Quantum Chemistry, 251–260.
© 1996 *Kluwer Academic Publishers.*

correlation energy being then introduced by second order many-body perturbation (UMP2) theory [20]. All electrons were always correlated, for we have shown [21] that core electrons play an important role in the calculation of hyperfine coupling constants. The most serious criticism to this approach would be that the wave function consisting of UHF orbitals does not represent a correct spin state of the molecular system under consideration. Since, however, all the computations reported in this study give a very low spin contamination ($0.75 < S^2 < 0.77$) we can expect quite accurate values of spin dependent properties.
Basis set effects were not in the ground of this study, so that the 6-311G** [22] basis set has been chosen as a compromise between reliability and computation times.

Isotropic Hyperfine coupling constants a_N are related to the spin densities $\rho(r_N)$ at the corresponding nuclei by

$$a_N = \frac{8\pi}{3} \frac{g_e}{g_0} g_N \beta_N \, \rho(r_N) \tag{1}$$

where g_e/g_0 is the ratio of the isotropic g value for the radical to that of the free electron, g_N and β_N are the nuclear magnetogyric ratio and nuclear magneton, respectively. In turn, the spin density at nucleus N can be calculated as the expectation value of the spin density operator over the electronic wave function

$$\rho_N = S_z^{-1} \langle \psi | \sum_{v=1} \delta(r_v - r_N) \, s_z(v) \, | \psi \rangle = \rho^\alpha(r_N) - \rho^\beta(r_N) \tag{2}$$

where the index v runs on all electrons, and S_z is the quantum number of the total electron spin (1/2 for radicals).

In the framework of the Born-Oppenheimer approximation, we can speak of a potential energy surface (PES) and of a "property surface", which can be obtained from electronic wave functions at different nuclear configurations. In this scheme, expectation values of observables (e.g. hyperfine coupling constants) are obtained by averaging the "property surface" on the nuclear wave functions. To proceed further, let us introduce a curvilinear path continuously describing the large amplitude motion (LAM) joining two (possibly equivalent) energy minima through a first order saddle point (SP). Next, the path is parametrized in terms of the signed arc length s in mass weighted (MW) cartesian coordinates. The only necessary condition on the path is that it must not contain any translational or rotational component. For the remaining f-1 internal degrees of freedom, {Qi} (which will be referred to as the small amplitude, SA,coordinates) the potential energy contributions are approximated to second order terms along the LA path. These local vibrational coordinates must be orthogonal to the path tangent, to translations and to infinitesimal rotations. In the adiabatic approach [23], the components of the SA coordinates in the space of the MW cartesian coordinates are the eigenvectors of the Hessian matrix from which translations, rotations and path tangent are projected out. The {Qi} are further assumed to adjust adiabatically to the motion along s, thus giving rise to the following effective potential:

$$V_{eff}(\mathbf{n}, s) = V_0(s) + \sum_{i=1}^{f-1} (n_i + \frac{1}{2}) \lambda_i(s) + \Delta V(s) \tag{3}$$

In the above equation $V_0(s) = V(s;Q=0)$, **n** is the array of conserved quantum numbers for the SA modes, and ΔV (neglected in this study) accounts for anharmonic effects and non orthogonality between the path tangent and the energy gradient [16,23]. In fact, the so called intrinsic reaction path (IRP) is always parallel to the gradient, so that the last contribution vanishes [23]. For intramolecular dynamics, however, the distinguished coordinate (DC) approach has the advantage of being isotope independent and well defined also beyond energy minima, while still retaining an almost negligible coupling between the gradient and the path tangent. This model corresponds to the construcion of the one-dimensional path through the optimization of all the other geometrical parameters at selected values of a specific internal coordinate. In the present context, the distinguished coordinate is the out-of-plane angle θ defined in Figure 1. Furthermore, the distance s along the path is set to zero at a suitable reference configuration (in the present case the planar structure where $\theta = 0°$).

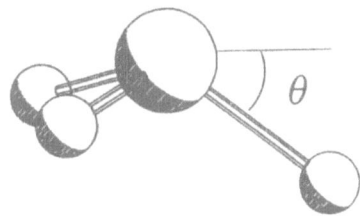

Figure 1: Definition of the out-of-plane angle θ

When the IRP is traced, successive points are obtained following the energy gradient. Because there is no external force or torque, the path is irrotational and leaves the center of mass fixed. Sets of points coming from separate geometry optimizations (as in the case of the DC model) introduce the additional problem of their relative orientation. In fact, the distance in MW coordinates between adjacent points is altered by the rotation or translation of their respective reference axes. The problem of translation has the trivial solution of centering the reference axes at the center of mass of the system. On the other hand, for non planar systems, the problem of rotations does not have an analytical solution and must be solved by numerical minimization of the distance between successive points as a function of the Euler angles of the system [16,24].

In the scenario just sketched, the large amplitude vibration along s is governed by the following equation:

$$\left((\widehat{T}_N(s) + V_{\textit{eff}}(\mathbf{n},s) \right) | j(s) \rangle = \varepsilon_j | j(s) \rangle \tag{4}$$

where $\widehat{T}_N$ is the kinetic energy operator and $|j(s)\rangle$ is a generic vibrational eigenstate with energy ε_j. The so-called vibrationally adiabatic zero curvature (VAZC) approximation is obtained neglecting the small couplings between the path tangent and the local vibrational coordinates appearing in the kinetic energy operator [23,25]. Since the arc length is measured in the space of mass weighted cartesian coordinates, we obtain a Schrödinger equation formally equivalent to that governing the motion of a particle with a unit mass in a one dimensional space. This model is intimately linked to the use of local vibrational basis functions centered at different points along the path. In our approach [16,26,27], cubic splines are used to interpolate the potential along the path and to generate a larger set of equispaced points on which cubic splines are also used as basis functions. This kind of treatment avoids any modelling of the ab-initio data and involves only analytical integrals. Although the size of the basis set is larger than the one necessary when employing Hermite, Morse or Gaussian functions, the spline approach remains competitive since the matrices to be diagonalized are banded with a constant width of 7. Furthermore, no new integrals are introduced by the computation of expectation values of observables (also represented by spline fittings), and the additional computational effort depends on the number of eigenstates to be taken into account, rather than on the dimension of the primitive spline basis set. The expectation value $< O >_j$ of a given observable in the eigenstate j corresponding to the eigenvalue ε_j is given by

$$< O >_j = <j(s)| O(s) |j(s)> \tag{5}$$

The temperature dependence of the observable is obtained by assuming a Boltzmann population of the vibrational levels, so that

$$< O >_T = \frac{\sum_j <j| O |j> exp\left[(\varepsilon_0 - \varepsilon_j)\right]/KT}{\sum_j exp\left[(\varepsilon_0 - \varepsilon_j)\right]/KT} \tag{6}$$

3. Results

Full geometry optimizations and calculations of harmonic force constants were performed at the UMP2/6-311G** level. Although this is not the main concern of this study, it is noteworthy that the relatively unexpansive theoretical treatment we have developped provides structural and spectroscopic parameters in close agreement with experiment (see Table 1). More precisely, the harmonic approximation seems quite adequate for CF_3, whereas strong anharmonicities affect the CH stretchings and the out-of-plane motion of CH_3. The wave number of this latter vibration is increased to 580 cm^{-1} (in much better agreement with the experimental value of 603 cm^{-1}) by our one-dimensional anharmonic treatment. Such a strong positive correction is in agreement with experimental estimates [29]. From another point of view, the two radicals are well suited to point out the influence of the shape of the potential well on vibrational effects: a simple well (CH_3) and a double-well with an high inversion barrier ($CF_3 \approx 120$ kJ mol^{-1}).

The influence of out-of-plane bending on geometrical parameters, electronic energy and coupling constants is shown in Figures 2-4. The linear relationship between the s coordinate and the θ angle is well evidenced in Figure 2a. We recall that in our approach, although the geometries in internal coordinates used to build the path are mass independent, the arc length s varies with the atomic masses, whereas the reduced mass governing the motion always remains unitary. The larger mass of fluorine versus hydrogen then explains the lower slope of the curve θ versus s for CF_3 than for CH_3. Also noteworthy is the increase of the CH and CF bond lengths upon inversion (Figure 2b).

Table 1: Geometrical parameters, wave numbers (cm^{-1}), IR intensities (km mol^{-1}, in brackets) and hyperfine coupling constants (G) of CH_3 and CF_3 radicals.

Parameter	CH_3		CF_3	
	Theor.	Exp.	Theor.	Exp.
CX	1.079	1.079	1.317	1.318
XCX	120.0	120.0	111.35	110.76
a_C (eq)	23.1		264.5	
$<a_C>$	37.3	38.4	263.9	271.6
a_X (eq)	-27.6		137.5	
$<a_X>$	-25.5	-23.1	136.9	142.4
v_1 (a_1)	3180 (0.0)	$\approx$3077	1123 (54.7)	1090
v_2 (a_1)	424 (82.8)	603	717 (16.2)	701
v_3 (e)	3373 (4.1)	3161	1302 (385.0)	1259
v_4 (e)	1447 (3.0)	1383	520 (2.2)	512

The symmetry assignment of vibrational states refers to C_{3v} point group. Experimental geometries and wave numbers are taken from [28,29] for CH_3 and [30] for CF_3. EPR parameters are taken from [31] for CH_3 at 96K and [32] for CF_3.at 77K.

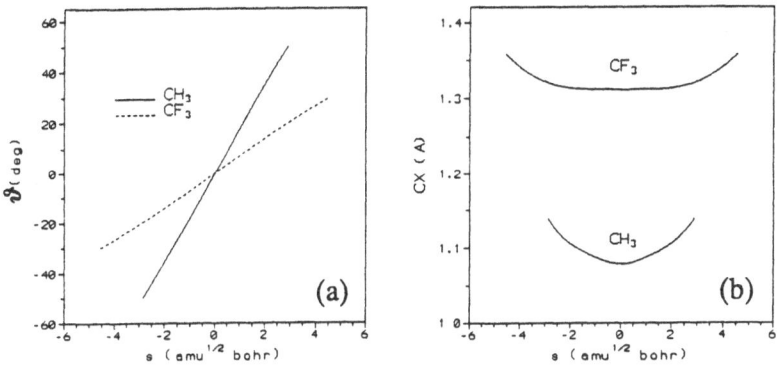

Figure 2: variation of the out of plane angle θ (a) and of the CX bond lengths (b) versus s

It is quite apparent (Figures 3,4) that the hyperfine constants of the central and terminal atoms in the two radicals are strongly influenced by the out-of-plane displacement. For the central atom, the coupling increases with $|\theta|$ and this is clearly related to a strong change in hybridization.

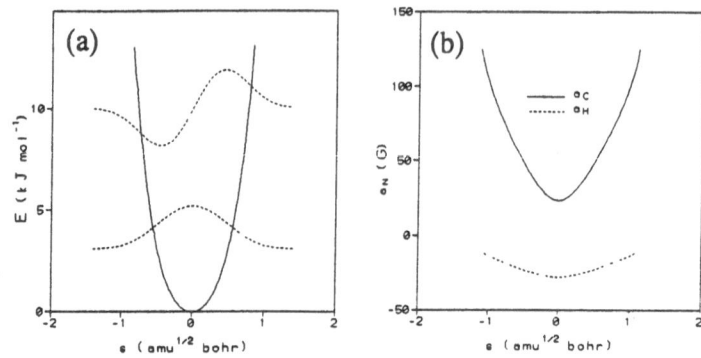

Figure 3: UMP2/6-311G** potential energy profile (a) and hyperfine coupling constants (b) of CH$_3$ versus s. Vibrational wave functions are normalized to 5.

The hyperfine coupling constants of the hydrogens increase smoothly (in absolute value) with inversion, while those of the fluorines show a more complex trend, reaching their maximum value around the equilibrium structure. At his stage, and at the repective equilibrium geometries, the couplings (Table 1) are far from experiment for CH$_3$, but closer to experiment for CF$_3$. More generally, the difference between computed and experimental values is inversely proportional to the height of the potential barrier, the effect being more pronounced at the central atom than at the α ones [33].

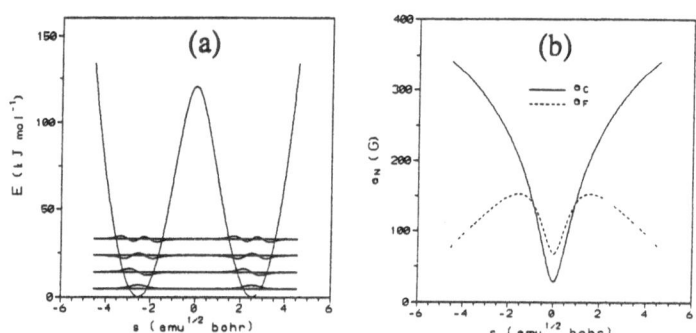

Figure 4: UMP2/6-311G** potential energy profile (a) and hyperfine coupling constants (b) of CF$_3$ versus s. Vibrational wave functions are normalized to 5.

4. Discussion

The similarity in the behaviour of coupling constants as a function of θ in both radicals allows to discuss vibrational averaging effects simply in terms of the potential governing the out-of-plane motion.

The ground vibrational wave function of planar systems (e.g. CH_3) is peaked at the planar structure. Vibrational averaging then changes the coupling constants toward values which would be obtained for an angle $\theta > \theta_{min}$ in a static description. The wave function of the ground vibrational state being symmetrically spread around $\theta = 0°$ introduces contributions of pyramidal configurations. This results in a noticeable increase of the absolute values of the coupling constants, which are minimal at planar structures (see Figure 3). Vibrational averaging then provides hyperfine coupling constants in close agreement with experiment. The effect is even more pronounced in the first excited vibrational state, whose wave function has a node at the planar structure and is more delocalized than the fundamental one, thus giving increased weight to pyramidal structures.

For radicals characterized by a double-well potential (e.g. CF_3) the vibrational effect acts in an opposite direction, bringing the coupling constants to values which would be obtained for $\theta < \theta_{min}$. The ground state vibrational wave function is now more localized inside the potential well, even under the barier, than outside. So it introduces more contributions of internal points. Vibrational effects, while still operative, are less apparent in this case since high energy barriers imply high vibrational frequencies with the consequent negligible population of excited vibrational states and smaller displacements around the equilibrium positions. This explains the good agreement between experimental and static theoretical computations.

Let us now turn to the second parameter, namely the shape of the "property surface". Around reference configurations, the dependence of the hyperfine coupling constants on the inversion motion is well represented by:

$$a(s) = a(s=0) + \left[\frac{\partial a}{\partial s}\right]_{s=0} s + \frac{1}{2}\left[\frac{\partial^2 a}{\partial s^2}\right]_{s=0} s^2 \qquad (7)$$

The average value of a can be written as:

$$<a(s)> = a(s=0) + \left[\frac{\partial a}{\partial s}\right]_{s=0} <s> + \frac{1}{2}\left[\frac{\partial^2 a}{\partial s^2}\right]_{s=0} <s^2> \qquad (8)$$

The mean and mean square values of the LA coordinate s represent the principal anharmonic and harmonic vibrational contributions, respectively [3].
In the case of a planar equilibrium structure, the lineaar term is absent since symmetry constraints impose that $\left[\frac{\partial a}{\partial s}\right]_{s=0} = <s> = 0$. Since, in our case, hyperfine coupling constants reach a minimum value at the planar reference structure (Figures 3b and 4b), the third term is always positive. Vibrational frequencies of this class of molecules are, of course small (Table 1), leading to large mean square amplitudes $<s^2>$, and consequently, to significant corrections to static values computed at the reference structure.
Unless Boltzmann averaging gives significant weight to vibrational states above the barrier, strongly pyramidal molecules like CF_3 can be effectively treated as systems governed by a single well potential unsymmetrically rising on the two sides of the minimum energy configuration. If we shift s so that now $s = 0$ at the equilibrium structure, the difference with the previous case resides in the presence of the linear term in Eq.(8). This is due to the

lack of any constraint on $\left[\frac{\partial a}{\partial s}\right]_{s=0}$ and $<s>$. Although the linear term contributes to $<a>$, it is, anyway, small and, since in our case $<s>$ and the first derivative of coupling constants have opposite signs (see appendix), it conterbalances the harmonic contribution. Thus the resulting correction on $<a>$ is small in all cases. this explains why the static results are very close to the dynamic ones.

5. Summary and conclusion

The results presented in the preceding sections call for the following general remarks.

i) As noticed in the earliest works on EPR [27,28], all the coupling constants increase, in absolute value, with the pyramidality at the radical center, the effect being always much prounounced at the radical cnter than at the surrounding atoms.

ii) Vibrational averagng of coupling constants is always operative, but can be masked by the compensation of effects related to the shape of the potential energy surface from one side, and of the "property surface" from the other.

iii) From a methodological point of view, standard polarized basis sets and limited CI are sufficient to compute hyperfine coupling constantsof localized π-radicals, if large amplitude vibrations are properly taken into account.

The most significant outcome of our study is that a qualitative understanding of vibrational averaging effects is possible along the line of reasoning developed above. This opens the opportunity for a more dynamically based analysis of EPR parameters for large non rigid radicals.

Acknowledgments

The work of V.B. and C.M. was sponsored by the Italian Research Council (CNR Comitato Informatica), whose support is gratefully acknowledged.

Appendix

Using second order perturbation theory [3], the mean and mean square values of the mass weighted coordinate s in the vibrational state $|j\rangle$ with quantum number j are explicitely given by:

$$<s>_j = \frac{-h}{8\pi^2 \omega^3} \frac{\partial^3 V(s)}{\partial s^3} (j + \frac{1}{2}) \tag{A1}$$

$$<s^2>_j = \frac{-h}{4\pi \omega} (j + \frac{1}{2}) \tag{A2}$$

where ω is the harmonic angular frequency

$$\omega = \frac{1}{2\pi} \left(\frac{\partial^2 V(s)}{\partial s^2}\right)^{1/2} \tag{A3}$$

In the above equation, h is the Planck constant, and c the speed of light. The mean values at the absolute temperature T are obtained from the same equations by the of $(j + 1/2)$ by

$$\coth\left(\frac{h\,\omega}{4\pi\,\mathrm{KT}}\right) \tag{A4}$$

where K is the Boltzmann constant. From one side, Eq. (A1) shows that $< s >$ and the cubic force constants $\frac{\partial^3 V(s)}{\partial s^3}$ have opposite signs. On the other side, Figure 4 shows that in the case of CF_3, $\frac{\partial^3 V(s)}{\partial s^3}$ and $\frac{\partial a(s)}{\partial s}$ have the same sign near the equilibrium structure. As a result, the linear term in Eq. (8) is negative, thus counterbalancing the positive quadratic term.

References

1. E. Fermi, Z. *Physik*, **60**, 320 (1930).
2. A. Abragam and M.H.L. Pryce, *Proc. Roy. Soc. (London)* **A205**, 135 (1951).
3. D. Papousek and M.R. Aliev, Molecular Vibrational-Rotational Spectra, Elsevier, Amsterdam (1982).
4. J.T. Hougen, P.R. Bunker, J.W.C. Johns, *J. Mol. Spectr.* **52**, 439 (1970).
5. W. Meyer, *J. Chem. Phys.* **51**, 5149 (1969).
6. S.Y. Chang, E.R. Davidson and G. Vincow, *J. Chem. Phys.* **52**, 5596 (1970).
7. T.A. Claxton and N.A. Smith, *Trans. Faraday Soc.* **66**, 1825 (1970).
8. V. Barone, J. Douady, Y. Ellinger, R. Subra and F. Pauzat, *Chem. Phys. Lett.* **65**, 542 (1979).
9. M. Peric, R. Runau, J. Romelt and S.D. Peyerimhoff, *J. Mol. Spectr.* **78**, 309 (1979).
10. Y. Ellinger, F. Pauzat, V. Barone, J. Douady and R. Subra, *J. Chem. Phys.* **72**, 6390 (1980).
11. D.M. Chipman, *J. Chem. Phys.* **78**, 3112 (1983).
12. P. Botschwinna, J. Flesh and W. Meyer, *Chem. Phys.* **74**, 321 (1983).
13. F. Pauzat, H. Gritli, Y. Ellinger and R. Subra, *J. Phys. Chem.* **88**, 4581 (1984).
14. F. Zerbetto and M.Z. Zgierski, *Chem. Phys.* **139**, 503 (1989).
15. V. Barone, P. Jensen and C. Minichino, *J. Mol. Spectr.* **154**, 252 (1992).
16. V. Barone and C. Minichino, *J. Chem. Phys.* (to be published).
17. V. Barone, C. Minichino H. Faucher, R. Subra and A. Grand, *Chem. Phys.Lett.* (in press).
18. M.J. Frisch, M. Head-Gordon, G.W. Trucks, J.B. Foresman, H.B. Schlegel, K. Raghavachari, M. Robb, J.S. Binkley, C. Gonzales, D.J. DeFrees, D.J. Fox, R.A. Whiteside, R. Seeger, C.F. Melius, J. Baker, R.L. Martin, L.R. Kahn, J.J.P. Stewart, S. Topiol and J.A. Pople, Gaussian 90, Gaussian Inc., Pittsburg (1990).
19. M.J. Frisch, G.W. Trucks, M. Head-Gordon, P.M.W. Gill, M.W. Wong, J.B. Foresman, B.G. Johnson, H.B. Schlegel, M.A. Robb, E.S. Replogle, R. Gomperts, J.L. Andres, K. Raghavachari, J.S. Binkley, C. Gonzales, R.L. Martin, D.J. Fox, D.J. DeFrees, J. Baker, J.P.P. Stewart and J.A. Pople, Gaussian 92, Gaussian Inc., Pittsburg (1992).

20.   C. Moller and M.S. Plesset, *Phys. Rev.* **46**, 618 (1934).
21.   Y. Ellinger, R. Subra, B. Levy and P. Millié, *J. Chem. Phys.* **62**, 10 (1975).
22.   W.J. Hehre, R. Ditchfield and J.A. Pople, *J. Chem. Phys.* **56**, 2257 (1972).
23.   W.H. Miller, N.C. Handy and J.E. Adams, *J. Chem. Phys.* **72**, 99 (1980).
24.   C. Zhixing, *Theor. Chim. Acta* **75**, 481 (1989).
25.   S.M. Colwell and N.C. Handy, *J. Chem. Phys.* **82**, 1281 (1985).
26.   B.W. Shore, *J. Chem. Phys.* **59**, 6450 (1973).
27.   P. Cremaschi, *Mol. Phys.* **40**, 401 (1980).
28.   G. Herzberg, Electronic Spectra and Electronic Structure of Polyatomic Molecules, Van-Nostrand, Princeton (1967).
29.   E. Hirota and C. Yamada, *J. Mol. Spectr.* **96**, 175 (1982).
30.   C. Yamada and E. Hirota, *J. Chem. Phys.* **78**, 1703 (1983).
31.   R.W. Fessenden, *J. Phys. Chem.* **71**, 74 (1967).
32.   R.W. Fessenden and R.H. Schuler, *J. Chem. Phys.* **43**, 2704 (1965).
33.   V. Barone, C. Minichino, A. Grand and R. Subra, to be published.

Ab–initio Calculations of Polarizabilities in Molecules: Some Proposals to this Challenging Problem

M. TADJEDDINE, J.P. FLAMENT
Ecole polytechnique, D.C.M.R., 91128 Palaiseau Cedex, France

1.Introduction

The polarizability expresses the capacity of a system to be deformed under the action of electric field : it is the first–order response. The hyperpolarizabilities govern the non linear processes which appear with the strong fields. These properties of materials perturb the propagation of the light crossing them; thus some new phenomenons (like second harmonic and sum frequency generation) appear, which present a growing interest in instrumentation with the lasers development. The necessity of prediction of these observables requires our attention.

The calculation of the static polarizability, α, is now well documented, actually common enough to give a test for the choice of the atomic basis sets in molecular calculations. On the other hand, few calculations concern the dynamic polarizability, i.e. when the $\hbar\omega$ energy of the electric field is no more zero but can vary and reach the electronic transition energies of the molecule. Computations are more complex; not only they must well describe the ground state in order to reproduce the static polarizability, but also the excited states (valence and Rydberg states) in order to give the resonance energies correctly.

The computation of these observables poses several problems :

- In the case of an electromagnetic perturbation, a first difficulty rises : the choice of the gauge. Indeed the gauge $(\vec{A}; U)$ is only a mathematical tool and the observables of interest (energies, susceptibilities...) must be gauge invariant; they are effectively if the computations use complete molecular bases. Our calculations, using bases unavoidably truncated, will be never gauge invariant. The discrepancies with respect to the gauge invariance is, in a way, a mesure of the quality of our computation, of the molecular basis set. In our calculations the gauge $(\vec{\mathcal{E}}.\vec{r})$ is used.

- Since we must restrict the number, N, of the molecular states used in the computations, what value have we to give to N ? And then, can we correct the obtained value for the polarizability in order to approximate to the exact value by evaluating the ignored terms ?

- The formula which gives the polarizability involves the excited states. As said before, it is necessary to be able to well describe them. The choice of the atomic

Y. Ellinger and M. Defranceschi (eds.), Strategies and Applications in Quantum Chemistry, 261–278.
© *1996 Kluwer Academic Publishers.*

basis set is essential : the bases used in usual calculations are not sufficient; we have to find suitable bases.

The first part of this paper responds to the first two problems through the calculation of the polarizability of CO (1). In this work, we bring our contribution to the three formal challenges enumerated by Ratner (2) in the special issue of *Int. J. Quant. Chem.* devoted to the understanding and calculation of the non linear optical response of molecules :

1. The frequency dependence is taken into account through a "mixed" time-dependent method which introduces a dipole–moment factor (i.e. a polynomial of first degree in the electronic coordinates) in a SCF–CI (Self Consistent Field with Configuration Interaction) method (3). The dipolar factor, ensuring the gauge invariance, partly simulates the molecular basis set effects and the influence of the continuum states. A part of these effects is explicitly taken into account in an extrapolation procedure which permits to circumvent the sequels of the truncation of the infinite sum–over– states.

2. The effects of electron correlation are investigted through the CIPSI (Configuration Interaction with Perturbatively Selected Configurations) calculations (4) of the molecular states.

3. The vibronic coupling features are evaluated in a perturbation treatment by taking account of temperature and electric field dependence (5).

The second part of this paper concerns the choice of the atomic basis set and especially the polarization functions for the calculation of the polarizability, α, and the hyperpolarizabiliy, γ. We propose field–induced polarization functions (6) constructed from the first– and second–order perturbed hydrogenic wavefunctions respectively for α and γ. In these polarization functions the exponent ζ is determined by optimization with the maximum polarizability criterion. These functions have been successfully applied to the calculation of the polarizabilities, α and γ, for the He, Be and Ne atoms and the H_2 molecule.

Throughout, atomic units will be used : $r_0 = 1\ bohr \simeq 5.29177 \times 10^{-11}m$; $E_H = 1\ Hartree \simeq 4.35975 \times 10^{-18}J$. The unit of the dipole moment is equal to $\mu_{au} = q_e r_0 \simeq 8.47836 \times 10^{-30}Cm$; the unit of the dipole polarizability is equal to $\alpha_{au} = q_e^2 r_0^2 E_H^{-1} \simeq 1.64878 \times 10^{-41}C^2 m^2 J^{-1}$ and that of the second hyperpolarizability to $\gamma_{au} = q_e^4 r_0^4 E_H^{-3} \simeq 6.23538 \times 10^{-65}C^4 m^4 J^{-3}$.

2.Calculation of the dynamic polarizability of CO : example of a mixed method

2.1. THEORY

2.1.1. *The sum–over–states approach*

The perturbation theory is the convenient starting point for the determination of the polarizability from the Schrödinger equation, restricted to its electronic part and the electric dipole interaction regime. The Stark Hamiltonian $-\vec{\mu}.\vec{\mathcal{E}}$ describes the dipolar interaction between the electric field $\vec{\mathcal{E}}$ and the molecule represented by its dipole moment $\vec{\mu}$. The perturbed molecular wavefunction is expanded in terms of the complete set of eigenfunctions $(|n\rangle, E_n)$ of the unperturbed molecular Hamiltonian H_0 and the components of the polarizability are given through an expansion over all electronic excited states. For a static external electric field,

$$\alpha_{uv}(0) = -2 \sum_{n\rangle 0} \frac{\langle 0|u|n\rangle \langle n|v|0\rangle}{E_0 - E_n} \tag{1}$$

where u and v run over the cartesian electronic coordinates x, y and z. For an oscillating electromagnetic field characterized by its pulsation ω

$$\vec{\mathcal{E}} = \sum_{\pm} \vec{\mathcal{E}}_0 e^{\pm i\omega t} \tag{2}$$

the dynamic polarizability is derived from the time–dependent perturbation theory :

$$\alpha_{uv}(\omega) = -\sum_{n\rangle 0} \left[\frac{\langle 0|u|n\rangle \langle n|v|0\rangle}{E_0 - E_n - \hbar\omega} + \frac{\langle 0|u|n\rangle \langle n|v|0\rangle}{E_0 - E_n + \hbar\omega} \right] \tag{3}$$

In such an expression, $\alpha_{uv}(\omega)$ must be read as the sum of two functions $(\alpha_{uv}(\omega) = -\epsilon''_{uv}(\omega) - \epsilon''_{uv}(-\omega))$, like

$$\epsilon''_{uv}(\omega) = \langle 0|u|\Phi_1^{v+}\rangle \tag{4}$$

where

$$|\Phi_1^{v+}\rangle = \sum_{n\neq 0} \frac{\langle n|v|0\rangle}{E_0 - E_n - \hbar\omega} |n\rangle = \sum_{n\neq 0} c_n^{v+} |n\rangle \tag{5}$$

The ket $|\Phi_1^{v+}\rangle$ and its counterpart $|\Phi_1^{v-}\rangle$ are calculated as a weighted sum over the $|n\rangle$ excited states; the weight of each state is well defined through its interaction with the $|0\rangle$ ground state by the $\vec{r}$ operator. The function $|\Phi_1\rangle = \sum_{\pm} |\Phi_1^{\pm}\rangle$ represents the first–order perturbed wavefunction whose knowledge is essential in the variation-perturbation treatment. Expression (5) has been proposed by Karplus and Kolker (7).

2.1.2. The polynomial approach

Previously, Kirkwood(8) had suggested another choice: he deduced the first–order perturbed wavefunction from the unperturbed one which was multiplied by a linear combination of the electronic coordinates, i.e. :

$$|\Phi_1^{v\pm}\rangle = g^{v\pm}(\vec{r})|0\rangle \tag{6}$$

with :

$$g^{v\pm}(\vec{r}) = \mathcal{E}_v \sum_u a_u^{v\pm} u \qquad (u = x, y, z) \tag{7}$$

$\mathcal{E}_v$ is the electric field component along v direction and $a_u^{v\pm}$ some constants. In this approach, the polarizability may be calculated very easily from the second-order perturbed wavefunction which is simply given by :

$$E^{(2)} = \langle 0|H^{(1)}|1\rangle \tag{8}$$

For a heteronuclear diatomique molecule of $C_{\infty v}$ symmetry (z being the molecular axis), it becomes :

$$E_{uu}^{(2)} = \sum_{\pm} a_u^{u\pm}(\langle u_i u_j \rangle - \langle u \rangle^2) \tag{9}$$

with :

$$\langle u \rangle = \langle 0 | \left(\sum_{i=1}^{n_e} u_i \right) | 0 \rangle \tag{10}$$

$$\langle u_i v_j \rangle = \langle 0 | \left(\sum_{i=1}^{n_e} u_i \right) \left(\sum_{j=1}^{n_e} v_j \right) | 0 \rangle \qquad (u \quad or \quad v = x, y, z) \tag{11}$$

where n_e is the electron number. As ω tends to zero, the constants $a_u^{u\pm}$ tend to

$$a_u^u = 2 \frac{\langle u_i u_j \rangle - \langle u \rangle^2}{n_e} \tag{12}$$

The normalization condition ($\langle 0 | \Phi_1 \rangle = 0$) imposes to move the origin to the center of electronic charge ($\langle u \rangle = 0, u = x, y, z$); thus, the polarizability may be written very simply in the limit of zero frequency :

$$\alpha_{uu} = 4 \frac{\langle u_i u_j \rangle^2}{n_e} \tag{13}$$

2.1.3. A mixed approach

The idea to combine a method only polynomial (Eq.6 with $g \neq 0$ and $c_n = 0$) with the SCF–CI procedure (Eq.5 with $g = 0$ and $c_n \neq 0$) has been initially developed for the calculation of magnetic observables (9) and later for the electric ones (10). Thus, the first–order perturbed wavefunction is given by :

$$|\Phi_1\rangle = \sum_{\pm} g^{\pm}(\vec{r})|0\rangle + \sum_{\pm, n \neq 0} c_n^{\pm}|n\rangle \tag{14}$$

and the component α_{uv} of the polarizability tensor becomes :

$$\alpha_{uv} = \sum_{\pm, w} a_w^{u\pm}\langle vw \rangle + \sum_{\pm, n} c_n^{u\pm}\langle 0|v|n \rangle \tag{15}$$

The calculations of the $a_w^{u\pm}$ and $c_n^{u\pm}$ constants lead to a system of linear equations similar to that of the SCF–CI method, but with three more lines and columns corresponding to the coupling of the polynomial function with the electric field perturbation. The methodology and computational details have already been discussed (1); we stress two points : the role of the dipolar factor, the nature and the number of the excited states to include in the summation.

2.2. DIPOLAR FACTOR

The dipolar factor $g(\vec{r})$ may be interpreted in terms of gauge invariance. The electric observables usually are calculated in the gauge $(\vec{A} = \vec{0}; U = -\vec{\mathcal{E}}.\vec{r})$. In the change to the gauge $(\vec{A}'; U')$, the Hamiltonian is transformed and the wavefunction $|\Psi\rangle$ becomes (11) :

$$|\Psi'\rangle = e^{i\chi(\vec{r},t)}|\Psi\rangle \qquad (16)$$

If the strength of the electric field is small enough, then :

$$|\Psi'\rangle \simeq (1 + i\chi(\vec{r},t))|\Psi\rangle \qquad (17)$$

As known (11), the gauge invariance is ensured if :

$$\chi(\vec{r},t) = \vec{r}.\vec{A}'(\vec{r},t) \qquad (18)$$

By omitting time–dependent terms, as in the preceding paragraph, the $|\Psi'\rangle$ function may be read as the sum of the unperturbed wavefunction $|\Psi\rangle$ and a term which is the product of this function by a linear combination of the electronic coordinates, i.e. the Kirkwood's $|\Phi_1\rangle$ function. Thus, the $g(\vec{r})$ dipolar factor ensures gauge–invariance.

But the role of the dipolar factor $g(\vec{r})$ in this mixed method is essential on the following point : its contribution in the α computation occurs in a complementary (and sometimes preponderant) way to that calculated only from the $|n\rangle$ excited states, the number of which is unavoidably limited by the computation limits. But before discussing their number, we have to comment the description of these states.

2.3. EXCITED STATES AND EXTRAPOLATION PROCEDURE

In a first approach, Rérat (10) described the $|n\rangle$ excited states of Eq.15 through Slater determinants, $|\phi_m\rangle$, constructed by monoexcitation of the ground state $|0\rangle$ through the $\vec{r}$ monoelectronic operator. By reason of orthogonality (deriving from $\langle 0|\Phi_1\rangle = 0$), all those necessary to the description of $|0\rangle$ were rejected. The lack of such determinants does not allow to have a good description of the excited states when they have a dominant configuration appearing also in $|0\rangle$. If this approach led to interesting static results with reduced basis sets, it could not reach the resonances correctly.

It is the reason for which the Slater determinants have been replaced by the $|n\rangle$ kets accounting for the true spectral states $|\Psi_n\rangle$ (1). These states have been computed independently by the CIPSI (4) program which treats the electronic correlation. Preliminary calculations of energies have been made by the standard CIPSI algorithm (4a) on small S subspaces of c.a. 400 determinants. Perturbation treatments involving larger subspaces (about 1000 for CO) have been achieved using the diagrammatic version of CIPSI (4b).

The quality of the $|\Psi_n\rangle$ states has been tested through their energy and also their transition moment. Moreover from the natural orbitals and Mulliken populations analysis, we have determined the predominant electronic configuration of each $|\Psi_n\rangle$ state and its Rydberg character. Such an analysis is particularly interesting since it explains the contribution of each $|\Psi_n\rangle$ to the calculation of the static or dynamic polarizability; it allows a better understanding in the case of the CO molecule : the difficulty of the calculation and the wide range of published values for the parallel component while the computation of the perpendicular component is easier. In effect in the case of CO :

- If the excited state is a valence state, without Rydberg character, its contribution to the polarizability may be important and sometimes essential. This is the case of the first $^1\Pi$ state.

- If the excited state presents an important Rydberg character, its contribution is very weak and is even negligible. For instance, this is the case for all the states $^1\Sigma^+$ states but the 7 and 8 ones.

Figure 1: Static parallel polarizability vs dipolar factor for CO. Solid(dashed) line with(without) dipolar factor, from Ref. 1

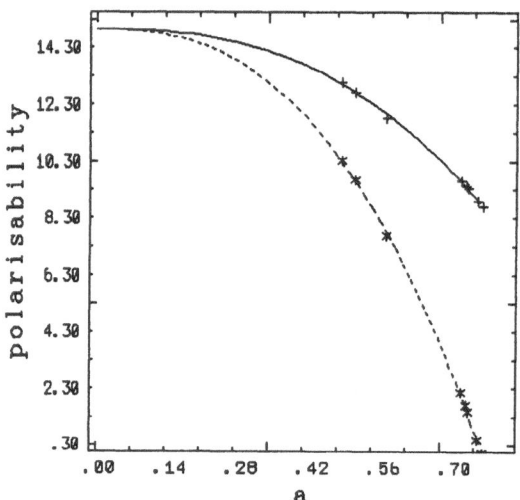

To summarize, if the low-lying states connected to the ground state by allowed dipole transition are not valence states but present a predominant Rydberg character, we have to introduce a lot of $|\Psi_n\rangle$ states; if not, the value of dynamic polarizability near the first resonance is poor.

To circumvent this difficulty, we have developed a procedure which allows us to reach an extrapolate value of α from a finite number N of low-lying true spectral states $|\Psi_n\rangle$.

We have shown that (1) :

1. Within the limits of a complete molecular basis set with exact $|\Psi_n\rangle$, the $a_w^{u\pm}$ coefficients of $g^{\pm}(\vec{r})$ tend to zero as N becomes infinite.

2. We can write an extrapolation formula of the form

$$\alpha_N - \alpha_N^{(0)} = ca_n^P \qquad (19)$$

where α_N and $\alpha_N^{(0)}$ are the polarizabilities calculated with N states, with and without the "polynomial" contribution. The value of the exponent p is determined by a least–square fit and then the extrapolated polarizability is obtained by a linear regression. In the case of the dynamic polarizability, this extrapolation is done separately for the cases $+\hbar\omega$ and $-\hbar\omega$.

Figure (1) gives an illustration of this extrapolation procedure for the calculation of the static parallel polarizability in CO. In this case the extrapolated value ($\alpha = 14.96$) was obtained with the following equation $\alpha_{zz} - \alpha_{zz}^{(0)} = 15.94a^{2.538}$.

It is important to underline two points :

- The extrapolation procedure rests upon the hypothesis of exact or very accurate eigenstates $|n\rangle$ which in practical calculations is seldom the case for the large molecules. The function $g(\vec{r})$ partly compensates the weakness of the atomic and molecular basis sets with the extrapolation procedure.

- This extrapolation has been obtained with a finite number N (usually less than 10) of spectral states lying under the first ionization potential; thus, the continuum is not taken into account explicitly in our calculations. It has been simulated through the $g(\vec{r})$ function and the extrapolation procedure as we are going to show it.

2.4.   CONTINUUM CONTRIBUTION

Table 1: Partial sums of discrete series for static polarizabilities of hydrogen (from Ref. 13)

N	$\alpha_{1s}^d(N)$	$\alpha_{1s}^{d,G}(N)$
0	0.000000	4.000000
1	2.959621	4.316847
10	3.639246	4.398945
100	3.662954	4.401826
200	3.663181	4.401853
300	3.663224	4.401858
400	3.663238	4.401860
500	3.663245	4.401861

Hydrogen atom, in its ground state, can be treated in an entirely analytic approach. The calculation of the second–order perturbed energy gives the well known values :

$$E_{1s}^{(2)} = -\frac{9}{4} \quad \text{and} \quad \alpha_{1s} = 4.5$$

for the static polarizability of the ground state. Since we have used exact analytic wavefunctions which are the eigenstates of the electronic Hamiltonian, the continuum contribution has been taken into account, i.e. α_{1s}^c.

On the other hand, the static polarizability can be calculated by a sum–over–states on the spectral states $|\Psi_n\rangle$; the discrete series $\alpha_{1s}^d(N)$ converge on a value defined by :

$$\alpha_{1s}^d(\infty) = \sum_{n \geq 2}^{N \to \infty} A_{n,1s} \tag{20}$$

Tanner and Thakkar (12) have obtained $\alpha_{1s}^d(\infty) = 3.66325789$. Then it is possible to deduce the continuum contribution $\alpha_{1s}^c = 0.8367$, i.e. about 18.6% of the total electronic polarizability.

In order to demonstrate the efficiency of the $g(\vec{r})$ function in the calculation of the polarizability, Rérat et al. (13) have carried out the calculation of the polarizability for the ground state of the hydrogen atom. This computation has been made with $(\alpha_{1s}^{d,G}(N))$ and without $(\alpha_{1s}^d(N))$ the dipolar factor, versus the N number of the spectral $|\Psi_n\rangle$ states involved in the calculation. The convergence of such series $\alpha_{1s}^{d,G}(N)$ and $\alpha_{1s}^d(N)$ leads to discrete values of 4.4018 and 3.6632 (i.e. the result of Tanner and Thakkar) corresponding respectively to 97.8% and 81.4% of the exact value. This result illustrates the fact that a large part of the continuum contribution is simulated through the use of the dipolar factor. Moreover the convergence of the series $\alpha_{1s}^{d,G}(N)$ is faster as we can see on table 1.

At last, the extrapolation procedure employed in that calculation gives the final $\alpha(N \to \infty)$ value to be 4.503, i.e. 0.07% above the exact static value of α.

Such a calculation with exact wavefunctions shows :

- the precise role and the rigorous contribution of the $g(\vec{r})$ function in particular for the continuum

- the efficiency of the extrapolation procedure to obtain accurate values.

2.5. VIBRONIC CORRECTIONS

The theoretical method, as developed before, concerns a molecule whose nuclei are fixed in a given geometry and whose wavefunctions are the eigenfunctions of the electronic Hamiltonian. Actually, the molecular structure is vibrating and rotating and the electric field is acting on the vibration itself. Thus, in a companion work, we have evaluated the vibronic corrections (5) in order to correct and to compare our results with experimental values.

In the particular case of diatomic molecules, the molecular geometry can be described by the reduced coordinate

$$\xi = \frac{R - R_e}{R_e} \tag{21}$$

where R is the internuclear distance; R_e, its equilibrium value in the electronic ground state. Energy and each component of the polarizability may be written as a power series in the reduced coordinate ξ around their equilibrium values :

$$E(\xi) = a_0\xi^2(1 + a_1\xi + ...) \tag{22}$$

$$\alpha(\xi) = \alpha_e + \alpha_e'\xi + \frac{1}{2}\alpha_e''\xi^2 + ... \tag{23}$$

where a_0 and a_1 are the well known Dunham constants and α'_e and α''_e the values of the first and second derivatives of the polarizability calculated at the equilibrium geometry.

By including the effect of the rotation (at a given T temperature, for the level $v = 0$) in a perturbation calculation, we have obtained (5) :

$$\Delta\alpha(T) = (\frac{kT}{a_0} - \frac{3}{4}ca_1)\alpha'_e + \frac{c}{4}\alpha''_e \qquad (24)$$

where we have introduced the dimensionless constant c like :

$$c = \frac{\hbar}{R_e}(2ma_0)^{-1/2} \qquad (25)$$

which characterizes the molecule (m is its reduced mass).

Moreover, for the observables depending on external electric field, its specific effect has to be investigated : the electric field induces new terms in the nuclear Hamiltonian, due to the change of equilibrium geometry and the nuclear motion perturbation. Pandey and Santry (14) has brought to the fore this effect and calculated the correction which only concerns the parallel component. It is represented by the following expression :

$$\Delta\alpha(\mathcal{E}) = \frac{1}{2a_0}\mu'^2_e \qquad (26)$$

where μ'_e is the value of the first derivative of the dipole moment calculated at the equilibrium geometry. On table 2 we have reported the results obtained for the

Table 2: Vibronic correction for the static polarizability of CO (from Ref. 5)

corrections	$\alpha_\perp$	$\alpha_\parallel$
$\Delta\alpha(T)$	0.032	0.136
$\Delta\alpha(\mathcal{E})$	0	0.456
$\Delta\alpha(T) + \Delta\alpha(\mathcal{E})$	0.032	0.592
α	11.26	15.52
exp.[a]	11.86	15.51

(a) from Ref. 15

static polarizability of CO : the correction for the perpendicular component may be neglected; for the parallel component, the vibronic correction mainly originates from the effect of the electric field cannot be neglected at all.

2.6. RESULTS FOR THE POLARIZABILITY OF CO

The quality of electronic calculations is confirmed by the very good agreement of the resonance energies for both α components if we compared to the experimental ones, as shown on table 3.

Table 3: Dynamic polarizability of CO from Ref. 1

discontinuity energy	$\alpha_\perp$	$\alpha_\parallel$
our work (1)	0.304	0.385
exp.[a]	0.296	0.396

(a) from Ref. 16

Moreover, the values obtained for the dynamic polarizability by varying the wavelength (until $\lambda \geq 3511\text{Å}$) are in good agreement with experiment (1). Table 4 resumes the results obtained for the static polarizability of CO :

1. On the first line, we have reported our results (1) obtained with the spectrocopic states, $|\Psi_n\rangle$, the dipolar factor $g(\vec{r})$ and the extrapolation procedure. In order to compare them with the experimental results (last line) we have corrected them by taking into account the vibronic coupling –temperature and electric field dependence– as developed before (second line). The parallel component, α_{zz}, is now in excellent agreement with experiment.

2. The two following lines present the results obtained later by Rérat et al. (17) : the method consists in adding one more term in the expression of $|\Phi_1\rangle$ given by Eq.14. He keeps the dipolar factor; from the summation on the spectroscopic states $|\Psi_n\rangle$, he retains only the first one of the symmetry of interest, thus there is no extrapolation procedure; on the other hand, he adds the Slater determinants $|\phi_m\rangle$ which contribute to the perturbation of the ground state by the operators $\vec{r}$ and $\vec{\nabla}$ and he takes into account the non–orthogonality of the zeroth and first–order perturbed wavefunctions. Their results show an improvement for both α components, in particular for anisotropy.

3. These results are compared with those obtained by Oddershede and Svendsen (18) using SOPPA or Sunil and Jordan (19) using MP4 or a coupled cluster approach, but without vibronic correction.

3.Determination of the polarization functions

In order to overcome the optimization process of the (hyper) polarizabilities calculations, we have been led to deeply study the perturbational and variational methods and in particular the variation–perturbation treatment introduced by Hylleras (20) since 1930. We will not develop here the theoretical framework of the recent study of N. El Bakali Kassimi (21). We propose criteria for generating adequate sets of polarization functions necessary to calculate (hyper) polarizabilities.
As our computations use the HONDO/8 program (22) which is based on the CPHF (Coupled Perturbed Hartree Fock) method (23) we begin by briefly recalling this method.

Table 4: Static polarizability components and anisotropy of CO

method	$\alpha_\perp$	$\alpha_\parallel$	$\alpha^{(a)}$	$\Gamma^{(b)}$
our work $(1)^{(c)}$	11.22	14.96	12.47	3.74
our work $(1)^{(c+e)}$	11.25	15.66	12.72	4.41
Ref. $17^{(d)}$	11.72	15.02	12.82	3.30
Ref. $17^{(d+e)}$	11.81	15.58	13.07	3.77
SOPPA$^{(f)}$			12.45	4.45
MP4 (SDTQ) $^{(g)}$	11.84	15.49	13.06	3.65
CCD $^{(g)}$	11.61	15.52	12.90	3.91
exp.$^{(h)}$	11.86	15.51	13.08	3.65

(a) $\alpha = (2\alpha_\perp + \alpha_\parallel)/3$ (b) $\Gamma = \alpha_\parallel - \alpha_\perp$

(c) calculation using dipolar factor and spectroscopic states with extrapolation procedure

(d) calculation using dipolar factor, spectroscopic states and Slater determinants

(e) with vibronic corrections

(f) from Ref. 18   (g) from Ref. 19   (h) from Ref. 15

3.1. THE CPHF METHOD

The variational theorem which has been initially proved in 1907 (24), before the birthday of the Quantum Mechanics, has given rise to a method widely employed in Quantum calculations. The finite–field method, developed by Cohen and Roothan (25), is connected to this method. The Stark Hamiltonian $-\vec{\mu}.\vec{\mathcal{E}}$ explicitly appears in the Fock monoelectronic operator. The polarizability is derived from the second derivative of the energy with respect to the electric field. The finite–field method has been developed at the SCF and CI levels but the difficulty of such a method is the well known loss in the numerical precision in the limit of small or strong fields. The latter case poses several interconnected problems in the calculation of polarizability at a given order, n :

- The strength of the field must not be so strong that higher order $(m)n$ effects come into play; and then, should we introduce the basis functions suited for order m to get a correct response of the system up to order n, even if we are concerned only with the nth–order ?

- The pointwise energies will be fitted by a Taylor espansion. What must be the order of the expansion ? How much points must be considered ? It is necessary to master the numerical techniques well.

By allowing the direct calculation of the successive derivatives (thus without resorting to any effective value of the field), the perturbation methods offers an elegant alternative. In the stationary perturbation theory, the CPHF is the most known. The CPHF originates in a perturbation development of the spin–orbitals and of their energies on the expansion of the electric field $\vec{\mathcal{E}}$. The polarizability (or the second hyperpolarizability) is derived from the second– (or fourth–) order perturbation energy. The CPHF is akin to the finite–field method on the point that they treat the bielectronic interactions in presence of the electric field in a self coherent way (26). On the other hand, it is basically different with respect to the use of variational principle : while the finite–field method variationaly treats the total energy in presence of the field, the CPHF, by using the perturbation development, allows variational approaches to the calculation of polarizabilities.

3.2. CHOICE OF TRIAL FUNCTION FOR THE POLARIZATION ORBITALS

In all the variational methods, the choice of trial function is the basic problem. Here we are concerned with the choice of the trial function for the polarization orbitals in the calculation of polarizabilities or hyperpolarizabilities. Basis sets are usually energy optimized but recently we can find in literature a growing interest in the research of adequate polarization functions (27).

By returning to the genuine meaning of the word "polarization", we propose polarization functions suited to the calculation of the electric property of interest : our polarization functions belong to the so–called field–induced ones (FIP) (28).

The foundation of our approach is the analytic calculations of the perturbed wavefunctions for a hydrogenic atom in the presence of a constant and uniform electric field. The resolution into parabolic coordinates is derived from the early quantum calculation of the Stark effect (29). Let us recall that for an atom, in a given Stark eigenstate, we have :

$$\alpha = -2E^{(2)} \tag{27}$$

$$\gamma = -24E^{(4)} \tag{28}$$

The calculated perturbed wavefunctions have been rewritten in terms of a combination of normalized Slater orbitals in real form. Ref. 6 gives a detailed illustration for the level $1s$.

At the beginning it is necessary to describe the unperturbed system very well, independently of the polarization functions : Let us assume that the unperturbed system is reasonably well described by using some finite set of basis functions $\{u_i^{(0)}\}$. As shown by Hirschfelder et al. (30) we only need the first–order perturbed function for α : $\{u_i^{(1)}\}$ and the second–order one for γ : $\{u_i^{(1)}\} \oplus \{u_i^{(2)}\}$.

We propose to construct the polarization functions from these perturbed wave functions. The genuine basis set $\{u_i^{(0)}\}$ has to be enriched by :

- the Slater orbitals (STO) which form $\{u_i^{(1)}\}$ in order to calculate α,

- the STO which form $\{u_i^{(1)}\} \oplus \{u_i^{(2)}\}$ in order to calculate γ.

Thus, by following the hydrogenic model, we know not only the kind of angular symmetry but also the value n of the quantum number of the suitable polarization functions. In the case of a true hydrogenic atom these STO appear in a given linear combination. To limit the size of the basis set, one could use an unique polarization orbital which would be the relevant combination (a contracted STO). In fact, the hydrogenic model does not apply exactly to any polyelectronic atom, so we let the coefficients of the combination vary freely, as new variational parameters, in the CPHF equations.

Moreover according to such a model, the same exponent ζ is used in each perturbed wavefunction, keeping the value of the non-perturbed wavefunction.

We propose to keep the same value of ζ in both polarization functions $\{u_i^{(1)}\}$ and $\{u_i^{(2)}\}$ but to dissociate their value from that of the zeroth order basis set $\{u_i^{(0)}\}$ which is taken in the literature once for all so as to describe the system for the best; let be ζ_0 this value. On the basis of the Hylleraas variation principle, we will determine the suitable value for the hydrogenic scale factor ζ in the polarization functions derived from $\{u_i^{(1)}\}$ and $\{u_i^{(2)}\}$ after optimization with respect to maximum polarizability.

Table 5: Polarization functions for α ($u^{(1)}$) and γ ($u^{(2)}$)

n	$|m|$	Stark eigenstate $u^{(0)}$	$u^{(1)(a)}$	$u^{(2)(a)}$
1	0	$|1s\rangle$	2p 3p	3s/3d 4s/4d 5s/5d
2	0	$\frac{1}{\sqrt{2}}[|2s\rangle \pm |2p_z\rangle]$	1s 2s/2p 3s/3d 4p/4d	4s/4f 5s/5p 5d/5f 6d/6f
2	1	$\frac{1}{\sqrt{2}}[|2p_x\rangle \pm i|2p_y\rangle]$	3d 4d	3s 4s 5s/5d
3	0	$-\frac{1}{\sqrt{3}}|3s\rangle + \sqrt{\frac{2}{3}}|3d_{z^2}\rangle$	2p 3p 4p/4f 5p/5f	3s/3d 4s/4d 5s/5d/5g 6s/6d/6g 7d/7g
3	0	$\frac{1}{\sqrt{3}}|3s\rangle \pm \frac{1}{\sqrt{2}}|3p_z\rangle + \frac{1}{\sqrt{6}}|3d_{z^2}\rangle$	1s 2s/2p 3s/3p/3d 4s/4p/4f 5p/5d/5f	5s/5g 6s/6p 6d/6f/6g 7d/7f/7g
3	1	$\frac{1}{\sqrt{2}}[|3p_u\rangle \pm |3d_{zu}\rangle]$ $u = x, y$	2p 3p/3d 4p/4f 5d/5f	5p/5g 6p/6d 6f/6g 7f/7g
3	2	$\frac{1}{\sqrt{2}}[|3d_{x^2-y^2}\rangle \pm |3d_{xy}\rangle]$	4p/4f 5p/5f	5s/5d/5g 6s/6d/6g 7d/7g

(a) STO with the pure spherical harmonics $Y_l^m(\theta, \phi)$

Table 5 presents the results for the first levels ($n = 1, 2, 3$). In this table, $u^{(1)}$ is the basis set to be added to $u^{(0)}$ for the calculation of α and $u^{(2)}$ to $u^{(0+1)}$ for γ. One must note that analytic expressions of $u^{(1)}$ and $u^{(2)}$ are developed over a series of monomials such as $r^{n-1}, r^{n-2}z, r^{n-3}z^2, r^{n-4}z^3$ and $r^{n-5}z^4$ (see Ref. 6). The first two monomials, r^{n-1} and $r^{n-2}z$ correspond exactly to ns and np orbitals. The others are combinations of nd/ns, nf/np and $ng/nd/ns$ respectively. Table 5 gives the orbitals with the pure spherical harmonics $Y_l^m(\theta, \phi)$. In programs using $6d$, $10f$ or $15g$ cartesian functions, only the nd, nf or ng need be given since they include the corresponding ns for the nd shells; np for nf and nd/ns for ng.

Applications to He and H_2 will give a deeper understanding of the determination of the polarization functions.

3.3. APPLICATION TO He AND H_2

Owing to their simplicity, the helium atom and the dihydrogen molecule have been the object of experiments (Ref. 31 for α and γ of He; Ref. 32 for α of H_2) and calculations, some of them near the Hartree–Fock limit (Ref. 33 for He and Ref. 34–36 for H_2). In order to test our polarization functions, we have taken the zeroth order basis set $\{u_i^{(0)}\}$ from the literature so as to describe the system best and our references values are the HF limit for any observable (E, α, γ).

3.3.1. *Helium*

The field–free atom has the configuration $1s^2$; the unperturbed wavefunction is described through a Huzinaga's CGTOs set (the 10 CGTOs one in Ref. 37). We use the $\{2p, 3p\}$ orbitals for $\{u_i^{(1)}\}$ and $\{3s/3d, 4s/4d, 5s/5d\}$ orbitals for $\{u_i^{(2)}\}$. The polarizability has been maximized with respect to the exponent ζ of the STOs of $\{u_i^{(1)}\}$ by using these polarization functions only; we have obtained : $\zeta = 1.38$. Then this value has been given also to the exponent of the STOs in $\{u_i^{(2)}\}$ at the second step of our calculations for the computation of γ.

Table 6: Effect of the polarization functions on the polarizabilities and hyperpolarizabilities of He ($\zeta_{opt}^{(a)} = 1.38$)

	E	α	γ	Ref.
with $u^{(0)(b)}$	−2.861669	0	0	6
with $u^{(0+1)(c)}$	−2.861669	1.3219	−5.4	6
with $u^{(0+1+2)(d)}$	−2.861669	1.3220	36.1	6
H–F limit	−2.86167	1.3222	36.0	33

(a) value of ζ of the polarization function for maximum polarizability

(b) $\{u^{(0)}\} = \{1s\}; 10\,CGTO$ from Ref. 37

(c) $\{u^{(1)}\} = \{2p, 3p\}$

(d) $\{u^{(2)}\} = \{3s/3d, 4s/4d, 5s/5d\}$

Table 6 clearly shows the effect of the polarization functions : the HF limits for energy, α and γ are reached at the first, second and third levels of calculation. Moreover, we have proved (6) that :

- the extension of the basis set does not produce further change, and

- if Table 6 shows the necessity of the d orbitals, any two of the three d shells give satisfactory values of γ.

This last result which will be verified with the following applications is a consequence of our choice for the polarization functions. In effect, the STOs have nodeless radial part and they all combine in phase in $\{u_i^{(2)}\}$ so that the resulting polarization function is also nodeless and can be approximately modeled by only one or two STOs with suitable exponents.

3.3.2. *Dihydrogen*

If the values published for α converge quite well (6.45 for $\alpha_\parallel$ and 4.5–4.6 for $\alpha_\perp$ in Ref. 38–40), nothing similar appears for γ components : 330 (39) $\rightarrow$ 687 (38) for γ_{zzzz}; 183 (39) $\rightarrow$ 214 (40) for γ_{xxzz}; 556 (38) $\rightarrow$ 704 (40) for γ_{xxxx}; such discrepancies exist, though there are actually p and d orbitals, required for α and γ calculations, in all the basis sets used. This evidences the extreme sensibility of γ to the quality of the wavefunction.

Mulliken (41) distinguishes two kinds of polarization. He calls "Coulomb polarization" what we are concerned with in this paper : the polarization produced by an electric field, and he calls "valence polarization" : a kind of polarization du to quantum–mechanical valence forces. In order to correctly describe the chemical bond in H_2, it is necessary to include the "valence polarization" function as soon as one calculates energy with the unperturbed function (i.e. the $2p$ orbital).

Table 7: Polarizabilities and hyperpolarizabilities of H_2 at $R_e = 1.4$ au, with optimization of ζ in polarization functions ($\zeta_{opt}^{(a)} = 1.1$) from Ref. 6

	E	$\alpha_\parallel$	$\alpha_\perp$	γ_{xxxx}	γ_{xxzz}	γ_{zzzz}
with $u^{(0)(b)}$	−1.133318	6.059	1.788			
with $u^{(0+1)(c)}$	−1.133326	6.453	4.606			
with $u^{(0+1+2)(d)}$	−1.133394	6.450	4.612	562	203	653

(a) value of ζ of the polarization function for maximum polarizability

(b) $\{u^{(0)}\} = \{1s, 2s, 2p\}$ with $\zeta_{1s} = 1.378, \zeta_{2s} = 1.176, \zeta_{2p} = 1.820$ from Ref. 35

(c) $\{u^{(1)}\} = \{2p, 3p\}$

(d) $\{u^{(2)}\} = \{3s/3d, 4s/4d, 5s/5d\}$

For this calculation we used the basis set $1s, 2s, 2p$ of Fraga and Ransil (35) which gives near HF limit quality for energy ($E_{HFL} = -1.13362957$ (34)). The polarization functions were derived from the $1s$ orbital only, like in He calculations. Their exponent was optimized using the maximum probability criterion ($\zeta_{opt} = 1.1$). Table 7 presents the obtained results.

Now with the $2p$ valence polarization, it is possible to partly describe the polarizability since the first step of calculation with the unperturbed wavefunction, especially the parallel component which is generally easier to calculate in CPHF. The optimized values of $\alpha_\parallel$ and $\alpha_\perp$ are excellent at the second step with $\{u_i^{(0)}\} \oplus \{u_i^{(1)}\}$.

Owing to the very large discrepancies in the data on γ, we have made new computations with other basis sets $\{u_i^{(0)}\}$ but with the same process; they converge at less than 4% from the previous ones, giving confidence in our results and in our procedure.

3.4. APPLICATION TO Be AND Ne

The two preceding applications showed that our hydrogenic model fits well with the helium atom and the dihydrogen molecule for the determination of the polarization functions except that their exponent ζ is different from ζ_0 which is the exponent of the genuine basis set $\{u_i^{(0)}\}$. It is obvious that the hydrogenic model will fit less and less as the atom will be described by more and more electrons.

Nevertheless our method of α and γ calculations has been successfully extended to Be and Ne atoms (21). Let us resume the principal results :

1. For less than 1% error for α, it is sufficient to "polarize" only the valence electrons in Be; the polarization of the $1s$ orbital leads to an α value within 0.1% of HFL value. Contrary to Be, the polarization of the inner shell is now absolutely negligible for Ne.

2. The transfer of exponent to the $\{u_i^{(2)}\}$ set leads to good values of γ : for Be : 4.05×10^4 instead of the HF limit 3.99×10^4 (42), and for Ne : 68 instead of 70 (42).

3. Contrary to the previous applications (He, H_2) we observe an increase of the values of α with the $\{u_i^{(2)}\}$ polarization function because $\{u_i^{(2)}\}$ contains p functions improving the first order wavefunction which, despite its size, was not at the HF limit.

4. At last, it is possible to still improve the results on γ by using two different values of the exponent :

   - ζ_1 by optimization of α for the STO of $\{u_i^{(1)}\}$
   - ζ_2 by optimization of γ for the STO of $\{u_i^{(2)}\}$

This last result is important for the generalization of our procedure to more complicated systems.

4. Conclusion

The computation of polarizabilities requires consideration of two complementary problems: the computational method and the basis set used.

We have first been concerned with the computational point of view. Through the calculation of the dynamic polarizability of CO, we have developed a method based on the conventional SCF–CI method, using the variational–perturbation techniques : the first–order wavefunction includes two parts (i) the traditional one, developed over the excited states and (ii) additional terms obtained by multiplying the zeroth–order function by a polynomial of first–order in the electronic coordinates. This dipolar factor makes an extrapolation procedure possible in critical cases (for example when the low–lying states are of Rydberg character as the $^1\Sigma^+$ states of CO). This calculation has shown the importance of the basis set and in particular the polarization functions necessary in such computations. We have studied this problem through the calculation of the static polarizability and even hyperpolarizability. The very good results of the hyperpolarizabilities obtained for various systems give proof of the ability of our approach based on suitable polarization functions derived from an hydrogenic model. Field–induced polarization functions have been constructed from the first– and second–order perturbed hydrogenic wavefunctions in which the exponent ζ is determined by optimization with the maximum polarizability criterion. We have demonstrated the necessity of describing the wavefunction the best we can, so that the polarization functions participate solely in the calculation of polarizabilities or hyperpolarizabilities.

References

1. M. Rérat, C. Pouchan, M. Tadjeddine, J.P. Flament, H.P. Gervais, and G. Berthier, *Phys. Rev.* **A43**, 5832, (1991)

2. M. Ratner, *Int. J. Quant. Chem.* **43**, 5, (1992)

3. E.N. Svendsen and T. Stroyer–Hansen, *Theoret. Chim. Acta* **45**, 53, (1977) H.F. Hameka and E.N. Svendsen, *Int. J. Quant. Chem.* **XI**, 129, (1977)

4. a) B. Huron, P. Rancurel and J.P. Malrieu, *J. Chem. Phys.* **58**, 5745, (1973); E. Evangelisti, J.P. Daudey and J.P. Malrieu , *Chem. Phys.* **75**, 91, (1983) b) R. Cimiraglia, *J. Chem. Phys.* **83**, 1746, (1985)

5. M. Tadjeddine, J.P. Flament, N.El Bakali Kassimi, H.P. Gervais, G. Berthier, M. Rérat and C. Pouchan, *J. Chim. Phys.* **87**, 989, (1990)

6. N. El Bakali Kassimi, M. Tadjeddine, J.P. Flament, G. Berthier and H.P. Gervais, *J. Mol. Struct. (THEOCHEM)* **254**, 177, (1992)

7. M. Karplus and H.J. Kolker, *J. Chem. Phys.* **39**, 1493, (1963)

8. J.G. Kirkwood, *Phys. Z.* **33**, 39, (1931)

9. J.P. Flament, H.P. Gervais and M. Rérat, *J. Mol. Struct. (THEOCHEM)* **151**, 39, (1987)

10. M. Rérat, *Int. J. Quant. Chem.* **36**, 169, (1989)

11. C. Cohen–Tannoudji, B. Diu and F. Laloe, Mécanique Quantique, Hermann, Paris, 1973; C. Cohen–Tannoudji, J. Dupont–Roc, G. Grynberg, Processus d'interaction entre photons et atomes, InterEditions/Editions du CNRS, Paris, 1988.

12. A.C. Tanner and A.J. Thakkar, *Int. J. Quant. Chem.* **24**, 345, (1983)

13. M. Rérat, M. Mérawa and C. Pouchan, *Phys. Rev.* **A45**, 6263, (1992)

14. P.K.K. Pandey and D.P. Santry, *J. Chem. Phys.* **73**, 2899, (1980)

15. quoted in Ref. 19

16. K.P. Huber and G. Herzberg, Molecular spectra and molecular structure IV: Constants of diatomiques molecules, Van Nostrand Reinhold, New York, 1979.

17. M. Rérat, M. Mérawa and C. Pouchan, *to be published*
    M. Mérawa, *Thèse*, Université de Pau et des Pays de l'Adour (1991)

18. J. Oddershede and E.N. Svendsen, *Chem. Phys.* **64**, 359, (1982)

19. K.K. Sunil and K. Jordan, *Chem. Phys. Letters* **145**, 377, (1988)

20. E. Hylleras, *Z. Phys.* **65**, 209, (1930)

21. N. El Bakali Kassimi, *Thèse*, Université de Paris VI (1992)

22. M. Dupuis, A. Farazdel, S.P. Karna and S.A. Maluendes, in MOTECC, Modern Techniques in Computational Chemistry, E. Clementi Ed. ESCOM, Leyden 1990).

23. A. Dalgarno, *Adv. Phys.* **11**, 281, (1962)

24. W. Ritz, *Jour. für die reine und angew. Math.* **135**, 1, (1907)

25. H.D. Cohen and C.C.J. Roothan, *J. Chem. Phys.* **43**, S34, (1965)

26. B. Champagne, J.G. Fripiat and J.M. André, *J. Chem. Phys.* **96**, 8330, (1992)

27. for a detailed bibliography, see Ref. 6.

28. G.D. Zeiss, W.R. Scott, N. Suzuki, D.P. Chong and S.R. Langhoff, *Mol. Phys.* **37**, 1543, (1979)

29. P.S. Epstein, *Phys. Rev.* **28**, 695, (1926)

30. J.O. Hirschfelder, W. Byers Brown and S.T. Epstein, *Adv. Quantum Chem.* **1**, 255, (1964)

31. L.L. Boyle, A.D. Buckingham, R.L. Disch and D.A. Dunmur, *J. Chem. Phys.* **45**, 1318, (1966); J.F. Ward and G.H.C. New, *Phys. Rev.* **185**, 57, (1969); T.M. Miller and B. Bedersen, *Adv. At. Mol. Phys.* **13**, 1, (1977)

32. N.J. Bridge and A.D. Buckingham, *Proc. R. Soc. London Ser.* A **295**, 334, (1966)

33. R.E. Sitter Jr. and R.P. Hurst, *Phys. Rev.* **A5**, 5, (1972)
    P.W. Fowler, *J. Chem. Phys.* **87**, 2401, (1987)

34. L. Laaksonen, P. Pyykkö and D. Sundholm, *Int. J. Quant. Chem.* **23**, 319, (1983)

35. S. Fraga and B.J. Ransil, *J. Chem. Phys.* **35**, 1967, (1961)

36. E.A. Mc Cullough Jr., *J. Chem. Phys.* **63**, 5050, (1975)
    D. M. Bishop and B. Lam, *J. Chem. Phys.* **89**, 1571, (1988)

37. R. Poirier, R. Kari and I.G. Csizmadia, Handbook of gaussian basis sets, Elsevier, Amsterdam, 1985.

38. P.W. Fowler and A.D. Buckingham, *Mol. Phys.* **67**, 681, (1989)

39. R.S. Watts and A.T. Stelbovics, *Chem. Phys. Letters* **61**, 351, (1979)

40. G. Maroulis and D.M. Bishop, *Chem. Phys. Letters* **128**, 462, (1986)

41. R.S. Mulliken, *J. Chem. Phys.* **36**, 3428, (1962)

42. T. Voegel, J. Hinze and F. Tobin, *J. Chem. Phys.* **70**, 1107, (1979)

Coupled Hartree-Fock Approach to Electric Hyperpolarizability Tensors in Benzene

P. LAZZERETTI, M. MALAGOLI and R. ZANASI
Università di Modena, Dipartimento di Chimica, via G. Campi 183, 41100 Modena, Italy

1. Introduction

In the presence of a static, spatially uniform electric field E_α, the electronic cloud of atomic and molecular systems gets polarized. The energy, W, can be written as a Taylor series [1–3]

$$
\begin{aligned}
W = {} & W_0 - \mathcal{M}_\alpha^{(0)} E_\alpha - \frac{1}{2}\alpha_{\alpha\beta} E_\alpha E_\beta \\
& -\frac{1}{6}\beta_{\alpha\beta\gamma} E_\alpha E_\beta E_\gamma - \frac{1}{24}\gamma_{\alpha\beta\gamma\delta} E_\alpha E_\beta E_\gamma E_\delta \\
& -\frac{1}{120}\delta_{\alpha\beta\gamma\delta\epsilon} E_\alpha E_\beta E_\gamma E_\delta E_\epsilon + \ldots,
\end{aligned}
\tag{1}
$$

where W_0 is the unperturbed energy, $\mathcal{M}_\alpha^{(0)}$ is the permanent electric dipole moment and the coefficients $\alpha_{\alpha\beta}$, $\beta_{\alpha\beta\gamma}$, etc. are known as (static) electric polarizabilities. Non-linear response of the system is rationalized via hyperpolarizabilities $\beta_{\alpha\beta\gamma}$, $\gamma_{\alpha\beta\gamma\delta}$ (sum over repeated Greek indices is implied), etc.. The total electric dipole moment of the molecule in the presence of the electric field is [1–3]

$$
\begin{aligned}
\mathcal{M}_\alpha = {} & -\frac{\partial W}{\partial E_\alpha} \\
= {} & \mathcal{M}_\alpha^{(0)} + \alpha_{\alpha\beta} E_\beta + \frac{1}{2}\beta_{\alpha\beta\gamma} E_\beta E_\gamma \\
& +\frac{1}{6}\gamma_{\alpha\beta\gamma\delta} E_\beta E_\gamma E_\delta + \frac{1}{24}\delta_{\alpha\beta\gamma\delta\epsilon} E_\beta E_\gamma E_\delta E_\epsilon + \ldots,
\end{aligned}
\tag{2}
$$

According to (1) and (2), the response tensors are defined

$$
\mathcal{M}_\alpha^{(0)} = -\left(\frac{\partial W}{\partial E_\alpha}\right)_{E\to 0},
\tag{3}
$$

$$
\alpha_{\alpha\beta} = -\left(\frac{\partial^2 W}{\partial E_\alpha \partial E_\beta}\right)_{E\to 0} = \left(\frac{\partial \mathcal{M}_\alpha}{\partial E_\beta}\right)_{E\to 0} = \left(\frac{\partial \mathcal{M}_\beta}{\partial E_\alpha}\right)_{E\to 0},
\tag{4}
$$

279

Y. Ellinger and M. Defranceschi (eds.), Strategies and Applications in Quantum Chemistry, 279–296.
© 1996 *Kluwer Academic Publishers.*

$$\beta_{\alpha\beta\gamma} = -\left(\frac{\partial^3 W}{\partial E_\alpha \partial E_\beta \partial E_\gamma}\right)_{E\to 0}$$

$$= \left(\frac{\partial^2 \mathcal{M}_\alpha}{\partial E_\beta \partial E_\gamma}\right)_{E\to 0} = \left(\frac{\partial^2 \mathcal{M}_\beta}{\partial E_\gamma \partial E_\alpha}\right)_{E\to 0} = \left(\frac{\partial^2 \mathcal{M}_\gamma}{\partial E_\alpha \partial E_\beta}\right)_{E\to 0}, \quad (5)$$

$$\gamma_{\alpha\beta\gamma\delta} = -\left(\frac{\partial^4 W}{\partial E_\alpha \partial E_\beta \partial E_\gamma \partial E_\delta}\right)_{E\to 0}$$

$$= \left(\frac{\partial^3 \mathcal{M}_\alpha}{\partial E_\beta \partial E_\gamma \partial E_\delta}\right)_{E\to 0} = \dots$$

$$= \left(\frac{\partial^3 \mathcal{M}_\delta}{\partial E_\alpha \partial E_\beta \partial E_\gamma}\right)_{E\to 0}. \quad (6)$$

Owing to permutational symmetry of the tensor indices, only

$$\prod_{i=1}^{r}\left(\frac{2}{i}+1\right) = \frac{3\cdot 4\cdot\dots(3+r-1)}{1\cdot 2\cdot\dots r} = \frac{1}{2}(r+2)(r+1) \quad (7)$$

components are distinct for a tensor of rank r appearing in (1). Thus the number of independent values which completely characterize the various tensors in eq. (1) is 3, 6, 10, 15, 21, ... respectively for $\mathcal{M}_\alpha^{(0)}$, $\alpha_{\alpha\beta}$, $\beta_{\alpha\beta\gamma}$, $\gamma_{\alpha\beta\gamma\delta}$, $\delta_{\alpha\beta\gamma\delta\epsilon}$.... Molecular point symmetry further reduces the number of linearly independent components, see, for instance, Refs. [4], [5]. For any tensor appearing in (1), denoted in general by $\mathcal{D}_{\alpha\beta\dots}$, let us rearrange its components as a column vector in cartesian space, i.e.,

$$\mathcal{D}_{\alpha\beta\dots} \equiv d_I. \quad (8)$$

If the basis set ϵ of unit vectors in cartesian 3-space transforms

$$\epsilon \to \epsilon' = \epsilon \mathbf{T}_\epsilon, \quad (9)$$

under an operation T, then the direct product matrix

$$\mathbf{T} = \mathbf{T}_\epsilon \times \mathbf{T}_\epsilon \times \dots, \quad (10)$$

can be introduced, so that

$$d_I \to d_I' = T d_I = \sum_J T_{IJ} d_J. \quad (11)$$

If T belongs to a group G and brings the physical system into self-coincidence, then the array of components will be stable under G, i.e.,

$$\{d_I'\} = \{d_I\}. \quad (12)$$

In addition one can always find a transformation leading to a symmetry adapted basis [4] $\bar{\epsilon}$, so that $\mathbf{T}$ is brought to the block diagonal form $\bar{\mathbf{T}}$ via the associated similarity transformation. The $\bar{\mathbf{T}}$ matrix can be written as a direct sum

$$\bar{\mathbf{T}} = \bar{\mathbf{T}}^{(\alpha,1)} \oplus \bar{\mathbf{T}}^{(\alpha,2)} \oplus \dots \bar{\mathbf{T}}^{(\alpha,n_\alpha)} \oplus \bar{\mathbf{T}}^{(\beta,1)} \oplus \bar{\mathbf{T}}^{(\beta,2)} \oplus \dots \bar{\mathbf{T}}^{(\beta,n_\beta)} \oplus \dots \quad (13)$$

where the different blocks of $\bar{\mathbf{T}}$ are classified according to the irreducible representation Γ_α, with frequency n_α, of G, and its μ-th appearence. Accordingly, in the new basis, the symmetry adapted tensor components are [4]

$$\bar{d}_I^{(\alpha,\mu)\prime} = \bar{d}_I^{(\alpha,\mu)}, \tag{14}$$

for every operation of the group and for each block (α, μ). This implies either that $\bar{d}_I^{(\alpha,\mu)} = 0$ or that, being invariant under the operations of G, it carries the one-dimensional totally symmetric representation, i.e., $\bar{T}_{II}^{(\alpha,\mu)} = 1$, $\bar{T}_{IJ}^{(\alpha,\mu)} = 0$. Thus, if the totally symmetric representation occurs m times in the direct product representation, then the tensor is fully determined by just m numbers. Therefore theoretical procedures for evaluating the higher-rank polarizability tensors appearing in (1) and (2) should efficiently exploit the symmetry properties of a given molecule to save computer effort. The number of independent parameters can be conveniently evaluated *a priori* via simple techniques based on symmetrized Kronecker products [4]. Tables reporting data for a number of groups are available [1].

Besides the elementary properties of index permutational symmetry considered in eq. (7), and intrinsic point group symmetry of a given tensor accounted for in eqs. (8)-(14), much more powerful group-theoretical tools [6] can be developed to speed up coupled Hartree-Fock (CHF) calculations [7–11] of hyperpolarizabilities, which are nowadays almost routinely performed in a number of studies dealing with non linear response of molecular systems [12–35], in particular at the self-consistent-field (SCF) level of accuracy.

The present paper is aimed at developing an efficient CHF procedure [6–11] for the entire set of electric polarizabilities and hyperpolarizabilities defined in eqs. (1)-(6) up to the 5-th rank. Owing to the $2n + 1$ theorem of perturbation theory [36], only 2-nd order perturbed wavefunctions and density matrices need to be calculated. Explicit expressions for the perturbed energy up to the 4-th order are given in Sec. IV.

A computer program for the theoretical determination of electric polarizabilities and hyperpolarizabilitieshas been implemented at the *ab initio* level using a computational scheme based on CHF perturbation theory [7–11]. Zero-order SCF, and first- and second-order CHF equations are solved to obtain the corresponding perturbed wavefunctions and density matrices, exploiting the entire molecular symmetry to reduce the number of matrix element which are to be stored in, and processed by, computer. Then $\alpha_{\alpha\beta}$, $\beta_{\alpha\beta\gamma}$ and $\gamma_{\alpha\beta\gamma\delta}$ tensors are evaluated. This method has been applied to evaluate the second hyperpolarizability of benzene using extended basis sets of Gaussian functions, see Sec. VI.

2. Solution of first-order CHF equation

The Hartree-Fock equations for the i-th element of a set containing occ occupied molecular orbitals $\{\phi_i\}$ in a closed shell system with $n = 2occ$ electrons are [8]

$$\left(\hat{F} - \epsilon_i\right)\phi_i - \sum_{j \neq i}^{occ}\phi_j\epsilon_{ji} = 0, \qquad \hat{F} = \hat{H} + \hat{G}, \tag{15}$$

where the orthonormality conditions are written

$$<\phi_i|\phi_j> = \delta_{ij}. \tag{16}$$

All the quantities appearing in (15) are expanded in powers of a formal perturbation parameter λ, which is finally put equal to unity, so that, for instance,

$$\phi_i = \phi_i^{(0)} + \phi_i^{(1)}\lambda + \phi_i^{(2)}\lambda^2 + \ldots. \qquad (17)$$

The matrix of Lagrange multipliers is usually chosen diagonal to zero order, so that

$$\epsilon_{ii}^{(0)} \equiv \epsilon_i^{(0)}, \qquad \epsilon_{ji}^{(0)} = 0, \quad j \neq i, \qquad (18)$$

and

$$\left(\hat{F}^{(0)} - \epsilon_i^{(0)}\right)\phi_i^{(0)} = 0, \qquad (19)$$

with

$$<\phi_i^{(0)}|\phi_j^{(0)}> = \delta_{ij}. \qquad (20)$$

To first order in λ,

$$\left(\hat{F}^{(0)} - \epsilon_i^{(0)}\right)\phi_i^{(1)} + \left(\hat{F}^{(1)} - \epsilon_i^{(1)}\right)\phi_i^{(0)} - \sum_{j \neq i}^{occ}\phi_j^{(0)}\epsilon_{ji}^{(1)} = 0, \qquad (21)$$

with

$$<\phi_i^{(0)}|\phi_j^{(1)}> + <\phi_i^{(1)}|\phi_j^{(0)}> = 0. \qquad (22)$$

Taking the Hermitian product with $\phi_i^{(0)}$ in eq. (21) one has

$$\epsilon_i^{(1)} \equiv \epsilon_{ii}^{(1)} = <\phi_i^{(0)}|\hat{F}^{(1)}|\phi_i^{(0)}>. \qquad (23)$$

Taking the product in eq. (21) with $\phi_k^{(0)}$, where k labels another occupied orbital in a non degenerate problem, i.e., $\epsilon_i^{(0)} \neq \epsilon_k^{(0)}$, using (19) and (20),

$$\left(\epsilon_k^{(0)} - \epsilon_i^{(0)}\right)<\phi_k^{(0)}|\phi_i^{(1)}> + <\phi_k^{(0)}|\hat{F}^{(1)}|\phi_i^{(0)}> - \epsilon_{ki}^{(1)} = 0. \qquad (24)$$

Owing to the arbitrary nature of the Lagrange multipliers, one can choose

$$\epsilon_{ki}^{(1)} = <\phi_k^{(0)}|\hat{F}^{(1)}|\phi_i^{(0)}>, \qquad (25)$$

so that the projection of the first-order i-th orbital on the subspace spanned by $occ-1$ occupied MO's $\{\phi_k^{(0)}, k \neq i\}$ vanishes. From the orthogonality condition (22) one has, for the i-th orbital,

$$2\Re<\phi_i^{(0)}|\phi_i^{(1)}> = 0. \qquad (26)$$

For real perturbations, e.g., in the presence of a static electric field, which is the case studied in the present paper, the zero- and first-order orbitals can always be chosen real, so that also

$$<\phi_i^{(0)}|\phi_i^{(1)}> = 0. \qquad (27)$$

At any rate, the projection of the first-order orbitals on the subspace of occupied $\{\phi_k^{(0)}\}$ is not needed within the McWeeny approach [7], where choice (25) is implicitly assumed; it is sufficient to calculate the projection $|\phi_i^{(1)}>_{VIR}$ on the subspace of virtual zero-order orbitals. Taking the Hermitian product with an unoccupied $\phi_k^{(0)}$ in eq. (21) one has

$$\left(\epsilon_i^{(0)} - \epsilon_k^{(0)}\right) <\phi_k^{(0)}|\phi_i^{(1)}> = <\phi_k^{(0)}|\hat{F}^{(1)}|\phi_i^{(0)}>, \tag{28}$$

and left-multiplying by the ket $|\phi_k^{(0)}>$ and summing over k gives

$$|\phi_i^{(1)}> \equiv |\phi_i^{(1)}>_{VIR} = M^{(i)}\hat{F}^{(1)}|\phi_i^{(0)}>, \tag{29}$$

where

$$M^{(i)} = \sum_k^{vir} \left(\epsilon_i^{(0)} - \epsilon_k^{(0)}\right)^{-1} |\phi_k^{(0)}><\phi_k^{(0)}| \tag{30}$$

is the Hartree-Fock propagator [10], [11] and the projector

$$P_{VIR} = \sum_k^{vir} |\phi_k^{(0)}><\phi_k^{(0)}| \tag{31}$$

is equivalent to the identity operator when acting on the subspace of virtual orbitals [7].

3. Solution of second-order CHF equation

We discuss a method to evaluate the second-order molecular orbitals appearing in eq. (17) consistent with the first-order computational scheme outlined in the previous section. In particular we take advantage of definition (30) to develop a compact approach explicitly oriented to numerical applications.
The second-order CHF equation for the i-th occupied orbital is

$$\left(\hat{F}^{(0)} - \epsilon_i^{(0)}\right)\phi_i^{(2)} + \left(\hat{F}^{(1)} - \epsilon_i^{(1)}\right)\phi_i^{(1)} + \left(\hat{F}^{(2)} - \epsilon_i^{(2)}\right)\phi_i^{(0)}$$
$$- \sum_{j\neq i}^{occ}\phi_j^{(1)}\epsilon_{ji}^{(1)} - \sum_{j\neq i}^{occ}\phi_j^{(0)}\epsilon_{ji}^{(2)} = 0, \tag{32}$$

where the orbitals satisfy the orthonormality condition to second order,

$$<\phi_i^{(0)}|\phi_j^{(2)}> + <\phi_i^{(1)}|\phi_j^{(1)}> + <\phi_i^{(2)}|\phi_j^{(0)}> = 0. \tag{33}$$

Taking in (32) the Hermitian product with $\phi_i^{(0)}$, and using eqs. (19), (21), and (22), one finds

$$\epsilon_i^{(2)} = <\phi_i^{(0)}|\hat{F}^{(2)}|\phi_i^{(0)}> - <\phi_i^{(1)}|\hat{F}^{(0)} - \epsilon_i^{(0)}|\phi_i^{(1)}>$$
$$+ 2\Re\left(\sum_j^{occ}<\phi_i^{(1)}|\phi_j^{(0)}>\epsilon_{ji}^{(1)}\right), \tag{34}$$

where the index $j = i$ can be omitted in the sum, in the present case of real perturbations, owing to eq. (27). Summing over i occupied, the last term in (34) vanishes.

Left-multiplying (32) by $\phi_k^{(0)}, k \neq i$ occupied in a non degenerate case, using (19), (21), and (33) one obtains

$$
\langle\phi_k^{(0)}|\hat{F}^{(2)}|\phi_i^{(0)}\rangle \quad - \quad \langle\phi_k^{(1)}|\hat{F}^{(0)}|\phi_i^{(1)}\rangle - \left(\epsilon_k^{(0)}\langle\phi_k^{(2)}|\phi_i^{(0)}\rangle + \langle\phi_k^{(0)}|\phi_i^{(2)}\rangle\epsilon_i^{(0)}\right)
$$

$$
+ \sum_j^{occ} \left(\epsilon_{kj}^{(1)}\langle\phi_j^{(0)}|\phi_i^{(1)}\rangle + \langle\phi_k^{(1)}|\phi_j^{(0)}\rangle\epsilon_{ji}^{(1)}\right) - \epsilon_{ki}^{(2)} = 0. \tag{35}
$$

This equation shows that second- and first-order Lagrangian multipliers are not independent, so that a specific selection of $\epsilon_{ij}^{(1)}$ will bias $\epsilon_{ij}^{(2)}$. Thus the choice (25) for the first-order Lagrangian multipliers makes the sum over j in eq. (35) vanish in the case of real perturbations, but, in general, there is no choice of $\epsilon_{ki}^{(2)}$ in eq. (35) which annihilates the projection of $|\phi_i^{(0)}\rangle$ on occupied $|\phi_k^{(0)}\rangle$ and $|\phi_i^{(0)}\rangle$.

However, using the McWeeny approach [7], it is sufficient to calculate only the projection $|\phi_i^{(2)}\rangle_{VIR}$ on the subspace of virtual zero-order orbitals in order to get the second hyperpolarizability tensor. This projection is evaluated via a procedure similar to the one used in solving the first-order equation (21). Taking in (32) the Hermitian product with the unoccupied $\phi_k^{(0)}$ and using (19), one finds

$$
\left(\epsilon_k^{(0)} - \epsilon_i^{(0)}\right)\langle\phi_k^{(0)}|\phi_i^{(2)}\rangle \quad + \quad \langle\phi_k^{(0)}|\hat{F}^{(2)}|\phi_i^{(0)}\rangle + \langle\phi_k^{(0)}|\hat{F}^{(1)}|\phi_i^{(1)}\rangle
$$

$$
- \sum_j^{occ}\langle\phi_k^{(0)}|\phi_j^{(1)}\rangle\epsilon_{ji}^{(1)} = 0, \tag{36}
$$

Multiplying on the left by the ket $|\phi_k^{(0)}\rangle$ and summing over k, one finds

$$
|\phi_i^{(2)}\rangle_{VIR} = M^{(i)}\hat{F}^{(2)}|\phi_i^{(0)}\rangle + |\theta_i^{(2)}\rangle, \tag{37}
$$

where

$$
|\theta_i^{(2)}\rangle = M^{(i)}\left(\hat{F}^{(1)}|\phi_i^{(1)}\rangle - \sum_j^{occ}|\phi_j^{(1)}\rangle\epsilon_{ji}^{(1)}\right) \tag{38}
$$

is available from the solutions (29) to the first-order eq. (21).

4. Computational scheme

We will now discuss an iterative scheme based on the CHF approach outlined in Sections II and III, using the McWeeny procedure [7] for resolving matrices into components, by introducing projection operators $\hat{R}_1$ and $\hat{R}_2$, with respect to the subspaces spanned by occupied and virtual molecular orbitals.

Expanding the occupied ϕ_i over an orthonormal atomic basis set χ of order m (which is assumed independent of the perturbation), one has

$$
\phi_i^{(0)} = \sum_{p=1}^{m}\chi_p c_{pi}^{(0)}, \quad \phi_i^{(1)} = \sum_{p=1}^{m}\chi_p c_{pi}^{(1)}, \quad \phi_i^{(2)} = \sum_{p=1}^{m}\chi_p c_{pi}^{(2)}. \tag{39}
$$

The projection operators have matrix representations

$$\mathbf{R}_1 \equiv \mathbf{R}^{(0)} = \sum_k^{\text{occ}} \mathbf{c}_k^{(0)} \mathbf{c}_k^{(0)\dagger} \tag{40}$$

and

$$\mathbf{R}_2 \equiv \left(1 - \mathbf{R}^{(0)}\right) = \sum_l^{\text{vir}} \mathbf{c}_l^{(0)} \mathbf{c}_l^{(0)\dagger}. \tag{41}$$

A perturbation expansion analogous to (17) holds for any matrix, for instance, the Fock matrix

$$\mathbf{F} = \mathbf{F}^{(0)} + \lambda \mathbf{F}^{(1)} + \lambda^2 \mathbf{F}^{(2)} + \dots, \qquad F_{pq}^{(i)} = <\chi_p | \hat{F}^{(i)} | \chi_q>, \tag{42}$$

and the density matrix

$$\mathbf{R} = \mathbf{R}^{(0)} + \lambda \mathbf{R}^{(1)} + \lambda^2 \mathbf{R}^{(2)} + \dots. \tag{43}$$

The first- and second-order coefficients $\mathbf{c}_i^{(1)}$ and $\mathbf{c}_i^{(2)}$ can also be resolved into projections on the subspaces of occupied and virtual molecular orbitals:

$$\mathbf{c}_i^{(1)} = \mathbf{o}_i^{(1)} + \mathbf{v}_i^{(1)}, \tag{44}$$

$$\mathbf{c}_i^{(2)} = \mathbf{o}_i^{(2)} + \mathbf{v}_i^{(2)}. \tag{45}$$

The projections $\mathbf{o}_k^{(1)}$, with k labeling another occupied orbital, vanish according to choice (25), and the projection $\mathbf{o}_i^{(1)}$ vanishes for a real perturbation, see eq. (27). Any matrix $\mathbf{A}$ can now be resolved into projection components [7] with respect to the occupied and virtual subspaces, that is,

$$\mathbf{A} = \mathbf{A}_{11} + \mathbf{A}_{12} + \mathbf{A}_{21} + \mathbf{A}_{22}, \qquad \mathbf{A}_{ij} \equiv \mathbf{R}_i \mathbf{A} \mathbf{R}_j, \quad i,j = 1,2. \tag{46}$$

For instance, the first-order density matrix can be written

$$\mathbf{R}^{(1)} = \mathbf{R}_{12}^{(1)} + \mathbf{R}_{21}^{(1)}, \tag{47}$$

where

$$\mathbf{R}_{21}^{(1)\dagger} = \mathbf{R}_{12}^{(1)} \equiv \mathbf{X} = \sum_k^{\text{occ}} \mathbf{c}_k^{(0)} \mathbf{v}_k^{(1)\dagger}. \tag{48}$$

The iterative scheme for the first-order coefficients $\mathbf{v}_i^{(1)}$ becomes

$$\mathbf{v}_i^{(1)} = \mathbf{M}^{(i)} \mathbf{F}^{(1)} \mathbf{c}_i^{(0)}, \tag{49}$$

$$\mathbf{M}^{(i)} = \sum_j^{\text{vir}} \left(\epsilon_i^{(0)} - \epsilon_j^{(0)}\right)^{-1} \mathbf{c}_j^{(0)} \mathbf{c}_j^{(0)\dagger}, \tag{50}$$

$$\mathbf{R}^{(1)} = \mathbf{X} + \mathbf{X}^\dagger, \tag{51}$$

$$\mathbf{F}^{(1)} = \mathbf{H}^{(1)} + \mathbf{G}^{(1)}, \qquad \mathbf{G}^{(1)} = \mathbf{G}(\mathbf{R}^{(1)}), \tag{52}$$

where

$$H_{pq}^{(1)} = <\chi_p|\hat{H}^{(1)}|\chi_q> \tag{53}$$

is used to start the iteration (49)-(52). In the case of electric perturbation, denoting by $r_{i\alpha}$ the coordinates of the i-th electron with charge $-e$,

$$\hat{H}^{(1)} = \hat{H}^{E_\alpha} E_\alpha, \qquad \hat{H}^{E_\alpha} = e\sum_i^n r_{i\alpha}. \tag{54}$$

To first order the repulsion matrix

$$G_{pq}^{(1)} = \sum_{rs} R_{sr}^{(1)}(<pr|qs> - <pr|sq>), \tag{55}$$

is obtained contracting the first-order density matrix with the two-electron integrals over the atomic basis,

$$<pr|qs> = e^2 \int d\tau_1 \int d\tau_2 \chi_p^*(1)\chi_r^*(2)r_{12}^{-1}\chi_q(1)\chi_s(2). \tag{56}$$

The second-order density matrix

$$\mathbf{R}^{(2)} = \sum_k^{occ} \left(\mathbf{c}_k^{(2)}\mathbf{c}_k^{(0)\dagger} + \mathbf{c}_k^{(1)}\mathbf{c}_k^{(1)\dagger} + \mathbf{c}_k^{(0)}\mathbf{c}_k^{(2)\dagger} \right) \tag{57}$$

is also resolved into four components according to (46),

$$\mathbf{R}_{ij}^{(2)} = \mathbf{R}_i\mathbf{R}^{(2)}\mathbf{R}_j, \quad i,j = 1,2, \tag{58}$$

that is,

$$\mathbf{R}^{(2)} = \mathbf{R}_{11}^{(2)} + \mathbf{R}_{12}^{(2)} + \mathbf{R}_{21}^{(2)} + \mathbf{R}_{22}^{(2)}, \tag{59}$$

$$\mathbf{R}_{11}^{(2)} = -\mathbf{X}\mathbf{X}^\dagger, \tag{60}$$

$$\mathbf{R}_{22}^{(2)} = \mathbf{X}^\dagger\mathbf{X}, \tag{61}$$

$$\mathbf{R}_{21}^{(2)\dagger} = \mathbf{R}_{12}^{(2)} \equiv \mathbf{Y} = \sum_k^{occ}\mathbf{c}_k^{(0)}\mathbf{v}_k^{(2)\dagger}. \tag{62}$$

Accordingly, only the projections $\mathbf{v}_k^{(2)}$ over the subspace of virtual orbitals are needed to compute the second-order density matrix.

The iterative scheme for the second-order coefficients, consistent with (49)-(52), is (in the case of electric perturbation there is no $\hat{H}^{(2)}$)

$$\mathbf{k}_i^{(2)} = \mathbf{M}^{(i)} \left(\mathbf{F}^{(1)}\mathbf{c}_i^{(1)} - \mathbf{X}^\dagger\mathbf{F}^{(1)}\mathbf{c}_i^{(0)} \right), \tag{63}$$

$$\mathbf{v}_i^{(2)} = \mathbf{M}^{(i)}\mathbf{F}^{(2)}\mathbf{c}_i^{(0)} + \mathbf{k}_i^{(2)}, \tag{64}$$

$$\mathbf{R}^{(2)} = -\mathbf{X}\mathbf{X}^\dagger + \mathbf{Y} + \mathbf{Y}^\dagger + \mathbf{X}^\dagger\mathbf{X} \tag{65}$$

$$\mathbf{F}^{(2)} \equiv \mathbf{G}^{(2)} = \mathbf{G}(\mathbf{R}^{(2)}). \tag{66}$$

The iteration starts with $v_i^{(2)} = k_i^{(2)}$. The second-order repulsion matrix $\mathbf{G}^{(2)}$ is defined analogously to (55). $\mathbf{M}^{(i)}$ and $\mathbf{X}$ matrices have been computed only once to solve the first-order CHF problem (i.e., to determine the polarizability α and the first hyperpolarizability β). $\mathbf{M}^{(i)}$ and $k_i^{(2)}$ are saved onto a file to be processed at each step of the iterative calculation (63)-(66): it seems worthy of notice that the present CHF algorithm, based on the Hartree-Fock propagator (30), is quite general, compact and suitable for efficient sequential determination of both first- and second-order perturbed orbitals. In addition, it can be easily extended to perturbations of higher order.

So far we have considered an orthonormal basis set χ. In actual calculations, employing non orthogonal sets of Gaussian functions with overlap matrix

$$\Delta_{pq} = <\chi_p|\chi_q>, \tag{67}$$

it is customary to orthogonalize according to the Löwdin procedure, i.e.,

$$\chi \rightarrow \chi \Delta^{-\frac{1}{2}}, \tag{68}$$

$$c_i^{(0)} \rightarrow \Delta^{\frac{1}{2}}c_i^{(0)}, \tag{69}$$

$$\mathbf{R}^{(0)} \rightarrow \Delta^{\frac{1}{2}}\mathbf{R}^{(0)}\Delta^{\frac{1}{2}}, \tag{70}$$

$$\mathbf{M}^{(i)} \rightarrow \Delta^{\frac{1}{2}}\mathbf{M}^{(i)}\Delta^{\frac{1}{2}}, \tag{71}$$

$$\mathbf{F}^{(0)} \rightarrow \Delta^{-\frac{1}{2}}\mathbf{F}^{(0)}\Delta^{-\frac{1}{2}}, \tag{72}$$

with similar equations for first- and second-order perturbed matrices. In the second-order iteration eqs. (63) and (65) are replaced by

$$k_i^{(2)} = \mathbf{M}^{(i)}\left(\mathbf{F}^{(1)}c_i^{(1)} - \Delta\mathbf{X}^\dagger\mathbf{F}^{(1)}c_i^{(0)}\right), \tag{73}$$

$$\mathbf{R}^{(2)} = -\mathbf{X}\Delta\mathbf{X}^\dagger + \mathbf{Y} + \mathbf{Y}^\dagger + \mathbf{X}^\dagger\Delta\mathbf{X} \tag{74}$$

for a non orthogonal basis.

The expression for the electronic contribution to electric dipole moment,

$$\mathcal{M}_\alpha^{(0)} = -2Tr[\mathbf{H}^{E_\alpha}\mathbf{R}^{(0)}], \tag{75}$$

is not affected by transformation (68)-(72), owing to the trace theorem. In addition, it can be shown that the iterative steps (49)-(52) are formally the same for a non orthogonal basis, as the formula for the polarizability

$$\alpha_{\alpha\beta} = -2Tr[\mathbf{H}^{E_\alpha}\mathbf{R}^{E_\beta}], \tag{76}$$

is also invariant under Löwdin orthogonalization. The overlap matrix appear only to third order in the expression for the first hyperpolarizability [10],

$$\beta_{\alpha\beta\gamma} = -4Tr[\mathbf{F}^{E_\alpha}(\mathbf{X}^{E_\beta\dagger}\Delta\mathbf{X}^{E_\gamma} - \mathbf{X}^{E_\gamma}\Delta\mathbf{X}^{E_\beta\dagger})] + [\beta\gamma\alpha] + [\gamma\alpha\beta], \tag{77}$$

where $[\beta\gamma\alpha]$ and $[\gamma\alpha\beta]$ are permutations of the expression in square brackets.

5. Symmetry transformations of second-order density

The electron density of a molecule in the presence of electric perturbation is a scalar field with perturbation expansion [6], [11]

$$
\begin{aligned}
P(\mathbf{r}) \;=\;& P^{(0)}(\mathbf{r}) + \sum_{\alpha} P^{\alpha}(\mathbf{r}) E_{\alpha} \\
&+ \frac{1}{2} \sum_{\alpha\beta} P^{\alpha\beta}(\mathbf{r}) E_{\alpha} E_{\beta} + \cdots .
\end{aligned} \tag{78}
$$

Relaxing the Einstein convention, sums over repeated Greek indices $\alpha = x, y, z$ are made explicit in this Section, to avoid misunderstanding whenever two couples of repeated indices α and β, with $\alpha \leq \beta$, appear in a formula, compare for (92) hereafter. Introducing a basis set χ of atomic functions, for the second-order term one defines the expansion

$$
P^{\alpha\beta}(\mathbf{r}) = 2 \sum_{pq=1}^{m} R_{pq}^{\alpha\beta} \chi_p(\mathbf{r}) \chi_q^*(\mathbf{r}). \tag{79}
$$

For any symmetry operator $T = T(\theta)$ (rewritten τ when operating on the domain of basis functions χ), for instance, the rotation-reflexion about the z-axis, with matrix representation

$$
\mathbf{T} = \begin{pmatrix} \cos\theta & -\sin\theta & 0 \\ \sin\theta & \cos\theta & 0 \\ 0 & 0 & \pm 1 \end{pmatrix}, \tag{80}
$$

over a basis set ϵ of Cartesian unit vectors, and belonging to a group G, one has

$$
\tau \chi_p(\mathbf{r}) = \chi_p(T^{-1}\mathbf{r}) = \sum_{q=1}^{m} \chi_q(\mathbf{r}) S_{qp}. \tag{81}
$$

In the transformed coordinate system,

$$
P^{\alpha'\beta'}(\mathbf{r}) = \sum_{\gamma\delta} T_{\alpha\gamma}^{-1} P^{\gamma\delta}(\mathbf{r}) T_{\beta\delta}^{-1} \tag{82}
$$

$$
P_{\text{trans}}^{\alpha'\beta'}(T\mathbf{r}) = \sum_{\gamma\delta} T_{\alpha\gamma}^{-1} P_{\text{trans}}^{\gamma\delta}(T\mathbf{r}) T_{\beta\delta}^{-1} = P^{\alpha\beta}(\mathbf{r}) \tag{83}
$$

$$
\begin{aligned}
\sum_{\alpha\beta} P^{\alpha\beta}(\mathbf{r}) T_{\lambda\alpha} T_{\mu\beta} &= \sum_{\alpha\beta\gamma\delta} T_{\lambda\alpha} T_{\alpha\gamma}^{-1} T_{\mu\beta} T_{\beta\delta}^{-1} P_{\text{trans}}^{\gamma\delta}(T\mathbf{r}) \\
= \sum_{\gamma\delta} \delta_{\lambda\gamma} \delta_{\mu\delta} P_{\text{trans}}^{\gamma\delta}(T\mathbf{r}) &= P_{\text{trans}}^{\lambda\mu}(T\mathbf{r}).
\end{aligned} \tag{84}
$$

Hence the transformation law for second-order density is

$$
P_{\text{trans}}^{\lambda\mu}(\mathbf{r}) = \sum_{\alpha\beta} T_{\lambda\alpha} T_{\mu\beta} P^{\alpha\beta}(T^{-1}\mathbf{r}) \tag{85}
$$

$$P^{\alpha\beta}(T^{-1}\mathbf{r}) = 2\sum_{pq=1}^{m} R_{pq}^{\alpha\beta}\chi_p(T^{-1}\mathbf{r})\chi_q^*(T^{-1}\mathbf{r}) = 2\sum_{pqrs=1}^{m} R_{pq}^{\alpha\beta}\chi_r S_{rp}\chi_s^* S_{sq}^* \tag{86}$$

Since the transformation belongs to the group G one has

$$P_{\text{trans}}^{\lambda\mu}(\mathbf{r}) = P^{\lambda\mu}(\mathbf{r}), \tag{87}$$

and the second-order density matrices transform according to

$$\mathbf{R}^{\lambda\mu} = \sum_{\alpha\beta} T_{\lambda\alpha} T_{\mu\beta} \mathbf{S}\mathbf{R}^{\alpha\beta}\mathbf{S}^\dagger. \tag{88}$$

Owing to permutational symmetry, at most six second-order matrices are independent. To account for point molecular symmetry let us introduce the symmetrized Kronecker square of $\mathbf{T}$, with matrix elements [4]

$$(\mathbf{T}_s^{[2]})_{\alpha\beta,\gamma\gamma} = T_{\alpha\gamma} T_{\beta\gamma}, \tag{89}$$

$$(\mathbf{T}_s^{[2]})_{\alpha\beta,\gamma\delta} = (T_{\alpha\gamma} T_{\beta\delta} + T_{\alpha\delta} T_{\beta\gamma}), \quad \gamma \neq \delta. \tag{90}$$

From eqs. (80), (89) and (90) one has

$$\mathbf{T}_s^{[2]} = \begin{pmatrix} \cos^2\theta & -2\sin\theta\cos\theta & 0 & \sin^2\theta & 0 & 0 \\ \sin\theta\cos\theta & \cos^2\theta - \sin^2\theta & 0 & -\sin\theta\cos\theta & 0 & 0 \\ 0 & 0 & \pm\cos\theta & 0 & \mp\sin\theta & 0 \\ \sin^2\theta & 2\sin\theta\cos\theta & 0 & \cos^2\theta & 0 & 0 \\ 0 & 0 & \pm\sin\theta & 0 & \pm\cos\theta & 0 \\ 0 & 0 & 0 & 0 & 0 & 1 \end{pmatrix} \tag{91}$$

Eventually one finds the final transformation law for the second-order density matrices $(\lambda \leq \mu)$,

$$\mathbf{R}^{\lambda\mu} = \sum_{\alpha\leq\beta} (\mathbf{T}_s^{[2]})_{\lambda\mu,\alpha\beta} \mathbf{S}\mathbf{R}^{\alpha\beta}\mathbf{S}^\dagger. \tag{92}$$

Hence, according to the present method, only the symmetry-distinct density matrices need to be computed.

Within our approach the entire molecular symmetry is exploited to increase the efficiency of the code in every step of the calculation. For a molecule belonging to a group G of order $|G|$, only $\approx \mathcal{N}^4/(8|G|)$ symmetry-distinct two-electron integrals over a basis set of $\mathcal{N}$ Gaussian atomic functions are calculated and processed at each iteration within SCF, first- and second-order CHF procedures. A skeleton Coulomb repulsion matrix $\widehat{\mathbf{G}}^{\alpha\beta}$ is obtained by processing the non-redundant list of unique two-electron integrals, then the actual repulsion matrices $\mathbf{G}^{\alpha\beta}, \alpha \leq \beta$, are obtained via the equation

$$\mathbf{G}^{\alpha\beta} = \sum_{T\in G} \left(\sum_{\gamma\leq\delta} (\mathbf{T}_s^{[2]})_{\alpha\beta,\gamma\delta}(\mathbf{S})^{-1}\widehat{\mathbf{G}}^{\gamma\delta}\mathbf{S}^{-1} \right). \tag{93}$$

This method turns out to be a major computer saver, as (i) the iterative steps become much faster, owing to the reduced number of integrals, and (ii) the occupancy of the mass storage gets smaller. Accordingly, one can afford large problems which would be otherwise intractable.

6. Fourth-order CHF energy

The CHF formulae (75)-(77) for the response tensors are established by expanding the Hartree-Fock energy [8],

$$W = 2Tr\mathbf{F}'\mathbf{R}, \qquad \mathbf{F}' = \mathbf{H} + \frac{1}{2}\mathbf{G}, \tag{94}$$

in powers of the electric field in the same way as (1). In the presence of multiple perturbations, the first-order perturbed core Hamiltonian can be written

$$\hat{H}^{(1)} = a\hat{H}^a + b\hat{H}^b + c\hat{H}^c + \ldots, \tag{95}$$

where the parameters a, b, c, etc., are related to the intensity of each perturbation. The expansion for the energy of a molecule becomes

$$\begin{aligned}
W = \ & W_0 + (aW^a + bW^b + cW^c + \ldots) \\
& + (a^2W^{a^2} + abW^{ab} + acW^{ac} + b^2W^{b^2} + bcW^{bc} + c^2W^{c^2} + \ldots) \\
& + (a^3W^{a^3} + a^2bW^{a^2b} + abcW^{abc} + \ldots) \\
& + (a^4W^{a^4} + a^3bW^{a^3b} + a^2b^2W^{a^2b^2} + a^2bcW^{a^2bc} + \ldots).
\end{aligned} \tag{96}$$

Comparing this general expression with the analogous expansion (1), one finds (no sum over repeated Latin indices)

$$W^{E_a} \equiv -\mathcal{M}_a^{(0)}, \tag{97}$$

$$W^{E_a^2} \equiv -\frac{1}{2}\alpha_{aa}, \qquad W^{E_aE_b} \equiv -\alpha_{ab}, \tag{98}$$

$$W^{E_a^3} \equiv -\frac{1}{6}\beta_{aaa}, \qquad W^{E_a^2E_b} \equiv -\frac{1}{2}\beta_{aab}, \qquad W^{E_aE_bE_c} \equiv -\beta_{abc}, \tag{99}$$

$$\begin{aligned}
W^{E_a^4} &\equiv -\frac{1}{24}\gamma_{aaaa}, \qquad W^{E_a^3E_b} \equiv -\frac{1}{6}\gamma_{aaab}, \\
W^{E_a^2E_b^2} &\equiv -\frac{1}{4}\gamma_{aabb}, \qquad W^{E_a^2E_bE_c} \equiv -\frac{1}{2}\gamma_{aacd}.
\end{aligned} \tag{100}$$

It should be noted that, in this notation, the order of the superscripts is irrelevant; $ab\ldots zW^{ab\ldots z}$ is the entire perturbed energy term linear in $ab\ldots z$ and there is no additional term with permuted $ab\ldots z$ indices.
The following terms appear in the expression of the 4-th order energy

$$\begin{aligned}
W^{a^4} = \ & Tr\mathbf{F}^{a^2}(\mathbf{R}_{11}^{a^2} + \mathbf{R}_{22}^{a^2}) + Tr\mathbf{F}^a(\mathbf{R}_{11}^{a^3} + \mathbf{R}_{22}^{a^3}) \\
& + 2Tr\mathbf{F}^{(0)}(-\mathbf{R}_{11}^{a^2}\mathbf{R}_{11}^{a^2} + \mathbf{R}_{22}^{a^2}\mathbf{R}_{22}^{a^2}),
\end{aligned} \tag{101}$$

$$\begin{aligned}
W^{a^3b} = \ & Tr\mathbf{F}^{a^2}(\mathbf{R}_{11}^{ab} + \mathbf{R}_{22}^{ab}) + Tr\mathbf{F}^{ab}(\mathbf{R}_{11}^{a^2} + \mathbf{R}_{22}^{a^2}) \\
& + Tr\mathbf{F}^a(\mathbf{R}_{11}^{a^2b} + \mathbf{R}_{22}^{a^2b}) + Tr\mathbf{F}^b(\mathbf{R}_{11}^{a^3} + \mathbf{R}_{22}^{a^3}) \\
& + 2Tr\mathbf{F}^{(0)}(-\mathbf{R}_{11}^{a^2}\mathbf{R}_{11}^{ab} - \mathbf{R}_{11}^{ab}\mathbf{R}_{11}^{a^2} + \mathbf{R}_{22}^{a^2}\mathbf{R}_{22}^{ab} + \mathbf{R}_{22}^{ab}\mathbf{R}_{22}^{a^2}),
\end{aligned} \tag{102}$$

$$W^{a^2b^2} = Tr\mathbf{F}^{a^2}(\mathbf{R}_{11}^{b^2} + \mathbf{R}_{22}^{b^2}) + Tr\mathbf{F}^{ab}(\mathbf{R}_{11}^{ab} + \mathbf{R}_{22}^{ab}) + Tr\mathbf{F}^{b^2}(\mathbf{R}_{11}^{a^2} + \mathbf{R}_{22}^{a^2})$$
$$+Tr\mathbf{F}^{a}(\mathbf{R}_{11}^{ab^2} + \mathbf{R}_{22}^{ab^2}) + Tr\mathbf{F}^{b}(\mathbf{R}_{11}^{a^2b} + \mathbf{R}_{22}^{a^2b})$$
$$+2Tr\mathbf{F}^{(0)}(-\mathbf{R}_{11}^{a^2}\mathbf{R}_{11}^{b^2} - \mathbf{R}_{11}^{b^2}\mathbf{R}_{11}^{a^2} - \mathbf{R}_{11}^{ab}\mathbf{R}_{11}^{ab}$$
$$+\mathbf{R}_{22}^{a^2}\mathbf{R}_{22}^{b^2} + \mathbf{R}_{22}^{b^2}\mathbf{R}_{22}^{a^2} + \mathbf{R}_{22}^{ab}\mathbf{R}_{22}^{ab}), \tag{103}$$

$$W^{a^2bc} = Tr\mathbf{F}^{a^2}(\mathbf{R}_{11}^{bc} + \mathbf{R}_{22}^{bc}) + Tr\mathbf{F}^{ab}(\mathbf{R}_{11}^{ac} + \mathbf{R}_{22}^{ac})$$
$$+Tr\mathbf{F}^{ac}(\mathbf{R}_{11}^{ab} + \mathbf{R}_{22}^{ab}) + Tr\mathbf{F}^{bc}(\mathbf{R}_{11}^{a^2} + \mathbf{R}_{22}^{a^2})$$
$$+Tr\mathbf{F}^{a}(\mathbf{R}_{11}^{abc} + \mathbf{R}_{22}^{abc}) + Tr\mathbf{F}^{b}(\mathbf{R}_{11}^{a^2c} + \mathbf{R}_{22}^{a^2c}) + Tr\mathbf{F}^{c}(\mathbf{R}_{11}^{a^2b} + \mathbf{R}_{22}^{a^2b})$$
$$+2Tr\mathbf{F}^{(0)}(-\mathbf{R}_{11}^{a^2}\mathbf{R}_{11}^{bc} - \mathbf{R}_{11}^{ab}\mathbf{R}_{11}^{ac} - \mathbf{R}_{11}^{ac}\mathbf{R}_{11}^{ab} - \mathbf{R}_{11}^{bc}\mathbf{R}_{11}^{a^2}$$
$$+\mathbf{R}_{22}^{a^2}\mathbf{R}_{22}^{bc} + \mathbf{R}_{22}^{ab}\mathbf{R}_{22}^{ac} + \mathbf{R}_{22}^{ac}\mathbf{R}_{22}^{ab} + \mathbf{R}_{22}^{bc}\mathbf{R}_{22}^{a^2}). \tag{104}$$

In the above formulae the projected (11) and (22) components of the density matrices are obtained from the series

$$\mathbf{X} = a\mathbf{X}_a + b\mathbf{X}_b + \cdots, \tag{105}$$

$$\mathbf{Y} = a^2\mathbf{Y}_{a^2} + b^2\mathbf{Y}_{b^2} + ab\mathbf{Y}_{ab}\cdots. \tag{106}$$

First- and second-order Fock Hamiltonians are given analogous expressions. The density corrections are given by

$$\mathbf{R}_{11}^{ab} = -(\mathbf{X}_a\mathbf{X}_b^{\dagger} + \mathbf{X}_b\mathbf{X}_a^{\dagger}), \tag{107}$$

$$\mathbf{R}_{22}^{ab} = \mathbf{X}_a^{\dagger}\mathbf{X}_b + \mathbf{X}_b^{\dagger}\mathbf{X}_a, \tag{108}$$

$$\mathbf{R}_{11}^{a^3} = -(\mathbf{X}_a\mathbf{Y}_{a^2}^{\dagger} + \mathbf{Y}_{a^2}\mathbf{X}_a^{\dagger}), \tag{109}$$

$$\mathbf{R}_{11}^{a^2b} = -(\mathbf{X}_a\mathbf{Y}_{ab}^{\dagger} + \mathbf{X}_b\mathbf{Y}_{a^2}^{\dagger} + \mathbf{Y}_{a^2}\mathbf{X}_b^{\dagger} + \mathbf{Y}_{ab}\mathbf{X}_a^{\dagger}), \tag{110}$$

$$\mathbf{R}_{11}^{abc} = -(\mathbf{X}_a\mathbf{Y}_{bc}^{\dagger} + \mathbf{X}_b\mathbf{Y}_{ac}^{\dagger} + \mathbf{X}_c\mathbf{Y}_{ab}^{\dagger} + \mathbf{Y}_{ab}\mathbf{X}_c^{\dagger} + \mathbf{Y}_{ac}\mathbf{X}_b^{\dagger} + \mathbf{Y}_{bc}\mathbf{X}_a^{\dagger}), \tag{111}$$

$$\mathbf{R}_{22}^{a^3} = \mathbf{X}_a^{\dagger}\mathbf{Y}_{a^2} + \mathbf{Y}_{a^2}^{\dagger}\mathbf{X}_a, \tag{112}$$

$$\mathbf{R}_{22}^{a^2b} = \mathbf{X}_a^{\dagger}\mathbf{Y}_{ab} + \mathbf{X}_b^{\dagger}\mathbf{Y}_{a^2} + \mathbf{Y}_{a^2}^{\dagger}\mathbf{X}_b + \mathbf{Y}_{ab}^{\dagger}\mathbf{X}_a, \tag{113}$$

$$\mathbf{R}_{22}^{abc} = \mathbf{X}_a^{\dagger}\mathbf{Y}_{bc} + \mathbf{X}_b^{\dagger}\mathbf{Y}_{ac} + \mathbf{X}_c^{\dagger}\mathbf{Y}_{ab} + \mathbf{Y}_{ab}^{\dagger}\mathbf{X}_c + \mathbf{Y}_{ac}^{\dagger}\mathbf{X}_b + \mathbf{Y}_{bc}^{\dagger}\mathbf{X}_a. \tag{114}$$

All of these formulae apply to the case of orthonormal basis sets [7]: corresponding expressions for the general case of metric Δ are easily obtained via similarity transformations, see, for instance, (70).

7. Second hyperpolarizability of benzene

The computational scheme outlined in Secs. IV and V has been applied to the calculation of the $\gamma_{\alpha\beta\delta\gamma}$ hyperpolarizability of benzene molecule, for which a number of *ab initio* studies have been already reported [15–18]. The computer program implementing the CHF algorithm has been checked with respect to corresponding finite-perturbation theory calculations [28]. For a number of molecular systems, belonging to several point groups, the results from calculations exploiting the full symmetry have been matched with corresponding ones, obtained by using lower subsymmetries for the same molecule [11], including C_1.

In particular the results of Ref. [16], obtained via a 4-31g polarized basis set, have been reproduced on an 486 IBM compatible PC, with a hard disk memory of 100 Mbyte. As a matter of fact, in that calculation, only 1 180 752 symmetry unique two-electron integrals $>1 \times 10^{-11}$ a.u. had to be stored within our method.

Five large basis sets have been employed in the present study of benzene; basis set I, which has been taken from Sadlej's tables [37], is a $(10s6p4d/6s4p)$ contracted to $[5s3p2d/3s2p]$, and contains 210 CGTOs. It has been previously adopted by us in a near Hartree-Fock calculation of electric dipole polarizability of benzene molecule [38]. According to our experience, Sadlej's basis sets [37] provide accurate estimates of first-, second-, and third-order electric properties of large molecules [39].

Basis sets II-V have been employed in estimating the Hartree-Fock limit of a number of second-order properties in the benzene molecule [40]. The primitive GTO sets range from $(11s7p2d/5s2p)$ to $(14s8p4d/8s3p)$, contracted respectively to $[6s5p1d/3s1p]$ and $[9s6p4d/6s3p]$. Although the exponents for the polarization functions of these basis sets were chosen in that paper to maximize the paramagnetic susceptibility, the extension of the basis sets (from 252 to 396 CGTO) guarantees a remarkable flexibility and excellent overall characteristics. The number $\mathcal{N}$ of symmetry unique two-electron integrals range from $\approx 11^6$ to $\approx 100^6$. The calculations have been carried out on a CONVEX C-220 and on an IBM 3090.

The ability of Sadlej basis sets [37] to provide reliable values of $\gamma_{\alpha\beta\delta\gamma}$ has been tested in a limited number of cases with encouraging results [11]. In the present work on benzene the Sadlej basis set yields theoretical estimates close to those obtained by Perrin *et al.* [16] and Karna *et al.* [17], but smaller than those reported by Augspurger and Dykstra [18]. The C-C bond distance retained in [18], however, is 1.397 Å, compared to 1.395 Åused by us, see Refs. [38] and [40].

The theoretical results provided by the large basis sets II-V are much smaller than those from previous references [15–18]: the present findings confirm that the second-hyperpolarizability is largely affected by the basis set characteristics. It is very difficult to assess the accuracy of a given CHF calculation of $\gamma_{\alpha\beta\gamma\delta}$, and it may well happen that smaller basis sets provide theoretical values of apparently better quality. Whereas the diagonal components of the electric dipole polarizability $\alpha_{\alpha\beta}$ are quadratic properties for which the Hartree-Fock limit can be estimated with relative accuracy a *posteriori*, e.g., via extended calculations [38], it does not seem possible to establish a variational principle for, and/or upper and lower bounds to, either $\beta_{\alpha\beta\gamma}$ and $\gamma_{\alpha\beta\gamma\delta}$.

As a matter of fact, the electric dipole polarizabilities obtained via basis sets II-IV are larger than those reported in Ref. 16 and Ref. 18: from Table 1 of Ref. 39 it can be seen that $\alpha_\perp$ from those basis sets ranges from 78.352 to 79.142 a.u., and that $\alpha_\parallel$

Table 1: Theoretical electric hyperpolarizability tensor of benzene molecule (in a.u.)[†].

Component	Basis sets					$PPMD^{(a)}$		$KTP^{(b)}$	$AD^{(c)}$
	I	II	III	IV	V	CHF	MP2		
γ_{xxxx}	12050	5200	6896	7593	7648	14118	19332	14164	19848
γ_{xxyy}	4017	1733	2299	2531	2549	4704	6402	4717	6612
γ_{xxzz}	6463	2432	3332	4590	4563	6498	8688	6457	9918
γ_{zzzz}	14422	4741	7241	10301	10320	12540	15708	12496	17142
$<\gamma>^{(d)}$	14482	5667	7792	9872	9793	15234	20388	15236	21948

[†] Conversion factors: 1 a.u.$=5.0366338\times10^{-40}$ e.s.u.$=6.2352583\times10^{-65}$ $C^4m^4J^{-3}$.

[a] Theoretical values from Ref. [16]

[b] Theoretical values from Ref. [17]

[c] Theoretical values from Ref. [18]

[d] The average value is defined by

$$<\gamma> = (1/5)[\gamma_{xxxx} + \gamma_{yyyy} + \gamma_{zzzz} + 2(\gamma_{xxyy} + \gamma_{yyzz} + \gamma_{zzxx})]$$

ranges from 42.305 to 45.284 a.u.. Sadlej basis sets give $\alpha_\perp =79.463$ and $\alpha_\parallel =45.448$ a.u., see Table 2 in Ref. [39]. These results are close to the estimated Hartree-Fock limits, $\alpha_\perp=79.4$ a.u., and $\alpha_\parallel=46.9$ a.u. [39]; accordingly they are much more accurate than those reported by Perrin et al. [16], i.e., $\alpha_\perp=74.22$ a.u. and $\alpha_\parallel=39.37$ a.u.. Our estimates are also more accurate than the best ones from Ref. 18, $\alpha_\perp=76.94$ a.u. and $\alpha_\parallel=41.36$ a.u..
These findings imply that our basis sets are definitely more reliable than those adopted in Ref. 16 and Ref. 18 for studying second-order electric properties. Accordingly, it seems quite difficult to understand that theoretical $\gamma_{\alpha\beta\gamma\delta}$ obtained via relatively small ad hoc basis sets are closer to the HF limit, if the same basis sets provide less accurate polarizabilities. This feature would mean that the problem of constructing suitable basis sets for the simultaneous evaluation of second-, third-, and fourth-rank electric properties of HF quality ought to be carefully reconsidered. Comparison with a few experimental values, obtained corresponding to different wavelengths [41–47], seems however to suggest that nuclear vibration [3] and electron correlation [15–18] play an important role. In particular, the correlation contributions estimated via second-order Moeller-Plesset techniques [16] are large. Accordingly, the present work confirms that CHF level of accuracy is insufficient to predict accurate hyperpolarizability of benzene molecule.
In any event, we are confident that the computational approach developed in this study, owing to its efficient use of molecular symmetry, can help develop large basis sets for first and second hyperpolarizabilities. An important aim would be that of estimating, at least at empirical level, Hartree-Fock limits for these quantities. To this end the use of basis sets polarized two times, according to the recipe developed by Sadlej [37], would seem very promising.

Acknowledgments

The authors wish to thank Dr. T. Prosperi for discussions. Financial support from the C.I.C.A.I.A. of the University of Modena, from the Italian M.U.R.S.T., and the Comitato Nazionale Scienze e Tecnologia dell'Informazione of the Italian C.N.R. is gratefully acknowledged.

References

1. A. D. Buckingham, *Adv. Chem. Phys.* **12**, 107 (1967); A. D. Buckingham and B. J. Orr, *Quart. Rev.* **21**, 195 (1967); M. P. Bogaard and B. J. Orr, in Molecular Structure and Properties, Int. Rev. Sci., Phys. Chem. Series 2, Vol. 2, A. D. Buckingham Ed., Butterworths, London 1975.

2. P. A. Franken and J. F. Ward, *Rev. Mod. Phys.* **35**, 23 (1963); J. F. Ward, *Rev. Mod. Phys.* **37**, 1 (1965); J. Orr and J. F. Ward *Molec. Phys.* **20**, 513 (1971); C. K. Miller, B. J. Orr, and J. F. Ward, *J. Chem. Phys.* **67**, 2109 (1977).

3. D. M. Bishop, *Rev. Mod. Phys.* **62**, 343 (1990); *J. Chem. Phys.* **86**, 5613 (1987); *ibid.* **90**, 3192 (1989); *ibid.* **95**, 5489 (1991); *Phys. Rev. Letters* **61**, 322 (1988); *Chem. Phys. Letters* **153**, 441 (1988).

4. R. McWeeny, Symmetry, Pergamon, Oxford, 1963.

5. S. J. Cyvin, J. E. Rauch, and J. C. Decius, *J. Chem. Phys.* **43**, 4083 (1965).

6. P. Lazzeretti and R. Zanasi, *Int. J. Quantum Chem. Phys.* **15**, 645 (1979).

7. R. McWeeny, *Phys. Rev.* **126**, 1028 (1962); G. F. Diercksen and R. McWeeny, *J. Chem. Phys.* **44**, 3554 (1966); R. McWeeny, *Chem. Phys. Letters* **1**, 567 (1968); J. L. Dodds, R. McWeeny, W. T. Raynes, and J. P. Riley, *Molec. Phys.* **33**, 611 (1977); J. L. Dodds, R. McWeeny, and A. J. Sadlej, *Molec. Phys.* **34**, 1779 (1977).

8. R. E. Wyatt and R. G. Parr, *J. Chem. Phys.* **41**, 514 (1964).

9. G. J. B. Hurst, M. Dupuis, and E. Clementi, *J. Chem. Phys.* **89**, 385 (1988).

10. P. Lazzeretti and R. Zanasi, *Chem. Phys. Letters* **39**, 323 (1976); *J. Chem. Phys.* **74**, 5216 (1981).

11. P. Lazzeretti, M. Malagoli, and R. Zanasi, *Z. Naturforsch.*, in the press

12. P. A. Christiansen and E. A. McCullough Jr., *Chem. Phys. Letters* **63**, 570 (1979).

13. J. Zyss and G. Berthier, *J. Chem. Phys.* **77**, 3635 (1982).

14. C. Daniel and M. Dupuis, *Chem. Phys. Letters* **171**, 209 (1990).

15. G. P. Das and D. S. Dudis, *Chem. Phys. Letters* **185**, 151 (1991).

16. E. Perrin, P. N. Prasad, P. Mougenot, and M. Dupuis, *J. Chem. Phys.* **91**, 4728 (1989).

17. S. P. Karna, G. B. Talapatra, and P. N. Prasad, *J. Chem. Phys.* **95**, 5873 (1991).

18. J. D. Augspurger and C. E. Dykstra, *Molec. Phys.* **76**, 229 (1992).

19. S. P. Karna, and M. Dupuis, *J. Comput. Chem.* **12**, 487 (1991).

20. J. Perez and M. Dupuis, *J. Phys. Chem.* **95**, 6525 (1991).

21. J. E. Bloor, *J. Molec. Struct. (THEOCHEM)* **234**, 173 (1991).

22. P. Chopra, L. Carlacci, H. F. King, and P. N. Prasad, *J. Phys. Chem.* **93**, 7120 (1989).

23. G. Maroulis, *J. Chem. Phys.* **94**, 1182 (1991); *ibid.* **96**, 6048 (1992); *Chem. Phys. Letters* **195**, 85 (1992).

24. G. Maroulis and A. J Thakkar, *J. Chem. Phys.* **93**, 4164 (1990).

25. M. Dory, L. Beudels, J. G. Fripiat, J. Delhalle, J. M. André, and M. Dupuis, *Int. J. Chem.* **42**, 1577 (1992).

26. B. L. Hammond and J. E. Rice, *J. Chem. Phys.* **97**, 1138 (1992).

27. G. Maroulis and D. M. Bishop, *Chem. Phys. Letters* **128**, 462 (1986); *Molec. Phys.* **58**, 273 (1986); *Theor. Chim. Acta* **69**, 161 (1986).

28. R. J. Bartlett and G. D. Purvis III, *Phys. Rev. A***20**, 1313 (1979).

29. L. Adamowicz and R. J. Bartlett, *J. Chem. Phys.***84**, 4988 (1986).

30. H. Sekino and R. J. Bartlett, *J. Chem. Phys.* **94**, 3665 (1991).

31. W. A. Parkinson and J. Oddershede, *J. Chem. Phys.* **94**, 7251 (1991).

32. G. J. M. Velders, J. M. Gillet, P. J. Becker, and D. Feil, *J. Phys. Chem.* **95**, 8601 (1991).

33. J. E. Rice, P. R. Taylor, T. J. Lee, and J. Almlöf, *J. Chem. Phys.* **94**, 4972 (1991).

34. G. Berthier, M. Defranceschi, P. Lazzeretti, G. Tsoucaris, and R. Zanasi, *J. Molec. Struct. (THEOCHEM)* **254**, 205 (1992).

35. S. P. Karna, P. N. Prasad, and M. Dupuis, *J. Chem. Phys.* **94**, 1171 (1991).

36. S. T. Epstein, The Variation Method in Quantum Chemistry, Academic, New York, 1974.

37. A. J. Sadlej, *Coll. Czech. Chem. Comm.* **53**, 1995 (1988).

38. P. Lazzeretti, M. Malagoli, and R. Zanasi, *Chem. Phys. Letters* **167**, 101 (1990).

39. P. Lazzeretti and J. A. Tossell, *J. Molec. Struct. (THEOCHEM)* **236**, 403 (1991).

40. P. Lazzeretti, M. Malagoli, and R. Zanasi, *J. Molec. Struct. (THEOCHEM)* **234**, 127 (1991).

41. P. Bogaard, A. D. Buckingham, and G. L. D. Ritchie, *Molec. Phys.* **18**, 575 (1970).

42. J. P. Hermann, *Opt. Comm.* **9**, 74 (1973).

43. M. D. Levenson and N. Bloembergen, *J. Chem. Phys.* **60**, 1323 (1974).

44. M. D. Levenson and N. Bloembergen, *Phys. Rev. B* **10**, 4447 (1974).

45. B. F. Levine and C. G. Bethea, *J. Chem. Phys.* **63**, 2666 (1975).

46. J. F. Ward and D. S. Elliot, *J. Chem. Phys.* **69**, 5438 (1978).

47. M-T. Zhao, M. Samoc, B. P. Singh, and P. N. Prasad, *J. Phys. Chem.* **93**, 7916 (1989).

Second Order Static Hyperpolarizabilities of Insaturated Polymers

D. HAMMOUTENE, G. BOUCEKKINE and A. BOUCEKKINE
Laboratoire de Chimie Théorique, U.S.T.H.B ,
BP 31 El Alia16111 Bab Ezzouar, Alger, Algérie

1. Introduction

In a previous work [1,2], we were interested in the calculation of second order hyperpolarizabilities of conjugated systems including substituted benzenes, pyridine N-oxydes and vinyl oligomers, in relation with non linear optical activity [3]. We showed that MNDO calculations were in good agreement with SCF ab initio results obtained using a double zeta basis set plus polarization and diffuse orbitals.

In this paper we present the hyperpolarizabilities, computed at the MNDO level, of different series of insaturated polymers, which are known to exhibit interesting chemical, mechanical or optical properties [4-16]. The influence of different structural factors, such as the lengthening of the polymeric chain, bond length alternation and conjugation should be investigated in order to help to the design of new active molecules.

2. Results and discussion

2.1. MOLECULES UNDER CONSIDERATION AND METHOD OF CALCULATION

Trans-polyenes $H-(-HC=CH-)_N-H$, trans-polyenynes $H-(HC=CH-C=C)_M-H$, cumulenes $H_2C=(C=C)_M=CH_2$ and polyynes $H-(C=C)_N-H$ have been studied (M=N-1). For centrosymmetric molecules, the first order hyperpolarizability β is equal to zero so that non linear effects are of second order nature . Furthermore, γ_{xxxx} (the x axis goes through the middle of the C-C bonds of the polyenes, or is the internuclear axis in the case of linear molecules) is the most important component of the second order γ hyperpolarizability tensor, the other components being negligible. Both γ_{xxxx} and the mean hyperpolarizability noted γ have been computed for the above mentioned polymers, the number of unit cells varying from N=1 to N=11.

The MNDO method [17] coupled with the finite perturbation (FP) technique [18], as implemented in the MOPAC5 [19] program has been used throughout this work.

The expression of the γ_{iiii} component as a numerical derivative, is the following:

Y. Ellinger and M. Defranceschi (eds.), Strategies and Applications in Quantum Chemistry, 297–311.
© 1996 *Kluwer Academic Publishers.*

$$\gamma_{iiii} = \frac{1}{\zeta_i^3} [\frac{1}{2} [\mu_i (2\zeta_i) - \mu_i (-2\zeta_i - \mu_i (2\zeta_i))]]$$

whereas the mean hyperpolarizability is given by:

$$\gamma = \frac{1}{5}(\sum_i \gamma_{iiii} + 2 \sum_{i<j} \gamma_{iijj})$$

μ_i (ξ_i) representing the value of the computed dipole moment when the numerical value ξ_i is given to the electric field strength.

2.2.   TRANS-POLYENES

These compounds have been the subject of several theoretical [7,11,13,20)] and experimental[21] studies. Ward and Elliott [20] measured the dynamic γ hyperpolarizability of butadiene and hexatriene in the vapour phase by means of the dc-SHG technique. Waite and Papadopoulos[7,11] computed static γ values, using a MacWeeny type Coupled Hartree-Fock Perturbation Theory (CHFPT) in the CNDO approximation, and an extended basis set. Kurtz [15] evaluated by means of a finite perturbation technique at the MNDO level [17] and using the AM1[22] and PM3[23] parametrizations, the mean γ values of a series of polyenes containing from 2 to 11 unit cells. At the ab initio level, Hurst et al.[13] and Chopra et al .[20] studied basis sets effects on γ_{xxxx} and γ. It appeared that diffuse orbitals must be included in the basis set in order to describe correctly the external part of the molecules which is the most sensitive to the electrical perturbation and to ensure the obtention of accurate values of the calculated properties.

The γ_{xxxx} and mean γ values computed at different theoretical levels are given in Table 1. We can see that the hyperpolarizability increases with the extension of the polymeric chain. It is worth noting that our MNDO values agree with the ab initio ones of Kurtz[15] but do not vary in a parallel direction to the CNDO results of Waite and Papadopoulos[7]. Note that the CNDO values are the closest to the experimental data for butadiene and hexatriene, but these latest data have been used to fit the CNDO parameters. Furthermore, the results of Hurst et al [13] show that the computed value of γ is very sensitive to any extension of the basis set. The MNDO calculations reach their best agreement with the more extended 6-31G+PD basis set. It is worth noting that a very good correlation exist between the calculated values of the two methods (coefficient of correlation equal to 0.998)[2].

Several authors have studied γ_{xxxx} which is very sensitive to the lengthening of the polymeric chain, as a function of the number N of unit cells, and have found a relationship of the form:

$$\gamma_{xxxx} = K. N^\lambda$$

where K and λ are parameters which can be evaluated using a least square method. We produce in Fig. 1, the variation curve of γ_{xxxx}. In Table 2, are given the computed λ values using different techniques.

Table 1: γ_{xxxx} and γ hyperpolarizability (in 10^{-36} esu) of transpolyenes $C_{2N}H_{2N+2}$

N	2	3	4	5	6	7	8	9	10	11
γ^a_{xxxx} (MNDO)	10.62	90.06	265.49	618.75	1223.73	2078.14	3123.11	4352.74	5828.97	7460.53
γa (MNDO)	2.20	17.54	53.19	122.38	243.82	413.85	625.73	867.78	1160.09	1483.08
γ^b (CNDO)	16.89	36.44	62.79	104.05	162.65	248.01				
γ^c (MNDO)	2.15	15.22	52.22	125.81	244.01	409.33	620.06	871.77	1159.29	1477.10
γ^c (AM1)	1.88	13.39	48.14	121.74	247.02	431.31	676.31	979.73	1336.24	1739.94
γ^c (PM3)	1.96	12.82	44.25	108.58	214.84	366.97	564.51	804.06	1080.72	1389.31
γ^d (ab initio) STO-3G	0.25	2.75	11.37	31.21	66.89	121.25	198.04	291.13	406.56	527.91
6-31G	0.55	4.99	20.60	57.90	128.28	240.64	408.58	621.45	899.35	1202.39
6-31G*	0.53	4.65	19.01	53.16	117.18	218.98	369.99			
6-31G +pd	7.50	17.74	41.53	90.13	174.63	304.86	493.13			
γ^e(exp)	13.80 (±0.78)	45.18 (±0.45)								

a : MNDO values (this work)
b : ref.[7,11]
c : ref.[15]
d : ref.[13]
e : experimental dynamic values taken from ref.[20]

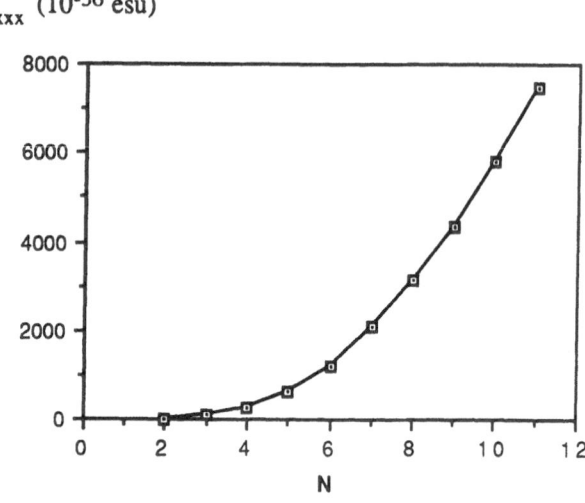

Figure 1: Variation of γ_{xxxx} with N in polyenes

We can see that our MNDO value $\lambda = 3.75$ is in better agreement with the ab initio results than with the empirical ones, $\lambda = 5 - 5.3$ [24,25].

Table 2: λ parameter values in the relation $\gamma_{xxxx} = KN^{\lambda}$

Method	λ	ref.
MNDO	3.75[a]	-
PPP	4.25	9
Free electron	5	24
Huckel	5.3	25
ab initio -CPHF	[3-4] [b]	13

[a] Our work ($K = 1.22 \ 10^{-36}$ esu)
[b] range of the values obtained with different basis sets and several
    Δr values ($\Delta r = r_{c-c} - r_{c=c}$)

2.3. TRANS POLYENYNES

Polydiacetylenes which constitute an important class of polenynic polymers can be synthetized photochemically in the solid state from substituted diacetylenes. Experimental studies have shown that polydiacetylenes exhibit χ^3 electrical susceptibilities similar to covalent semi-conductors' ones[24-28] either in the solid state[29] or in solution[30].

However, this kind of compounds has not been extensively studied theoretically[12,15]. At the semi-empirical level , we point out the MNDO calculations of Williams[12] concerning chains of less than 14 carbon atoms, and the INDO computations of Kirtman[15] related to polymeric chains containing up to 60 carbons. To our knowledge no ab-initio evaluation of γ has been done for polyenynes. In Table 3, are given our γ_{xxxx} and γ values, obtained using the MNDO method, and Williams' ones[12].

As in the polyenes case, we can see that these values increase with the lengthening of the polymeric chain. We observe the good agreement between our values and those obtained by Williams[12] for the three first compounds of the series, whereas for the higher polymers our values are lower. After checking of the calculations, it seems to us that an error could have occur in Williams' computations.

The variation of γ_{xxxx} in the polyenynes as a function of N (Fig. 2) follows the relationship

$$\gamma_{xxxx} = 0.94 \ 10^{-36} \ . \ N^{3.44}$$

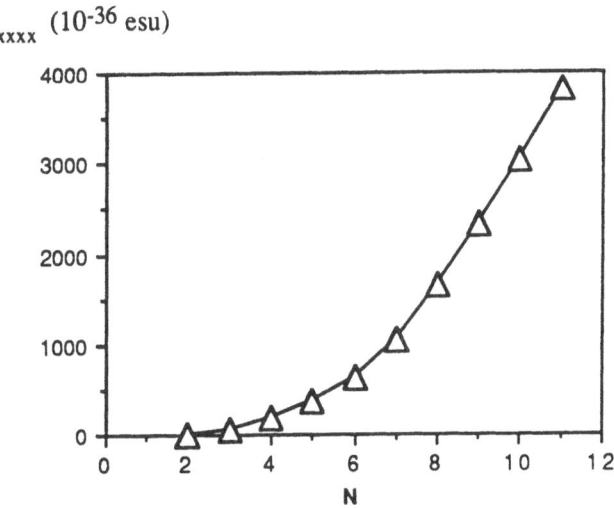

Figure 2: Variation of γ_{xxxx} with N in polyenynes

2.4.   CUMULENES

The geometry of these compounds is very different from the usual conjugated structures which generally exhibit bond alternation. For this reason, cumulenes possess a great structural and electronic homogeneity. Very few theoretical studies have been carried out on these polymers. However, we note the non empirical calculations of Chopra et al.[20] at the SCF level using a 3-21G basis set, of the γ hyperpolarizability of the first cumulenes (N = 2,3,4). On another hand, Beratan et al.[31] carried out tight binding computations to evaluate γ_{xxxx}/N of higher cumulenes, up to 80 carbon atoms.

Table 3: γ_{xxxx} and γ hyperpolarizability (in 10^{-36} esu) of polyenynes

N	2	3	4	5	6	7	8	9	10	11
γ_{xxxx}[a]	6.12	59.04	194.49	357.70	631.17	1044.95	1646.67	2328.47	3040.60	3818.86
γ[a]	1.24	11.47	38.27	70.50	123.42	206.81	324.24	460.74	599.74	758.51
γ[b]	1.2	11.2	38.1	130.1	347.3	594.4	-	-	-	-

a : MNDO values (this work)
b : MNDO values from ref.[12]

Table 4: γ_{xxxx} and γ hyperpolarizability (in 10^{-36} esu) of cumulenes

N	2	3	4	5	6	7	8	9	10	11
γ_{xxxx}[a]	-1.76	6.57	47.09	161.67	539.41	1463.88	2724.27	5489.22	9858.37	17721.64
γ_{xxxx}[b]	-4.70	-26.74	-111.31	-	-	-	-	-	-	-
γ[a]	-0.11	1.76	10.12	33.37	109.36	302.45	547.12	1101.45	1984.17	3555.31

a : MNDO values (this work)
b : The ab initio values of ref.[24] have been multiplied by 6 in order to obtain the same γ definition

In Table 4, are given our γ_{xxxx} MNDO values and those of Chopra et al.[20]. The mean hyperpolarizability increases in a non linear manner with the extension of the polymeric chain. The MNDO results are, C_4H_4 excepted, of a different sign and lower in absolute value than the Chopra et al values. We believe that this discrepancy is due to the fact that the basis set used in the ab initio calculation does not include diffuse orbitals which are necessary to describe correctly this kind of electric properties[32-34]. Furthermore, we see that γ_{xxxx} varies as a function of N, according to the following relationship (Fig. 3):

$$\gamma_{xxxx} = 0.01 \ N^{6.02}$$

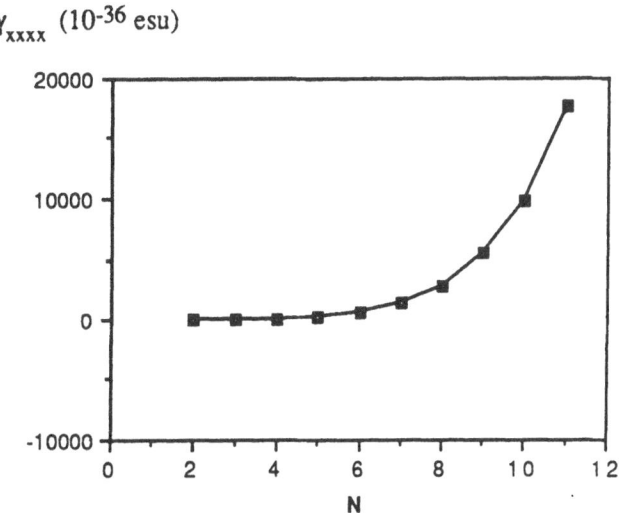

γ_{xxxx} $(10^{-36}$ esu$)$

Figure 3: Variation of γ_{xxxx} with N in cumulenes

2.5. POLYYNES

These monodimensional compounds, rich in π electrons, have been the object of several experimental[35] and theoretical work[20,35,36]. Perry et al.[35], using a powder SHG technique, have studied diaryl polyynes and have shown that some of them exhibit second order hyperpolarizabilities of very high magnitude. On another hand, Jameson and Fowler[36] carried out ab initio calculations in order to study basis sets effects on the electrical properties of acetylene and diacetylene. Furthermore, Chopra et al [20], then Maroulis and Thakkar [37] have been interested in the influence of the lengthening of the polymeric chain on these properties, and studied polyynes up to 8 carbon atoms. Beratan et al.[31] carried out tight binding calculations on high polyynes. Our MNDO results and the values obtained by the above mentioned authors are given in Table 5.

We note that, as previously, γ and γ_{xxxx} increase with the lengthening of the polymeric chain. It is worth noting that the MNDO results vary in a parallel manner with the ab initio

Table 5: γ_{xxxx} and γ hyperpolarizability values (in 10^{-36} esu) in some polyynes $C_{2N}H_2$

N	2	3	4	5	6	7	8	9	10	11
γ_{xxxx} [a] (MNDO)	3.85	30.09	114.62	321.27	533.90	905.69	1455.05	2135.09	2985.53	3901.40
γ_{xxxx} [b](ab initio) STO-3G	1.14	8.64	30.54							
3-21G	2.52	16.98	58.74							
6-31G	2.94									
6-31G*	2.70									
3-21Gsp	10.02									
3-21G+pdd	9.42									
γ_{xxxx} [c] (ab initio)	9.84									
γ_{xxxx} [d,e,f](ab initio)	10.17[d]	32.02[e]	76.78[f]							
γ [a] (MNDO)	0.96	6.33	23.34	64.77	109.91	183.18	293.76	432.15	601.14	785.16
γ [c] (ab initio)	4.92									
γ [d,e,f] (ab initio)	5.78[d]	10.30[c]	18.18[f]							

a : this work; b: ref.[20] ; c: ref.[36] ; d,e,f: ref.[37]

values which have been obtained at the SCF level using basis sets including diffuse orbitals. As previously, a $\gamma_{xxxx} = K\,N^\lambda$ relationship exists between γ_{xxxx} and N, and the λ MNDO value equal to 3.97 is close to the result of Maroulis and Thakkar ($\lambda = 3.0$). The corresponding graph has been reported in Figure 4.

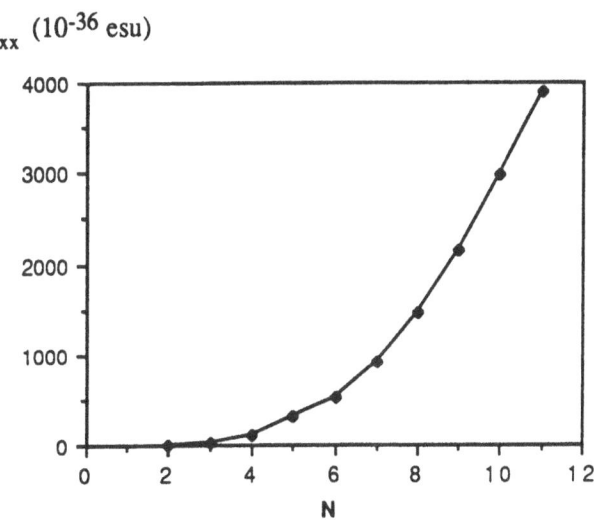

γ_{xxxx} (10^{-36} esu)

Figure 4: Variation of γ_{xxxx} with N in polyynes

2.6. COMPARISON OF THE γ_{xxxx} EVOLUTION IN THE FOUR SERIES OF POLYMERS

In Table 6, are given the MNDO γ_{xxxx} values of the four series of polymers, polyenes (A), polyenynes (B), cumulenes (C) and polyynes (D).
In Figure 5, the variation curve of γ_{xxxx}/N as a function of N, is plotted.

As can be seen in Table 6 and Figure 5, up to $N = 5$, γ_{xxxx} varies approximately as follows:
$$\gamma_{xxxx}(C) < \gamma_{xxxx}(D) < \gamma_{xxxx}(B) < \gamma_{xxxx}(A)$$
and $\gamma_{xxxx}(B) = 1/2\ (\gamma_{xxxx}(A) + \gamma_{xxxx}(D))$, for $N < 4$.
For $N>5$, and more particularly for $N=7$, we observe an important increase of γ_{xxxx} (C) relatively to $\gamma_{xxxx}(B)$ and $\gamma_{xxxx}(D)$, whereas $\gamma_{xxxx}(A)$ still remains of the higher magnitude.

Table 6: MNDO γ_{xxxx} hyperpolarizability (in 10^{-36} esu) of the four studied oligomers

N	2	3	4	5	6	7	8	9	10	11
Polyenes (A)	10.62	90.06	265.69	618.75	1223.73	2078.14	3123.11	4352.74	5828.97	7460.53
Polyenynes(B)	6.12	59.04	194.49	357.70	631.17	1044.95	1646.67	2328.47	3040.60	3818.86
Cumulenes (C)	-1.76	6.57	47.09	161.67	539.41	1463.88	2724.27	5489.22	9858.37	17721.64
Polyynes (D)	3.85	30.09	114.62	321.27	533.90	905.69	1455.05	2135.09	2985.53	3901.40

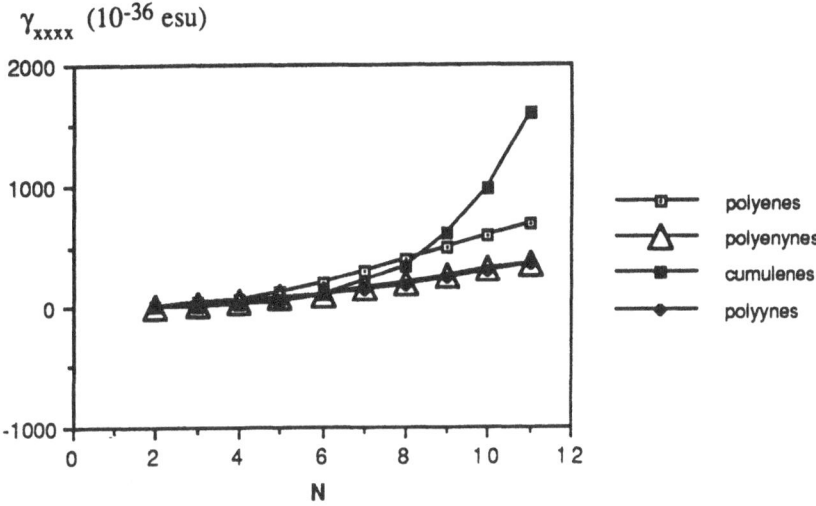

Figure 5: Variation of γ_{xxxx}/N with N in the 4 series of oligomers

Beyond N=9, γ_{xxxx}(C) becomes higher than γ_{xxxx}(A). In general, the relative classification of the studied polymers hyperpolarizabilities does not follow the increase in the number of π electrons. An explanation can be found, if we consider the two important factors which are the lengthening of the polymeric chain and bond alternation. In Table 7, are given the MNDO optimized lengths L, of the studied oligomers. The variation of L as a function of N, is plotted in Figure 6. We can see that for any N value, we have approximately the classification:

$$L(D) < L(C) < L(B) < L(A).$$

For the lower polymers (N<5) one has the same classification of L and γ_{xxxx} for polyenes and polenynes.

Beyond N=7, the longitudinal γ_{xxxx} hyperpolarizability of cumulenes increases and becomes higher than the corresponding polyenes values for 8<N<11. In order to explain this result, we computed a bond alternation index Δr defined as the mean value of the differences of the bond lengths of consecutive C-C bonds. The values obtained are given in Table 8. The variation of Δr as a function of N is plotted in Figure 7, for the four series of polymers. These curves indicate that Δr is the smallest in the cumulenes family and that its value is negligible beyond N=7. The regular geometry of these compounds is certainly at the origin of their hyperpolarizability exaltation, particularly beyond N=9. This result is in agreement with the work of André et al.[38] on polyenes, who showed that bond alternation reduces the magnitude of electrical polarizabilities and hyperpolarizabilities

Furthermore, it is worth noting that up to N=11 (Figure 5) no saturation of the γ tensor is observed. Beratan et al.[31] estimate that no less than forty unit cells are necessary to reach such a saturation in the case of cumulenes.

L (Å)

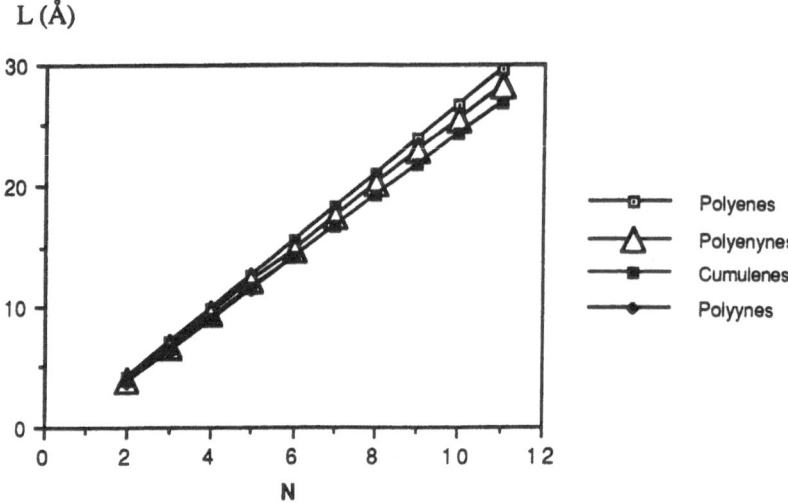

Figure 6: Variation of the length L with N in the 4 series of oligomers

Bond
alternation (Å)

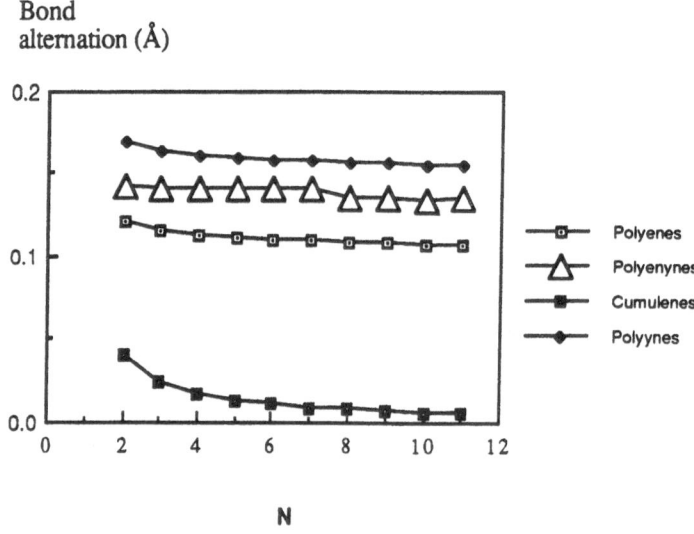

Figure 7: Variation of the bond alternation index with N in the 4 series of oligomers

Table 7: Chain length L (in A°) in the four oligomer series

N	2	3	4	5	6	7	8	9	10	11
Polyenes (A)	4.15	6.97	9.79	12.61	15.44	18.25	21.07	23.89	26.72	29.53
Polyenynes(B)	3.96	6.72	9.34	12.10	14.72	17.48	20.12	22.89	25.51	28.28
Cumulenes (C)	3.89	6.44	8.99	11.55	14.10	16.70	19.20	21.75	24.30	26.85
Polyynes (D)	3.77	6.33	8.90	11.47	14.03	16.60	19.17	21.73	24.30	26.87

Table 8: Δr (A°) mean variation

N	2	3	4	5	6	7	8	9	10	11
Polyenes (A)	0.121	0.115	0.112	0.110	0.109	0.109	0.108	0.108	0.107	0.107
Polyenynes(B)	0.142	0.141	0.141	0.141	0.141	0.141	0.135	0.135	0.134	0.135
Cumulenes (C)	0.040	0.024	0.017	0.013	0.011	0.009	0.008	0.007	0.006	0.006
Polyynes (D)	0.169	0.163	0.160	0.159	0.158	0.157	0.156	0.156	0.155	0.155

3. Conclusions

The MNDO method combined with a finite perturbation technique has been used for the computation of the static hyperpolarizabilities γ and γ_{xxxx} of four series of insaturated oligomers, including polyenes, polyenynes, cumulenes and polyynes. The MNDO results are in good agreement with the available ab initio values obtained at the SCF level using extended basis sets including diffuse orbitals. This study permits to confirm the ability of the semi-empirical MNDO method to give reliable values for second order hyperpolarizability with a very small computing time. The computed electrical properties increase in a non linear manner with the lengthening of the polymeric chain. Actually we note a relationship between γ_{xxxx} and the number of unit cells N of the form $\gamma_{xxxx} = K N^{\lambda}$. The γ_{xxxx} hyperpolarizabilities of polyenes are two to three times higher than those of polyynes, because the effect of the polyenes length and the weaker bond alternation compensate for the richest but localized π system of polyynes. Polyenynes hyperpolarizabilities values are generally intermediate between those of polyenes and polyynes. Higher cumulenes exhibit the greatest hyperpolarizabilities.This is due to the quasi-inexistence of bond alternation in their structure.

References

1.   D. Hammoutene, G. Boucekkine, A. Boucekkine, G. Berthier and A. Le Beuze, *Molecular Engineering,* **1**, 333 (1992).
2.   D. Hammoutene, G. Boucekkine, A. Boucekkine and G. Berthier, *J. Mol. Struct.* (submitted).
3.   J. Zyss, *J. Chem. Phys.* **70** , 3333 (1979); J. Zyss and G. Berthier, *J. Chem. Phys.* **77**, 3635 (1982).
4.   K. C. Rustagi and J. Ducuing, *Opt. Commun.* **10**, 258 (1974).
5.   H.F. Hameka, *J. Chem. Phys.* **67**, 2935 (1977);
        E.F. McIntyre and H. F. Hameka, *J. Chem. Phys.* **68**, 3481 (1978).
6.   O. Zamani-Khamiri, E.F. McIntyre and H.F. Hameka, *J. Chem. Phys.* **72**, 5906 (1980); O. Zamani-Khamiri and H.F. Hameka, *J. Chem. Phys.* **73**, 5693 (1980).
7.   M.G. Papadopoulos, J. Waite and C.A. Nicolaides, *J. Chem. Phys.*, **77**, 2527 (1982).
8.   B. Kirtman, *Chem. Phys. Lett.* **143**, 81 (1988).
9.   C.P. DeMelo and R. Silbey, *J. Chem. Phys.* **88** , 2558 (1988); ibid. *J. Chem. Phys.* **88**, 2567 (1988).
10.   B.M. Pierce, *J. Chem. Phys.* **91**, 791 (1989); ibid., *Spec. Publ.-R. Soc. Chim.* **69**, 48 (1989).
11.   M.G. Papadopoulos and J. Waite,*J. Mol. Struct.* **202**, 121 (1989).
12.   G.R.J. Williams, *J. Mol. Struct.* **153**, 185 (1987).
13.   G.J.B. Hurst, M. Dupuis and E. Clementi, *J. Chem. Phys.* **89**, 385 (1988).
14.   B. Kirtman, *Int. J. Quantum Chem.* **36**, 119 (1989).
15.   H.A. Kurtz, *Int. J. Quantum Chem. Quant. Chem. Symp.* **24**, 791 (1990).
16.   J.M. Andre and J. Delhalle, *Chem. Rev.* **91**, 843 (1991).
17.   M.J.S. Dewar and W. Thiel, *J. Am. Chem. Soc.* **99**, 4899 (1977); ibid., *J. Am. Chem. Soc.* **99**, 4907 (1977); *Theor. Chim. Acta* **46**, 89 (1977).
18.   H.D. Cohen and C.C.J. Roothaan, *J. Chem. Phys.* **43**, S34 (1965).
19.   J.J.P. Stewart, *Q.C.P.E.* n° 455 (1983); MOPAC version 5.0 (1989).

20. P. Chopra, L. Carlacci, H.F. King and P. N. Prasad, *J. Phys. Chem.* **93**, 7120 (1989).
21. J.F. Ward and D.S. Elliott, *J. Chem. Phys.* **69**, 5438 (1978).
22. M.J.S. Dewar, E.G. Zoebisch, E.F. Healy and J.J.P. Stewart, *J. Am. Chem. Soc.* **107**, 3902 (1985).
23. J.J.P. Stewart, *J. Comput. Chem.*, **10**, 209 (1989); ibid., **10**, 221 (1989).
24. G.M. Carter, M.K. Thakur, Y.J. Chen and J.V. Hryniewicz, *Appl. Phys. Lett.* 457 (1985).
25. A.J. Heeger, D. Moses and M. Sinclair, *ACS Symp. Ser.,* **346,**372 (1987).
26.   W.M. Dennis, W. Blau and D.J. Bradley, *Opt. Eng.* **25**, 538 (1986).
27. M. Sinclair, D. Moses, A.J. Heeger, K. Vilhelmsson, B. Valk and M. Salour, *Solid State Commun.* **61**, 221 (1987).
28. P.A. Chollet, F. Kajzar and J. Messier, *Mater. Sci.* **13**, 39 (1987).
29. C. Sauteret, J.P. Hermann, R. Frey, F. Pradere, J. Ducuing, R.H. Baughman and R.R. Chance, *Phys. Rev. Lett.* **36**, 956 (1976).
30. M.L. Shand and R.R. Chance, *J. Chem. Phys.* **69**, 4482 (1978).
31. D.N.Beratan, M.A. Lee, D.W. Allender and S. Risser, <u>Liquid crystal chemistry, physics and applications</u> *SPIE* **1080**, p101 (1989).
32. H.J. Werner and W. Meyer, *Mol. Phys.* **31**, 855 (1976).
33. P. Lazzeretti and R. Zanasi, *Chem. Phys. Lett.* **39**, 323 (1976); ibid., *J. Chem. Phys.* **74**, 5216 (1981).
34. A. J. Sadlej, *Chem. Phys. Lett.* **47**, 50 (1977).
35. J.W.Perry, A.E. Stiegman, S.R. Marder, D.R. Coulter, D.N. Beratan, D.E. Brinza, F.L. Flavetter and R.H. Grubbs, <u>Non linear optical properties of organic materials</u> , *SPIE*, **971**, p17 (1988).
36. C.J. Jameson and P.W. Fowler, *J. Chem. Phys.* **85**, 3432 (1986).
37. G. Maroulis and A.J. Thakkar, *J. Chem. Phys.* **95** ,9060 (1991).
38. J.M. Andre, C. Barbier, V.P. Bodart and J. Delhalle, <u>Non linear optical properties of organic molecules and crystals</u>, Vol. 2, D.S. Chemla and J. Zyss Ed. Academic Press, p137 (1987).

An ab initio Study of the Magnetic Properties of the Isoelectronic Series BeH⁻, BH, CH⁺ and MgH⁻, AlH, SiH⁺

An ab initio Study of the Magnetic Properties of the Isoelectronic Series BeH^-, BH, CH^+ and MgH^-, AlH, SiH^+

M. NAIT ACHOUR, A. BOUCEKKINE and R.LISSILLOUR
Laboratoire de Chimie Théorique, Institut de Chimie, U.S.T.H.B.,BP 31 El-Alia 16111 Bab-Ezzouar, Alger, Algérie and Laboratoire de Chimie Théorique, Université de Rennes 35042 Rennes Cedex, France

1. Introduction

The magnetic properties of the BH molecule have been theoretically studied by several authors [1-6], because this compound is supposed to exhibit a temperature-independent paramagnetism. Its isoelectronic molecules, BeH^- and CH^+ have been the subject of similar studies [3-5]. Recently Fowler and Steiner [5] emitted the hypothesis that an isolated BeH^- anion is weakly diamagnetic, whereas it could become paramagnetic under some conditions of environment. Furthermore, the calculation of the magnetic susceptibility of the AlH molecule which is of particular interest because its electronic structure is similar to that of BH, has shown that this compound is weakly diamagnetic [7]. In this work, we plan to re-examine more systematically the magnetic properties of BeH^- by mean of SCF ab initio calculations including several sets of diffuse orbitals, and to extend the study to the MgH^- and SiH^+ molecules which are the AlH analogs.

2. Calculations and discussion

2.1. METHOD OF CALCULATION

First of all, let us point out that electron correlation effects on second order magnetic properties (susceptibilities, screening constants) were investigated by several authors [6,8], and that it was found that calculations at the Hartree-Fock level give reliable results for these properties. Actually, it is well known that computed SCF diamagnetic susceptibilities, using large basis sets, agree excellently with the corresponding experimental values. We retained, for our part, to employ at the SCF level, London field-dependent atomic orbitals (the so-called gauge invariant atomic orbitals: GIAO) [9] which ensure the origin independence of the calculated magnetic susceptibilities. The London approach has been extended and widely used to study the magnetic properties of conjugated molecules between 1951 and 1953 by G. Berthier et al. [10-13]. At the ab initio level, it was shown [3,4] that the calculated magnetic susceptibilities using the London functions in a triple-zeta basis set supplemented by one eccentric polarization function (a s-type bond function) are very close to those obtained using very large field-independent basis set near the Hartree-

313

Y. Ellinger and M. Defranceschi (eds.), Strategies and Applications in Quantum Chemistry, 313–318.
© 1996 *Kluwer Academic Publishers.*

Fock limit. It has also been pointed out by Wolinski et al. [14] that the use of GIAO's permits to reduce the basis set dimension when evaluating screening constants. All the details of the method of calculation are given in references [15-19], so they will not repeated here.

2.2. RESULTS FOR BeH⁻

The starting point is our previously performed calculations [3] using the Huzinaga basis set [20] (9s) for Be and (4s) for H, triple-zeta contracted, supplemented by the three 2p orbitals proposed for Be by Ahlrichs and Taylor [21] with exponents equal to 1.2 , 0.3 and 0.05 respectively. This initial basis set, noted I, includes one s-type bond-function the exponent of which is equal to 0.5647. Several sets of diffuse orbitals have then been added to this basis I. Their corresponding exponents were determined by downward extrapolation from the valence basis set, using the Raffenetti [22] and Ahlrichs [21] procedure. Three supplementary basis sets noted II, III and IV containing respectively one, two and three extra diffuse orbitals, have thus been constituted. The corresponding exponents α of these supplementary diffuse orbitals are reported in Table 1.

Table 1: Diffuse exponents of H and Be

Basis set	H	Be	
	α_s	α_s	α_p
II	0.022725	0.016778	0.010920
III	0.022725	0.016778	0.010920
	0.004801	0.004842	0.002229
IV	0.022725	0.016778	0.010920
	0.004801	0.004842	0.002229
	0.001014	0.001397	0.000455

In Table 2, we have reported (in 10^{-6} erg.G^{-2}.mol^{-1}) the $\chi_{//}$ and $\chi_{\perp}$ principal components, respectively parallel and perpendicular to the internuclear axis, the mean valu $\chi = 1/3(\chi_{//} + 2\chi_{\perp})$ and the anisotropy $\Delta\chi = \chi_{//} - \chi_{\perp}$ of the BeH⁻ molecule susceptibility tensor, obtained using each of the previously defined basis sets. The internuclear distance R has been taken equal to 2.5 a.u.

As we can see,the diffuse orbitals play a dramatic part in the description of the magnetic properties of BeH⁻ : not less two sets of these orbitals (basis set III) are necessary to obtain an accurate and converging value of the susceptibility.The BeH⁻ anion should be diamagnetic and its mean susceptibility is of the order of -2.10^{-6}erg.G^{-2}.mol^{-1}. Note that the use of a single set of supplementary diffuse orbitals is not sufficient to bring to light this magnetic property.

Table 2: BeH⁻ magnetic susceptibility

Basis set	$\chi_{//}$	$\chi_\perp$	χ	$\Delta\chi$
I	-23.84	38.17	7.50	-62.01
II	-35.31	21.23	2.38	-56.54
III	-39.00	16.48	-2.01	-55.48
IV	-39.15	16.35	-2.15	-55.50

We carried out a second calculation for BeH⁻ at an internuclear distance of 2.67 a.u. corresponding to the minimum of energy. The mean magnetic susceptibility χ obtained value, which is equal to $-0.86.10^{-6}$ erg.G^{-2}.mol^{-1}, agrees more closely to the value calculated by Fowler and Steiner [5], at the same internuclear distance, which is equal to $-0.50 \ 10^{-6}$ erg. G^{-2}.mol^{-1}.

2.3. RESULTS FOR AlH, SiH⁺ AND MgH⁻

The Veillard basis set [23] (11s,9p) has been used for Al and Si, and the (11s,6p) basis of the same author has been retained for Mg. However, three p orbitals have been added to this last basis set, their exponents beeing calculated by downward extrapolation. The basis sets for Al, Si and Mg have been contracted in a triple-zeta type. For the hydrogen atom, the Dunning [24] triple-zeta basis set has been used. We have extended these basis sets by mean of a s-type bond function. We have optimized the exponents α and locations d of these eccentric polarization functions, and the internuclear distance R of each of the studied molecules. These optimized parameters are given in Table 3.

Table 3: Optimized parameters

Compound	R(Å)	α	d(Å)
MgH⁻	1.81	0.3	0.905
AlH	1.65	0.3	0.825
SiH⁺	1.52	0.5	0.76

The optimization of the geometry leads to a good agreement with experiment for AlH (R_{exp} = 1.648 Å [25]). For MgH⁻ a previously calculated internuclear distance [25] using a 6-31G* basis set, is equal to 1.863 Å which is not very close to our value for this molecule. Note, however, that in the particular case of MgH⁻, the energy presents a flat minimum between 1.79 Å and 1.82 Å, its variation beeing of the order of 10^{-5} a.u in this interval.

For MgH⁻, we have extended the previously defined basis set, noted I, by means of one and two sets of diffuse orbitals, the exponents of which have been computed by downward extrapolation. These basis sets are reported in Table 4.

Table 4: Diffuse orbital exponents for Mg and H

Basis set	H	Mg	
	α_s	α_s	α_p
I	-	-	0.2175
			0.0764
			0.0268
II	0.022725	0.08424	0.009430
III	0.022725	0.08424	0.009430
	0.004801	0.01983	0.003197

In Table 5, we show the calculated magnetic susceptibilities for AlH and SiH$^+$, and the Lipscomb et al. [7] obtained values for AlH using an extended basis set of field independent Slater type orbitals.

Table 5: Magnetic susceptibilities (10^{-6}erg.G^{-2}.mol^{-1}) of AlH and SiH$^+$

Compoud	$\chi_{//}$	$\chi_\perp$	χ	$\Delta\chi$
AlH	-21.85	8.63	-1.53	-30.48
AlH [a]	-21.87	8.88	-1.37	-30.75
SiH$^+$	-15.32	28.06	13.60	-43.38

[a] ref. [7]

Our results are in close agreement with the values obtained by Lipscomb et al. [7]; this confirm the validity of our approach. Note that, AlH is predicted to be weakly diamagnetic but SiH$^+$ should be paramagnetic.

The results for the MgH$^-$ molecule are given in Table 6.

Table 6: MgH$^-$ magnetic susceptibilities (10^{-6} erg.G^{-2}.mol^{-1})

Basis set	$\chi_{//}$	$\chi_\perp$	χ	$\Delta\chi$
I	-38.42	9.54	-6.44	-47.96
II	-51.38	-7.98	-22.44	-43.40
III	-51.56	-7.26	-22.02	-44.30

Our results indicate that the basis set I cannot describe correctly MgH⁻ and that the magnetic susceptibility of this anion is strongly depending on the inclusion of diffuse orbitals in the basis set. We notice that the basis set II permits to obtain reliable results, its further extension by extra diffuse functions (basis set III) leading approximately to the same results. MgH⁻ should be diamagnetic, and its mean susceptibility is of the order of -22. 10^{-6} erg.G^{-2}.mol^{-1}.

It should be noted that diffuse functions which are necessary for a good description of the magnetic properties of anions, have been found needless when computing the susceptibilities of the neutral molecules.

In Table 7, we reported the mean magnetic susceptibilities χ of the BH and CH⁺ molecules, obtained by G. Berthier et al. [3], using the same SCF ab initio method, employing a triple-zeta basis set augmented by a s-type bond function. We produce also in this table, our values for BeH⁻ and those of the AlH, SiH⁺ and MgH⁻ series for comparison. We note, that the magnetic susceptibilities exhibit the same features in the two analogous series of molecules, namely that diamagnetism decreases when the heavy atom nuclear charge increases.

Table 7: Mean magnetic susceptibilities χ (10^{-6} erg.G^{-2}.mol^{-1})

Compound	χ	χ	$\Delta\chi$
BeH⁻	-0.86	-22.02	MgH⁻
BH	18.72	-1.53	AlH
CH⁺	32.07	13.60	SiH⁺

3. Conclusion

The good agreement between our results and those obtained by Lipscomb and al. [7], permits to think that the calculated susceptibility values for SiH⁺ and MgH⁻ are accurate. We observe also, that SiH⁺is predicted to be paramagnetic as its CH⁺ counterpart [3], whereas the AlH and MgH⁻ molecules are diamagnetic. We could confirm the weakly diamagnetic character of the BeH⁻ molecule whose susceptibility is strongly dependent on the introduction of several diffuse orbitals in the basis set. In both series of compounds, diamagnetism decreases with the increase of the heavy atom nuclear charge.

References

1. R.M. Stevens, R.M. Pitzer and W.N. Lipscomb, *J. Chem. Phys.* **42**, 3666 (1965)
2. R.A. Hegstrom and W.N. Lipscomb, *J. Chem. Phys.* **45**, 2378 (1966).
3. G. Boucekkine-Yaker, A. Boucekkine, A. Zaucer and G. Berthier, *Int. J. Quant. Chem..* **23**, 365 (1983).
4. G. Boucekkine-Yaker, A. Boucekkine, and G. Berthier, *Int. J. Quant. Chem..* **18**, 369 (1984).

5.   P.W. Fowler and E. Steiner, *Mol. Phys.* **74**, 1147 (1991).
6.   M. Iwai and A. Saika, *Int. J. Quant. Chem.*. **24**, 623 (1983).
7.   E. A. Laws, R.M. Stevens and W.N. Lipscomb, *J. Chem. Phys.* **54**, 4269 (1971)
8.   M. Iwai and A. Saika, *J. Chem. Phys.* **77**, 1951 (1982).
9.   F. London, *J. Phys. Radium* **8**, 397 (1937).
10.   G. Berthier, M. Maillot and B. Pullman, *J. Phys. Radium* **12**, 717 (1951).
11.   G. Berthier, M. Maillot, A. Pullman and B. Pullman, *J. Phys. Radium* **13**, 15 (1952).
12.   M. Maillot, G. Berthier and B. Pullman, *J. Phys. Radium* **12**, 652 (1951).
13.   M. Maillot, G. Berthier and B. Pullman, *J. Phys. Radium* **50**, 176 (1953).
14.   K. Wolinski,J.F. Hinton and P. Pulay, *J. Am. Chem. Soc.* **112**, 8251 (1990).
15.   M. Zaucer and A. Azman, *Croat. Chem. Acta* **47**, 17 (1975).
16.   R. Ditchfield, *Mol. Phys.* **27**, 789 (1974).
17.   M. Zaucer, D. Pumpernik, M. Hladnik and A. Azman, *Z. Nathurforsch A* **32**, 411 (1977)
18.   G. Boucekkine-Yaker, L. Brunet and G. Berthier, *J. Chim. Phys.* **84**, 671 (1987).
19.   A. Boucekkine, G. Boucekkine-Yaker, M. Nait Achour and G. Berthier, *J. Mol. Struct.* (Theochem) **166**, 109 (1988).
20.   S. Huzinaga, *J. Chem. Phys.* **42**, 1293 (1965).
21.   R.Ahlrichs and P.R.Taylor, *J. Chim. Phys.* **78**, 315 (1981).
22.   R.C. Raffenetti, *J. Chem. Phys.* **58**, 4452 (1973).
23.   A.Veillard, *Theoret..Chim. Acta* **12**, 405 (1968).
24.   T.H. Dunning Jr., *J. Chem. Phys.* **53**, 2830 (1970).
25.   G.W. Spitznagel, T. Clark, P.von Ragué Schleyer and W.J. Hehre, *J.Comput. Chem.* **8**, 1109 (1987).

CI Calculations of Miscellaneous Spectroscopic Observables for the PN X¹Σ, A¹Π and ¹Δ States

G. de BROUCKERE
University of Amsterdam, Department of Physics and Astronomy, Valckenierstraat 65,
1018 XE Amsterdam, The Nederlands

1. Introduction

1.1 PN X$^{1}\Sigma^{+}$ STATE

The PN molecule whose ground state electronic configuration is

$$(X^{1}\Sigma^{+})\ 1\sigma^{2}2\sigma^{2}3\sigma^{2}4\sigma^{2}1\pi^{4}5\sigma^{2}6\sigma^{2}7\sigma^{2}2\pi^{4}$$

first investigated by Curry et al [1], has attracted a considerable amount of interest among experimentalists due to the availability of high quality optical spectra. The presence of sharp spectral lines made it possible to determine many of the rotational-vibrational spectroscopic constants, such as r_{e},k_{e}, $\omega_{e}x_{e}(y_{e})$, B_{e},α_{e} with high accuracy. Some of these constants were subsequently refined by Wyze et al [2] using high resolution microwave spectroscopy by means of which several pure rotational transitions were also measured.
In recent years PN has been among the growing number of first and second row molecular species observed from a variety of astronomical sources, including Orion(KL), W51M, SgrB2. Several pure rotational transitions have been unambiguously identified [3].It is of primary importance to assist astrophysicists in identifying potential interstellar species that as many spectroscopic constants as possible be available in order to recognize the measured spectral lines. As a large number of small phosphorus-containing compounds have either not been detected experimentally or have gaps in the known spectroscopic constants, theory might be able to fill in the missing information.

1.2. LOW-LYING PN A $^{1}\Pi$ AND $^{1}\Delta$ STATES

The best characterized excited state of PN remains to be, without contest, the A¹Π state yet experimentally detected in 1933 [1]. During the next fifty years no new excited state findings were reported for PN. However during the last decade a revival of experimental as well as theoretical interest has lead to a reexamination of both X¹Σ⁺ and A¹Π states. Aside from improving the accuracy of certain spectroscopic constants' values, a few perturbing states interacting selectively with some A¹Π low lying vibrational levels (i.e. v'=0,1,2,3) have been characterized [4]. These states are ³Δ, ¹Δ and ³Σ⁻ , all arising from

319

Y. Ellinger and M. Defranceschi (eds.), Strategies and Applications in Quantum Chemistry, 319–332.
© 1996 *Kluwer Academic Publishers.*

the ($...1\pi^4 2\pi^3 7\sigma^2 3\pi^1$) orbital occupancy. For example, the perturbation on v'=1 in the $A^1\Pi$-$X^1\Sigma^+$ bands, first noticed by Curry et al [1] , was due to nearby low rovibrational levels of a $^3\Sigma^-$ state, while the $^3\Delta$ state was shown to perturb the level v'=0. Whereas a few spectroscopic constants had been determined for the perturbed $A^1\Pi$ vibrational states by these perturbing states a similar analysis had never been performed for the $^1\Delta$ state although this state should perturb in the same proportions [4] certain low lying $A^1\Pi$ vibrational levels. For the sake of completeness, let us mention a new $^1\Sigma^+ \rightarrow X^1\Sigma^+$ transition [5] has been reported arising from the $2\pi \rightarrow 3\pi$ excitation which should interact with high lying $A^1\Pi$ vibrational levels as well as a set of four new excited states, i.e. two $^1\Pi$s and two $^1\Sigma^+$s, but an absolute vibrational assignment for these states was not possible [6]. The technology of astrophysical measurements on excited states for molecules such as PN is still in its infancy and, to the best of our knowledge, no results have been reported.

2. Procedure

2.1.   POTENTIAL CURVES OF THE X $^1\Sigma$, A $^1\Pi$ and $^1\Delta$ STATES

Details of the extended triple zeta basis set used can be found in previous papers [7,8]. It contains 86 cartesian Gaussian functions with several d- and f-type polarisation functions and s,p diffuse functions. All cartesian components of the d- and f-type polarization functions were used. CI wave functions were obtained with the MELDF suite of programs [9]. Second order perturbation theory was employed to select the most energetically double excitations, since these are typically too numerous to otherwise handle. All single excitations, which are known to be important for describing certain one-electron properties, were automatically included. Excitations were permitted among all electrons and the full range of virtuals.

All three states were described by a single set of SCF molecular orbitals based on the occupied canonical orbitals of the $X^1\Sigma^+$ state and a transformation of the canonical virtual space known as "K-orbitals" [10] which , among other properties, approximate the set of natural orbitals. Transition moments within orthogonal basis functions are easier to derive.

For the X state the composition of the reference space was obtained by performing two Hartree-Fock single and double excitations (HFSD-CI) calculations at two typical internuclear distances, i.e. R_e (equilibrium geometry) and about $3R_e$, and adding to the HF configuration all those configurations whose coefficients in either of these CIs were $\geq$ |0.03| . The resulting list of 44 configurations constituted the occupancy of the reference space for Multi-Reference Single and Double CI (MRSD-CI) calculations in the region $2.61a_0$-$5.53a_0$. The energy threshold value, hereafter referred to as ETHRESH, used in the perturbation theory selection procedure of the configurations was set equal to $3.5 \times 10^{-7} E_h$.

For the $A^1\Pi$ state, considerably more configurations contribute the above threshold coefficient and in order to increase the value of the sum of the squares of the CI coefficients in the reference space (Σc_i^2) with respect to that obtained in either of the HFSD-CIs, two MRSD-CI calculations at the above internuclear distances were next performed, keeping in either case all configurations with expansion coefficients $c_i \geq$ |0.03|.

An avoided crossing did occur at ~$3R_e$ which, however, was absent at $5.53a_0$ [8].Thus the combined space spanned by all configurations whose coefficients equalled or exceeded $|0.03|$ in either of the MRSD-CIs (at R_e and $5.53R_e$) was chosen for the reference space to describe the Π potential function in the $2.61a_0$ - $5.53a_0$ region. This space spanned 51 configurations. ETHRESH was set equal to $5.0 \times 10^{-7} E_h$. This value as the above one should ensure that the overwhelming majority of the correlation energy was recovered via the variational CI calculation.

For the $^1\Delta$ state, this procedure led to still much larger reference spaces and larger CI wave functions because of the still larger number of configurations possessing expansion coefficients $\geq |0.03|$. In order to keep the calculations tractable, this threshold was set equal to 0.043 while lowering the value of ETHRESH to $4.0 \times 10^{-7} E_h$. This reference space was spanned by 28 configurations. In light of the primary goal for the $^1\Delta$ state of being able to compute properties in the neighbourhood of the potential curve's minimum *with acceptable accuracy*, this wavefunction should still be adequate. This will be illustrated by comparing our lifetime results with those of CO obtained by a different CI approach as well as a set of spectroscopic constants with those obtained by similar CI calculations based on another algorithm.

Estimates of the energy contributions from higher than double excitations out of the reference space were obtained by means of one form of the "Davidson correction" [11,7]. More details can be found in references [7,8].

3. Results

3.1.   PN X $^1\Sigma^+$ STATE

3.1.1. *Potential energy curve; one-electron properties; spectroscopic constants*

The potential energy curve including the Davidson correction is shown in Figure 1. Among the calculated one-electron properties [7] only a few ones did show sizeable correlation effects (Table 1). The calculated and experimental values of the electric dipole moment are unexpectedly yet in good agreement although multiple bonded systems are known to require the use of a large number of higher angular momentum basis functions [12]. The rather large theoretical difference found for the $^{14}N_7$ nuclear coupling remains at the present time an open question: if the experimental value is accurate, this difference is due to a too small value of the electric field gradient. It should be pointed out this observable is neither easy to measure with precision nor to compute accurately, especially for multiple bonded systems involving third period atoms.

3.1.2. *Spectroscopic constants*

Selected spectroscopic constants (Table 2), derived by the well-known Dunham polynomial fit expansion method [13,14] were calculated from both the variational MRSD-CI energies and the estimated full CI energies, i.e. including the Davidson correction. In general, the effects of the Davidson correction appear to be small. A very good agreement, up to the second decimal, is obtained in case of r_e leading to an exceedingly small theoretical deviation (0.007%) on α_e. The effect of the correction on the latter observable is sizeable (24%) and decreases the agreement with experiment. The observables k_e, ω_e, $\omega_e x_e$ - in contrast with r_e, B_e, α_e - are closer to experiment at the full CI approximation, leading also to a subtantial lowering of the theoretical discrepancy for the anharmonicity whereas the relative error on the fundamental frequency amounts merely to 1.5%.

Table 1. Calculated and measured one-electron properties of the PN X $^1\Sigma^+$ state

	SCF	HF SD-CI	MR SD-CI	EXP
SCF Energy	-395.1719			
CI Energy		-395.5497	-395.5838	
Est.full CI En		-395.6048	-395.6120	
Generated	1	94,735	14,625,064	
Selected	1	15,562	93,076	
$\mu_z{}^*$	-10.76	-9.93	-9.70 P$^+$- N$^-$	I9.176±0.002I
$\Theta_{//}{}^*$	-8.22	-8.64	-8.88	
$<q_{zz}(N)>$	-1.3398	-1.1641	-1.0899	
$eq_{zz}(N)Q/h$	-3.239	-2.814	-2.635	5.1728±0.0005

Unless otherwise noted, the phosphorus atom is placed at the origin of the reference frame, with nitrogen pointing in the positive z direction

Energy(a.u.); $\mu_z(10^{-30}$C m); $\Theta_{//}(10^{-40}$C m^2); $eq_{zz}(N)Q/h$(MHz); $q_{zz}(N)(10^{22}$Vm^{-2}) $Q=+0.01.10^{-28}m^2$[16]

Entries
'Generated'   Total number of spin and symmetry adapted configurations
'Selected'   Number of spin and symmetry adapted configurations selected by second-
               order perturbation theory and treatedvariationally
'*'   Property calculated with respect to the center of mass.
               For the dipole moment, the polarity has not been measured experimentally.
               The sense found is the same as that of McLean and Yoshimine [17] and the
               agreement between observation and the MR SD-CI result leaves little doubt
               on the correctness of this sense.
'EXP'   Experimental values, see Raymonda and Klemperer [18]
'Est.full CI En' Estimated full-CI energy [11]

These results have been compared with those issued from smaller CI calculations based on another algorithm [15]. Except for the anharmonicity, our results are generally closer to experiment. Another instructive comparison has been made with CASSCF results on singly bonded CS$^+$ [19] using a comparable basis set: for r_e, B_e and α_e the theoretical deviations are quite comparable, the CASSCF result on the anharmonicity being however much closer to experiment. The size of the H-CI matrix handled in the latter methodology (350,000 - 385,000) is not commensurable with that in the present study (93,000 -

110,000). Therefore, it appears that the overall agreement obtained for a variety of spectroscopic constants is comparable for the two methods while the present method allows us to use a more compact wavefunction. It should also be noted that a good CI description of a triple bonded system involving a third period atom is much harder to achieve. It can be concluded that the shape of the theoretical potential energy curve reflects its experimental counterpart with acceptable accuracy in the interatomic region of interest.

Table 2. Values of some spectroscopic constants for the PN X $^1\Sigma^+$ state

	A	B	EXPT
E_{min} (au)	-395.58390	-395.61219	
E_0 (au)	-395.58070	-395.60912	
r_e (a$_0$)	2.8320	2.8421	2.817583
r_e (Å)	1.4986	1.5040	1.490866
k_e .(10^{-5} dyne/cm)	11.430	10.485	10.144
ω_e (cm^{-1})	1418.3	1358.4	1337.24
E (ZP)^f	701.9	675.5	
$\omega_e x_e$(cm^{-1})	29.58	14.71	6.983
$\omega_e y_e$ (cm^{-1})	0.90	0.09	0.007
B_e (cm^{-1})	0.7783	0.7728	0.7865
α_e (cm^{-1})	0.00554	0.00685	0.005536

In column A , use is made of the variational MRSD-CI energies. In column B, these energies are corrected for higher excitations (see text)

Entries

'E_{min}'   Energy of the minimum of the potential curve
'E_0 '   Energy corresponding to vibrational quantum number v=0

'E(ZP)'   Zero point energy $=0.5\omega_e - 0.25\omega_e x_e + 0.125\omega_e y_e$
            The Dunham polynomial fit expansion of the theoretical curves involves polynomials of 9th and 10th degree leading to the data in columns 'A' and 'B' respectively.

'f'   Using the variational CI energies and correcting these for higher order excitations, the radial Schrödinger equation solutions for E(ZP) are equal to 712.1cm^{-1} and 667.4cm^{-1} respectively with r_{min}=1.4171Å and r_{max}=2.93Å.

3.1.3. *Pure rotational and vibrorotational transitions; spontaneous radiative lifetimes*

Pure rotational transitions, vibrorotational transitions and spontaneous radiative lifetimes have been derived by solving numerically [20] the one-dimensional radial part of the Schrödinger equation for the single X state preceded by constructing an interpolation curve - using tensioned spline functions [21] - of this state adopting the data including the Davidson correction. For the calculations of lifetimes within a single state ,in contrast to two distinct states , the vibrational/rovibrational decays occuring via a cascade mode, the electric dipole moment function replaces the usual electric dipolar transition moment function required in case of transitions between states. More details can be found in [7,8] as well as the references quoted therein.

A remarkable agreement with experiment had been found for pure rotational excitations (Table 3). The calculated values are systematically a little smaller because the theoretical internuclear distance is slightly larger than the experimental one, i.e. r_e=1.490866 Å (expt), r_e=1.5040 Å (theor.) and this deviation affects the parameter values (α_e, B_e) leading to our previous observation. A slight deviation with respect to the expected effect of the anharmonicity on the vibrorotational transitions is observed in contrast to the same effect noted on the pure rotational excitations.

To the best of our knowledge, pure rotational transition calculations for the PN X state are reported for the first time.

Table 3. Pure rotational and rovibrational transition energies

(v',v")	(0,0)	(1,0)	(1,1)	(2,1)	(2,2)	(3,2)	(3,3)	(4,3)	(4,4)
(j'.j")									
(2,1)	3.03	1590.22	2.94	1636.25	2.85	1655.73	2.78	1669.72	2.72
	(3.13)*		(3.11)						
(2,3)	4.55	1588.61	4.41	1634.69	4.28	1654.24	4.17	1668.27	4.07
	(4.69)*		(4.66)		(4.63)				
(3,4)	6.07	1586.95	5.88	1633.09	5.70	1652.70	5.56	1666.79	5.43
	(6.26)		(6.22)		(6.17)		(6.13)		(6.08)
(4,5)	7.58	1585.24	7.35	1631.45	7.13	1651.13	6.94	1665.27	6.79
	(7.83)*		(7.77)		(7.72)		(7.66)		

Experimental values (in parentheses) from microwave spectroscopy [2] and interstellar measurements [3] (upper index *)
(v',v"), (j',j") vibrational and rotational quantum numbers for upper and lower levels respectively

The calculated lifetimes (Table 4) are several powers of ten larger then those corresponding to usual electric dipolar transitions (10^{-7}-10^{-8} s). They constitute therefore true predictions which require special techniques of measurements that were available only in recent years.

Molecular lifetimes corresponding to pure vibrational (v =0-4) and rovibrational (v =0-4, j=1-5) levels - derived for the first time - were found to be eight to ten powers of ten larger than those corresponding to an electric dipolar $^1\Pi \rightarrow ^1\Sigma$ transition ($^1\Pi$ state above $^1\Sigma$ state). In the latter case large transition moments for non-orthogonal off-diagonal vibrational states are also responsible for the resulting magnitude of these lifetimes. In this case, the off-diagonal elements of the dipole moment for orthogonal vibrational states are several powers of ten smaller compared to their transition moment counterparts, leading to much smaller values of the transition probabilities and Einstein spontaneous emission coefficients and therefore very large lifetimes. A very important rotational effect on the lifetime for v'=0, becoming rather weak for v'=1 and being not existent for higher v's considered in this study is noted.

Table 4. Molecular spontaneous vibrational/rovibrational lifetimes, $\tau_{v'}$ (s)

v'	τ_1	τ_{1a}	τ_2	τ_3	τ_4
0	∞	107,661	18,998	6,929.12	1099.97
1	39.514	41.711	41.565	41.212	39.619
2	1.721	1.738	1.738	1.737	1.734
3	1.112	1.125	1.125	1.124	1.122
4	0.706	0.713	0.713	0.713	0.712

$\tau_1 : j'=j''=0$ $\tau_{1a} : j'=0, j''=1$ $\tau_2 : j'=1, j''=0,2$ $\tau_3 : j'=2, j''=1,3$ $\tau_4 : j'=4, j''=3,5$

3.2.   PN A ¹Π and ¹Δ EXCITED STATES

3.2.1. *Potential energy curves for the A ¹Π and ¹Δ states. Selected one-electron properties. Spectroscopic constants*

MR SD-CI potential curves including the Davidson corrections are presented in Figure 1.

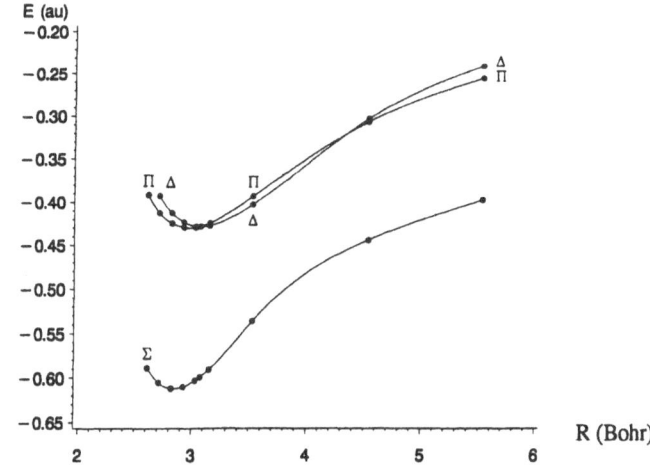

Fig.1. Potential energy curves including the Davidson correction
of the X ¹Σ, A ¹Π and ¹Δ states of PN

The correlation effects on the electric dipole moment (Table 5) are much larger at the MR SD-CI level for the $A^1\Pi$ state while the HF SD-CI algebraic values differ slightly for the two states, the qualities of the respective reference spaces being comparable[8].We find the following magnitudes' sequence, $\mu_z(X^1\Sigma^+)>\mu_z(^1\Delta)>\mu_z(A^1\Pi)$, among the dipole moments and presently it cannot be verified by experiment. MR SD-CI correlation effects noticed on the electric quadrupole moment are also larger for the $A^1\Pi$ state. These effects on the nuclear nitrogen couplings' z component for the Π state appear to be negligible with respect to their HF SD-CI homologues which is generally quite unusual in case of the ground state. The absolute value of the $^{14}N_7$ coupling of the Π state, 2.554MHz, is smaller than that for the $X^1\Sigma^+$ state, 2.635MHz, whereas the $^1\Delta$ state nuclear coupling appears to be the smallest : this sequence seems plausible as the electric field gradient at the nitrogen nucleus should diminish as one goes to higher energy molecular excited states.

Table 5. Calculated one-electron properties of the PN A $^1\Pi$ state at the experimental equilibrium geometry and $^1\Delta$ state at the theoretical equilibrium geometry

	A $^1\Pi$ STATE		$^1\Delta$ STATE	
	HF SD-CI	MR SD-CI	HF SD-CI	MR SD-CI
CI Energy	-395.3084	-395.3818	-395.3311	-395.3719
Est.full CI En	-395.5366	-395.4293	-395.5109	-395.4284
Generated	217,808	18,071,450	438,753	6,314,582
Selected	29,798	158,005	43,708	106,695
$\mu_z{}^*$	-6.90	-5.09	-6.86	-6.55
$Q_{//}{}^*$	4.87	4.10	-2.33	-2.46
$<q_{xx}(N)>$	-1.6319	-1.3992	+0.3937	+0.4233
$eq_{xx}(N)Q/h$	-3.9464	-3.3838	+0.9522	+1.0236
$<q_{zz}(N)>$	+1.0614	+1.0563	-0.7875	-0.8466
$eq_{zz}(N)Q/h$	+2.5663	+2.5540	-1.9045	-2.0473

For the A $^1\Pi$ state the experimental equilibrium geometry $r_e=2.92284a_0$, while the corresponding theoretical value for the $^1\Delta$ state $r_e=3.0711a_0$

Energy (a.u.); $\mu_z(10^{-30}$ Cm); $Q_{//}(10^{-40}$ Cm^2); $q_{zz}(N),q_{xx}(N)$ $(10^{22}$ Vm$^{-2})$;
$eq_{zz}(N)Q/h$ (MHz) and analoguously for the other components.
Entries
'Generated'   Total number of spin and symmetry adapted single and double excitations
'Selected'   Number of spin and symmetry adapted configurations selected by second-
                    order perturbation theory and treatedvariationally
'*'   Property calculated with respect to the center of mass
'Est.full CI En' Estimated full-CI energy [11]

Several spectroscopic constants derived with and without the Davidson correction (Table 6) show little differences except for k_e, α_e and $\omega_e x_e$, the overall agreement with experiment being satisfactory. As for the X state, the Davidson correction tends to reflect experiment sometimes better, e.g. for $\omega_e x_e$, B_e, whereas the discrepancy with respect to experiment is sizeably reduced for α_e.

Table 6. Values of spectroscopic constants for the PN A ¹Π and ¹Δ bands

	A ¹Π BAND			¹Δ BAND	
	A	B	EXP	A	B
E_{min} (a.u.)	-395.38241	-395.42984		-395.37192	-395.42830
E_0 (a.u.)	-395.37997	-395.42742		-395.36941	-395.42580
r_e, (a_0)	2.9754	2.9730	2.92284	3.0734	3.0711
r_e (Å)	1.5745	1.5732	1.5467	1.6264	1.6252
other		1.566			1.622
k_e (10^{-5}dyne/cm)	6.585	6.460	6.899	6.947	6.946
ω_e (cm⁻¹)	1076.5	1066.3	1103.09	1105.7	1105.7
other		1071.7			967.9
E(ZP) (cm⁻¹)	535.6	530.5		549.7	548.8
$\omega_e x_e$ (cm⁻¹)	10.43	10.72	7.222	12.41	16.17
other		11.40			1.97
$\omega_e y_e$ (cm⁻¹)	-0.10	-0.12		-0.17	-0.30
B_e (cm⁻¹)	0.7051	0.7063	0.73071	0.6609	0.6619
other		0.713			0.664
α_e (cm⁻¹)	0.00076	0.00183	0.00663	0.00390	0.00298
other		0.0075			0.0070
T_e (cm⁻¹)		40,007.6	39,805		40,345.5
other		40,328			41,619.16

A seventh degree polynomial fit to the theoretical curves has been used (columns A and B). In column A use is made of the variational MR SD-CI energies while in column B estimated full-CI energies [11] are utilised

Entries

'E_{min}'	Energy of the minimum of the potential curve
'E_0 '	Energy corresponding to vibrational quantum number v=0
'E(ZP)'	Zero point energy $=0.5\omega_e - 0.25\omega_e x_e + 0.125\omega_e y_e$
'EXP'	Experimental values for the A¹Π band [22]
'other'	see [15]

A deviation of merely 0.5% is found for the experimental $T_e(^1\Pi)$ value. Most of these results for both states compared favourably with those obtained by Grein and Kapur [15] using smaller basis sets and another CI algorithm also based on single and double excitations out of a multi-reference space. For $\omega_e x_e$ and T_e of the $A^1\Pi$ state our results (with the Davidson correction) are closer to the experimental values.

We find the $^1\Delta$ state to lie at merely 338cm^{-1} above the $A^1\Pi$ state. The N_2 electronic spectra exhibit strong similarities with PN and, in particular, the same sequence is observed, the corresponding states being separated by only 188.65cm^{-1} [22] which should make our PN value very likely to be observed.

3.2.2 *Spontaneous radiative lifetimes for the A $^1\Pi$ - X$^1\Sigma^+$, $^1\Delta$ - A $^1\Pi$ bands and corresponding lifetimes for the A $^1\Pi$ and $^1\Delta$ states*

Spontaneous radiative lifetimes via electric dipole transitions for the $A^1\Pi \rightarrow X^1\Sigma^+$ and $^1\Delta \rightarrow A^1\Pi$ transitions are summarized in Table7 for several sets of (v',v") values. Experimentally, the detection of the so-called Hanle signal which consists of measuring the change of fluorescence intensity in a transition band as the magnetic sublevels separate in a variable external magnetic field provides a direct measure of the upper state lifetime [23] which has been performed only for v'=0. This value agrees reasonably with its theoretical counterpart. A lack of a still better agreement with experiment might be related to the presence of the very nearby $^1\Delta$ state perturbing selectively the $A^1\Pi$s v'=0 level, no experimental analysis by high resolution spectroscopy [4] or through the Hanle signal measurement [23] being known for the $^1\Delta$ state (see also section 1.2). Whereas for the first transition band no rotational dependence for (j',j")$\neq$ 0 is noticed and the lifetimes increase with the vibrational quantum number, a small but clear-cut rotational effect does exist for v'=0 - 2 in case of the second system. Moreover these lifetimes decrease with *v'* (except for v'=4) in contrast with the previous system. These effects are related to the evolution of the respective transition moments as function of the internuclear distance [8]. The large difference in magnitude among the lifetimes of both systems are linked to the corresponding values of the Einstein coefficients [8]. These results have been compared with those issued from a theoretical study of the same transition in CO [24], isoelectronic in the valence shell's electrons: it appears that our PN lifetimes are of the same order of magnitude as those in CO. Because a monotonic decrease in the CO lifetimes from low v' up to high-lying vibrational levels has been predicted too which were found consistent with experiment for the latter levels, we believe our PN lifetime values should be of the correct order of magnitude.

A lifetimes' comparison for each excited state (Table 8) shows that they are smaller for the $A^1\Pi$ state except for v'=0. The vibrational decays occur by means of cascade processes. Rotational effects (v'=1 - 2 ; j'=1 - 2) appear to be even more intense than for the corresponding transition bands and in either case these effects disappear for j'=4. For both states the lifetimes decrease with upward v' values. For the $^1\Delta$ state a comparison of our results with those on CO shows that the PN $^1\Delta$ lifetimes are larger by an amount similar to that noted for the PN $^1\Delta \rightarrow A^1\Pi$ transition. Unfortunely no conclusive lifetime measurements for the CO $^1\Delta$ state seem to exist nowadays such that no useful informations can be invoked from the CO molecule to subtantiate our corresponding lifetimes for the PN $^1\Delta$ state.

Comparing the vibrational $^1\Delta$ lifetimes issued from both decay mechanisms (Tables 7 - 8), it is readily seen that the electric dipolar transition decay is always slightly favoured. A similar conclusion holds for the A$^1\Pi$ state but, as expected, the vibrational transition probabilities are much larger for the dipolar decay which lead to much smaller vibrational lifetimes with respect to those via the cascade mode of decay, the differences amounting to five to six powers of ten (Table 7 - 8).

Table 7. Spontaneous radiative lifetimes $\tau_{v'}$ j' (s) of the $^1\Delta$ and A $^1\Pi$ bands via
$^1\Delta\rightarrow$A $^1\Pi$ and A $^1\Pi\rightarrow$X $^1\Sigma^+$ electric dipolar transitions

	A$^1\Pi\rightarrow$X$^1\Sigma^+$				$^1\Delta\rightarrow$A$^1\Pi$			
τ	τ_1	τ_{1a}	τ_2	τ_4	τ_0	τ_1	τ_2	τ_3
v'								
0	9.424 -7	4.712 -7	4.713 -7	4.715 -7	6.569	3.307	3.310	3.315
		(2.3±0.7) -7[a]						
1	9.923 -7	4.955 -7	4.956 -7	4.957 -7	2.962 -1	1.487 -1	1.492 -1	1.497 -1
					(3.4 -2)			
2	1.029 -6	5.146 -7	5.147 -7	5.147 -7	1.048 -2	5.253 -3	5.258 -3	5.265 -3
					(5.2 -3)			
3	1.084 -6	5.421 -7	5.421 -7	5.421 -7	4.126 -3	2.065 -3		2.066 -3
					(1.9 -3)			
4	1.144 -6	5.725 -7	5.725 -7	5.725 -7	4.362 -3	2.182 -3		2.182 -3
					(1.0 -3)			
5	1.231 -6	6.159 -7	6.159 -7	6.157 -7	3.463 -3	1.733 -3		1.736 -3
					(6.6 -4)			

Columns 2 - 5 ($^1\Pi\rightarrow^1\Sigma$): τ_1: j'=j"=0   Columns 6 - 9 ($^1\Delta\rightarrow^1\Pi$): τ_0 :j'=j"=0
τ_{1a}:j'=1,j"=0,1,2 τ_1:j'=2,j"=1,2,3
τ_2:j'=2,j"=1,2,3 τ_2:j'=3,j"=2,3,4
τ_4:j'=4,j"=3,4,5 τ_3:j'=4,j"=3,4,5

Values are given as (mantissa exponent); for example 9.424 -7 reads 9.424 .10^{-7}
a see [23]
In parentheses, CO theoretical results [24]

Table 8. Spontaneous radiative lifetimes (s) of the A $^1\Pi$ and $^1\Delta$ bands. Cascade decay

v'	A $^1\Pi$				$^1\Delta$			
	τ_1	τ_{1a}	τ_2	τ_4	τ_1	τ_{1a}	τ_2	τ_4
0	∞	166,252	45,705	9.058 3	∞	4.420 4	1.487 4	7.316 3
1	0.404	2.428 -1	2.022 -1	2.023 -1	1.515 (8.3 -2)	9.472 -1	7.577 -1	7.580 -1
2	1.871 -1	1.124 -1	9.369 -2	9.374 -2	8.025 -1 (3.8 -2)	5.015 -1	4.013 -1	4.014 -1
3	1.114 -1	6.693 -2	5.575 -2	5.579 -2	5.364 -1 (2.5 -2)	3.352 -1	2.682 -1	2.686 -1
4	7.656 -2	4.600 -2	3.832 -2	3.836 -2	3.895 -1 (2.0 -2)	2.435 -1	1.948 -1	1.950 -1
5	5.699 -2	3.424 -2	2.852 -2	2.854 -2	2.977 -1 (1.6 -2)	1.861 -1	1.489 -1	1.489 -1

Columns 2 - 5 (A $^1\Pi$ results) τ_1:j'=j''=0   Columns 6 - 9 ($^1\Delta$ results) τ_1:j'=j''=0

τ_{1a}:j'=1,j''=1,2 τ_{1a}:j'=2,j''=2,3

τ_2:j'=2,j''=1,2,3 τ_2:j'=3,j''=2,3,4

τ_4:j'=4,j''=3,4,5 τ_4:j'=4,j''=3,4,5

Values are given as (mantissa exponent); for example 2.428 -1 reads 2.428 .10⁻¹

In parentheses, CO theoretical results [24]

3.2.3. *Miscellaneous spectroscopic observables*

The emission spectrum observed by high resolution spectroscopy for the A $^1\Pi$ - X $^1\Sigma^+$ vibrational bands [4] has been very well reproduced theoretically for several low-lying vibrational quantum numbers and the spectrum for the $^1\Delta$ - A$^1\Pi$ vibrational bands has been theoretically derived for low vibrational quantum numbers to be subjected to further experimental analysis [8]. Related Franck-Condon factors for the latter and former transition bands [8] have also been derived and compared favourably with semi-empirical calculations [25] performed for the former transition bands. Pure rotational, vibrational and rovibrational transitions appear to be the largest for the X ground state followed by those for the A$^1\Pi$ and $^1\Delta$ states respectively [8]. Whereas accurate data on pure rotational excitations were available for the X ground state [2] with which a fairly good theoretical agreement was obtained (Table 3), no such data exist for these excited states. Because the three states have been treated identically in solving the corresponding one-dimensional Schrödinger equations, the qualities of the respective reference spaces being rather close [7,8] and noting the algorithm provides exact solutions within the numerical procedure utilized, we believe the theoretical data should also be verified by accurate microwave measurements. The experimental trend shown by the expectation values of the electric dipole moment over a few vibrational functions is well reflected theoretically for the X state [7]. Different patterns over several vibrational quantum numbers for this observable are also predicted for each excited state [8] which, to-date, cannot be experimentally verified.

It is a pleasure for the author of being invited to contribute to this book as a tribute to Gaston Berthier who taught him in the late sixties at 'Ecole Normale Supérieure (rue Lhomond, Paris)' how to use a particular molecular orbital formalism, developped in his group, for a study on transiton metal complexes. This has been the beginning of a fruitful collaboration over the years.

This work was performed as part of the research programme of the 'Foundation for Fundamental Research on Matter (FOM)' which is financially supported by the 'Netherlands Organization for the Advancement of Scientific Research (NWO)'.

References

1.   V.J. Curry, L and G. Herzberg, *Z.Phys.* **86**, 348 (1933).
2.   F.C. Wyze, E.L. Manson and W. Gordy, *J.Chem.Phys.* **57**, 1106 (1972).
3.   L.M. Ziuris, *Astrophys.J.* **321**, L81 (1987).
4.   S.N. Ghosh, R.D. Verma and J. VanderLinde,*Canad.J.Phys.* **59**, 1640 (1981).
      R.A. Gottscho, R.W. Field and H. Lefebvre-Brion, *J.Mol.Spectr.* **70**, 420(1978).
5.   B. Coquart and J.C. Prudhomme, *J.Mol.Spectrosc.* **87**, 75 (1981).
6.   R.D. Verma, S.N. Ghosh and Z. Iqbal, *J.Phys.B.* **20**, 3961 (1987).
7.   G. de Brouckère, D. Feller, J.J.A. Koot and G. Berthier, *J.Phys.B.* **25**, 4433 (1992).
8.   G. de Brouckère, D. Feller and J.J.A. Koot , *J.Phys.B,* in press.
9.   L. McMurchie, S. Elbert, S. Langhoff and E.R. Davidson "MELDF Suite of Programs" substantially modified by D. Feller, R.J. Cave, D. Rawlings, R. Frey, R. Daasch,
      L. Nitzche, P. Phillips, K. Iberle, C. Jackels and E.R. Davidson (1990/1991).
10.   D. Feller and E.R. Davidson, *J.Chem.Phys.* **74**, 3977 (1981).
11.   E.R. Davidson and D.W. Silver, *Chem.Phys.Lett.* **52**, 403 (1977).
12.   S.R. Langhoff, C.W. Bauschlicher and P.R. Taylor, *Chem.Phys.Lett.* **180**, 88 (1991).
13.   J.L. Dunham, *Phys.Rev.* **41**, 721 (1932).
14.   The suite of programs VIBROT (Vibration-rotation analysis program for diatomic molecules) was written by T.H. Dunning, Molecular Science Research Center, Pacific Northwest Laboratory, Richland, 1978/1979, and modified by R. Eades (1983) and D. Feller (1990).
15.   F. Grein and A. Kapur,*J.Mol.Spectrosc.* **99**, 25 (1983).
16.   G.H. Fuller, *J.Phys.Chem.Ref.Data* **5**, 835 (1976).
17.   A.D. McLean and M. Yoshimine, *IBM J.Res.Dev.* **12**, 206 (1968).
18.   J. Raymonda and W. Klemperer,*J.Chem.Phys.* **55**, 232 (1971).
19.   M. Larsson, *Chem.Phys.Lett.* **117**, 331 (1985).
20.   The suite of programs INTENSITY was written by W.T. Zemke and W.C. Stwalley, Department of Chemistry, Wartburg College, (1978)
21.   The program SPLINE was written by K.A. Kaiser, Data processing, Southern Illinois University, (1978).

22.   K.P. Huber and G. Herzberg, <u>Molecular Spectra and Molecular Structure. IV. Constants of Diatomic Molecules</u>, Van Nostrand Reinhold, New York (1979). G. Herzberg, <u>Molecular Spectra and Molecular Structure. I. Spectra of Diatomic Molecules</u> (p.154), Van Nostrand, New York (1961).
23.   M.B. Moeller, M.R. McKeever and S.J. Silvers, *Chem. Phys.Lett.* **31**, 398 (1975)
24.   M.E. Rozenkrantz and K.Kirby, *J.Chem.Phys.* **90**, 6528 (1989).
25.   M.B. Moeller and S.J. Silvers, *Chem. Phys.Lett.* **19**, 78 (1973).

Theoretical Treatment of State-Selective Charge Transfer Processes. N^{5+} + He as a Case Study

M.C. BACCHUS-MONTABONEL
Laboratoire de Spectrométrie Ionique et Moléculaire (URA CNRS n°171),
Université Lyon I, 43 Bd du 11 Novembre 1918, 69622 Villeurbanne, France

1. Introduction

Charge transfer recombination of multiply charged ionic species in collision with neutral atoms or molecules is an important process in astrophysical plasmas [1,2] and controlled nuclear fusion research [3]. From an experimental point of view, these reactions have been extensively studied in recent years using a wide variety of techniques (VUV spectroscopy [4-11], energy gain spectroscopy [12,13], electron spectrometry [14,15]). Much attention has also been paid to the interpretation of the electron capture processes using model potential methods [16-18] which allow generally a fair description of the phenomena in the case of closed shell systems, or ab initio methods [19-24] necessary for the study of open-shell systems as for example low-charged ions or metastable states.

Recently, we have developed a full theoretical treatment of electron capture processes involving an ab initio molecular calculation of the potential energy curves and of the radial and rotational couplings followed, according to the collision energy range concerned, by a semi-classical [21-23] or quantal [24] collision treatment.

As a test case, we report in this paper the study of the N^{5+} + He collision. This work has been undertaken in connection with photon spectroscopy experiments regarding the electron capture for the reactions

$$N^{5+}(1s^2)^1S + He(1s^2)^1S \rightarrow N^{4+} + He^+ \quad \text{single-electron capture (SC)}$$
$$\text{or} \quad N^{3+} + He^{2+} \quad \text{double-electron capture (DC)}$$

at collision energies in the range [10-100 keV] [4,7].

In accordance with previous investigations [8,9], these experiments have shown a quite different behaviour for $N^{5+}(1s^2)$ than for other multicharged ions such as the isoelectronic ion O^{6+} [10]. The single-electron capture process has been shown to be dominant on the n = 3 levels and in particular on the 3s level for collision energies lower than 50 keV. A high probability of double capture has also been observed characterized by an intense peak at $\lambda = 76.5$ nm attributed to the $2s^2\,^1S \rightarrow 2s\,2p\,^1P\ N^{3+}$ transition [4,5,7]. Furthermore, for this system, metastable levels can be populated by foil excitation and may thus be present in the incident beam. Experimental measurements have been performed for the

333

Y. Ellinger and M. Defranceschi (eds.), Strategies and Applications in Quantum Chemistry, 333–348.
© 1996 *Kluwer Academic Publishers.*

$N^{5+}(1s2s)^3S + H_2$ and $N^{5+}(1s2s)^3S + He$ collisions at 60 keV from photon spectroscopy [6] and at 51 keV from electron spectroscopy [14].

A complete theoretical treatment of the $N^{5+} + He$ collision should therefore take into account, first the single-electron capture process from the ground state entry channel $N^{5+}(1s^2)^1S + He(1s^2)^1S$ and also from the metastable level $N^{5+}(1s2s) + He(1s^2)$ in order to take care of the fraction of metastable N^{5+} ion in the incident beam, as well as the double-electron capture process from both ground and metastable N^{5+} ions.

The single-electron capture process from the ground state $N^{5+}(1s^2)^1S + He(1s^2)^1S$ is the easiest one to handle and also the most important one. The capture being dominant on the $n = 3$ levels, and the effect of spin-orbit coupling being of negligible importance for electron capture in the energy range of interest, we have determined the potential energy curves corresponding to the entry channel $^1\Sigma^+$ $\{N^{5+}(1s^2)^1S + He(1s^2)^1S\}$ and all the $^1\Sigma^+$, $^1\Pi$ and $^1\Delta$ states corresponding to the $\{N^{4+}(1s^2,3l) + He^+(1s)\}$ configuration.

The consideration of the collision involving the metastable ion $N^{5+}(1s2s) + He(1s^2)$ requires the calculation of much higher levels. The work has been undertaken in tight connection with experimental investigations in order to reduce the number of states involved in the molecular calculation. From an experimental point of view, it is assumed that only the triplet metastable state $N^{5+}(1s2s)^3S$ will be involved in the collision because of the shorter lifetime of the singlet state [25] with respect to the time-of-flight of the ions from their production to the collision cell. Besides, among the doublet and quartet states $N^{4+}(1s2snl)^{2,4}L$ produced in the single-electron capture process, the doublet states are rapidly autoionized, when the quartet states are metastable with respect to Coulomb autoionization and then only transitions involving these quartet states may be observed.

As in the collision of the ground state ion $N^{5+}(1s^2)$ on a He target, the main process has been shown experimentally to be the core-conserving single-electron capture on the $n = 3$ levels [6,14] with a small amount of capture on the $n = 4$ levels. The transfer-excitation process corresponding to a single-electron capture and an excitation of the core leading to $N^{4+}(1s2pnl)$ states has also been observed, with a dominance of the capture on the $3l$ levels [6,10]. In view of all these experimental findings we have thus considered the collisional processes

$$N^{5+}(1s2s)^3S + He(1s^2)^1S \rightarrow N^{4+}(1s2s3l)^4L + He^+(1s)^2S$$
$$\rightarrow N^{4+}(1s2p3l)^4L + He^+(1s)^2S$$

and calculated the potential energy curves for the $^3\Sigma^+$ entry channel and the $^3\Sigma^+$ and $^3\Pi$ states of the $\{N^{4+}(1s2s3l)^4L + He^+(1s)^2S\}$ configuration as well as the $^3\Sigma^+$ and $^3\Pi$ states corresponding to the $\{N^{4+}(1s2p3s) + He^+(1s)\}$ configuration which could partly account for the transfer-excitation process.

For the double-electron capture $N^{5+}(1s^2) + He(1s^2) \rightarrow N^{3+}(1s^2 2s2p) + He^{2+}$ a very large energy gap separates the two potential energy curves involved in the process, thus many molecular curve crossings may be important and the population of the $N^{3+}(2s2p) + He^{2+}$ level should probably come from a cascade effect. In practice, a complete treatment including all the potential energy curves is impossible [18,26].

As this process has been shown experimentally [5] to be dominant on the $N^{3+}(2s2p)$ + He^{2+} level, we have performed a calculation including the entry channel $^1\Sigma^+$ {$N^{5+}(1s^2)$ + $He(1s^2)$}} and the $^1\Pi$ and $^1\Sigma^+$ states corresponding to the {$N^{3+}(2s2p)$ + He^{2+}} configuration. We have also considered the {$N^{4+}(2l)$ + $He^+(1s)$} states which are energetically very close to the double capture states. Such a calculation could certainly hardly provide quantitative results, but it could give some qualitative information on the behaviour of the collisional system.

It has otherwise been shown experimentally [5] that the double-electron capture occurs mainly from the $N^{5+}(1s^2)$ + $He(1s^2)$ ground state. We have thus neglected the contribution of the $N^{5+}(1s2s)$ metastable ion.

The electron capture processes are driven by non-adiabatic couplings between molecular states. All the non-zero radial and rotational coupling matrix elements have therefore been evaluated from ab initio wavefunctions.

2. Computational method

The potential energy curves have been determined by ab initio calculations with configuration interaction according to the CIPSI algorithm [27]. The SCF calculation has been performed by means of the Psondo program [28,29] for the electronic configuration $1\sigma^2$ $2\sigma^2$ of NHe^{5+}. From a molecular point of view, the $N^{5+}(1s^2)$ + He and the $N^{5+}(1s2s)$ + He collisional systems have to be considered separately.

For the ground state system $N^{5+}(1s^2)$ + He, compact configuration interaction (CI) spaces have been used in the calculation (about 100-150 determinants) with a threshold $\eta = 0.01$ for the contribution to the perturbation. According to the deep energy difference between 1σ and 2σ, the 1σ molecular orbital has been frozen in the CI procedure. The basis of atomic functions used in the calculation [21] is a 9s5p3d basis of gaussian functions for nitrogen and a 4s1p basis for helium optimized from the 6-311G** basis of Krishnan et al. [30]. Diffuse functions have been added — 2s2p1d for nitrogen and 1p for helium — and optimized with respect to the experimental data by means of a one-configuration calculation for the excited states of N^{4+} and He^+ respectively.

This basis leads to a reasonable agreement with experiment [31] for a large number of atomic levels of nitrogen (Table 1). For the determination of the couplings between the states involved in the double electron capture process, a less expanded basis set has been chosen — 8s4p3d basis set for nitrogen and 3s1p basis set for helium — leading to shorter computation time while saving a fairly reasonable agreement with experiment.

The calculation performed for the metastable $N^{5+}(1s2s)$ + He system has necessitated somewhat larger CI spaces (200-250 determinants) in order to reach the same perturbation threshold $\eta = 0.01$, the 1σ molecular orbital being not frozen for this calculation. The basis of atomic orbitals has been also expanded to a 10s6p3d basis of gaussian functions for nitrogen reoptimized on $N^{5+}(1s^2)$ for the s exponents and on $N^{4+}(1s^22p)$ for the p exponents and added of one s and one p diffuse functions [22]. For such excited states, no experimental data are available, so the most diffuse functions have been optimized with respect to highly accurate atomic data of Chung [32] taking into account relativistic correction terms. The results are reported on Table 2 and show a rather reasonable agreement, the accuracy is of course somewhat poorer than for the ground state $N^{5+}(1s^2)$ but we are dealing with much more excited states. The comparison with relativistic atomic calculations gives besides an insight over the importance of relativistic terms which seem to be quite negligible with respect to the rate of accuracy reached in such calculations.

Table 1. Comparison of calculated atomic levels with experimental data (Bashkin and Stoner [31]) (in a.u.).

	Experiment	9s5p3d / 4s1p		8s4p3d / 3s1p	
		N	He	N	He
$N^{5+}(1s^2)$	6.448	6.429		6.424	
$N^{4+}(3d)$	5.057	5.061		5.065	
$N^{4+}(3p)$	5.027	5.024		5.001	
$N^{4+}(3s)$	4.928	4.902		4.898	
$N^{4+}(2p)$	3.216	3.190		3.193	
$N^{4+}(2s)$	2.849	2.822		2.822	
$N^{3+}(2s3d)$	1.956	1.972		1.976	
$N^{3+}(2s3p)$	1.844	1.848		1.843	
$N^{3+}(2s3s)$	1.773	1.771		1.771	
$N^{3+}(2p3p)$	1.073	1.096		1.096	
$N^{3+}(2s2p)$	0.596	0.619		0.619	
$N^{3+}(2s^2)$	0	0		0	
$He^+(1s)$	0.904	0.878		0.887	
$He(1s^2)$	0	0		0	

The evaluation of the radial coupling matrix elements between molecular states of the same symmetry

$$g_{KL}(R) = < \psi_K \mid \frac{\partial}{\partial R} \mid \psi_L >$$

has been performed by means of the finite difference technique [33]

$$g_{KL}(R) = \lim_{\Delta \to 0} \frac{1}{\Delta} < \psi_K(R) \mid \psi_L(R+\Delta) >$$

For reasons of numerical accuracy, we have performed a three-point differentiation using calculations at R-Δ and R+Δ with a parameter $\Delta = 0.0012$ a.u.. The origin of the electronic coordinates has been generally taken at the N nucleus in order to eliminate the non-vanishing coupling terms at long-range. The importance of possible translation effects has nevertheless been estimated in the case of the metastable $N^{5+}(1s2s)$ + He system by performing the calculations of $g_{KL}(R)$ using both the N and He nuclei as the origin of electronic coordinates.

Table 2. Comparison of present calculated atomic levels with experimental data [31] and other calculations [32] (in a.u.).

	This calculation	K.T. Chung non relativistic	K.T. Chung relativistic	Experiment
$N^{4+}(1s2s3d)^4D$	-30.832782	-30.817929	-30.837461	
$N^{4+}(1s2s3p)^4P$	-30.864605	-30.891621	-30.911482	
$N^{4+}(1s2s3s)^4S$	-30.992015	-31.013330	-31.033845	
$N^{4+}(1s2p3d)^4F$	-30.594964	-30.586870	-30.604060	
$N^{4+}(1s2p3p)^4D$	-30.729597	-30.649589	-30.667166	
$N^{4+}(1s2p3s)^4P$	-30.798135	-30.722802	-30.740776	
$N^{5+}(1s2s)^1S$	15.642948			15.675432
$N^{5+}(1s2s)^3S$	15.400174			15.435593
$N^{5+}(1s^2)^1S$	0.0			0.0

The rotational coupling matrix elements between Σ-Π and Π-Δ states have been evaluated analytically by use of the L_+ and L_- operators.

3. Molecular results

3.1. GROUND STATE SYSTEM $N^{5+}(1s^2)^1S + He(1s^2)^1S$

The potential energy curves of the $^1\Sigma^+$, $^1\Pi$ and one $^1\Delta$ states involved in the single- and double-electron capture processes are displayed in Fig. 1. The $^1\Pi$ and $^1\Delta$ potential energy curves show no evidence of avoided crossings, but three avoided crossings appear in the range [6.0-9.0 a.u.] between the entry channel and the $^1\Sigma^+$ $\{N^{4+}(1s^2,3l) + He^+\}$ states of single-electron capture and at about 9.0 a.u. between the $^1\Sigma^+$ $\{N^{4+}(2s) + He^+(1s)\}$ and $^1\Sigma^+$ $\{N^{3+}(2s2p) + He^{2+}\}$ states.

The asymptotic energy values obtained by a configuration interaction calculation at 25 a.u. corrected by the coulombic repulsion term (the $1/R^4$ term has been neglected) are seen to be in quite good agreement with experiment (Table 3).

The main features of the radial coupling matrix elements are presented in Fig. 2. In correspondence with the avoided crossings between the $^1\Sigma^+$ potential energy curves of single-electron capture, sharp peaked functions appear at respectively 6.35, 7.50 and 8.30 a.u.. They are approximately 1.23, 2.53 and 12.21 a.u. high and respectively 0.75, 0.50 and less than 0.10 a.u. wide at half height.

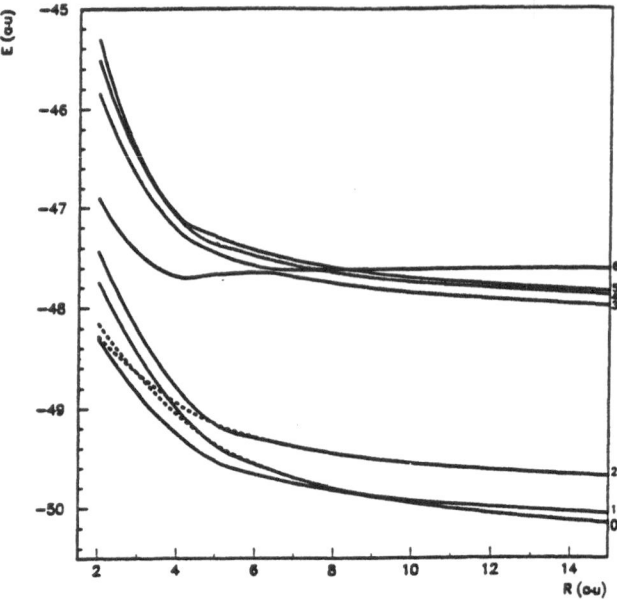

Fig. 1. Potential energy curves of the $^1\Sigma^+$, $^1\Pi$, $^1\Delta$ states of NHe^{5+}.
—— $^1\Sigma^+$ states, ---- $^1\Pi$ states, ···· $^1\Delta$ state.
0: Σ and Π states dissociating to $\{N^{3+}(2s2p) + He^{2+}\}$
1: Σ state dissociating to $\{N^{4+}(2s) + He^+(1s)\}$
2: Σ and Π states dissociating to $\{N^{4+}(2p) + He^+(1s)\}$
3: Σ state dissociating to $\{N^{4+}(3s) + He^+(1s)\}$
4: Σ and Π states dissociating to $\{N^{4+}(3p) + He^+(1s)\}$
5: Σ, Π and Δ states dissociating to $\{N^{4+}(3d) + He^+(1s)\}$
6: Σ state dissociating to $\{N^{5+}(1s^2) + He(1s^2)\}$.

Table 3. Comparison of experimental data [31] with theoretical energies of $^1\Sigma^+$ states obtained by a CIPSI calculation at 25 a.u. corrected by the coulombic repulsion term (in a.u.).

	Theoretical energies	Experimental energies
$N^{5+}(1s^2) + He(1s^2)$	2.708	2.694
$N^{4+}(3d) + He^+(1s)$	2.220	2.208
$N^{4+}(3p) + He^+(1s)$	2.181	2.177
$N^{4+}(3s) + He^+(1s)$	2.073	2.078
$N^{4+}(2p) + He^+(1s)$	0.370	0.367
$N^{4+}(2s) + He^+(1s)$	0	0

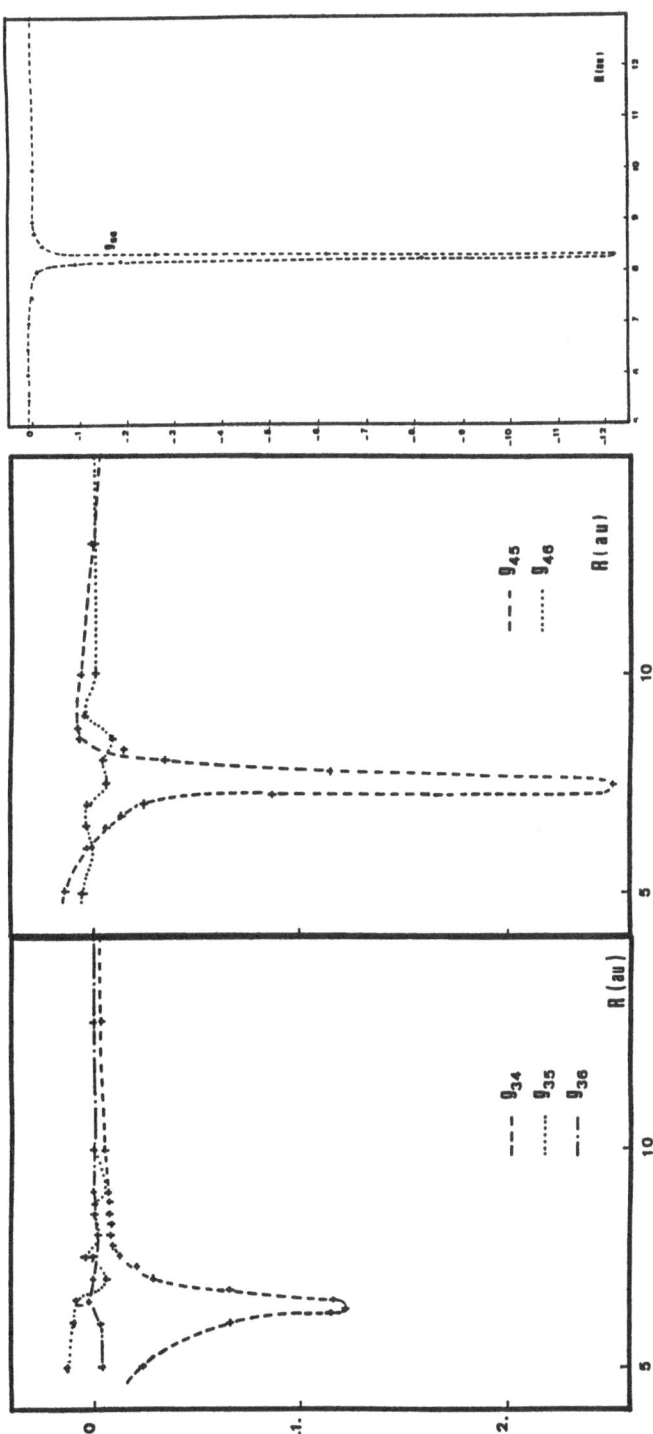

Fig. 2. a, b, c. Non-adiabatic radial coupling matrix elements for the $^1\Sigma^+$ states of single-electron capture.

The rotational coupling matrix elements between $^1\Sigma^+$ and $^1\Pi$ states of NHe^{5+} are displayed in Fig. 3. At large internuclear distances, rotational couplings are seen to be almost equal to 1.0 a.u. for $^1\Pi$ and $^1\Sigma^+$ states, corresponding to the same configuration, i.e. {N^{4+}(3p) + He$^+$(1s)} and {N^{4+}(3d) + He$^+$(1s)}.

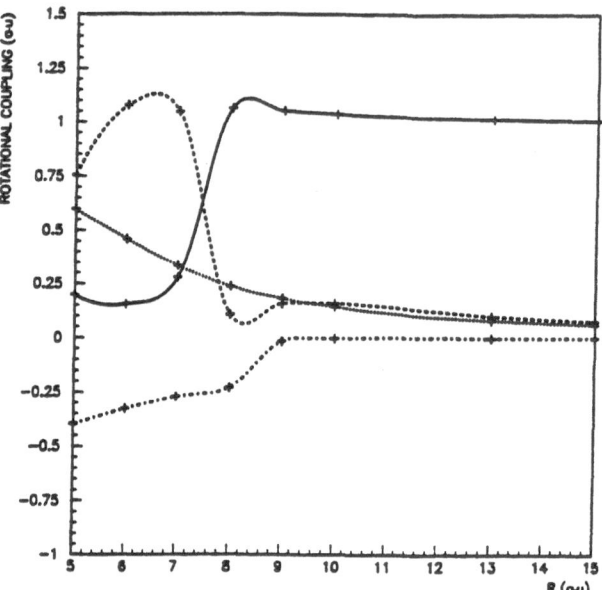

Fig. 3.   Rotational coupling matrix element between the $^1\Pi$ {N^{4+}(3p) + He$^+$(1s)} state and the $^1\Sigma^+$ states of single-electron capture.

······ g_{73}, ——— g_{74}, ----- g_{75}, — · — · g_{76}.

3.2.   METASTABLE SYSTEM N^{5+}(1s2s)^{3}S + He(1s)^{1}S

The potential energy curves of the $^3\Sigma^+$ and $^3\Pi$ states are presented in Fig. 4. They show four avoided crossings in the range [5.0-10.0 a.u.] between the entry channel, the state corresponding to {N^{4+}(1s2p3s) + He$^+$} and the three states of single-electron capture {N^{4+}(1s2s3l) ^{4}L + He$^+$}.

In relation with these avoided crossings, the radial coupling matrix elements present sharp peaks at respectively 5.4, 6.6, 7.55 and 9.5 a.u. (Fig. 5). We may notice that these radial couplings are almost insensitive to the choice of the origin of electronic coordinates. The most sensitive one is the g_{23} function at short internuclear distance range, but we may expect weak translational effects for such electron capture processes dominated by collisions at large distance of closest approach.

With regard to the results obtained for the ground state system N^{5+}(1s^2) + He, all the crossings are shifted towards shorter internuclear distances and have a lower height. The main remark concerns, however, the presence of an avoided crossing between the entry channel and the $^3\Sigma^+$ state corresponding to the $\{N^{4+}(1s2p3s) + He^+\}$ configuration which explains immediately the possibility of a transfer-excitation process for the $N^{5+}(1s2s)+He$ collision; such a process was not observed with the ground state.

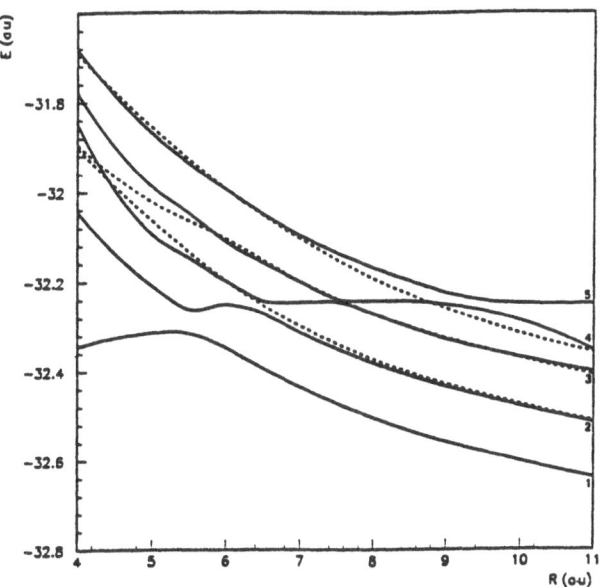

Fig. 4.   Potential energy curves for the $^3\Sigma^+$ and $^3\Pi$ states of NHe^{5+} metastable ion.
        —— $^3\Sigma^+$ states, ---- $^3\Pi$ states.
        1: Σ state dissociating to $\{N^{4+}(1s2s3s) + He^+(1s)\}$
        2: Σ and Π states dissociating to $\{N^{4+}(1s2s3p) + He^+(1s)\}$
        3: Σ and Π states dissociating to $\{N^{4+}(1s2s3d) + He^+(1s)\}$
        4: Σ and Π states dissociating to $\{N^{4+}(1s2p3s) + He^+(1s)\}$
        5: Σ state dissociating to $\{N^{5+}(1s2s)^3S + He(1s^2)\}$.

4. Collision dynamics

4.1. SINGLE-ELECTRON CAPTURE PROCESS FROM THE GROUND STATE $N^{5+}(1s^2)^1S + He(1s^2)^1S$

This is, beyond all doubt, the most important process and the only one which has been already tackled with theoretically. Nevertheless, the prediction given by the classical overbarrier transition model is not correct for this collision [9] and the modified multichannel Landau-Zener model developed by Boudjema et al. [34] cannot explain the experimental results   for collision velocities higher than 0.2 a.u.. With regard to the collision energy range, we have thus performed a semi-classical [35] collisional treatment

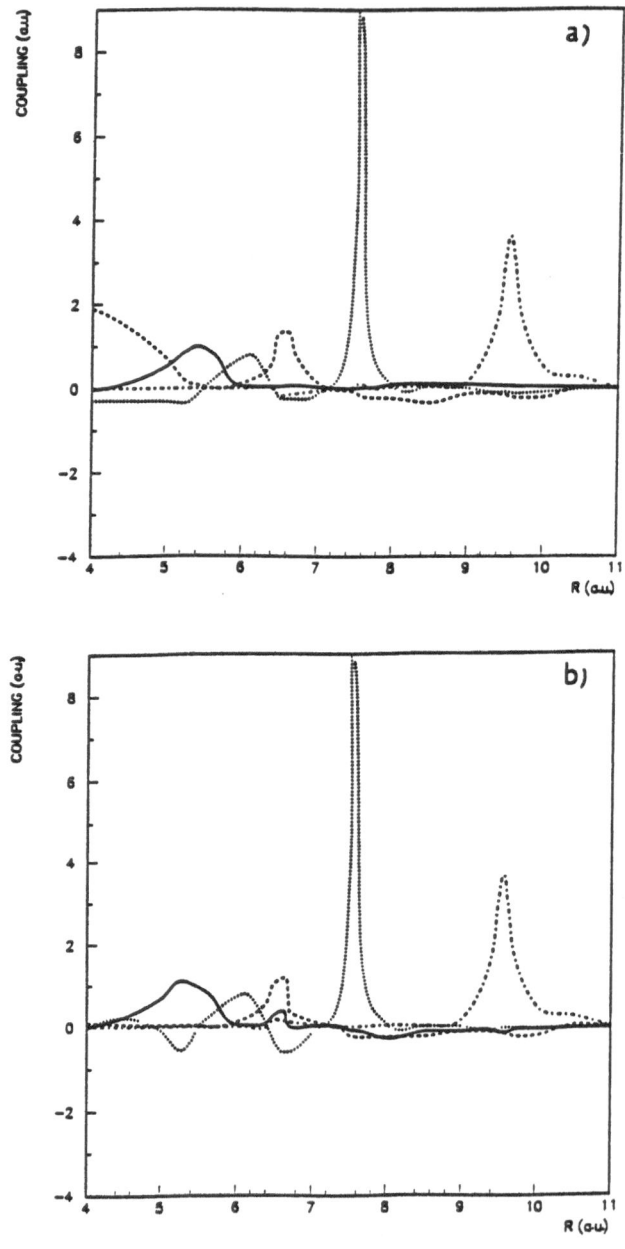

Fig. 5.a, b. Non adiabatic radial coupling matrix elements between $^3\Sigma^+$ states.
a) Origin N. b) Origin He.
—— g_{12}, ---- g_{23},···· g_{34}, — · — g'_{45} ($g_{45} \times 10$).

of the single-electron capture process using the ab initio molecular data. The $^1\Delta$ state has been neglected in the calculation. The strong radial couplings between the entry channel and the $^1\Sigma^+$ states dissociating to $\{N^{4+}(1s^2,3l) + He^+(1s)\}$ have been fitted by Lorentzian shape functions, while the other couplings and the potential energies have been fitted by spline cubic functions.

The partial cross-sections on the n = 3 levels are displayed in Table 4 and Fig. 6 and show a fairly good overall agreement with the experimental results of Cotte et al. [4,7] and Dijkkamp et al. [9]. From a numerical point of view, the error bar has been estimated experimentally to ±30% by Cotte et al. [4,7] and to ±5% by Dijkkamp et al. [9]. Theoretically, the error bar could be evaluated to about ±20%, the main difficulty arising in the determination of the sharp radial couplings.

Table 4. Single-electron capture cross-sections on the n = 3 levels (in 10^{-16} cm^2). (For comparison with Dijkkamp results, the collision energy is given in parenthesis when different from ours).

Collision energy (keV)		Theoretical calculation	Cotte et al. [4]	Dijkkamp et al. [9]
10	σ_{3s}	7.6	7.1	9.4
	σ_{3p}	2.2	2.0	3.2 (12.5 keV)
	σ_{3d}	0.6	0.8	1.6
25	3s	5.8	5.8	6.5
	3p	2.1	2.4	3.1
	3d	1.6	1.5	1.9
35	3s	4.8	5.3	5.2
	3p	2.2	2.4	3.0 (37.5 keV)
	3d	2.1	1.8	2.5
50	3s	4.5	5.0	4.6
	3p	2.8	2.5	3.1
	3d	3.2	2.5	3.0
75	3s	3.2		3.5
	3p	3.0		3.1
	3d	4.2		3.8
100	3s	2.4		2.9
	3p	2.6		3.1
	3d	5.6		4.4

Our calculations reproduce the variation of the partial cross-sections with the collision energy, in particular the variation of the $\{N^{4+}(3d) + He^+\}$ level which could not be interpreted by the Landau-Zener model [34]. This feature seems to be driven at high energy by the rotational coupling (Table 5), the $^1\Pi$ levels showing a preponderant contribution to the cross-section at 100 keV, especially for the states of the $\{N^{4+}(3d) + He^+\}$ configuration.

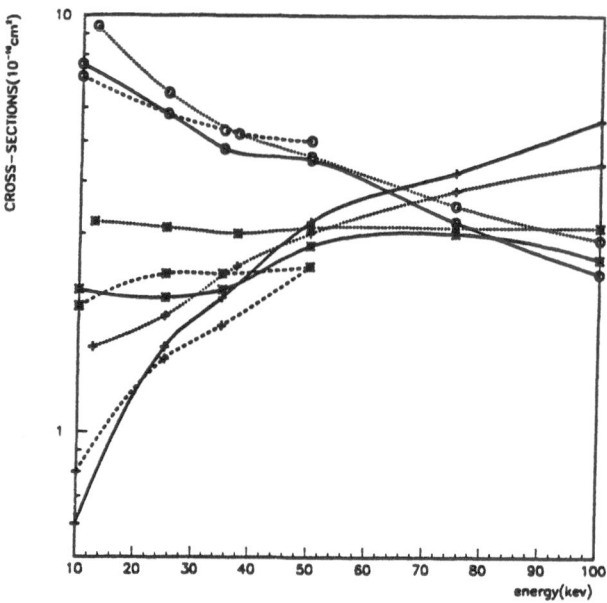

Fig. 6.   Partial cross-sections of single-electron capture.
——— This calculation, ----- Cotte et al. [4,7], ····· Dijkkamp et al. [9].

4.2.   SINGLE-ELECTRON CAPTURE PROCESS FROM THE METASTABLE $N^{5+}(1s2s)^3S + He(1s^2)^1S$

In consideration with the experimental data available, the collision dynamics has been performed for two energies, 60 and 50 keV, by means of a semi-classical method using the EIKONXS program [36]. As seen in Table 5, the contribution of the Π levels coupled by rotational couplings appears to be quite negligible over the contribution of the Σ levels for collision energies up to 50 keV. The collisional treatment has thus been performed with $^3\Sigma^+$ states only. Two calculations have been undertaken: one with the entry channel and the three states of single-electron capture $\{N^{4+}(1s2s3l)^4L + He^+(1s)^2S\}$ and one including besides the transfer-excitation state. The partial cross-sections of capture are presented in Table 6 and compared with the experimental results of Bouchama and Druetta at 60 keV [6]. Taking into account the experimental error bar, which is at least of 30% in view of the weakness of the observed lines, the accordance appears to be quite good. This result gives even confidence in the experimental results which remain particularly difficult to analyse. Besides, the theoretical partial cross-sections are affected by less than 20% by changing the origin of electronic coordinates, which is far behind the experimental error bar.

Table 5. Values of the single-electron capture cross-sections for the $^1\Sigma^+$ and $^1\Pi$ states (in 10^{-16} cm^2).

Collision energy (keV)	σ_{3s}	σ_{3p}		σ_{3d}	
	$^1\Sigma^+$	$^1\Sigma^+$	$^1\Pi$	$^1\Sigma^+$	$^1\Pi$
10	7.64	0.63	10^{-3}	2.19	4.10^{-3}
50	4.50	2.76	2.10^{-3}	3.22	0.01
100	2.43	0.80	1.85	1.42	4.22

Table 6. Calculated values of the partial cross-sections of capture at 60 keV. Comparison with experiment [6] (in 10^{-16} cm^2).

	O origin	He origin	Experiment ($\pm 30\%$)
σ_{3s}	6.65	7.78	7
σ_{3p}	4.75	3.92	3
σ_{3d}	4.77	2.83	2
σ_{TE}	2.57	2.22	1.7
(N^{4+}(1s2p3s) + He$^+$)			

A comparison of the partial cross-sections of capture at 50 keV, for the collision with He of the ground state N^{5+}(1s^2)^{1}S and the metastable N^{5+}(1s2s)^{3}S is given in Table 7. About the same values are obtained for both systems, with a slightly higher value of σ_{3d} for the metastable. This shows, a posteriori, that neglecting the fraction of metastable — which is often done when no informations are available — should not lead to too high an experimental error bar.

Table 7. Comparison of the capture partial cross-sections with N^{5+}(1s^2) and N^{5+}(1s2s)^{3}S at 50 keV (in 10^{-16} cm^2).

	N^{5+}(1s2s)^{3}S + He(1s^2) our calculation		N^{5+}(1s^2) + He(1s^2) our calculation	experiment	
	O origin	He origin	O origin	ref. [4]	ref. [9]
σ_{3s}	4.85	3.37	4.5	5.0	4.6
σ_{3p}	3.78	3.57	2.8	2.5	3.1
σ_{3d}	4.33	3.12	3.2	2.5	3.0

4.3. DOUBLE-ELECTRON CAPTURE PROCESS FROM THE GROUND STATE $N^{5+}(1s^2)^1S + He(1s^2)^1S$

A quantitative treatment of this process should be "formidable" as stated by Crandall [8]. We tried just to get an insight into a possible interpretation of the phenomena.

For that, we performed two semi-classical calculations in the 10-50 keV energy range: first a three-channel calculation including the entry channel and the $^1\Sigma^+$ and $^1\Pi$ states corresponding to the $\{N^{3+}(1s^22s2p) + He^{2+}\}$ configuration, and a six-channel calculation including besides the $^1\Sigma^+$ and $^1\Pi$ states corresponding to $\{N^{4+}(n=2l) + He^+(1s)\}$.

Assuming the contribution of the potential energy curves which have not been taken into account to be almost constant with the collision energy, such calculations could provide a relative estimate of the variation of the double capture cross-sections with the collision energy. The results presented in Fig. 7 seem to be coherent with this hypothesis and to corroborate a cascade effect for the double electron capture process.

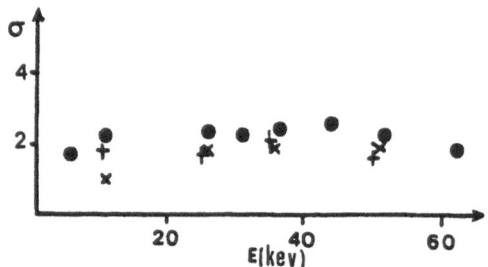

Fig. 7. Comparison of experimental and theoretical values of the cross-section for the double capture process.
o experimental cross-sections (in 10^{-16} cm^2) [5], Ỹ three-channel calculation, + six-channel calculation (arbitrary units).

5. Concluding remarks

This work provides an accurate and complete treatment of the collision of a multicharged ion on a neutral atom target — here the helium atom — taking into account both ground and metastable states.

As far as the molecular calculation is concerned, the use of an ab initio method is necessary for an adequate representation of the open-shell metastable $N^{5+}(1s2s)$ + He system with four outer electrons. The CIPSI configuration interaction method used in this calculations leads to the same rate of accuracy as the spin-coupled valence bond method (cf. the work on CH^{3+} by Cooper et al. [19] or on NH^{4+} by Zygelman et al. [37]).

For the collision dynamics, a semi-classical method is quite accurate for intermediate energies about 50 keV. Note the use of a more recent collision program using the propagation method for the $N^{5+}(1s2s)$ + He metastable system.

The agreement shown between calculations and experiment gives confidence both in the theoretical method used here, and in the analysis of experimental spectra, in particular in the case of metastable state. Furthermore, the interpretation of the transfer-excitation process is straight forward from the knowledge of the potential energy curves of the collisional system.

Acknowledgements

I gratefully thanks M. Druetta for very fruitful discussions all over this work.

References

1. P. Petitjean, C. Boisson and D. Péquignot, *Astron. Astrophys.* **240**, 433 (1990)
2. H.W. Moos, S.T. Durrance, T.E. Skinner, P.D. Feldman, J.L. Bertaux and M.C. Feston, *Astrophys. J. Lett.* **275**, L 19 (1983).
3. R.J. Fonck, D.S. Darrow and K.P. Jaehnig, *Phys. Rev. A* **29**, 3288 (1984).
4. P.H. Cotte, M. Druetta, S. Martin, A. Denis, J. Désesquelles, D. Hitz and S. Dousson, *Nucl. Instrum. Methods B* **9**, 743 (1985).
5. T. Bouchama, J. Désesquelles, M. Druetta, M. Farizon and S. Martin, *J. Phys. B: At. Mol. Phys.* **20**, L 457 (1987).
6. M. Druetta, T. Bouchama and S. Martin, *Nucl. Instrum. Methods B* **40/41**, 50 (1989); T. Bouchama and M. Druetta, private communication.
7. P.H. Cotte, Thesis, Lyon, France (1984).
8. D.H. Crandall, *Phys. Rev. A* **16**, 958 (1977).
9. D. Dijkkamp, D. Ciric, E. Vlieg, A. De Boer and F.J. De Heer, *J. Phys. B: At. Mol. Phys.* **18**, 4763 (1985).
10. Y.S. Gordeev, D. Dijkkamp, A.G. Drentje and F.J. De Heer, *Phys. Rev. Lett.* **50**, 1842 (1983).
11. A. Brazuk, D. Dijkkamp, A.G. Drentje, F.J. De Heer and H. Winter, *J. Phys. B: At. Mol. Phys.* **17**, 2489 (1984).
12. K. Okuno, H. Tawara, T. Iwai, Y. Kaneko, M. Kimura, N. Kobayashi, A. Matsumoto, S. Ohtani, S. Tagaki and S. Tsurubuchi, *Phys. Rev. A* **28**, 127 (1983).
13. P. Roncin, M. Barat, M.N. Gaboriaud, L. Guillemot and H. Laurent, *J. Phys. B: At. Mol. Opt. Phys.* **22**, 509 (1989).
14. A. Bordenave-Montesquieu, P. Benoît-Cattin, A. Gleizes, S. Dousson and D. Hitz, *J. Phys. B: At. Mol. Phys.* **18**, L 195 (1985).
15. R. Mann, F. Folkmann and H.F. Beyer, *J. Phys. B* **14**, 1161 (1981).
16. M. Gargaud, J. Hanssen, R. Mc Carroll and P. Valiron, *J. Phys. B* **14**, 2259 (1981).
17. J. Hanssen, R. Gayet, C. Harel and A. Salin, *J. Phys. B* **17**, L 323 (1984).
18. C. Harel, H. Jouin and B. Pons, *J. Phys. B: At. Mol. Opt. Phys.* **24**, L 425 (1991).
19. D.L. Cooper, M.J. Ford, J. Gerratt and M. Raimondi, *Phys. Rev. A* **34**, 1752 (1986).

20. M. Kimura and R.E. Olson, *Phys. Rev. A* **31**, 489 (1985); *J. Phys. B* **18**, 2729 (1985).
21. M.C. Bacchus-Montabonel, *Phys. Rev. A* **36**, 1994 (1987); *Phys. Rev. A* **40**, 6088 (1989).
22. M.C. Bacchus-Montabonel, *Phys. Rev. A* **46**, 217 (1992).
23. M.C. Bacchus-Montabonel, C. Courbin, R. Mc Carroll, *J. Phys. B: At. Mol. Phys.* **24**, 4409 (1991).
24. M. Gargaud, M.C. Bacchus-Montbonel, R. Mc Carroll, *J. Chem. Phys.* (to be published).
25. C.D. Lin, W.A. Johnson and A. Dalgarno, *Phys. Rev. A* **15**, 154 (1977).
26. T.P. Gozdanov, R.K. Janev and V. Yu Lazur, *Phys. Scr.* **32**, 64 (1985).
27. B. Huron, J.P. Malrieu and P. Rancurel, *J. Chem. Phys.* **58**, 5745 (1973).
28. M. Dupuis, J. Rys and H.F. King, Hondo 76, Q.C.P.E. Program, n° 338 (1976).
29. J.P. Daudey, Atelier pour le calcul des états excités des molécules, Technical Note, Toulouse, France, 1990.
30. R. Krishnan, J.S. Binkley, R. Seeger and J.A. Pople, *J. Chem. Phys.* **72**, 650 (1980).
31. S. Bashkin and J.O. Stoner, Atomic Energy Levels and Grotrian Diagrams, North-Holland, Amsterdam, 1975.
32. K.T. Chung (private communication).
33. M.C. Bacchus-Montabonel, R. Cimiraglia, and M. Persico, *J. Phys. B: At. Mol. Phys.* **17**, 1931 (1984).
34. M. Boudjema, P. Benoît-Cattin, A. Bordenave-Montesquieu and A. Gleizes, *J. Phys. B: At. Mol. Phys.* **21**, 1603 (1988).
35. M. Aubert and C. Lesech, *Phys. Rev. A* **13**, 632 (1976).
36. R.J. Allan, Technical Memorendum, Daresbury Laboratory, 1990.
37. B. Zygelman, D.L. Cooper, M.J. Ford, A. Dalgarno, J. Gerratt and M. Raimondi, *Phys. Rev. A* **46**, 3846 (1992).

An Ab initio Study of the Lowest $^{1,3}\Sigma^+$ States of BH. Quasi Diabatic Curves and Vibronic Couplings

M. PERSICO,
Dipartimento di chimica e chimica industriale, Università di Pisa, v. Risorgimento 35, I-56126 Pisa, Italy
R. CIMIRAGLIA,
Dipartimento di chimica, Università di Ferrara, v. Borsari 46, I-44100 Ferrara, Italy
F. SPIEGELMANN,
Laboratoire de physique quantique, UA 505 CNRS, Université Paul Sabatier, 118 route de Narbonne, F-31062 Toulouse, France

1.Introduction

The electronic $^1\Sigma^+$ states and potential energy curves of BH have been studied in the past at considerably different levels of detail and accuracy. Of the $X^1\Sigma^+$ ground state, four vibrational levels have been observed [1-3], therefore only the lowest part of the RKR curve is known [3-4]. A dissociation energy of 28760 $\pm$150 cm^{-1} has been derived from the break-off of rotational lines for the $A^1\Pi$ state, assuming that it corresponds to the top of a barrier in the potential energy curve of that state [2,5]. However, the above assumption has been criticised [6], and a recent analysis of the tunnelling lifetimes in the $A^1\Pi$ state yields $D_e = 29420 \pm 30$ cm^{-1} [7], in good agreement with the most complete theoretical determinations [8,9]. The whole potential curve for the $X^1\Sigma^+$ state has been determined by several theoretical methods [10-12], almost at the same level of accuracy as the calculations of D_e.
Four vibrational levels were observed also for the $B^1\Sigma^+$ and the $C^1\Sigma^+$ states and only one for the $E^1\Sigma^+$ state [1,2]. Ab initio calculations have shown that the $B^1\Sigma^+$ potential energy curve has two minima [11,13-15], but the vibrational states which should be accomodated in the outer minimum have never been observed. The double minimum is due to a Rydberg-ionic curve crossing, which should affect also the shape of the potential energy curves of the next excited states, C and $E^1\Sigma^+$. However, only the $C^1\Sigma^+$ curve has been determined [15], and for a range of internuclear distances such that the expected avoided crossings could not be detected.
The aim of this work is to obtain the four lowest $^1\Sigma^+$ curves and wavefunctions of BH at the same level of accuracy and to bring out the interplay of ionic, Rydberg and valence states at energies and internuclear distances which were not previously investigated. We have therefore made use of a method, already put forward by us [16,17] to determine at once quasi-diabatic and adiabatic states, potential energy curves and approximate nonadiabatic couplings. We have analogously determined the first three $^3\Sigma^+$ states, of which only the lowest had been theoretically studied

349

Y. Ellinger and M. Defranceschi (eds.), Strategies and Applications in Quantum Chemistry, 349–365.

[14,18,19]. The determination of electronic quantities is preliminary to the study of molecular dynamics and radiative properties.

In section 2 we describe the method employed and in sections 3 and 4 we comment upon the electronic energy curves and wavefunctions.

2.Method

We have performed ab initio calculations employing a gaussian atomic basis set similar to that of Jaszunski et al [11], but slightly superior for the presence of more diffuse s and p functions; these were necessary in order to represent higher Rydberg terms of the B atom and the H⁻ anion formed upon dissociation of the ionic state. The s and p exponents were taken from van Duijneveldt's compilation [20]. His 13s,8p basis for the B atom was contracted with the scheme 6111111,311111; two diffuse s and one p function were added, with exponents 0.036, 0.014 and 0.015, respectively; three d functions (not including the s combination of the cartesian components) were added, with exponents close to Jaszunski's: 1.0, 0.4, 0.12; the final basis had 10s,7p,3d functions on B. For H we used van Duijneveldt's 9s basis, contracted with the scheme 51111 and supplemented with one more s (exponent 0.0165) and three p (1.0, 0.4, 0.135) functions; the exponents of the most diffuse s and p functions were such as to minimize the full CI energy of H⁻; the basis for H was thus 6s,3p. The computed electron affinity of H is 5735 cm^{-1}, to be compared with the experimental value of 6081 cm^{-1} [21]. The closed shell SCF energy of BH at the experimental equilibrium distance ($R = 2.336$ bohr) is -25.131322 a.u., while the HF limit is -25.131647 a.u. [22].

SCF-CI calculations were performed at 20 different internuclear separations, from 1.2 bohr to $+\infty$. The lowest separate atom states are B(^{2}P, 2p) and H(^{2}S); therefore, in order to have a homolytic dissociation and three degenerate 2p orbitals on B we have adopted the closed shell Fock hamiltonian with fractional occupation [23]: one electron was placed in the 3σ orbital, correlating with H(1s) at infinite separation, and 1/3 each in the 4σ and 1π orbitals correlating with B(2p).

Electron correlation was treated by the CIPSI multi-reference perturbation algorithm ([24,25] and refs. therein). The Quasi Degenerate Perturbation Theory (QDPT) version of the method was employed, with symmetrisation of the effective hamiltonian [26], and the Møller-Plesset baricentric (MPB) partition of the C.I. hamiltonian. The zeroth order states $\left|\eta^{(0)}\right\rangle$ were quasi-diabatic, as described by Cimiraglia et al [16]. The adiabatic states $\left|\psi^{(0)}\right\rangle$ are obtained by diagonalisation and then submitted to a unitary transformation:

$$\left|\eta^{(0)}\right\rangle = \left|\psi^{(0)}\right\rangle C^{(0)\dagger} \qquad (1)$$

The $C^{(0)}$ transformation matrix is determined by the requirement that the sum of square overlaps $\sum_I \left\langle \eta_I^{(0)} | R_I \right\rangle^2$ be maximized. The R_I are simple reference wavefunctions chosen so as to represent the diabatic states of interest. Their quality does not affect the accuracy to which the adiabatic energies and wavefunctions are determined, but only the degree of "diabaticity" of the corresponding $\eta_I^{(0)}$ functions. The QDPT is the natural complement of such a method, in that it allows a perturbed effective hamiltonian matrix **H** in the quasi-diabatic basis to be constructed. The $\eta_I^{(0)}$ wavefunctions themselves are perturbed to first order:

$$|\eta_I\rangle = \left|\eta_I^{(0)}\right\rangle + \left|\eta_I^{(1)}\right\rangle \tag{2}$$

and the final adiabatic energies and states are obtained by diagonalisation of the **H** matrix:

$$\mathbf{H\,C} = \mathbf{C\,E} \tag{3}$$

$$|\psi_K\rangle = \sum_I C_{IK}\,|\eta_I\rangle \tag{4}$$

The **C** matrix, the columns of which, $\mathbf{C}_K$, are the eigenvectors of **H**, is normally not too different from the $\mathbf{C}^{(0)}$ matrix defined above. However, the QDPT treatment, applied either to an adiabatic or to a diabatic zeroth-order basis, is necessary in order to prevent serious artefacts, especially in the case of avoided crossings [27]. The preliminary diabatisation makes it easier to interpolate the matrix elements of the hamiltonian and of other operators as functions of the nuclear coordinates and to calculate the nonadiabatic coupling matrix elements:

$$G_{KL}(R) \equiv \left\langle \psi_K \left| \frac{\partial}{\partial R} \right| \psi_L \right\rangle \tag{5}$$

$$T_{KL}(R) \equiv \left\langle \psi_K \left| \frac{\partial^2}{\partial R^2} \right| \psi_L \right\rangle \tag{6}$$

(see appendix); moreover, one can apply theoretical or empirical corrections to the diabatic energies, in order to improve the location of the avoided crossings and the general shape of the adiabatic curves; finally, the simple inspection of the diabatic curves and of the $\mathbf{C}_K$ vectors helps in bringing out the nature of the adiabatic states and the relevant physical effects.

The construction of quasi-diabatic states often meets with the so called intruder state problem. This can be stated as follows: a given set of adiabatic states, contiguous in the energy ordering, is not spanned by the same basis of (approximately) diabatic states for all nuclear geometries. Such is the case under study in this work. The first four singlet and triplet Σ^+ states of BH at very large internuclear distances originate from $H(^2S,1s)$ and

1) $B(^2P, 2p)$, ground state;
2) $B(^2S, 3s)$, Rydberg excitation $2p \to 3s$ (excitation energy, $40040\ cm^{-1}$);
3) $B(^2D, 2s2p^2)$, valence excitation $2s \to 2p$ ($47857\ cm^{-1}$);
4) $B(^2P, 3p)$, Rydberg excitation $2p \to 3p$ ($48614\ cm^{-1}$).

We may tentatively identify the diabatic references $R_1 \ldots R_4$ with antisymmetrized products of these B and H atomic states. The Rydberg series converges to the first ionisation level ($66928\ cm^{-1}$); the electron affinity of H brings the energy of the ionic configuration B^+H^- to $60847\ cm^{-1}$. Six more excited states of B lie under this level [28]; the attractive Coulomb potential of B^+H^- crosses all the flat curves originated from such states, and also those of $B(^2S, 3s)$, $B(^2D, 2s2p^2)$ and $B(^2P, 3p)$, at large internuclear distances ($R > 10$ bohr). As a result, the four lowest $^1\Sigma^+$ states, for intermediate R, are linear combinations of three neutral and one ionic state, approximately orthogonal to R_4. To circumvent the intruder state problem, we change the definition of the fourth reference state, following a procedure already described [17]. For the singlet states, R_4 will be the linear combination:

$$R_4 = \lambda_1[B(^2P, 3p)\ H(^2S)] + \lambda_2[B^+H^-] \tag{7}$$

with coefficients depending on the internuclear distance: λ_1 and λ_2 are determined at the same time as the $C^{(0)}$ matrix, and by the same maximum overlap criterium, $\sum_I \langle \eta_I^{(0)} | R_I \rangle^2 = max$. As a result, the associated $|\eta_4\rangle$ state is less strictly diabatic, or diabatic only in certain ranges of R, but some interesting properties still hold: for instance, the $\langle \eta_I | \frac{\partial}{\partial R} | \eta_4 \rangle$ matrix elements should be small or negligible as asserted in the appendix.

The situation for the $^3\Sigma^+$ states is similar, because of the autoionizing state $H^-(^3S, 1s2s)$, lying about 13800 cm^{-1} above the ground state of the H^- anion according to our full CI calculations. The $B^+H^-(^3S)$ state contaminates the fourth $^3\Sigma^+$ at short R. However, we have chosen to study only the first three $^3\Sigma^+$ states, which are approximately spanned by $B(^2P, 2p)\ H(^2S)$, $B(^2S, 3s)\ H(^2S)$ and, depending on the internuclear distance, $B(^2D, 2s2p^2)\ H(^2S)$ or $B(^2P, 3p)\ H(^2S)$. Therefore, a linear combination of the latter two states gives $|R_3\rangle$, while $|R_1\rangle$ and $|R_2\rangle$ are defined as for the singlets.

The orbitals employed to build the reference configurations were taken from atomic SCF calculations for the appropriate states of B, B^+, H and H^-; only for the valence orbitals of the Rydberg states, $B(^2S, 3s)\ H(^2S)$ and $B(^2P, 3p)\ H(^2S)$, we have preferred to perform RHF calculations on the BH^+ cation.

The CIPSI zeroth-order subspace was selected by performing preliminary calculations at different R for four $^1\Sigma^+$ and three $^3\Sigma^+$ states; at medium and large distances ($R > 4$ bohr), however, we determined five $^1\Sigma^+$ and four $^3\Sigma^+$ states, in order to guarantee a correct description of the ionic singlet and of both the $B(^2D, 2s2p^2)\ H(^2S)$ and $B(^2P, 3p)\ H(^2S)$ triplets, too. The same subspace, union of the subspaces resulting from selections at different R, was employed in all the final calculations. We adopted a modified version [25] of the three classes CIPSI algorithm originally proposed by Evangelisti et al [29]. The zeroth-order wavefunctions and energies are determined by diagonalisation in a subspace of 1287 determinants, (selection threshold = 0.018); of these, 396 determinants (selection threshold = 0.038) originate single and double excitations in the perturbation step.

The matrix elements of the electric dipole and of the $\frac{\partial}{\partial R}$ operators were determined for the perturbed η_I wavefunctions. The finite differences technique was applied to evaluate $\langle \eta_I | \frac{\partial}{\partial R} | \eta_J \rangle$, with $\Delta R = 0.005$ bohr (see [16] and refs. therein). All the matrix elements in the $|\eta\rangle$ basis were interpolated by cubic splines: notice that the direct interpolation of the adiabatic matrix elements would require many more ab initio calculations, especially in the proximity of the avoided crossings. The adiabatic states and energies are then determined by diagonalisation for any desired value of R (eqs. 3 and 4) and the matrix elements of other operators are transformed from the diabatic to the adiabatic basis; this procedure allows an approximate evaluation of the second derivative nonadiabatic couplings, $\langle \psi_K | \frac{\partial^2}{\partial R^2} | \psi_L \rangle$ (see appendix).

The first few bound vibrational states in each potential were determined by the Numerov algorithm [30], in conjunction with a technique, similar to Hajj's [31], which allows to change both the starting and the ending point of the numerical integration according to the features of each vibrational state: this is advantageous especially when dealing with double minimum problems and with bound states approaching the dissociative limit.

3.Singlet electronic states and energies

The ground state non relativistic electronic energy of BH at equilibrium distance can be evaluated in -25.2845 ± 0.00015 a.u., following the same procedure as Meyer and Rosmus [10], but using more accurate values for D_e [7] and the relativistic correction [32]; the uncertainty on the total arises from the experimental determination of the dissociation energy. Given the HF limit, one evaluates the correlation energy as 0.1528 a.u. We obtained a MPB total energy of -25.26125 a.u. (about 85% of the correlation energy accounted for), while the variational (zeroth-order) CI gives -25.18465 a.u. These values may be compared with some previous results: full CI on a DZ + polarisation basis set, -25.22763 a.u. [33]; CEPA with two different basis sets, -25.22588 a.u. [10] and -25.25512 a.u. [12]; CASSCF, -25.22413 a.u. [11]; MP4, -25.21498 a.u. [34]; QCISD, -25.25358 [8].

The main features of the singlet adiabatic energy curves are summarized in table 1. To avoid any ambiguity in the comparison with other results, we have given the first four vibrational levels for each electronic state, rather than the vibrational constants. Some of the data from previous works have been determined by spline interpolation of the potentials and Numerov integration of the vibrational equations. The MPB vibrational level spacings for the X, B and C states are too low by about 15, 30 and 65 cm^{-1}, respectively; the CASSCF results of Jaszunski et al 1981, limited to the X and B states, are better under this respect. On the other hand, we have the best results as concerns the dissociation energies, D_e, the transition energies of BH, T_e, and those of the B atom, $T_{R=\infty}$. Notice that the R_e, $R_{barrier}$ and ΔE values for the barrier and the outer minimum of the B state, shown in the first column of table 1, are not true experimental data: they belong to the hybrid potential of Luh and Stwalley [4], obtained by rescaling the theoretical results of Jaszunski et al [11]. An accurate treatment of the electron correlation is mandatory in order to evaluate such quantities, because of the wide differences in electronic structure between the states under consideration; however, in our opinion, the residual deviations from the experimental data are mainly due to deficiencies of the atomic basis set: in fact, CIPSI calculation for the isolated B atom and the B^+, H^- ions, with enlarged zeroth order spaces, did not improve the results.

Our potential energy curves too need some rescaling, in order to give, as far as possible, the correct ordering of the vibronic states belonging to different electronic terms, and the right position of the avoided crossings occurring at large R. This is best done by considering the diabatic energies, H_{ii}, because a constant or smoothly varying correction is not applicable to the adiabatic curves, at least in those regions where they undergo abrupt avoided crossings. The diagonalisation of a corrected diabatic hamiltonian matrix yields modified adiabatic energies and states, according to eqs.(3) and (4). Therefore, the nature of the electronic states and their dependence on R must be taken here into consideration. Between $R = 10$ and 25 bohr, the ionic potential curve of coulombic shape crosses $B(^2S, 3s)$, $B(^2D, 2s2p^2)$ and $B(^2P, 3p)$ (from now on, we shall indicate the neutral states with the corresponding atomic term of boron, for simplicity). Within this range of distances we were able to determine five diabatic

Table 1. Spectroscopic constants for the $^1\Sigma^+$ states of BH.
Distances in bohr and energies in cm^{-1}.

		Experimental[a]	MPB	+scaling	CASSCF[d]	NO-CI[e]
$X^1\Sigma^+$	R_e	$2.3289 \pm .0004^b$	2.323	2.312	2.329	2.357
	D_e	29420 ± 30^c	28794	29427	28400	28751
	$E_{vib}(v=0)^f$	1172	1157	1178	1170	1458
	$E_{vib}(v=1)^f$	3442	3411	3471	3436	4144
	$E_{vib}(v=2)^f$	5615	5576	5675	5607	6459
	$E_{vib}(v=3)^f$	7696	7632	7768	7684	8539
$B^1\Sigma^+$	$R_{e,inner}$	$2.2983 \pm .0005^b$	2.293	2.288	2.298	2.331
	$D_{e,inner}$	17120 ± 30	16054	17131	15164	—
	$R_{e,outer}$	5.830^b	5.756	5.702	5.834	—
	$\Delta E_{outer-inner}$	4001^b	4618	5542	4437	—
	$R_{barrier}$	3.832^b	3.805	3.807	3.826	3.770
	$\Delta E_{barrier-inner}$	10097^b	9633	10085	9679	10315
	T_e^g	52336	52405	52336	52512	50479
	$T_{R=\infty}^h$	40040	39665	40040	39328	—
	$E_{vib}(v=0)^f$	1181	1161	1176	1183	1408
	$E_{vib}(v=1)^f$	3429	3383	3429	3420	3968
	$E_{vib}(v=2)^f$	5502	5441	5521	5472	6114
	$E_{vib}(v=3)^f$	7378	7270	7389	7327	—
$C^1\Sigma^+$	R_e	2.2921	2.296	2.312	—	2.326
	D_e	22000 ± 30	20407	22004	—	21070
	T_e^g	55281	56706	55280	—	59038
	$T_{R=\infty}^g$	47857	48319	47857	—	55395
	$E_{vib}(v=0)^f$	1223	1189	1165	—	1443
	$E_{vib}(v=1)^f$	3588	3491	3435	—	4115
	$E_{vib}(v=2)^f$	5846	5698	5626	—	6439
	$E_{vib}(v=3)^f$	8000	7798	7720	—	8498
$E^1\Sigma^+$	R_e	2.300	2.259	2.257	—	—
	D_e	16220 ± 30	15263	16233	—	—
	T_e^i	61808 ± 30	62020	61808	—	—
	$T_{R=\infty}^h$	48614	48489	48614	—	—

[a] Experimental data from Huber and Herzberg [5], if not otherwise specified; vibrational energies for the X and B states from Luh and Stwalley [4], and for the C state from Johns et al [2]. [b] Refs. [3,4]. [c] Refs. [7-9]. [d] Ref. [11]. [e] Ref. [15]. [f] Vibrational energies from the bottom of the potential energy curve. For the B curve, only the vibrational states localized in the inner minimum are considered. Angular momentum J=0. [g] Transition energy from the X state, minimum to minimum. [h] Transition energy at the separated atom limit (atomic transitions of the B atom). [i] Transition energy as in (g), from T_{0-0} and a preliminary estimate of the zero point energy in the C state.

and adiabatic singlets, because here the intruder state problem, as stated in the previous section, is absent (the next ionic-Rydberg crossing is expected at $R = 36$ bohr, with the B(^{2}D, 3d) state). In the five states calculations, the reference wavefunctions were the same as described above, but in place of the linear combination of eq.(7) we had the two separate components B(^{2}P, 3p) H(^{2}S) and B$^+$H$^-$. Such a treatment enables us to describe all the aforesaid crossings: however, small errors in the transition energies and ionisation potential of the B atom, or in the electron affinity of H, bring about large displacements of the curve crossings along the R coordinate, because of the reduced slope of the ionic curve in this region. We have therefore applied the following constant corrections Δ_2 to the calculated diabatic energies H_{22} to H_{55}, i.e. to the transition energies for the B(^{2}S, 3s) , B(^{2}D, 2s2p^2) , B(^{2}P, 3p) and B$^+$H$^-$ states, respectively: $\Delta_2 = +0.001709, -0.002878, +0.001344, +0.00270$ a.u. The same corrections have been applied in the four states treatment, taking thoroughly into account that the fourth diabatic state gradually transforms from neutral B(^{2}P, 3p) to ionic between 15 and 20 bohr. Although the corrections to the diabatic potentials do not change abruptly in the crossing regions, they may depend smoothly on the molecular geometry, as a consequence of specific deficiencies of basis set or correlation treatment. Therefore, for $R \leq 2.3$ bohr, about the equilibrium distance for all the states, we have applied a different set of corrections, Δ_1. The Δ_1 values were adjusted so to reproduce the experimental dissociation and transition energies, D_e and T_e, within 1 cm^{-1}. For the $E^1\Sigma^+$ state T_e is unknown and has been estimated from T_{0-0} and a preliminary computed value of the zero point energy. In conclusion we have: $\Delta_1 = -0.00014, -0.00276, -0.01305, -0.00341$ a.u., respectively for the B(^{2}P, 2p) , B(^{2}S, 3s) , B(^{2}D, 2s2p^2) and B(^{2}P, 3p) states; the ionic state, as we shall see, does not contribute to the first four adiabatic states in the equilibrium region. Between $R_1 = 2.3$ and $R_2 = 6$ bohr, we connect smoothly the Δ_1 and Δ_2 corrections:

$$\Delta H_{II} = \frac{1}{2}(\Delta_1 + \Delta_2) + \frac{3}{2}\frac{R - R^*}{R_2 - R_1}(\Delta_2 - \Delta_1) - 2\left(\frac{R - R^*}{R_2 - R_1}\right)^3 (\Delta_2 - \Delta_1) \quad (8)$$

where $R^* = (R_1 + R_2)/2$.

The third column of table 1 contains the spectroscopic constants of the four $^1\Sigma^+$ states, based on the corrected adiabatic potentials. Apart from the D_e, T_e and $T_{R=\infty}$ energies, which are bound to coincide with the experimental values, we observe a remarkable improvement in the computed vibrational levels of the $B^1\Sigma^+$ state. The ΔE and R values here presented for the barrier and outer minimum of this potential curve are probably the best estimates available to date. The vibrational levels belonging to the $C^1\Sigma^+$ state are much less accurately determined, the energy differences $\Delta E(calc. - exp.)$ ranging from -58 to -280 cm^{-1}. The only previous determination of this potential energy curve leads to overestimate the same levels by 200-600 cm^{-1} (the same holds for the lowest electronic states, and may be partly due to a poor interpolation, i.e. to an insufficient number of points on the potential curves). No comparison with previous determinations is possible for the $E^1\Sigma^+$ state.

Figure 1 shows the corrected adiabatic energies for the $^1\Sigma^+$ states, $E_K(R)$, which come out practically identical from the four states and five states treatments (the same holds for the nonadiabatic coupling and dipole moment matrix elements). Figure 2 shows the corrected diabatic energies, $H_{ii}(R)$, according to the four states treatment; only the upper portions of the B(^{2}P, 3p) and ionic curves in the range $10 < R < 25$

Figure 1. Adiabatic energies, $^1\Sigma^+$ states.

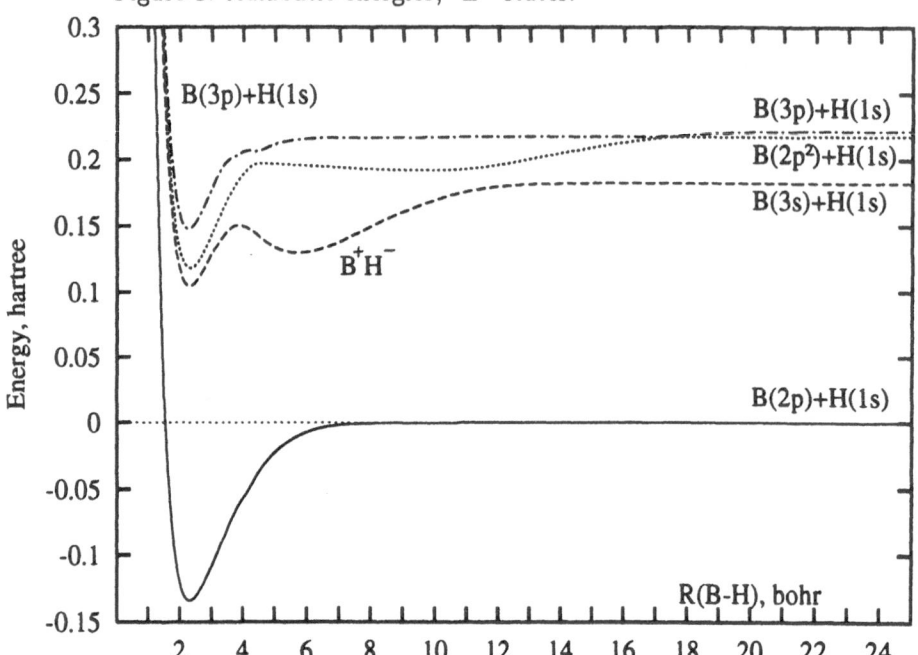

Figure 2. Diabatic energies, $^1\Sigma^+$ states.

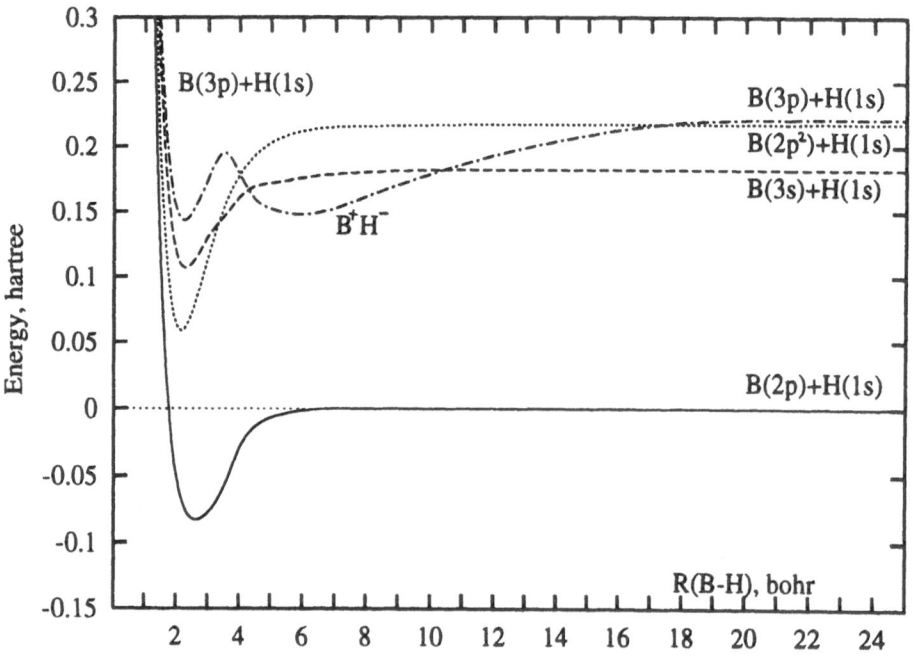

bohr have been taken from the five states treatment. All the diabatic states here considered are attractive. The two Rydberg states, $B(^2S, 3s)$ and $B(^2P, 3p)$, have a well depth of about 17000 cm^{-1}, approximately the same as in the BH^+ ground state $(D_{00} = 15730~cm^{-1}$ [5]). The $B(^2D, 2s2p^2)$ state is more attractive, about 34600 cm^{-1}, and has the shortest equilibrium bond length; this can be rationalized, if one considers that in the $B(^2S, 3s)$ and $B(^2P, 3p)$ states a single electron bond is formed between the $B(2p)$ and the $H(1s)$ orbitals, whereas in the $B(^2D, 2s2p^2)$ state a two electron bond is formed between the $B(2s)$ and the $H(1s)$ orbitals. As a result, the $B(^2D, 2s2p^2)$ state crosses $B(^2S, 3s)$ around $R = 3.2$ bohr. The minima of the neutral species are placed between 2.2 and 2.6 a.u., whereas the ionic state has a minimum around $R = 6$ bohr, in the region where the other curves are almost flat. This is due to the large ionic radius of H^-: for the unperturbed anion, an extrapolation from X-ray electron density data gives $R_{ion} = 4.9$ bohr [35]. As a consequence, the ionic curve crosses the three neutral excited ones, both in the long range with its outer limb, and around $R = 4$ bohr with its inner limb.

The resulting adiabatic ground state is substantially $B(^2P, 2p)$ at all distances, with a non negligible contribution of B^+H^- between 4 and 7 bohr, and a stronger mixing with $B(^2D, 2s2p^2)$ at shorter distances. The H_{13} interaction between $B(^2P, 2p)$ and $B(^2D, 2s2p^2)$ is rather large (0.105 a.u. at $R = 2.3$ bohr, whereas all other off diagonal matrix elements are at least 6 times smaller): this makes the ground state minimum deeper and tighter than the corresponding diabatic one, and displaces the $B(^2D, 2s2p^2)$ state to higher energies, so that it ends up being the second excited state, $C^1\Sigma^+$, instead of the first.

The $B^1\Sigma^+$ state is characterized by the double minimum. It dissociates as $B(^2S, 3s)$ + $H(^2S)$, but it becomes mainly ionic in the outer potential well. The crossing of the diabatic curves can be located at $R = 10.4$ bohr, but the minimum energy gap, $\Delta E_{23} \simeq 3600~cm^{-1}$) and the maximum of $G_{23}(R)$ (see fig. 3) occur at $R = 11.5$ bohr; the shift is due to the decrease of $H_{24} \simeq \Delta E_{23}$ with increasing the internuclear distance. Anyway, the crossing region extends over a range of a few bohr, because of the rather large ratio of H_{24} versus the slope difference, $\frac{\partial}{\partial R}(H_{22} - H_{44})$. Around the maximum in the $B^1\Sigma^+$ curve the ionic contribution almost disappears: the B state is again substantially $B(^2S, 3s)$ down to the equilibrium distance. This description agrees with those based on Natural Orbitals, by Pearson et al [14], and on MCSCF, by Jaszunski et al [11].

The next state, $C^1\Sigma^+$, dissociates as $B(^2D, 2s2p^2)$ $H(^2S)$. It undergoes an abrupt change to a combination of B^+H^- and $B(^2P, 3p)$ at $R = 17.5$ bohr. Here the minimum energy gap is much smaller, $\Delta E_{34} \simeq 50~cm^{-1}$, and the $G_{34}(R)$ function consistently more sharply peaked (see fig.3). In this range of distances (10 to 20 bohr), the ionic state has much larger interaction matrix elements H_{ij} with $B(^2S, 3s)$ and $B(^2P, 3p)$ than with $B(^2D, 2s2p^2)$; this is due to the diffuse charge distribution of the Rydberg states and of H^- and is reflected on the very different features of the avoided crossings here described. Around 11 bohr the $C^1\Sigma^+$ state switches gradually to $B(^2S, 3s)$, still keeping a partial ionic character. This state too presents a double well potential, but the outer minimum is shallower and very flat: $D_e \simeq 5740~cm^{-1}, R_e \simeq 9.6$ bohr and a barrier of about 1290 cm^{-1} which separates it from the inner minimum. The barrier is located at about 4.5 bohr, where the mixed $B(^2D, 2s2p^2)$ /ionic character disappears, in the range of about 0.5 bohr. For shorter distances the $C^1\Sigma^+$ state is substantially $B(^2D, 2s2p^2)$.

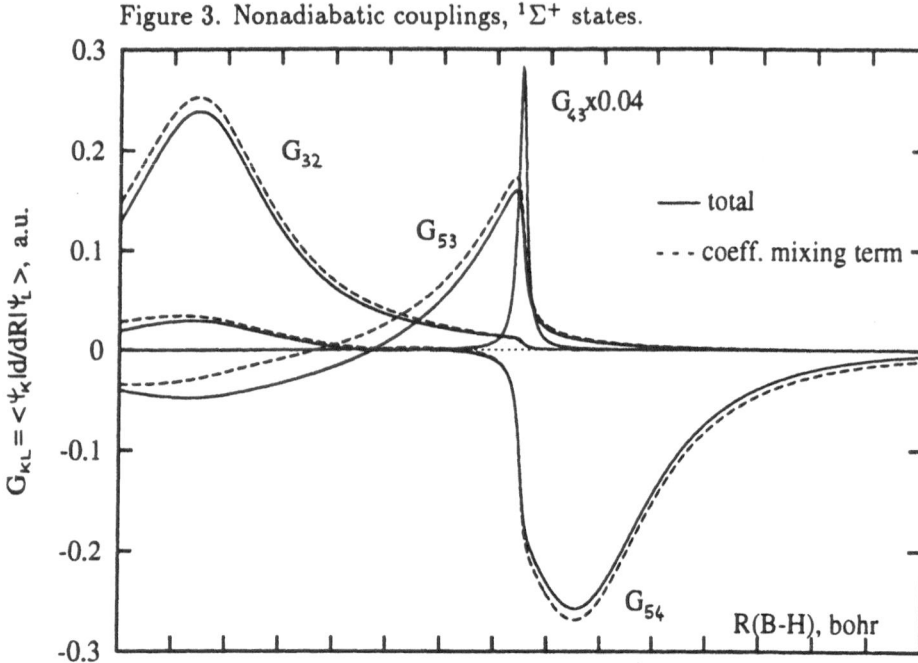

Figure 3. Nonadiabatic couplings, $^1\Sigma^+$ states.

The last state here considered, $E^1\Sigma^+$, dissociates as B(^{2}P, 3p) H(^{2}S) . It under-goes two avoided crossings, one broad between 15 and 20 bohr, with the ionic state, and one very narrow, at $R = 17.5$ bohr, involving the partially ionic and the B(^{2}D, 2s2p^2) state, whence the peculiar shape of the $G_{35}(R)$ and G_{45} coupling func-tions. The E state remains B(^{2}D, 2s2p^2) , with a flat potential energy curve, down to $R \simeq 4.5$ bohr. Around $R = 4$ bohr the ionic state crosses again B(^{2}D, 2s2p^2) and B(^{2}P, 3p) , so that at shorter distances the $E^1\Sigma^+$ state is substantially B(^{2}P, 3p) . Although the above considerations about the diabatic/adiabatic relationships make sense and are confirmed by the shape of the $G_{KL}(R)$ and dipole moment functions (see fig.4), one should keep in mind that the identification of the adiabatic states with linear combinations of VB-like wavefunctions still contains some degree of arbitrari-ness; this appears at two distinct levels. First, as in pure VB treatments, the reference functions are not perfectly orthogonal among themselves, therefore one cannot rig-orously attach a different meaning to each of them. The overlaps generally increase when shortening the R distance; in particular, the 3s Rydberg orbital of boron over-laps with the diffuse 1s of H$^-$, therefore the overlap between R_2 (B(^{2}S, 3s)) and R_4 (combination of B(^{2}P, 3p) and B$^+$H$^-$) reaches a maximum of about 0.4 around $R = 6$ bohr, where the weight of the ionic structure in R_4 is largest; also, the $\langle R_1|R_3\rangle$ overlap is large at short distances, about 0.5 at $R = 2.3$ bohr, due to the superposition of the B(2p) and H(1s) orbitals. The second source of arbitrariness is the greater or lesser ability of the reference functions R_I to span the adiabatic subspace $|\psi_1\rangle \ldots |\psi_N\rangle$: one can take $S = \sqrt{\sum_I^N \langle R_I|\eta_I\rangle^2 /N}$ as a measure of the adequacy of the reference set. In our case, the values of S range from a quite satisfying 0.96 at large R to an acceptable 0.86 at short R.

Figure 4: Dipole moments, $^1\Sigma^+$ states.

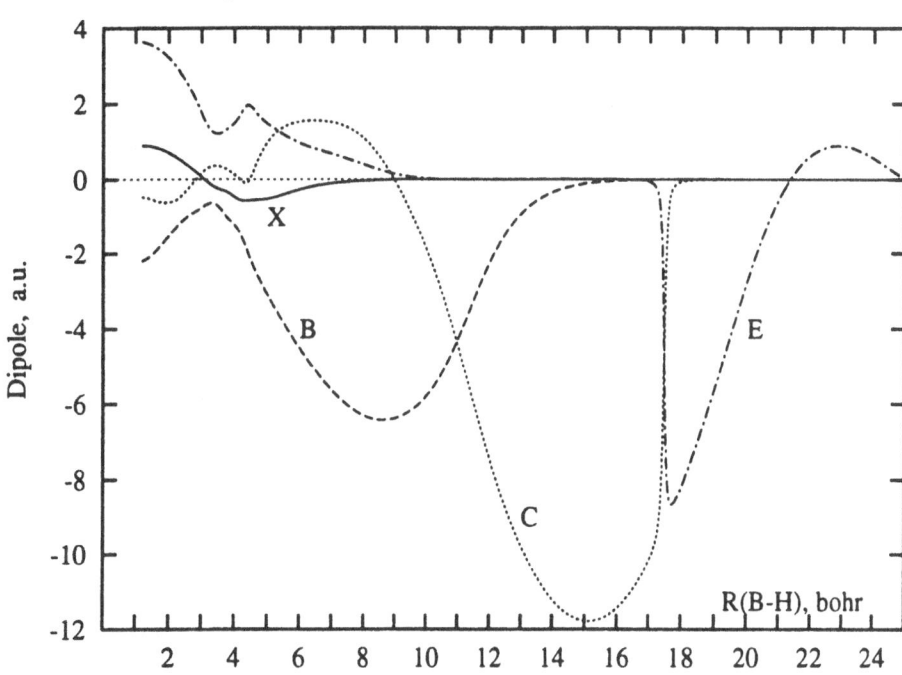

The success of the diabatisation procedure can be ultimately judged by computing the $\langle \eta_I |\frac{\partial}{\partial R}| \eta_J \rangle$ matrix elements, as we have done in this work. Fig. 3 shows the most important couplings, obtained from the five state treatment at large distances. The state mixing term alone provides a very good approximation of the coupling functions (eq. (18)). The contribution containing the $\langle \eta_I |\frac{\partial}{\partial R}| \eta_J \rangle$ functions, second term in the r.h.s. of eq. (17), is largely negligible. In figs. 5 and 6 we show the coupling functions for the first four singlets, at shorter distances. Here, the state mixing term does not approximate the exact couplings with uniform accuracy: in general, the agreement is much better for large couplings, such as those obtained in the proximity of avoided crossings, than for small couplings, between states which are well separated in energy (notice the different scales of figs. 3, 5 and 6). In practice, the diabatisation procedure is most useful and the approximate couplings are most accurate, in those cases where electronic transition probabilities are higher, and the "exact" evaluation of couplings is more difficult (see also ref. [17]).

In a recent work [36], Bak et al presented a new method to obtain first-order corrected radial couplings from MCSCF wavefunctions and applied it to three $^1\Sigma^+$ states of

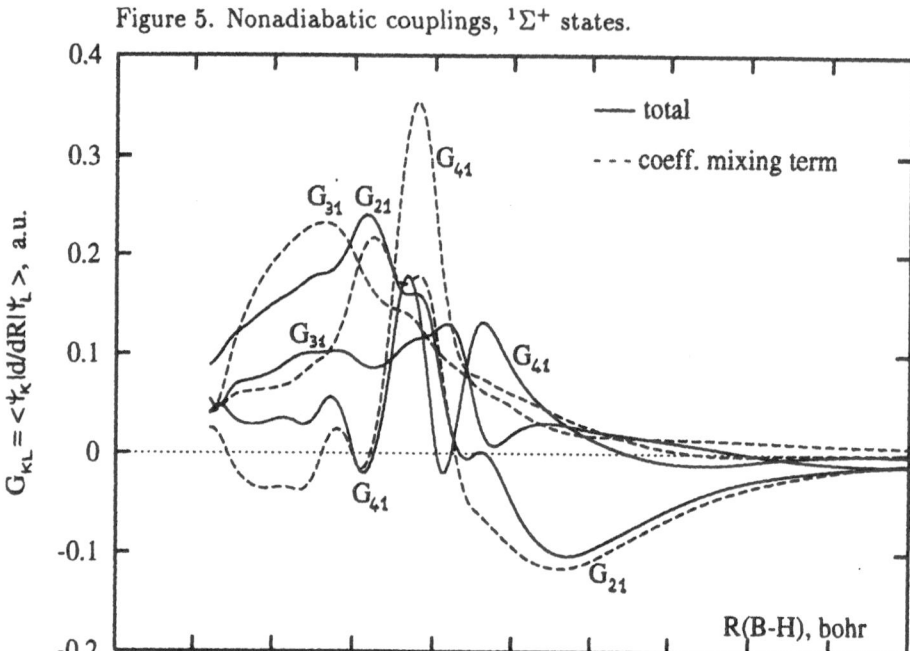

Figure 5. Nonadiabatic couplings, $^1\Sigma^+$ states.

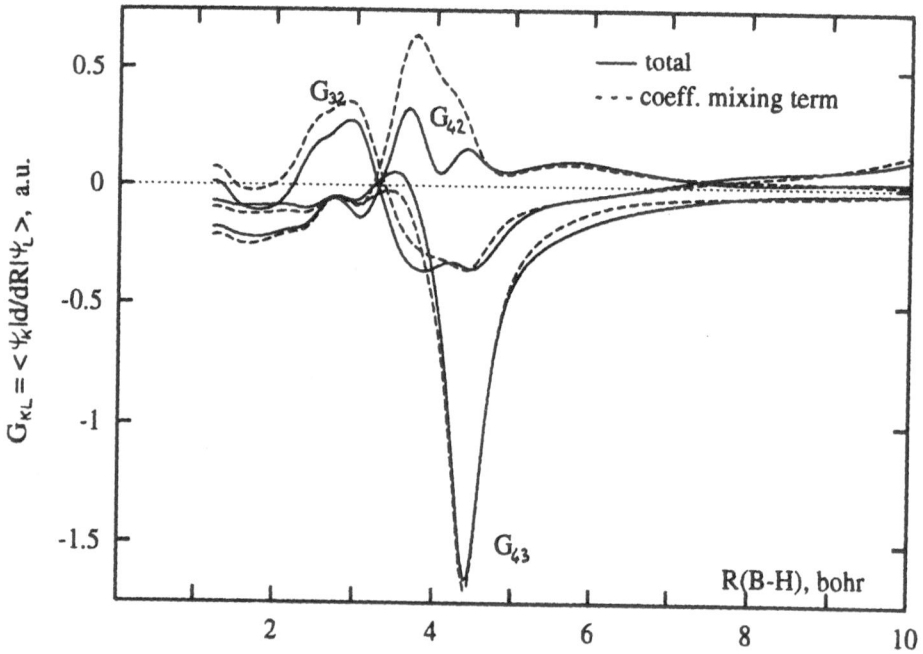

Figure 6. Nonadiabatic couplings, $^1\Sigma^+$ states.

BH: their results are in good agreement with ours. A diabatisation procedure partly related to ours has been put forward by Gadéa and Pélissier: see ref.[37] for a recent application.

4. Triplet states

We have computed diabatic and adiabatic energies and wavefunctions for three triplet states, originating from the asymptotes: $B(^2P, 2p)$ $H(^2S)$, $B(^2S, 3s)$ $H(^2S)$ and $B(^2D, 2s2p^2)$ $H(^2S)$. As already mentioned, the first two diabatic reference states simply coincide with the asymptotic wavefunctions, while $|R_3\rangle$ is a linear combination of $B(^2D, 2s2p^2)$ $H(^2S)$ and $B(^2P, 3p)$ $H(^2S)$. At large R we have almost pure $B(^2D, 2s2p^2)$, below $R = 5$ bohr we have $B(^2P, 3p)$: the transition takes place gradually around $R = 6$. The reason is that $B(^2D, 2s2p^2)$ $H(^2S)$ is repulsive, in the triplet spin coupling. On the contrary, $B(^2D, 2s2p^2)$ $H(^2S)$ and $B(^2S, 3s)$ $H(^2S)$ are binding, in analogy with the corresponding singlet and the BH^+ core. Two more avoided crossings are quite evident in fig. 7: the repulsive $B(^2P, 2p)$ $H(^2S)$ curve crosses both $B(^2S, 3s)$ $H(^2S)$ and $B(^2P, 3p)$ $H(^2S)$. The adiabatic curves alone, also shown in fig. 7, may not be interpreted so easily, because of the closeness of the two crossings both in the distance and in the energy scale.

Fig. 8 shows the nonadiabatic coupling functions between $^3\Sigma^+$ states, which are completely dominated by the double crossing feature. The coefficient mixing term is a very good approximation of the total matrix elements, thus confirming the considerations already put forward with regard to the singlets.

The lowest adiabatic state is completely dissociative. The second and third are bound, but an efficient predissociation of the associated vibronic states can be predicted, on the basis of the strong couplings and small energy gaps in the region of the minima.

5. Conclusions

In this work we have carried out a thorough investigation on the lowest $^{1,3}\Sigma^+$ states of the BH molecule. In order to best elucidate the physical nature of the electronic states, we have resorted to a quasi-diabatic description within the CIPSI-QDPT method. Such technique is particularly adequate to characterize the behaviour of both the electronic curves and the nonadiabatic coupling functions in the presence of narrowly avoided crossings, a situation which is very common when considering the dissociation of a molecule in electronically excited states. In such cases, the diabatisation technique here applied is able to yield approximate adiabatic coupling functions in very good agreement with exactly evaluated ones. An application of our results, concerning the time evolution of vibronic states, line widths, radiative and predissociative lifetimes, is in progress.

6. Appendix

In this section we give the relations between the nonadiabatic coupling matrix elements in the quasi-diabatic and adiabatic representations. We do not obtain simple

Figure 7. Diabatic and adiabatic energies, $^3\Sigma^+$ states.

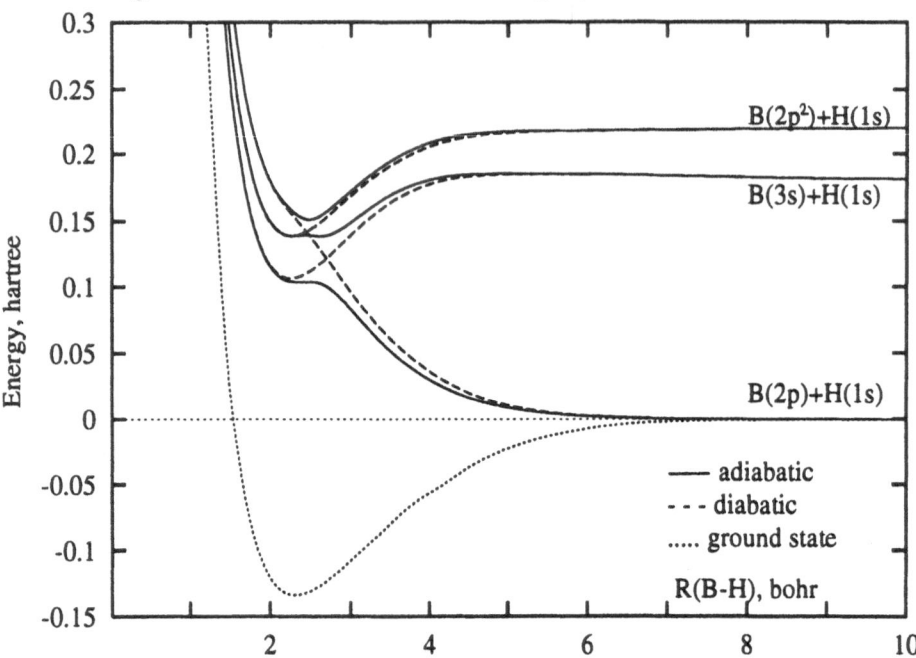

Figure 8. Nonadiabatic couplings, $^3\Sigma^+$ states.

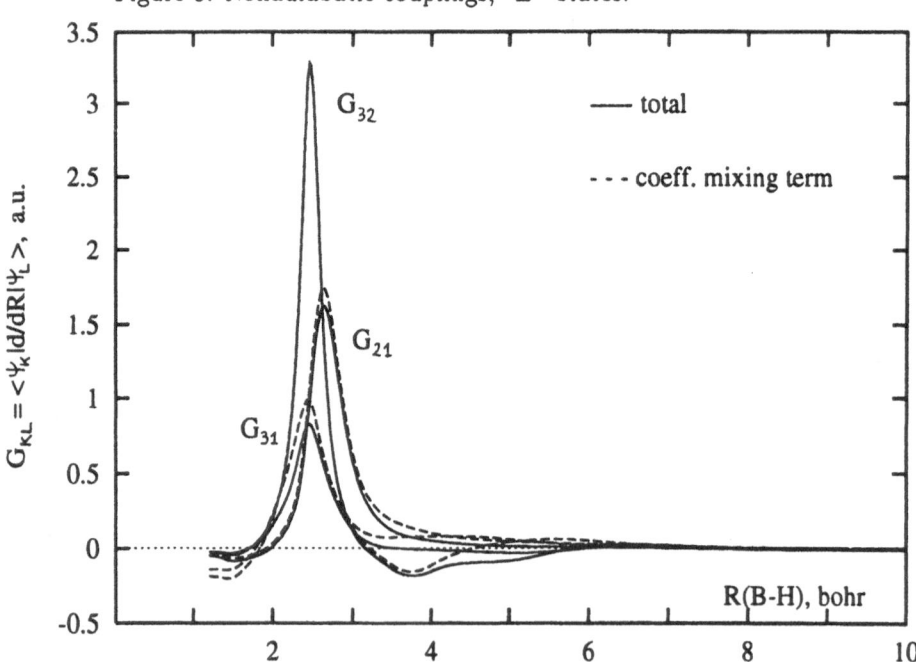

matrix transformation equations, because the differential operators, $\frac{\partial}{\partial R}$ and $\frac{\partial^2}{\partial R^2}$, also operate on the transformation coefficients. We have, by definition:

$$|\psi\rangle = |\eta\rangle \, \mathbf{C} \tag{9}$$

$$\mathbf{H} = \langle \eta \, |\hat{\mathcal{H}}_{el}| \, \eta \rangle \tag{10}$$

$$\mathbf{H} \, \mathbf{C}_K = E_K \, \mathbf{C}_K \tag{11}$$

where $\mathbf{C}_K$ is a column of the $\mathbf{C}$ matrix. The same equations hold for zero-order or perturbed matrices and wavefunctions. The nonadiabatic coupling matrices in the two representations are:

$$\mathbf{G} = \left\langle \psi \left| \frac{\partial}{\partial R} \right| \psi \right\rangle \tag{12}$$

$$\mathbf{G}_D = \left\langle \eta \left| \frac{\partial}{\partial R} \right| \eta \right\rangle \tag{13}$$

$$\mathbf{T} = \left\langle \psi \left| \frac{\partial^2}{\partial R^2} \right| \psi \right\rangle \tag{14}$$

$$\mathbf{T}_D = \left\langle \eta \left| \frac{\partial^2}{\partial R^2} \right| \eta \right\rangle \tag{15}$$

We shall make use of the Hellmann-Feynman type formula:

$$\mathbf{C}_K^\dagger \frac{\partial \mathbf{C}_L}{\partial R} = (E_L - E_K)^{-1} \mathbf{C}_K^\dagger \frac{\partial \mathbf{H}}{\partial R} \mathbf{C}_L \tag{16}$$

for $K \neq L$. Notice that $\mathbf{C}_K^\dagger \frac{\partial \mathbf{C}_K}{\partial R} = 0$. In practical applications, this equation allows to substitute the numerical differentiation of the coefficients with that of the hamiltonian matrix. There are two advantages:
1) it is easier to impose the hermiticity constraint $H_{IJ} = H_{JI}^*$ than the orthonormality constraint $\mathbf{C}_K^\dagger \mathbf{C}_L = \delta_{KL}$;
2) the H_{IJ} elements normally are smooth functions of R; on the contrary, the C_{IK} coefficients may undergo rapid variations in the proximity of avoided crossings, thus making the numerical differentiation more difficult.
From eqs.(9,12,13), the relation between $\mathbf{G}$ and $\mathbf{G}_D$ follows:

$$G_{KL} = \mathbf{C}_K^\dagger \frac{\partial \mathbf{C}_L}{\partial R} + \mathbf{C}_K^\dagger \mathbf{G}_D \mathbf{C}_L \tag{17}$$

The "state mixing" term, the first in the r.h.s., usually dominates, at least in the presence of avoided crossings. Its determination reduces to a simple problem of interpolation of the H_{IJ} matrix elements, according to eq.(16). The second term corresponds, for large R, to the electron translation factor (see for instance [38]). This term depends on the choice of the reference frame: that is, for baricentric frames, it depends on the isotopic masses. It contains the $\mathbf{G}_D$ matrix, which may be determined by numerical differentiation of the quasi-diabatic wavefunctions [16]: this calculation is more demanding, especially in the case of many internal coordinates. It is therefore interesting to adopt the approximation:

$$G_{KL} \simeq \mathbf{C}_K^\dagger \frac{\partial \mathbf{C}_L}{\partial R} = (E_L - E_K)^{-1} \mathbf{C}_K^\dagger \frac{\partial \mathbf{H}}{\partial R} \mathbf{C}_L \tag{18}$$

The G_{KL} functions obtained by this equation are labelled "coefficient mixing term" in figs. 3, 5, 6 and 8.

The second derivative matrix elements give rise to three terms:

$$T_{KL} = \mathbf{C}_K^\dagger \frac{\partial^2 \mathbf{C}_L}{\partial R^2} + 2\, \mathbf{C}_K^\dagger \mathbf{G}_D \frac{\partial \mathbf{C}_L}{\partial R} + \mathbf{C}_K^\dagger \mathbf{T}_D \mathbf{C}_L \tag{19}$$

Only the first one is a pure state mixing term. The Hellmann-Feynman expression of the two terms containing coefficient derivatives involves a bit of algebra. For $K \neq L$ we have:

$$\begin{aligned}
&\mathbf{C}_K^\dagger \frac{\partial^2 \mathbf{C}_L}{\partial R^2} + 2\, \mathbf{C}_K^\dagger \mathbf{G}_D \frac{\partial \mathbf{C}_L}{\partial R} = \\
&2(E_L - E_K)^{-2} \left[\mathbf{C}_K^\dagger \frac{\partial \mathbf{H}}{\partial R} \mathbf{C}_K - \mathbf{C}_L^\dagger \frac{\partial \mathbf{H}}{\partial R} \mathbf{C}_L \right] \mathbf{C}_K^\dagger \frac{\partial \mathbf{H}}{\partial R} \mathbf{C}_L + \\
&2\mathbf{C}_K^\dagger \left[(E_L - E_K)^{-1} \frac{\partial \mathbf{H}}{\partial R} + \mathbf{G}_D \right] \left[\sum_M (E_L - E_M)^{-1} \mathbf{C}_M \mathbf{C}_M^\dagger \right] \frac{\partial \mathbf{H}}{\partial R} \mathbf{C}_L + \\
&(E_L - E_K)^{-1} \mathbf{C}_K^\dagger \frac{\partial^2 \mathbf{H}}{\partial R^2} \mathbf{C}_L
\end{aligned} \tag{20}$$

The summation runs over all $M \neq K, L$. The diagonal matrix elements are:

$$\begin{aligned}
&\mathbf{C}_K^\dagger \frac{\partial^2 \mathbf{C}_K}{\partial R^2} + 2\, \mathbf{C}_K^\dagger \mathbf{G}_D \frac{\partial \mathbf{C}_K}{\partial R} = \\
&2\mathbf{C}_K^\dagger \sum_M \left[(E_M - E_K)^{-1} \frac{\partial \mathbf{H}}{\partial R} + \mathbf{G}_D \right] (E_K - E_M)^{-1} \mathbf{C}_M \mathbf{C}_M^\dagger \frac{\partial \mathbf{H}}{\partial R} \mathbf{C}_K
\end{aligned} \tag{21}$$

The calculation of the $\mathbf{T}_D$ matrix, involving second derivatives of the electronic wavefunctions, is more expensive and subject to numerical inaccuracy than that of $\mathbf{G}_D$. A simple approximation for the third term in eq.(19) is based on a partial expansion of the identity operator in terms of the diabatic basis:

$$\begin{aligned}
\mathbf{C}_K^\dagger \mathbf{T}_D \mathbf{C}_L &= \mathbf{C}_K^\dagger \frac{\partial \mathbf{G}_D}{\partial R} \mathbf{C}_L - \mathbf{C}_K^\dagger \left\langle \frac{\partial \boldsymbol{\psi}}{\partial R} \middle| \frac{\partial \boldsymbol{\psi}}{\partial R} \right\rangle \mathbf{C}_L \\
&\simeq \mathbf{C}_K^\dagger \left[\frac{\partial \mathbf{G}_D}{\partial R} \mathbf{C}_L + \mathbf{G}_D^2 \right] \mathbf{C}_L
\end{aligned} \tag{22}$$

Notice that the usually dominant term, the one containing $\frac{\partial \mathbf{G}_D}{\partial R}$, is not approximated.

References

1. S. H. Bauer, G. Herzberg and J. W. C. Johns, *J. Mol. Spectrosc.*, **13**, 256 (1964)

2. J. W. C. Johns, F. A. Grimm and R. F. Porter, *J. Mol. Spectrosc.*, **22**, 435 (1967)

3. F. S. Pianalto, L. C. O'Brien, P. C. Keller and P. F. Bernath, *J. Mol. Spectrosc.*, **129**, 348 (1988)

4. W.-T. Luh and W. C. Stwalley, *J. Mol. Spectrosc.*, **102**, 212 (1983)

5. K. P. Huber and G. Herzberg, <u>Constants of diatomic molecules</u>, van Nostrand, Princeton, 1979

6. O. Gustafsson and M.Rittby, *J. Mol. Spectrosc.*, **131**, 325 (1988)

7. M. Persico, *Mol. Phys.* , in press.

8. L. A. Curtiss and J. A. Pople, *J. Chem. Phys.* , **90**, 2522 (1989)

9. J. M. L. Martin, J. P. François and R. Gijbels, *J. Chem. Phys.* , **91**, 4425 (1989)

10. W. Meyer and P. Rosmus, *J. Chem. Phys.* , **63**, 2356 (1975)

11. M. Jaszunski, B. O. Roos and P.-O. Widmark, *J. Chem. Phys.* , **75**, 306 (1981)

12. P. Botschwina, *Chem. Phys. Letters* , **129**, 279 (1986)

13. J. C. Browne and E. M. Greenawalt, *Chem. Phys. Letters* , **7**, 363 (1970)

14. P. K. Pearson, C. F. Bender and H. F. Schaefer III, *J. Chem. Phys.* , **55**, 5235 (1971)

15. S. A. Houlden and I. G. Csizmadia, *Theoret. Chim. Acta* , **35**, 173 (1974)

16. R. Cimiraglia, J.-P. Malrieu, M. Persico and F. Spiegelmann, *J. Phys. B*, **18**, 3073 (1985)

17. M. Persico, in Spectral Line Shapes, vol.3, p.587, F. Rostas ed., de Gruyter, Berlin, 1985

18. J. F. Harrison and L. C. Allen, *J. Mol. Spectrosc.*, **29**, 432 (1969)

19. R. J. Blint and W. A. Goddard III, *Chem. Phys.* , **3**, 297 (1974)

20. F. B. van Duijneveldt, *I.B.M. Technical Report* RJ 945, n.16437, San José, 1971

21. H. Hotop and W. C. Lineberger, *J. Phys. Chem. Ref. Data*, **14**, 731 (1985)

22. L. Laaksonen, P. Pyykkö and D. Sundholm, *Chem. Phys. Letters* ,**96**, 1 (1983)

23. N. C. Baird and R. F. Barr, *Theoret. Chim. Acta* , **36**, 125 (1974)

24. B. Huron, J.-P. Malrieu and P. Rancurel, *J. Chem. Phys.* , **58**, 5745 (1973)

25. R. Cimiraglia and M. Persico, *J. Comp. Chem.*, **39**, 39 (1987)

26. F. Spiegelmann and J.-P. Malrieu, *J. Phys. B*, **17**, 1235 (1984)

27. F. Spiegelmann and J.-P. Malrieu, *J. Phys. B*, **17**, 1259 (1984)

28. S. Bashkin and J. O. Stoner, Atomic Energy Levels and Grotrian Diagrams, vol.1, North-Holland, Amsterdam, 1975

29. S. Evangelisti, J.-P. Daudey and J.-P. Malrieu, *Chem. Phys.* , **75**, 91 (1983)

30. B. Numerov, *Publ. Observatoire Central Astrophys. Russ.* **2**, 188 (1933)

31. F. Y. Hajj, *J. Phys. B*, **13**, 4521 (1980)

32. D. Sundholm, *Chem. Phys. Letters* , **149**, 251 (1988)

33. R. J. Harrison and N. C. Handy, *Chem. Phys. Letters* , **95**, 386 (1983)

34. J. A. Pople and P. v. R. Schleyer, *Chem. Phys. Letters* , **129**, 279 (1986)

35. O. Johnson, *Inorg. Chem.*, **12**, 780 (1973)

36. K. L. Bak, P. Jørgensen, H. J. Aa. Jensen, J. Olsen and T. Helgaker, *J. Chem. Phys.* , **97**, 7573 (1992)

37. A. Boutalib and F. X. Gadéa, *J. Chem. Phys.* , **97**, 1144 (1992)

38. J. B. Delos and W. R. Thorson, *J. Chem. Phys.* , **70**, 1774 (1979)

Magnesium Photoionization: a K-Matrix Calculation with GTO Bases

R. MOCCIA

*Dipartimento di Chimica e Chimica Industriale, Università di Pisa,
Via Risorgimento 35, I-56126 Pisa, Italy*

P. SPIZZO

*Istituto di Chimica Quantistica ed Energetica Molecolare del CNR, Via
Risorgimento 35, I-56126 Pisa, Italy*

1.Introduction

The theoretical study of the molecular photoionization processes is an active field of the current research. The calculation of the photoionization cross sections may be advantageously done by resorting to the use of L^2 bases. In fact, these bases are used by many powerful and available codes devised to treat bound-state problems, which may be adapted to consider also the electronic continuum. If only the integral cross section is required, the calculations are greatly simplified by methods, like the Stieltjes imaging (1), that allow to obtain these quantities without a detailed knowledge of the continuum wavefunctions. But if more detailed quantities are wanted, like the branching ratios, the differential cross sections, the structure due to narrow resonances etc., a more detailed knowledge of the continuum wavefunctions is necessary. In the atomic case, there exist efficient L^2 techniques for the continuum properties and may be applied in connection with the powerful CI packages that are currently available. Their application to molecular systems is thwarted by the requirement that the variational functions must be accurate in a region far away from the nucleus. Thus, they should be expanded upon bases of very diffuse orbitals, like the STO or STOCOS ones (2). These bases are very cumbersome for molecular calculations and only in few cases (usually the hydrides) one may tradeoff the problems in the multicenter bielectronic integrals for the well-known shortcomings of a monocentric expansion. To take a full advantage of the current packages for molecular structure, it is clearly necessary to develop an L^2 technique capable of extracting the continuum properties from the comparative short-range representations allowed by the GTO bases, which are an almost obligatory choice for molecules.

Unfortunately, in the molecular systems the theoretical predictions for the already formidable electronic problem cannot be checked fairly against the experimental data, since the nuclear motions may play major effects. From here the need to check these methods in calculations on atomic systems, where accurate theoretical and comparable experimental reference data are already available.

Y. Ellinger and M. Defranceschi (eds.), Strategies and Applications in Quantum Chemistry, 367–378.
© 1996 *Kluwer Academic Publishers.*

In previous works, we have developed efficient L^2 techniques which have been applied to various atomic systems using STOCOS bases (3–5) and then to Helium with GTO bases (6). This last calculation has shown the capability of our K-matrix technique to obtain the continuum properties with GTO bases. As a matter of fact, accurate results were obtained also in the energy regions of the autoionizing states, where it is necessary to recover the interactions between diffuse discrete states and a continuum. The present paper applies this method to Magnesium and shows that it deals effectively also with other technical difficulties that are encountered in molecular calculations, e.g. the orthogonality to the inner shells and the strong short-range deviations from the Coulomb potential.

Unless otherwise specified, all quantities are expressed in atomic units.

2.Method of calculation

The present method is an extension of the K-matrix technique pioneered by Fano (7). Our previous works have discussed thoroughly its general aspects, the discretization procedure (4,8) and the implementation upon the short-range GTO bases (6), so only a concise description will be given here.

The K-matrix method is essentially a configuration interaction (CI) performed at a fixed energy lying in the continuum upon a basis of "unperturbed functions" that (at the formal level) includes both discrete and continuous subsets. It turns the Schrödinger equation into a system of integral equations for the K-matrix elements, which is then transformed into a linear system by a quadrature upon a finite L^2 basis set.

In the present implementation, the unperturbed functions are not subject to any orthogonality constraint nor are required to diagonalize any model hamiltonian. This freedom yields a faster convergence of the variational expansion with the basis size and allows to obtain the phaseshift of the basis states without the analysis of their asymptotic behaviour.

For conciseness, throughout this article it is understood that all the states and manifolds have well defined symmetry, so the corresponding labels $(L, M_L, S, M_S,$ parity) and projectors are omitted wherever this is possible without ambiguities.

2.1. THE K-MATRIX UNPERTURBED BASIS STATES

The formal basis employed in the K-matrix calculation includes the relevant partial wave channel (pwc) subspaces plus a "localized channel" (lc) of discrete functions. These last are usual CI states and their inclusion in the basis allows to efficiently reproduce the autoionizing states and the correlation effects.

In the atomic case, the pwc's are defined by the ion level I and the l value of the electron partial wave, i.e. the formal pwc subsets span the tensor product of the ion level states times the one-electron l-wave manifold. In the following, the subspaces will be indexed with greek letters; a subspace index $\alpha = I_\alpha, l_\alpha$ will designate explicitly an open pwc subspace, while an index β an arbitrary subspace. The lc subspace will be numbered 0 and $\hat{Q}_\beta$ will denote the projector in the subspace β.

The formal basis employed for the pwc β is made by the eigenfunctions $|\Phi_{\beta E}\rangle = N|\Phi_\beta^+ W_{\beta\epsilon}\rangle$ of the channel hamiltonian $\hat{H}_\beta = \hat{Q}_\beta \hat{H} \hat{Q}_\beta$, i.e. the hamiltonian projected in the *pwc* subspace β. The Φ_β^+ are coupled products of the ion states times the angular and spin functions of the outer electron (the so-called channel functions in the close-coupling jargon) and N is a normalization factor arising from the lack of strong-orthogonality of the outer orbital to the ion states. The radial functions $W_{\beta\epsilon}$ behave asymptotically as standing shifted Coulomb waves and the *pwc* basis functions may be indexed by the corresponding energy $E = E_I^+ + \epsilon$. For ease of writing, the same notation is employed for both discrete and continuous eigenfunctions, normalized to unity and $\delta(E - E')$ respectively.

Since these formal bases, which are supposed to describe the true continuum background, will be represented upon finite L^2 sets, all the quantities which must be interpolated from these representations (i.e. matrix elements and phaseshifts) must be smooth functions of the energy index: this requires a suitable redefinition of the channel hamiltonian $\hat{H}_\beta$ if this supports narrow shape resonances.

Using GTO bases, it cannot be expected that the variational representations of the electron waves are sufficiently accurate far outside the so-called "molecular region", i.e. the rather limited region of space where the potential clearly deviates from the asymptotic Coulomb form. Therefore the phaseshifts of the *pwc* basis states cannot be obtained from the analysis of their long-range behaviour, as was done in previous works with the STOCOS bases. In the present approach, this analysis may be avoided since the K-matrix technique allows to determine, by equation [3] below, the phase-shift difference between the eigenfunctions of $\hat{H}_\beta$ and the auxiliary basis functions $|\Phi_{\beta E}^{(C)}\rangle = N|\Phi_\beta^+ W_{\beta\epsilon}^{(C)}\rangle$, where $W_{\beta\epsilon}^{(C)}$ are the bound and continuum radial eigenfunctions in the Coulomb field. This requires variational representations accurate only where the potential felt by the outer electron is different from that of a pure Coulomb field and therefore this phaseshift difference builds up. Unless in special cases, the auxiliary basis $|\Phi_{\beta E}^{(C)}\rangle$ cannot be an orthonormal one.

2.2. THE K-MATRIX METHOD

If at the energy E there are n open channels, one may define n linearly independent trial functions $|\{\Psi_E\}\rangle = (|\Psi_{1E}\rangle, ..., |\Psi_{nE}\rangle)$ of the form

$$|\Psi_{\alpha E}\rangle = |\Phi_{\alpha E}\rangle + \sum_{\beta'} \fint dE' \, |\Phi_{\beta'E'}\rangle \, P\frac{1}{E - E'} \, K_{\beta'E',\alpha E} \tag{1}$$

where P denotes the principal value of the integral and $\fint$ the summation over the discrete part and the integration over the continuous one of the subsets. The expansion coefficients $K_{\beta'E',\alpha E}$ are determined by imposing $\langle \Phi_{\beta'E'}|\hat{H} - E|\Psi_{\alpha E}\rangle = 0$ and this leads to the system of coupled integral equations

$$K_{\beta'E',\alpha E} - \sum_{\beta''} \fint dE'' \, V_{\beta'E',\beta''E''}(E) \, P\frac{1}{E - E''} \, K_{\beta''E'',\alpha E} = V_{\beta'E',\alpha E}(E) \tag{2}$$

$$V_{\beta'E',\beta''E''}(E) = \langle \Phi_{\beta'E'}|\hat{H} - E|\Phi_{\beta''E''}\rangle - (E' - E)\delta_{\beta'\beta''}\delta(E' - E'')$$

The delta-function addendum removes the divergences from these matrix elements and allows their representation upon L^2 bases. When the *pwc* basis functions are the eigenfunctions of the projected hamiltonian $\hat{Q}_\beta \hat{H} \hat{Q}_\beta$, the intrachannel matrix elements $V_{\beta E', \beta E''}(E)$ are identically zero, but this is not required to apply the method and indeed it must not hold for the auxiliary basis $|\Phi_{\beta E}^{(C)}\rangle$.

The K-matrix on the energy-shell $\mathbf{K}(E)$, defined by $K_{\alpha'\alpha}(E) = K_{\alpha'E, \alpha E}$, is a real symmetric matrix on the real energy axis. As discussed in (4), it is related to the scattering matrix and contains the quantities needed to analyze the resonances.

It should be noted that the integral equations [2] determining the elements $K_{\beta'E', \alpha E}$, derived as an energy-variational problem, correspond also to the stationary condition of the variational functional proposed by Newton (9). Thus the K-matrix elements obeying equation [2] guarantee a stationary value for the K-matrix on the energy shell.

When only one channel is open, the phaseshift $\delta_\alpha(E)$ of $|\Psi_{\alpha E}\rangle$ is related to that $\delta_{\alpha 0}(E)$ of the unperturbed basis function $|\Phi_{\alpha E}\rangle$ by

$$\delta_\alpha(E) = \delta_{\alpha 0}(E) - \tan^{-1}[\pi K_{\alpha\alpha}(E)] \tag{3}$$

This relation allows, as said above, to obtain the phaseshifts of the basis functions $|\Phi_{\beta E}\rangle$ by a single-channel K-matrix calculation on the basis $|\Phi_{\beta E}^{(C)}\rangle$, whose non-Coulomb phaseshifts are zero by construction.

The real K-matrix variational wavefunctions $|\Psi_{\alpha E}\rangle$ satisfy

$$\langle \{\Psi_{E'}\}|\hat{H} - E''|\{\Psi_E\}\rangle = (E - E'')\left[1 + \pi^2 \mathbf{K}(E)^2\right] \delta(E - E')$$

so that two sets of complex orthonormal eigenfunctions may be obtained by:

$$|\{\Psi_E^{(\pm)}\}\rangle = (|\Psi_{1E}^{(\pm)}\rangle, ..., |\Psi_{nE}^{(\pm)}\rangle) = (|\Psi_{1E}\rangle, ..., |\Psi_{nE}\rangle)\left[1 \pm i\pi \mathbf{K}(E)\right]^{-1} \mathbf{\Delta}^{(\pm)}(E)$$

where $\Delta_{\alpha'\alpha}^{(\pm)}(E) = \delta_{\alpha\alpha'} e^{\pm i \xi_{\alpha E}}$ and $\xi_{\alpha E}$ is the total phaseshift of $|\Phi_{\alpha E}\rangle$. The states $|\Psi_{\alpha E}^{(-)}\rangle$ obey the boundary conditions suitable for photoionization processes, since they contain an outgoing Coulomb wave (with zero phaseshift) only in the channel α.

As discussed in (4), the K-matrix has a pole at energies near a resonance and this yields a convenient method for the analysis of the narrow autoionizing states. The matrix representation of equation [2] upon a finite L^2 basis may be in fact recast in the form (4)

$$\mathbf{K}_{\alpha E} = \mathbf{P}^{-1}(E)\left[\mathbf{P}^{-1}(E) - \mathbf{V}(E)\right]^{-1} \mathbf{V}_{\alpha E}$$

where $\mathbf{P}(E)$ is an almost diagonal matrix arising from the integration of $P/(E - E')$ times the polynomials that interpolate the K- and V-matrix elements over the basis grid. Across a narrow resonance, the real symmetric matrix $[\mathbf{P}^{-1}(E) - \mathbf{V}(E)]$ should be a smooth function of the energy and one of its eigenvalues must change sign. Its inverse may be therefore approximated, quickly and accurately, using the smallest (in modulus) eigenvalues and their eigenvectors, which may be linearly interpolated across the resonance. The blocked and almost-diagonal matrix $\mathbf{P}(E)$ may be easily inverted, so it is possible to sweep the resonance profile with a great saving of CPU time.

This approach proved accurate and convenient for the analysis of the narrow resonances; the results presented in this work have been obtained without employing this trick because the limited dimensions of this single-channel problem are easily handled by the standard method.

2.3. THE BASIS FUNCTIONS

The present calculations are at the frozen-core level; the inclusion of a phenomeno-
logical core potential is straightforward and has been avoided here only because it
will complicate the comparison of the results on GTO and STOCOS bases. All the
basis functions have the form $|1s^2 2s^2 2p^6 \Phi_v\rangle$, where $|1s^2 2s^2 2p^6\rangle$ is a single Slater de-
terminant built with the SCF orbitals for the ground state; for simplicity the core
will be hereafter omitted. The strong-orthogonality to a closed shell SCF core does
not cause any loss of generality and has been imposed for computational ease on the
valence group functions $|\Phi_v\rangle$. No other orthogonality constraint is imposed on the
basis functions, in particular the waves are not strong-orthogonal to the ion states nor
the *pwc* basis functions are mutually orthogonal. Each of these conditions, beside
slowing the convergence of the variational expansion, would inhibit the phaseshift
determination by the K-matrix calculation on the auxiliary basis.

The core orbitals have been expanded upon an $11s/5p$ GTO basis and the group
energy of the core is -198.747 against the value -198.823 obtained with a $4s/3p$
STO basis. As well known, very long gaussian expansions are needed to obtain inner
orbitals of near-HF (SCF limit) quality, which is instead reached by relatively short
STO expansions. The GTO core employed here yields an higher total energy, but
generates a more attractive potential and hence more negative attachment energies
for the outer electrons.

The $3s$ and $3p$ orbitals of Mg^+ have been expanded upon all the GTO employed
for the inner orbitals plus other 5 GTO whose orbital exponents were optimized for
them.

For the description of the $3s\epsilon l$ continua and of the higher $3pnl$ resonances, the bases
include a large number of configurations of the forms $|3s\chi_{nl}\rangle$, $|3p\chi_{nl}\rangle$, where $3s, 3p$
are the lowest one-electron states in the field of the SCF core.

The localized basis function for the set 0 (ls) are usual frozen-core valence-shell CI
states; all the bound states involved in the present calculations are also described at
this level.

2.4. THE REPRESENTATION OF THE PARTIAL WAVE CHANNELS

The L^2 representations $|\phi_{\beta j}\rangle = N|\Phi_\beta^+ w_{\beta j}\rangle$ of the unperturbed states $|\Phi_{\beta E}\rangle$ (uppercase
letters are used for the formal basis and lowercase ones for their L^2 representations)
have been obtained by diagonalizing the electrostatic hamiltonian over the $|3s\chi_{nl}\rangle$
basis configurations mentioned above.

The variational representations $w_{\beta j}^{(C)}$ of the regular Coulomb waves have been obtained
by diagonalizing the Coulomb hamiltonian upon the same orbital basis employed to
expand the waves $w_{\beta j}$. The following discussion refers explicitly to the states $|\Phi_{\beta E}\rangle$
but apply equally well to this ones.

The basis orbitals employed for the electron wave have the form

$$R_{\xi l m}(\underline{r}) = N_{\xi l} \; r^l \, P_3(\xi, r^2) \; e^{-\xi r^2} \, Y_{lm}(\theta, \phi)$$

where $P_3(\xi, r^2)$ is a third-degree polynomial in r^2 with coefficients depending on ξ.
The present work is mainly aimed at demonstrating the ultimate accuracy of the
method, so the sequence of the orbital exponents ξ and the coefficients of $P_3(\xi, r^2)$
were determined to yield a large number of variational states in the lowest continuum

with the least redundance of the metric. In atomic calculations, the rather expensive introduction of these polynomial factors appears to be fully justified only in rather special cases. Indeed, here one deals generally with a rather limited number of open channels and may therefore employ a large number of basis orbitals, so a low metric redundance is the only practical advantage. As a matter of fact, results of comparable quality were obtained in preliminary calculations without this factor. The molecular calculations, instead, require the proper consideration of many partial wave channels, so the choice of the above parameters may be used to minimize the number of basis functions.

The variational pwc states obtained with these bases include accurate representations for the lowest bound states of the channel hamiltonian, broad "wavepackets" in the higher Rydberg region, a number of narrow wavepackets in the lowest continuum and again broad wavepackets at higher energy. In this context, narrow wavepacket means a variational state whose wave is almost exact, i.e. $C_{\beta j} w_{\beta j} \approx W_{\beta \epsilon}$, inside a sufficiently large sphere. When, as in the present case, the channel hamiltonians $\hat{H}_\beta$ do not support shape resonances, the energy-normalization constants of these narrow wavepackets may be fairly well approximated from the energy spacings, $C_{\beta j}^2 \approx 2/(E_{\beta, j+1} - E_{\beta, j-1})$.

The present method does not involve the analysis of the long-range behaviour of the states, so its application requires only that the narrow wavepackets are accurate inside the molecular region. By equation [3], the phaseshifts of these states may be determined through a K-matrix calculation on the auxiliary basis, so it is assumed that the narrow wavepackets might be continued outside the molecular region as shifted Coulomb waves.

The K-matrix calculations may be obviously performed only at energies inside the range covered by the narrow wavepackets, which should allow to interpolate the matrix elements and the phaseshifts of the channel basis functions. The contributions to the integrals in equations [1,2] from the region of the narrow wavepackets are obtained by interpolating the integrands on the grid supplied by the variational basis. Those from the high-energy regions, which should be small and weakly energy dependent, are instead approximated by summing the contributions of the broad wavepackets.

The L^2 basis for the discretized K-matrix calculations contains the narrow wavepackets in their "energy-normalized" form $C_{\beta j} |\phi_{\beta j}\rangle$ and all the other variational states with unit normalization.

The calculations performed with gaussian bases have been checked employing also the STOCOS bases which include, beside Slater and Hydrogenic orbitals, a large number of STOCOS functions

$$R_{\xi l m}(\underline{r}) = N_{\xi k l} \, r^l e^{-\xi r} \cos(kr)$$

These are probably the most efficient L^2 bases for calculations in the electronic continua of atomic systems and furnish reliable reference data for the GTO calculations.

3.Results

The present work is essentially concerned with the comparison of the results obtained with the GTO and STOCOS bases: a thorough comparison of the STOCOS results with the experimental data has been given in (3).

As noted above, the field exerted by the present GTO core on the outer electrons

Table 1. Positions of a few levels with respect to the double and single ionized ground states.

state	E−E(Mg^{2+})			E−E(Mg$^+$)		
	GTO	STO	exp. (10)	GTO	STO	exp. (10)
3s ^{2}S^e	−0.54253	−0.54182	−0.55254			
3p ^{2}P^o	−0.38498	−0.38440	−0.38975	+0.15755	+0.15742	+0.16279
3s^2 ^{1}S^e	−0.82025	−0.81884	−0.83355	−0.27772	−0.27702	−0.28099
3s4s ^{1}S^e	−0.62501	−0.62402	−0.63533	−0.08248	−0.08220	−0.08278
3s3p ^{1}P^o	−0.66255	−0.66168	−0.67384	−0.12002	−0.11987	−0.12129
3s4p ^{1}P^o	−0.59835	−0.59754	−0.60878	−0.05583	−0.05572	−0.05615

Table 2. Oscillator strengths for the first transitions from the ground state.

transition	GTO/LG	GTO/VG	STO/LG	STO/VG	exp. (11)
3s^2 ^{1}S^e − 3s3p ^{1}P^o	1.761	1.745	1.761	1.746	1.75−1.83
3s^2 ^{1}S^e − 3s4p ^{1}P^o	0.1068	0.1119	0.1154	0.1129	0.107
3s^2 ^{1}S^e − 3s5p ^{1}P^o	0.0262	0.0244	0.0252	0.0245	0.022
3s^2 ^{1}S^e − 3s6p ^{1}P^o	0.0100	0.0089	0.0092	0.0090	0.008
3s^2 ^{1}S^e − 3s7p ^{1}P^o	0.0049	0.0042	0.0044	0.0043	0.004

Table 3. Wave energies and widths of autoionizing states. $a(-b)$ means $a \cdot 10^{-b}$.

state	GTO (E,Γ)		STOCOS (E,Γ)	
3p^2 ^{1}S^e	0.0272	1.62(−3)	0.0273	1.51(−3)
3p4s ^{1}P^o	0.0732	1.27(−2)	0.0713	1.40(−2)
3p5s ^{1}P^o	0.1158	4.10(−3)	0.1153	3.91(−3)
3p6s ^{1}P^o	0.1326	1.84(−3)	0.1321	1.70(−3)
3p7s ^{1}P^o	0.1410	9.89(−4)	0.1406	9.03(−4)
3p3d ^{1}P^o	0.1060	5.75(−5)	0.1061	8.87(−5)
3p4d ^{1}P^o	0.1280	2.69(−5)	0.1277	2.53(−5)
3p5d ^{1}P^o	0.1384	9.59(−6)	0.1381	1.02(−5)

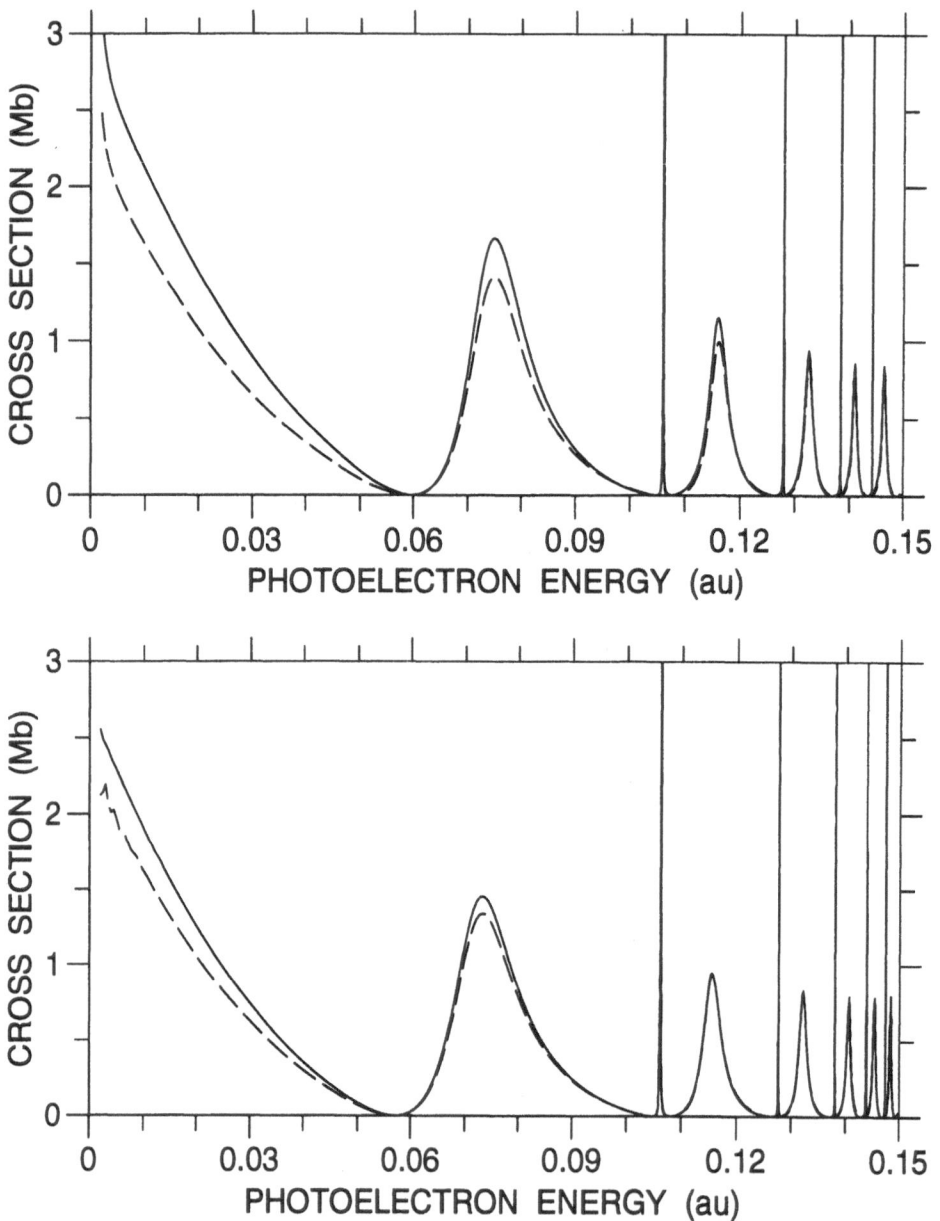

Fig.1 Ground state photoionization cross section. Full line: LG results, broken line: VG results.
Upper figure: GTO basis calculation up to the 3p6d/3p8s resonances.
Lower figure: STOCOS basis calculation up to the 3p7d/3p9s resonances.

is slightly more attractive than that of a near-HF core and the attachment energies for the outer electrons are slightly more negative than those obtained in valence-analogous calculations with STO bases. This partially compensates the error due to the intershell correlation, so, by a fortunate cancellation of effects, the GTO calculations yield valence energies in better agreement with experiment than those carried out with STO bases; the most significant positions are reported in table 1. Also the transition energies, of course, turn out slightly better with the GTO basis than with the STOCOS one: the $3s^2 - 3s3p$ transition is predicted at 0.15770 and 0.15715 with the GTO and STOCOS bases against the experimental value 0.15970.

In spite of the relatively poor description of the core, the present GTO calculations yield accurate oscillator strengths: the data for the lowest-energy transitions from the Mg ground state are given in table 2 and compare well with the STOCOS results. The experimental values given in this table represent a less relevant comparison, due to their uncertainties and/or normalizations upon theoretical results (3). All the transition probabilities have been computed with both the length gauge (LG) and the velocity gauge (VG) forms of the dipole operator and the gauge invariance of the results is only slightly worse than that achieved with the STOCOS bases.

The properties of the $^1P^o$ continuum and the ground state photoionization cross section have been studied from the $3s$ threshold up to a wave energy of about 0.150. From this energy to the $3p$ ionization threshold at about 0.157, the crowding of the resonances of the $3pns$ and $3pnd$ series makes hopeless further variational calculations. The quality and regularity of the present results, however, allow to extrapolate safely the properties of this region, e.g. by fitting formulae based on the quantum defect theory.

In addition, we have investigated the broad $3p^2$ $^1S^e$ resonance, which lies rather close to the ionization threshold and represents therefore a stringent test for the capabilities of the method in the delicate low-energy region.

The positions and the widths for the above autoionizing states have been obtained from the analysis of the scattering matrix as described in (4); the results of the present GTO and STOCOS calculations are reported in table 3. On the whole, the GTO and STOCOS results compare quite well, with a significant difference only for the width of the $3p3d$ resonance. As discussed in (3), a frozen-core calculation of this kind reproduces accurately the position of the levels $3snl, 3pnl$ $(n \geq 4)$ with respect to their parent ion. Indeed, the calculated wave energies of the $3pns$ and $3pnd$ resonances compare well with their experimental counterparts when corrected for the intershell energy difference 0.0054 between the $3s$ and $3p$ ion levels. Instead, the position of the $3p^2$ $^1S^e$ resonance cannot be reproduced very accurately with respect to neither of these thresholds, since its intershell correlation energy is intermediate between theirs. The GTO and STOCOS calculations yield values in good agreement for this resonance, although the predicted width is somewhat larger than the experimental value $1.27 \cdot 10^{-3}$ (12).

The ground state photoionization as calculated with the GTO and STOCOS bases is given in figure 1. The agreement of the two calculations is satisfactory, apart in the critical region of the very low wave energies. It is interesting to note that the correlation effects in this continuum are so strong that the cross section to the unperturbed pwc basis functions (which are the best frozen-core single-configuration approximations) is far from being the coarse average of that to the correlated states: it amounts to 10 Mb at the threshold and would lie off the figure in most of the energy range.

The phaseshifts for the $^1P^o$ manifold are reported in figure 2 from the ionization threshold up to the photoelectron energy 0.1, across the lowest broad resonance

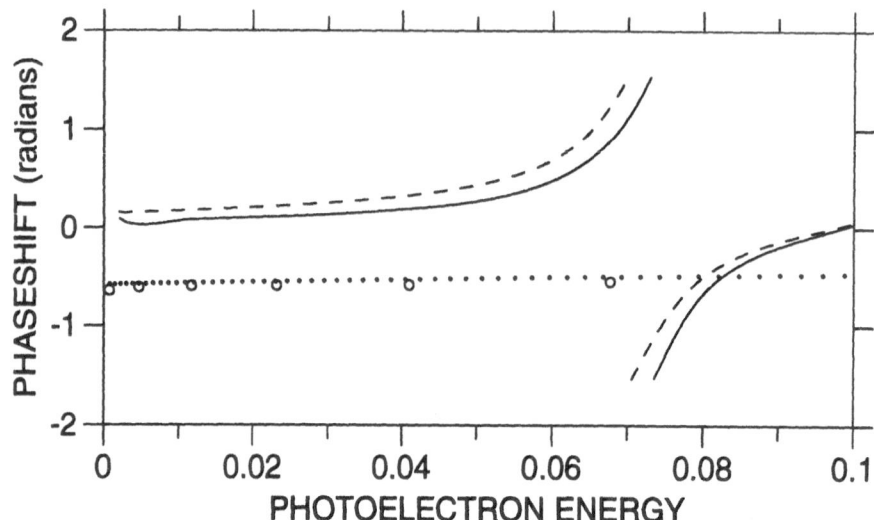

Fig.2 p-wave phaseshifts. Full line: GTO results, broken line: STOCOS
results, circles: GTO basis states, points: STOCOS basis states.

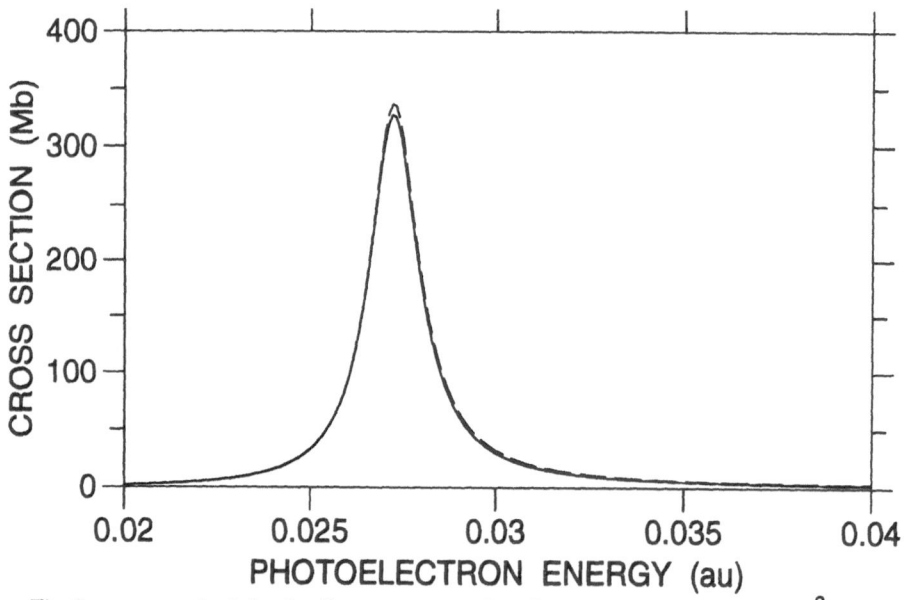

Fig.3 s-wave photoionization cross section from 3s3p across the $3p^2$
resonance. Full line: LG results, broken line: VG results.

($3p4s$). The phaseshifts of the unperturbed STOCOS states plotted in this figure (points) have been calculated by fitting their asymptotic behaviour, but nearly identical results have been obtained by the single-channel K-matrix step. Only this last technique has been employed for the GTO basis; the calculated phaseshifts (circles) agree closely with the STOCOS results and demonstrate the ability of our method to obtain these quantities from moderately diffuse bases. The difference in the phaseshifts for the K-matrix states in the GTO and STOCOS calculations is partially due to the different wave energy of the resonance (about 0.002 lower with the STOCOS basis).

The $s-$wave contribution to the photoionization from the $3s3p\ ^1P^o$ level is plotted in figure 3 and shows a quite satisfactory gauge invariance. Its peak value is in excellent agreement with that yielded by our previous STOCOS calculations, 346 Mb (3).

4.Conclusions

It may be concluded that a method based on the K-matrix technique may be conveniently adapted to calculate the continuum properties using variational L^2 basis functions that are accurate only inside the "molecular region". This means that the calculations may be carried out upon GTO bases, which allow the extension of the proposed method to molecular systems, as already checked for H_2 (13).

Acknowledgment

One of the authors (R.M.) gratefully acknowledges the financial help of the Progetto Finalizzato Chimica Fine of the CNR.

References

1. P.W. Langhoff and C.T. Corcoran, *J. Chem. Phys.* 61, 146 (1974); I. Cacelli, V. Carravetta, R. Moccia and A. Rizzo, *J. Phys. Chem.* 92, 979 (1988); I. Cacelli, V. Carravetta, R. Moccia and A. Rizzo, *J. Chem. Phys.* 89, 7301 (1988)

2. I. Cacelli, R. Moccia and V. Carravetta, *Chem. Phys.* 90, 313 (1984); I. Cacelli, V. Carravetta, A. Rizzo and R. Moccia, *Physics Reports* 205, 283 (1991)

3. R. Moccia and P. Spizzo, *J. Phys.* B21, 1121, 1133 and 1145 (1988), *Phys. Rev.* A39, 3855 (1989)

4. R. Moccia and P. Spizzo, *Phys. Rev.* A43, 2199 (1991) and references therein

5. R. Moccia and P. Spizzo, *J. Phys.* B23, 3557 (1990); *Il Nuovo Cimento* 13D, 757 (1991)

6. R. Moccia and P. Spizzo, *Can. J. Chem.* 70, 513 (1992)

7. U. Fano, *Phys. Rev.* 124, 1866 (1961)

8. I. Cacelli, V. Carravetta and R. Moccia, *J. Chem. Phys.* 85, 7038 (1986)

9. R.G. Newton, Scattering Theory of Waves and Particles, Springer-Verlag, New York, 1982

10. S. Bashkin and J.O. Stoner, Atomic Energy Levels and Grotrian Diagrams, North Holland, Amsterdam, 1975

11. C.J. Mitchell, *J. Phys.* B8, 25 (1975); L. Lundin, B. Engman, J. Hilke and I. Martinson, *Phys. Scr.* 8, 274 (1973); F.M. Kelly and M.S. Mathur, *Can. J. Phys.* 56, 1122 (1978)

12. D.J. Bradley, C.H. Dugan, P. Ewart and A.F. Purdie, *Phys. Rev.* A13, 1416 (1976); R.E. Bonanno, C.W. Clark and T.B. Lucatorto, *Phys. Rev.* A34, 2082 (1986)

13. I. Cacelli, R. Moccia and A. Rizzo, *J. Chem. Phys.* (in press)

Investigation of Photochemical Paths by a Combined Theoretical and Experimental Approach

F. MOMICCHIOLI, I. BARALDI, A. CARNEVALI and G. PONTERINI
Dipartimento di Chimica, Università di Modena, Via Campi 183, I-41100, Modena, Italy

1. Introduction

The attempt to set the wide field of photochemical reactivity within the framework of quantum chemical methods is relativity recent, if one considers what has occured in other fields like, for instance, electronic molecular spectroscopy. In the common view, the origin of theoretical photochemistry dates back to two papers, published in 1972, which first provided a general description of biradical-like species by elucidating their electronic structure in terms of the basic 3x3 CI model [1], and bringing to light their rôle in organic photochemistry [2]. From then on Salem, Michl and others, starting essentially from analysis of state correlation diagrams, have introduced several new theoretical concepts, such as *avoided crossing, funnel, twisted intramolecular charge transfer* (TICT) state, etc., which are now currently used to rationalize a variety of photophysical and photochimical behaviours (for comprehensive descriptions, see ref. [3-6]). By accurate calculations on simple model systems [4,5], the above mentioned concepts have been shown to have fairly general validity, so they can be seen as real supports for the modern theory of organic photochemistry.

On the other hand, when dealing with photochemistry of large molecules in dense media (e.g organic dyes in liquid solution) the application of the above interpretative framework faces two serious problems. The first one is that construction of potential energy surfaces of the ground state (S_0) and some low-lying electronic excited states (at least S_1 and T_1) cannot be fulfilled by the same ab initio extended-CI procedures used for the model compounds. Thus, one should have resort to all-valence-electron NDO (neglect of differential overlap) methods which yet, in their current formulations, have proved more or less inadequate to build ground and excited state potential surfaces, particularly along paths leading to conformational rearrangements or formation of photoreaction intermediates (radical-like structures) (for detailed analysis, see ref. [7-10]. The second problem comes from the fact that the potential surfaces for the isolated molecule, even if they were rightly calculated, are in principle poorly representative of the photochemical behaviour in solution phase where energy minima and barriers may be substantially affected by solvent polarity and viscosity [6,11-16].

Y. Ellinger and M. Defranceschi (eds.), Strategies and Applications in Quantum Chemistry, 379–399.
© 1996 *Kluwer Academic Publishers.*

Starting from such considerations, during the last decade we have searched for an effective approach to the theoretical understanding of the condensed-phase photochemistry of large organic molecules. Our study had two major components: 1) formulation of a new model Hamiltonian of the all-valence-electron type capable of providing, in the case where it is adequately solved, qualitatively correct descriptions of the reaction paths in both the ground and lowest excited states, and 2) implementation of photophysical and photochemical measurements devised to minimize the degree of arbitrariness inherent in any comparison between the results of molecular quantum-mechanical calculations and the experimental observations in condensed phase. As for point 1), our studies led us to publish in the early 1980s two modified INDO-based methods, namely C INDO [9] (limited to prediction of ground state properties) and CS INDO [10] (capable of reliably handling both ground and excited state properties), subsequently applied to the study of electronic spectra and *cis-trans* (ground and excited state) isomerizations in a variety of conjugated systems: diarylethylenes [17-20], polyphenyls [21], binaphthyls [22,23], cyanines [24,25], diphenylmethane dyes [26], donor-acceptor-type stilbene derivatives [27,28], etc. . Point 2) was fulfilled by setting up in our laboratory a complete equipment for both stationary and time-resolved nanosecond spectroscopy as well as by access to picosecond spectroscopic apparatuses of external laboratories.

The remainder of the present article is divided into two parts. The first one reviews the main points of our combined theoretical-experimental approach. The second one reports an application of it to the study of the mechanism and dynamics of *trans-cis* photoisomerization of bisdimethylaminopentamethine cyanine (BMPC)[#1].

2. Combining theory and experiment

2.1. CONSTRUCTION OF MOLECULAR WAVEFUNCTIONS AND POTENTIAL SURFACES: THE CS INDO MODEL

CS INDO [10] (as well as the parent C INDO [9]) shares the same basic idea as the PCILO scheme [29,30]: to exploit the conceptual and computational advantages of using a basis set of hybrid atomic orbitals (AOs) directed along, or nearly, the chemical bonds.

In the PCILO scheme the hybrid AOs, determined according to Del Re's method [31], are used to construct a basis of molecular orbitals (MOs) localized on the bonds and lone pairs, and from this to build a configuration basis set constituted by a fully localized determinant, representing the chemical formula, and the excited configurations obtained by utilizing the antibonding orbitals. On this basis, two different perturbative CI procedures were developed in successive times for the ground state [29,30] and the excited states [32,33]. In short, the early one is a perturbation expansion up to the third order for the fully localized determinant, taken as the zeroth order ground state wavefunction. As is well known [34], the CNDO version of such PCILO scheme

---

[#1] This study starts from previous work carried out by three of us (F.M., I.B., G.P.) in collaboration with Berthier [24,25], and may be considered as the logical pursuance of it.

overcomes, at least partly, the most striking failures of the ordinary CNDO-SCF procedure as far as the conformational predictions are concerned. The PCILO-CNDO method for the excited states, as proposed by Langlet and Malrieu [32,33], is based on a second-order perturbation treatment of multiconfigurational zeroth-order wavefunctions determined by variational CI on a proper basis of local single excitations. A similar procedure, using the CIPSI method [35] for a better choice of the zeroth-order wavefunctions, was applied to study *cis-trans* photoisomerism in styrene [36] and s-*trans*-1,3-pentadiene [37] and emphasized the usefulness of the excitonic scheme in interpreting photoreaction mechanisms.

In conclusion, from the scanty reported applications the CIPSI-PCILO-CNDO procedure stands as an interesting investigation tool in photochemistry. However, with especial reference to large conjugated systems, the PCILO-CNDO scheme has some limitations arising from both the very model (e.g. arbitrariness sometimes arising in the choice of the zeroth-order localized structure [30,38]) and the CNDO parametrization (e.g. underestimation of the internal rotation barriers [9,34] and, at the same time, large overestimation of transition energies [30,36,37]), the latter defects being retained at the INDO level of approximation [39].

Thus, we resolved to reconsider the delocalized NDO-type MO-SCF techniques and explore the possibility, if any, of decidedly improving their predictive capabilities through the use of hybridi:.ed AO basis sets. It is well known that the main defect of CNDO and INDO SCF procedures, making them hardly usable to predict conformations and rotational barriers of conjugated molecules [9,34,40], is an anomalous stabilization of geometries with perpendicular arrangement of π subsystems. This failure originates in a large overestimation of σ-π (hyperconjugative) interactions [7-9] which is traceable in turn to the fact that CNDO-INDO methods adopt averaged β_{AB}^0 bonding parameters (i.e. depending only on the nature of atoms A and B) in order to satisfy all invariance conditions [41]. In fact, the use of averaged β_{AB}^0 parameters causes inadequate differentiation of the resonance integrals ($\beta_{\mu\nu} = \beta_{AB}^0 S_{\mu\nu}$) corresponding to the various types of interactions (σ-σ, π-π, σ-π,...) occurring in conjugated systems. Of course, the correct differentiation might be approximately restored by introducing specific screening constants ($k_{\sigma\sigma}$, $k_{\pi\pi}$, $k_{\sigma\pi}$,....), but the realization of this simple idea requires the characters of the interactions to be unequivocally identifiable in any context. In planar geometries, the belonging of the basis AOs to π or σ systems is fixed by symmetry, so proper screenings can be introduced for σ-σ and π-π interactions using pure Slater orbitals (e.g. with p_z orbitals forming the π system), as is done, for instance, in CNDO/S [42] and INDO/S [43] methods. However, in out-of-plane twisted conformations the use of pure Slater orbitals does not make it possible to discriminate between σ and π symmetries (now definable only within each planar subsystem), and hence the CNDO-INDO/S procedure for the evaluation of $\beta_{\mu\nu}$ integrals becomes ineffective. On the other hand, by switching from the usual STO valence set to a set of hybrid AOs retaining σ or π (local) symmetry in both planar and non-planar geometries, the resonance integrals can be made to correspond to chemically well-defined interactions and can therefore be specifically reparametrized according to a formula like

$$\beta_{\alpha\beta} = \beta^0_{AB} k_{i(\alpha)j(\beta)} S_{\alpha\beta} \qquad\qquad i,j=\sigma,\pi \qquad\qquad (1)$$

where $S_{\alpha\beta}$ is the overlap integral between the hybrid orbitals χ_α (atom A) and χ_β (atom B) and $k_{i(\alpha)j(\beta)}$ is a screening factor depending on the nature of the involved hybrids. There is ample evidence [9,17,44] that the INDO SCF procedure transformed according to this scheme (C INDO) can provide predictions comparable to those of minimal-basis-set ab initio SCF calculations for conformations and rotational barriers of conjugated molecules in the ground state.

Starting from the C INDO scheme, in a second step we derived a new version of the method (CS INDO, where C and S stand for conformations and spectra) [10] capable of correctly handling electronic spectra and excited state potential surfaces, yet preserving the quality of the C INDO predictions as far as the ground state properties are concerned. The relevant changes incorporated into CS INDO are: 1) re-modelling of the screening effects for $\beta_{\alpha\beta}$ integrals, 2) scaling down of the electron repulsion integrals (γ_{AB}) according to one of the current "spectroscopic" parametrizations (e.g. Pariser-Parr, Ohno-Klopman, Mataga-Nishimoto), 3) use of a new formula for core-core repulsions (E^{CR}_{AB}) self-fitting the adopted γ_{AB} parametric function. Moreover, rather extended - properly selected CIs are performed as needed for a correct excited state treatment.

At the present stage of development [45] the essential steps of the CS INDO CI procedure can be summarized as follows:

1) A basis set of hybrid AOs, having σ, π or n character, is prepared by the Slater s,p valence set according to the maximum overlap criterion [31,46]. An extension to d orbitals is in progress.

2) The selection of the screening factors for $\beta_{\alpha\beta}$ integrals is made assuming $k_{\sigma\sigma}=1$ and obtaining the best values of $k_{\sigma\pi}$, $k_{\pi\pi}$, $k_{\sigma n}$ by fitting rotational barriers, $\pi\to\pi^*$ transition energies and $n\to\pi^*$ transition energies, respectively. The remaing factors ($k_{n\pi}$, k_{nn}) are then deduced by a simple proportionality criterion. It being understood that $k_{\sigma\sigma}$ is kept equal to one ("pivot" parameter) the optimized values of the other screening factors are slightly dependent on the adopted CI scheme (see point 4). The introduction of three extra-parameters related to chemically distinct interactions and controlling distinct molecular properties, is the main feature of the CS INDO method.

3) Apart from minor exceptions all other parameters are given the same values as in standard INDO (electronegativities, Slater-Condon parameters, bonding parameters) or CNDO/S (one-centre repulsion integrals) methods. Two-centre repulsion integrals are usually evaluted by the Ohno-Klopman formula.

4) With reference to our main target (large conjugated molecules), CI calculations are expanded on the subspace of the configurations generated within a restricted MO basis set encompassing all π and π^* type orbitals which, in CS INDO, are easily identifiable in non-planar conformations, too. From this common starting point CI calculations are

performed by purely variational or mixed variational-perturbative (CIPSI-type [35,47]) approaches. In both cases a rather limited number of representative configurations (e.g. the monoexcited configurations "localized" on the characteristic chromophores) form a privileged reference set for the construction of the CI matrix.

The results of the above cited applications [18-28,45] have clearly shown that CS INDO method is fairly successful in combining equally satisfactory predictions of electronic spectra and potential surfaces (especially along internal rotation pathways) of conjugated molecules, a goal never reached by other NDO-type procedures. CS INDO shares, at least partly, the interpretative advantages of the CIPSI-PCILO-CNDO procedure [32,33,36,37], coming from using the same hybrid AO basis sets, but improves its predictive capabilities as far as spectroscopic and photochemical properties are concerned.

The advantages of CS INDO with respect to all other NDO-type procedures derive, in summary, from the resonance integrals being made to depend on the nature (in the chemical sense) of the interacting orbitals. This implies, in principle, loss of the hybridizational invariance [41], but the practical disadvantages are slight since in general the hybridization is nearly determined by the molecular structure and, when ambiguity arises (e.g. with nonbonding hybrids of heteroatoms), hybridization may well be fixed according to an energetic criterion [44]. A more serious deficiency of CS INDO, inherent in the INDO approximations, concerns the description of those Coulombic interactions where the angular dependence plays an important rôle (e.g. lone-pair interactions). This defect, due to the use of spherically averaged two-centre electron-repulsion integrals (required to preserve rotational invariance), could be partly overcome by switching to the NDDO level of approximation [48,49], but the benefit would not balance the difficulty of working out a complete re-parametrization of NDDO for the excited states.

From the foregoing considerations, taken as a whole, the CS INDO CI method appears to be a suitable tool to try to explore ground and excited state potential surfaces of large conjugated molecules. On the other hand, such systems are commonly studied in solution, so one must face the extra problem of the possible solvent effects (see next section) on the mechanism and dynamics of photochemical processes. Of course, no effects of the solvent viscosity can be explicitly treated within the framework of the electronic theory of photochemistry. However, leaving aside specific solute-solvent interactions, the dielectric solvent effects can be conveniently evaluated by theoretical models treating the solvent as a continuum, essentially the reaction field and the virtual charge (solvaton) models [3]. Quantum chemical SCF treatments incorporating the dielectric effects of the solvent have been developed for both models [50-52]. Such direct quantum chemical approaches are certainly advisable for studies limited to the ground state, but they are hardly practicable in photochemistry where the solvent effects on the ground state and some lowest excited states are to be evaluated at the same time. Thus we limit ourselves to calculating state by state the solvation energy (E_{sol}) of a solute molecule using its electrical properties as obtained in the isolated-molecule approximation. Following the conclusions of a recent study [27], where the two continuum models have been comparatively tested, we evaluate E_{sol} according to the solvaton model which, after Hedrich et al. [53], can be conveniently expressed as:

$$E_{sol} = -0.5(1-\epsilon^{-1})\left[\sum_A Q_A^2 / R^{eff} + \sum_A \sum_{B \neq A} Q_A Q_B / (r_{AB} + R^{eff})\right] \qquad (2)$$

where ϵ is the static dielectric constant of the solvent, R^{eff} is an average effective atomic radius, r_{AB} is the distance between the centres of the atoms A and B, and Q_A (Q_B) is the net charge on the atom A (B) derived from the CI treatment of the isolated molecule. As pointed out in ref. [27], this "microscopic" model is superior to the "macroscopic" ones based on the global dipole moment of the solute, since it is able to take into account local solute-solvent interactions. This is especially important in large molecules, where high local net charges may well occur in spite of a small global dipole moment.

2.2. PRODUCTION OF SELECTED PHOTOPHYSICAL AND PHOTOCHEMICAL DATA

As said above, our theoretical tools are especially effective for studying photoisomerizations (in a generalized sense) of conjugated non-rigid systems. Such processes usually involve large amplitude motions of rather bulky groups, so that coupling of these motions with solvent drag is often strong. Furthermore, in many cases, marked changes of the electrical properties, related to separation or localization of charge, take place along the reaction coordinate (e.g. sudden-polarization [54] and TICT-formation [6] phenomena). As a consequence, coulombic interactions with solvent molecules may deeply affect the potential governing the internal motion of a solute molecule. A fruitful experimental approach should provide selected and organized experimental information on both the spectroscopy of the species and the dynamic of the processes involved in the photoisomerization.

This purpose is achieved in our laboratory carrying out photostationary (steady-state fluorometry and photolysis) and time-resolved (time-correlated fluorescence single-photon counting, laser or conventional flash photolysis) experiments, at variable temperature and in solvents with different polarities and viscosities. The results may consist as well in the absorption and emission spectra of a number of transients (e.g. triplets and photoisomers) as in lifetimes, quantum yields and rate constants. The analysis of the temperature dependences of the latter affords the preexponential factors (A) and activation energies (E_a) of the processes under study in the solvent employed. The repetition of the variable temperature measurements in several solvents of very different dielectric constants, yet similar and low viscosities, will make the "pure" medium-polarity effects on the kinetic parameters emerge and will help to check the validity of a theoretical model [55] as well as to verify the reliability of the calculated solvation energies. However, in order that a more quantitative check of the calculated potential barriers may be possible, activation energies must be cleared of the contributions arising from solvent frictional effects. Such "intramolecular" (but for electrostatic solute-solvent interactions) activation energies are obtained as the slopes of isoviscosity plots made with data measured in a series of homologous solvents. Within a certain accuracy, the procedure itself provides a check of the reasonableness of the assumptions on which it rests [16].

Steady-state experiments can also be designed within the same kind of strategy. As an example, we can cite recent works [25,45], where the results of a quantitative analysis of the resolved absorption spectra of a number of *trans* and *cis* isomers of cyanine dyes were compared with calculated oscillator strengths and transition energies so as to propose the identification of the observed phototropic species as well defined *cis* isomers.

Starting from the results of such a theoretical-spectroscopic investigation on BMPC [25], in the next section we report a typical application of the above outlined approach, in which kinetic measurements as functions of the solvent properties have been prompted by theoretical considerations and the experimental results are used in turn to analyse critically the calculated potential energy curves.

3. An example: the *trans→cis* photochemical and *cis→trans* thermal back isomerization of BMPC

Photoisomerism of BMPC (Scheme 1) has already been investigated by us [25] through the comparison between the calculated spectra of the *trans* and the two mono-*cis* (2-3 and 3-4 *cis*) isomers, on the one hand, and the experimentally resolved spectra of the

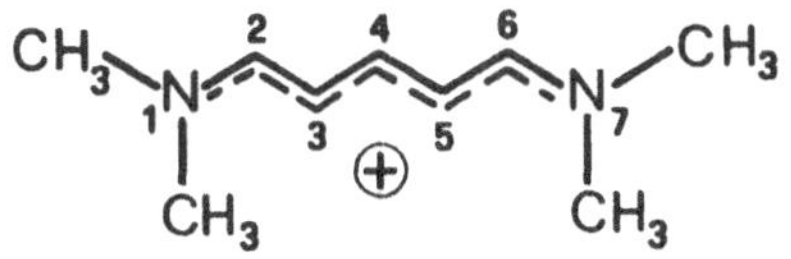

BMPC

Scheme 1

stable and photochemically produced forms, on the other. It was concluded that irradiation of the stable (all-*trans*) form results in the formation of a single isomer with *cis*-planar structure and, by considerations on the intensity relations between ^1A-^1B and ^1A-$^1C_2$ (*cis* peak) transitions, the phototropic species was assigned as the 3-4 *cis* isomer. Both these findings appeared to be in agreement with the predictions of an early theoretical study based on simple MO correlation diagrams as well as on explicit potential-energy curve (CS INDO CI) calculations [24] for unsubstituted pentamethine cyanine (PC). Ref.s [24,25], taken together, disproved previous theoretical interpretations [56,57] according to which cyanine photoisomers should correspond to twisted ground state conformations (resulting from a 90°-rotation around one of the C-C bonds) strongly stabilized by electrostatic solute-solvent interactions. In view of such a striking interpretative contrast, and considering that calculations of ref. [24] had been carried out for a model system (PC) neglecting any solvent effect, we decided to tackle the dynamic aspects of BMPC photoisomerization in solution more thoroughly through a proper sequence of calculations and experiments.

As a first step of this work, we wanted to verify if the predictions of the previous theoretical study on the prototype system (PC) [24] were valid for BMPC, as well. For this aim we theoretically analysed the pathways leading from *trans* to (2-3 and 3-4) mono-*cis* isomers and explored, in addition, the possibility of concerted isomerizations at two C-C bonds (taking the 2-3, 4-5 double isomerization as an example).

Let us first recall some basic characteristics of the cyanine isomerization mechanism, as emerging from simple MO correlation diagrams like those of Fig. 1. In

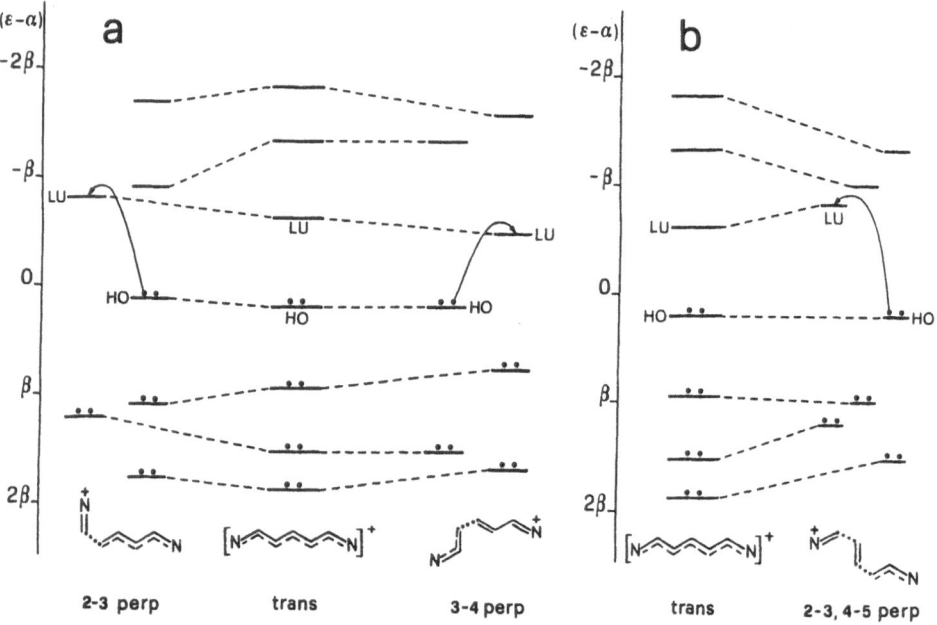

Figure 1. Hückel π-MO correlation diagrams for *planar→perp* twisting of pentamethine cyanine about the 2-3 and 3-4 bonds (a), and simultaneous twisting about 2-3 and 4-5 bonds (b). For the *perp* forms the orbitals belonging to the different π fragments are indicated and the CT nature of the HOMO-LUMO excitation is emphasized.

reference to Fig. 1a, it is evident that isomerization at one C-C bond involves formation of a twisted (*perp*, θ=90°) intermediate characterized by two decoupled π subsystems having even and odd numbers of π centres. Due to the π-MO localization, at θ=90° the cationic charge is borne by the even (polyenic) fragment in the ground state and shifts to the odd (polymethinic) fragment upon HOMO-LUMO excitation. A substantially similar situation arises with two-bond isomerizations (Fig. 1b) even if the scheme is made more complex because of the presence of three mutually orthogonal π subsystems at θ=90°. In summary, Fig. 1 emphasizes the fact that the foreseeable photoisomerization intermediates have TICT-like nature (related to full localization of the positive charge, instead of charge separation as in the classic TICT donor-acceptor systems, e.g. DMABN [6]). As previously argued [24], the CT nature of the *perp* forms points to some other basic properties: 1) quasi-degeneracy of the S_1 and T_1 (essentially HOMO-LUMO) states, 2) low efficiency of the S_1-T_1 intersystem crossing, 3) possible occurrence of intramolecular minima at the *perp* S_1 conformations.

Since the negligible contribution of intersystem crossing to the S_1 radiationless decay of cyanine-like systems has been firmly established [26,58,59], henceforth we will consider only the properties of the S_1 and S_0 potential surfaces involved in the

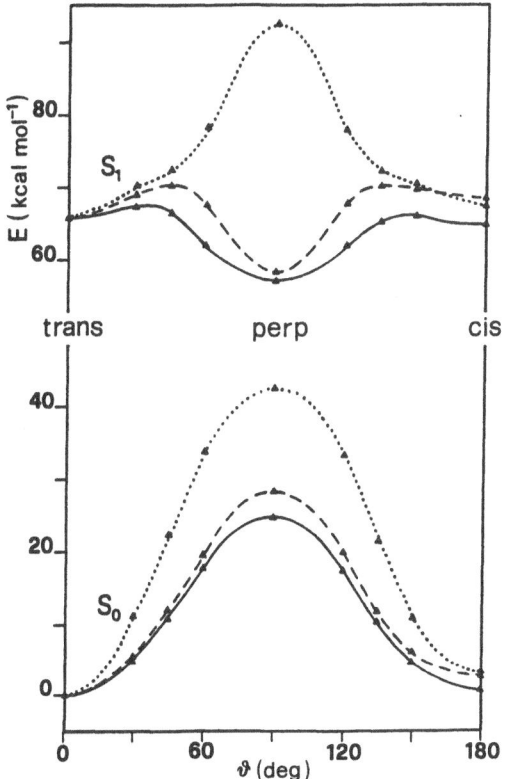

Figure 2. CS INDO CI potential energy curves describing *trans-cis* isomerization of BMPC about the 3-4 bond (——), the 2-3 bond (– – –), and both the 2-3 and 4-5 bonds (·······) in the ground (S_0) and lowest singlet-excited (S_1) states. Full triangles (▲) indicate the calculated points; energies are referred to the *trans* planar ground state. Almost all elements of the SCF calculation, including integrals, structural parameters and hybridization of the AOs, were as in ref. [25]. Apart from a bond angle optimization around the 2-3 *cis* sterically hindered form, no geometry relaxation was allowed for during twisting motion. S_0 and S_1 state energies were calculated by a second-order perturbative CI starting from monoconfigurational zero-order wavefunctions ([HOMO²] and [HOMO,LUMO]) and using a MO subset (19 HOMOs + 21 LUMOs) encompassing all π and π^* type orbitals. The characteristic CS INDO screening parameters were given the values $k_{\pi\pi}=0.60$, $k_{\sigma\pi}=0.75$, $k_{\sigma\sigma}=1$, which fit in with the adopted CI scheme (see ref. [24]).

trans→cis photoisomerization and the *cis→trans* thermal back isomerization, respectively. The results of the CS INDO CI calculations for the 2-3 and 3-4 one-bond and 2-3, 4-5 two-bond isomerizations are represented in Fig. 2. The concerted two-bond *trans→cis* isomerization is predicted to be hindered by high barriers at θ=90° in both the ground (42.4 kcal mol⁻¹) and the excited state (26.6 kcal mol⁻¹). This result does not explain the very short fluorescence lifetime (τ≤20 ps for BMPC⁺,ClO₄⁻ in octanol at room temperature [57]; for τ in 1:1 ethanol-methanol mixture see later) nor the aptitude of BMPC and other streptocyanines for photoisomerization [25]. We consider this as sufficient proof to rule out definitely simultaneous isomerizations at two C-C bonds. On the other hand, the potential curves for both 2-3 and 3-4 single isomerizations are in keeping with the observed photophysical and photochemical behaviours. The main aspect is the presence of pronounced *perp* minima in S₁ which can be easily reached from the directly excited S₁-*trans* form by overcoming of more or less little barriers (4.5 and 1.9 kcal mol⁻¹ for the isomerizations around 2-3 and 3-4 links, respectively) and through which decay to the ground state *trans* and *cis* isomers may rapidly occur (the barriers for thermal back isomerization of 2-3 *cis* and 3-4 *cis* isomers being 25.4 and 24.2 kcal mol⁻¹, respectively). From the comparison between the calculated barriers the formation of the 3-4 *cis* isomer appears to be decidedly favoured in agreement with both the CS INDO CI calculations on PC [24] and the analysis of the spectrum of the phototropic form of BMPC in methanol solution [25].

In conclusion almost all the experimental observations concerning photophysics and photochemistry of BMPC in alcoholic solutions appear to be fairly accounted for by

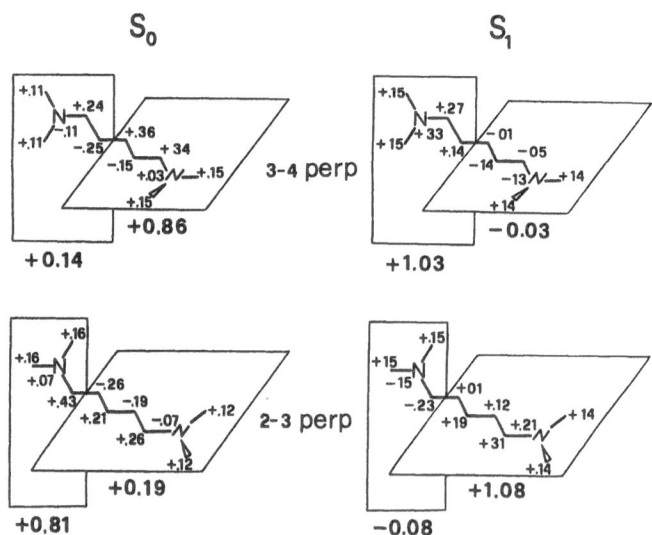

Figure 3. Charge distributions in the S₀ and S₁ states of the 2-3 and 3-4 *perp* forms (θ=90°) of BMPC, as obtained by single configuration (ground state SCF and [HOMO,LUMO]) wavefunctions. Only the overall charge is reported for the methyl substituents. The global net charges of the perpendicular subunits, emphasizing localization of the charge, are also indicated.

calculations performed in the "free space" approximation. Of course, this requires supplementary investigations since, in principle, one should expect both the photoinduced *trans→cis* and the thermal *cis→trans* isomerization dynamics to be influenced by solvent polarity, owing to the charge localization phenomenon in the S_1 and S_0 *perp* forms (Fig. 3), as well as by solvent viscosity, as has been observed with many other polymethine cyanines [16].

As for the theoretical treatment, we could only try to include the electrostatic solute-solvent interactions and, in fact, we corrected the electronic potential energies for the solvation effects by simply adding E_{sol} as calculated according to the solvaton model [eq. (2)]. The resulting potential curves are to be seen as effective potentials at equilibrium, i.e. reflecting orientational equilibrium distributions of the solvent dipoles around the charged atoms of the solute molecule. In principle, the use of potentials thus corrected involves the assumption that solvent equilibration is more rapid than internal rotation of the solute molecule. Fig. 4 points out the effects produced on the potential

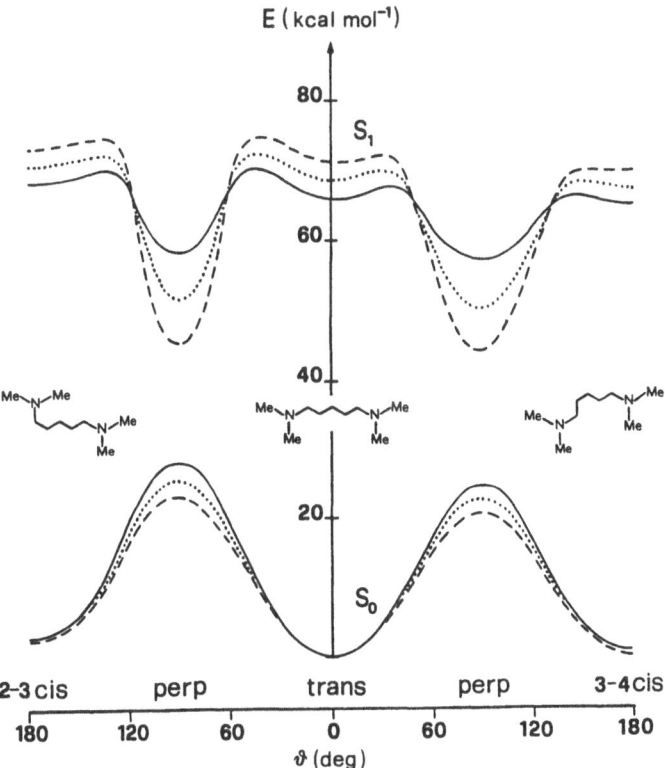

Figure 4. Potential energy curves for the S_0 and S_1 states of BMPC as a function of the angle of rotation about the 2-3 and 3-4 bonds : (———) in the gas phase (see Fig. 2), (·······) in a weakly polar solvent ($\varepsilon=2$), (– – –) in a highly polar solvent ($\varepsilon=40$). The energy of the molecule in solution was obtained by adding up its energy in the gas phase and E_{sol} as calculated by eq. 2 (the parameter R^{eff} was taken equal to 4 a.u. according to ref. [27]).

energy curves by two solvents of very different dielectric constants ($\varepsilon=2$ and 40) when using the net charges derived from monoconfigurational descriptions of S_0 (SCF determinant) and S_1 ([HOMO,LUMO] configuration) states. Owing to the fact that the cation charge begins to localize at highly twisted conformations, solvation manifests itself as a potential energy lowering (much more marked in S_1 than in S_0) around the *perp* forms ($60° \leq \theta \leq 130°$). As a consequence, the small barriers impeding *trans→perp* conversion in S_1, located at low values of θ ($30° \leq \theta \leq 45°$), appear to be very little affected by the solvent polarity, while the barriers to *trans-cis* isomerization in S_0, located at $\theta=90°$, undergo a significant, yet not very large, decrease when the solvent polarity increases. Anyway, some 50% of the limit polarity effect is obtained on passing from the gas phase ($\varepsilon=1$) to a solvent of low dielectric constant ($\varepsilon=2$). In order to check the internal consistency of the predicted solvation effects we recalculated E_{sol} using the atomic charges obtained by rather extended CI wavefunctions (Fig. 5). Not surprisingly, appreciable changes were found only for the excited state and were confined to a smaller

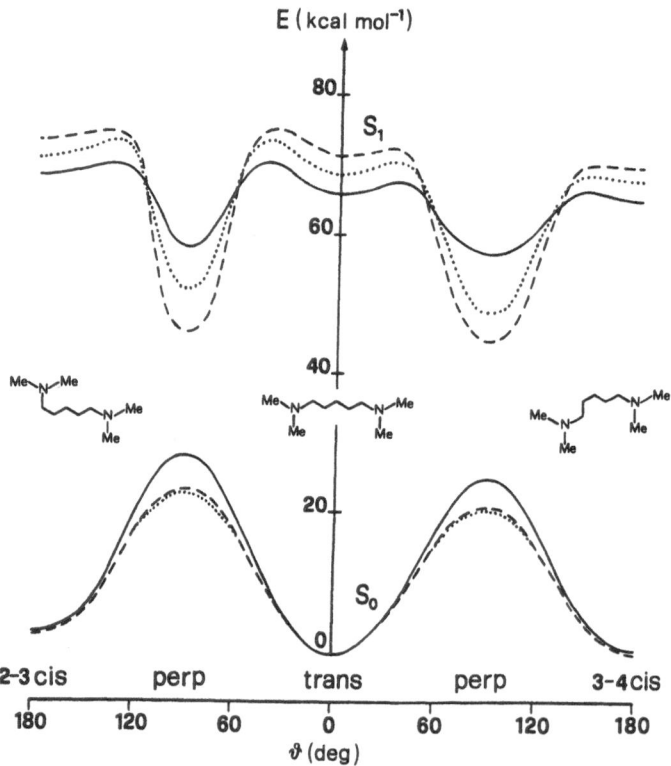

Figure 5. Effects of an highly polar solvent ($\varepsilon=40$) on the potential energy curves describing *trans-cis* isomerizations at the 2-3 and 3-4 C–C bonds of BMPC when using charge distributions obtained by single configuration (– – –) or extended CI (·······) wavefunctions . For the other details see caption of Fig. 4.

stabilization of the *perp* form. In summary, Fig.s 4,5 indicate, in agreement with the results of Fig. 2, that BMPC photoisomerization proceeds preferably through rotation around the 3-4 bond and involves overcoming of a small barrier (between 1 and 2 kcal mol^{-1}) almost independent of the solvent polarity. On the other hand, not very large yet significant polarity effects are expected for the barrier hindering 3-4 *cis→trans* back isomerization in S_0, which is of ≈24 kcal mol^{-1} in the gas phase and goes down to ≈20 kcal mol^{-1} in a highly polar solvent (ε=40).

The validity of the above conclusions rests on the reliability of theoretical predictions on excited state barriers as low as 1-2 kcal mol^{-1}. Of course, this required as accurate an experimental check as possible with reference to both the solvent viscosity effects, completely disregarded by theory, and the dielectric solvent effects. As for the photoisomerization dynamics, the needed information was derived from measurements of fluorescence lifetimes (τ) and quantum yields (ϕ$_F$) on solution of BMPC$^+$,ClO$_4$$^-$. Leaving out solvents of very low dielectric constant, where extensive formation of ion pairs may occur [60], the observed photophysical properties are confidently referable to the unperturbed BMPC cation. Figure 6 shows the temperature dependence of the

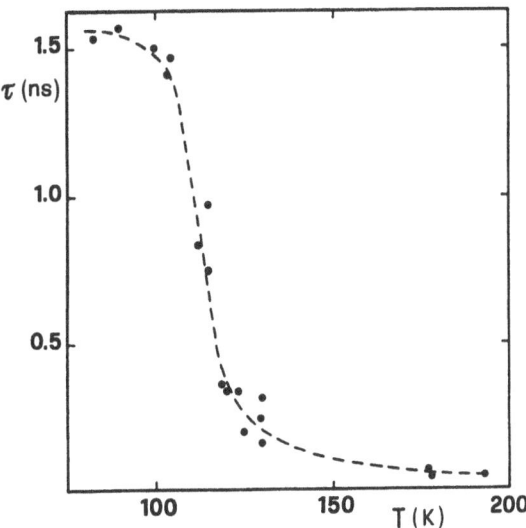

Figure 6. Temperature dependence of the fluorescence lifetime of BMPC in 1:1 ethanol-methanol. Measurements were carried out at the LENS laboratory of Florence by a picosecond apparatus using as an excitation source (at 380 nm) a dye laser pumped by a frequency-doubled cw Nd-YAG laser and recording the fluorescence time profiles by a streak camera. Since the overall instrumental response time was 75-80 ps, decays with τ≥200 ps, observed at T≤130 K, were analyzed without deconvolution. At 177, 178 and 193 K, the lifetimes were roughly estimated as τ≈(FWHM^2-77^2)$^{1/2}$, where FWHM was the width at half maximum of the decay. Because of the rather high sample absorbances (A$_{max}$≈2), self absorption may have reduced the lifetimes to some extent.

fluorescence lifetime of BMPC in 1:1 EtOH-MeOH mixture; τ is about 1.6 ns at temperatures lower than 100 K and drops suddenly upon heating the sample: $\tau \approx 200$ ps at $T \approx 130$ K. Further heating causes a much slower lifetime shortening. Such a behaviour is parallel to that of the solvent viscosity, which undergoes a steep change between 100 and 130 K due to matrix melting [61]. This indicates that solvent friction considerably affects the photoisomerization dynamics of BMPC and poses the very question whether the photoisomerization of BMPC is a barrierless process, so that its kinetics would be affected by temperature only through the variation of the solvent viscosity. The results reported in Fig. 7 clearly show that the answer to this question is "no". The fluorescence quantum yields of BMPC in several linear alcohols at 298 K increase with solvent viscosity much more slowly than when measurements are carried out in ethanol at different temperatures. It is apparent that, in ethanol, a temperature lowering causes an increase of the quantum yield not only indirectly, by increasing the solvent viscosity to values comparable with those of the longer-chain alcohols, but also in a direct way: the photoisomerization of BMPC in alcohols features a significant intramolecular barrier.

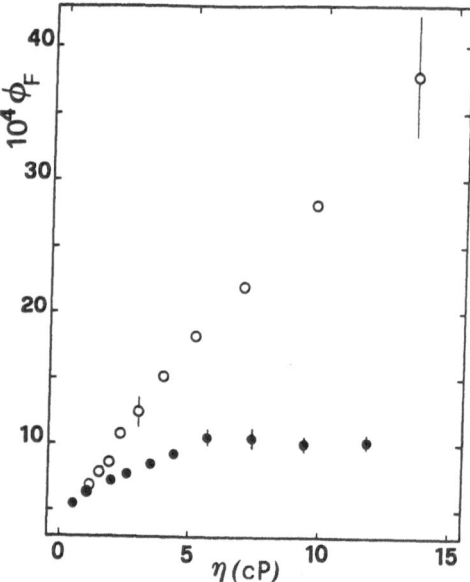

Figure 7. Dependence of the fluorescence quantum yield of BMPC on solvent viscosity: (•) in linear alcohols, from methanol to decanol, at 25°C, (o) in absolute ethanol between 200 and 298 K. The quantum yields were measured on optically thin samples ($A_{max} \leq 0.2$). The value in ethanol, 5.7×10^{-4}, was determined relative to quinine sulfate in 0.5 mol l^{-1} H_2SO_4 ($\phi_F = 0.55$ [62]) and 9,10-diphenylanthracene in cyclohexane ($\phi_F = 0.90$ [63]). It was then used as a reference for the determinations in the other alcohols.

Such a qualitative conclusion is supported by the observation that the room-temperature fluorescence spectrum of BMPC in alcohols (Fig. 8) is a good mirror image of the corresponding absorption band: its width at half maximum (2100 cm^{-1}) is only slightly larger than that of the absorption band (1850 cm^{-1}), the Stokes shift is small (1650 cm^{-1}) and there is no evidence of the long tail extending to the red which is often observed in the case of barrierless torsion in S_1 (e.g., crystal violet [64]). This indicates that fluorescence is emitted from a narrow collection of S_1 conformations centred around a minimum not much shifted with respect to the S_0 one, in keeping with the existence of an intramolecular barrier to photoisomerization[#2]. Evidently, this is in qualitative agreement with the theoretical predictions (Fig.s 4,5).

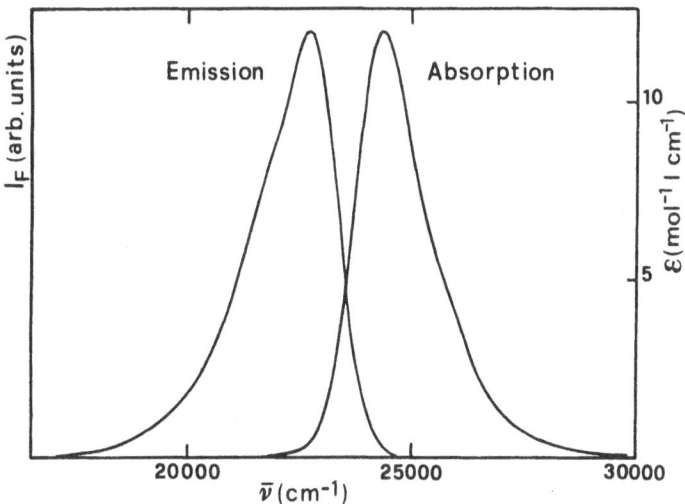

Figure 8. Lowest-energy absorption band and corrected fluorescence spectrum of BMPC in 1:1 ethanol-methanol at 23°C.

In order to go further into the experimental check we constructed Arrhenius plots of the fluorescence quantum yield of BMPC in a few solvents (methanol, ethanol, propanol, hexanol and methylene chloride), all of which showed good linearity. The activation energies and A/k_F ratios, calculated from the slopes and intercepts of those plots, are collected in Table 1. The smooth increase of both parameters in the alcohol series is mainly associated with the increase of solvent viscosity. On the other hand, decrease of the solvent dielectric constant from 32.7 (methanol) to 8.9 (dichloromethane) causes a small but significant increase of the activation energy; also, this increase is probably somewhat compensated by the decrease of the viscous-flow

---

[#2] The oscillator strength of the longest wavelength absorption band of BMPC (1.1, [25]) is very similar to those of two previously studied carbocyanines (DOC and DTC) [45] so that we can expect that, for BMPC as well as for DOC and DTC, the radiative constant (k_F) is equal to 2-3×10^8 s^{-1}. Combining this value with the fluorescence quantum yield of BMPC in methanol, $\phi_F = 5.3 \times 10^{-4}$, we can estimate its room-temperature fluorescence lifetime to be $\approx$ 2 ps.

activation energy of the latter solvent with respect to the former one[#3]. However, in view of its slight size this change is not contrary to the theoretically predicted "insensitiveness" of the *trans→perp* conversion barrier to solvent polarity.

Table 1. Activation energies and ratios of the preexponential factors to the radiative rate constants (A/k_F) for the photoisomerization of BMPC in several solvents. Solvent dielectric constants at room temperature, ε [65], and viscous flow activation energies, E_η [66], are shown too.

Solvent	ε	E_a (kcal mol^{-1})	E_η (kcal mol^{-1})[#]	10^{-5} (A/k_F)[*]
CH$_3$OH	32.7	1.94 ± 0.10	2.46	0.5 ± 0.1
C$_2$H$_5$OH	24.6	2.17 ± 0.14	3.06	0.7 ± 0.1
C$_3$H$_7$OH	20.3	2.58 ± 0.19	4.11	1.0 ± 0.2
C$_6$H$_{13}$OH	13.3	2.84 ± 0.24	5.28	1.3 ± 0.5
CH$_2$Cl$_2$	8.9	2.12 ± 0.14	1.60	0.7 ± 0.1

[#] A similar trend is shown by the values of η (cP) at 298 K: 0.55, 1.10, 1.98, 4.66, 0.42.
[*] Preexponential factors, all of the order of 10^{13} s^{-1}, can be estimated using k_F= 2–3×10^8 s^{-1} (see footnote 2).

With the aim of getting a quantitative evaluation of the "intramolecular" activation energy for the photoisomerization of BMPC in alcohols, a parameter which can be directly compared with calculated barriers, isoviscosity plots were drawn at 2, 6 and 10 cP using data obtained in methanol, ethanol, propanol and hexanol (Fig. 9). In all cases, the hexanol points lie slightly above the lines drawn through the points of the three shorter alcohols. This is probably a manifestation of the saturation of viscosity effects emphasized by Fig. 7: as the size of the alcohol molecule increases, the microscopic friction felt by the isomerizing solute is less and less adequately described by the solvent shear viscosity. Therefore, in spite of the fact that, due to the very low barrier and associated frequency, an almost diffusive reaction dynamics is expected [16], shear viscosity only provides a rough description of the frictional interaction between the twisting solute and the solvent molecule. This is confirmed by the finding that the slopes of the isoviscosity plots, determined omitting the hexanol points, decrease with viscosity. The "intramolecular" activation energies obtained at 2, 6 and 10 cP were equal to 1.24, 1.08 and 0.98 kcal mol^{-1}, respectively. In conclusion, because of the approximations in our analysis, related to the problematic use of shear viscosity as a measure of solvent friction, we can only provide an estimate of the "intramolecular"

[#3] Comparison of these results with those found for DOC and DTC , whose activation energies in dichloromethane were equal or even smaller than in methanol [55], indicates that the effect of solvent polarity on the photoisomerization barrier, although still small, is more pronounced for the open-chain cyanine BMPC than for the carbocyanines.

activation energy for the photoisomerization of BMPC in short-chain linear alcohols: $E=1.1\pm0.2$ kcal mol^{-1}. This finding strongly supports the theoretical prediction that photoisomerization of BMPC proceeds efficiently by twisting about the 3-4 bond in the S_1 surface. In fact, the S_1 calculated barriers to *trans*→3-4 *perp* conversion in highly polar solvents ($\varepsilon=40$) ranged between 1.10 kcal mol^{-1} and 1.75 kcal mol^{-1} (Fig. 5) according to whether the atomic charges derived from single configuration or extended-CI S_1 wavefunctions were used to evaluate E_{sol} (the calculated barrier in the isolated molecule being 1.85 kcal mol^{-1}).

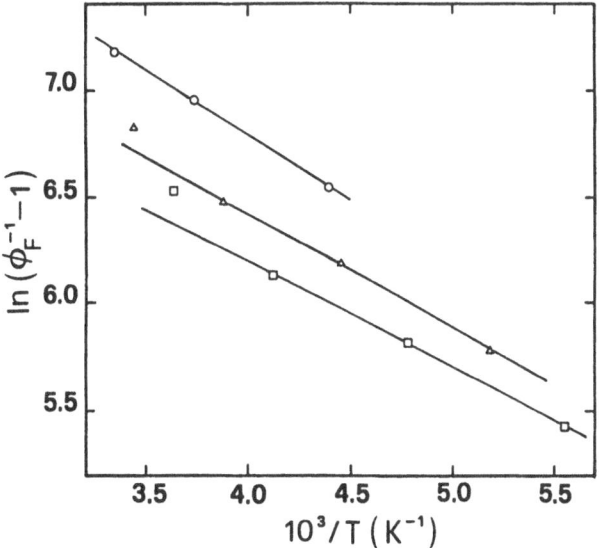

Figure 9. Isoviscosity plots for the photoisomerization kinetics of BMPC in methanol, ethanol, propanol and hexanol (points from right to left on each line). Viscosities: (o) 2 cP, (△) 6 cP, (◻) 10 cP.

Finally, the dependence of the *cis*→*trans* back isomerization kinetics of BMPC in the ground state on solvent polarity was investigated by measuring spectrophotometrically the rate constant of this process in methanol, dichloromethane, chlorobenzene and toluene. These solvents have similar room-temperature viscosities (from 0.45 to 0.8 cP) and viscous-flow activation energies (from 1.60 to 2.51 kcal mol^{-1} [66]). Because of this, and of the probably small relative frictional contribution to the overall activation energy connected with the high curvature of the potential function at the barrier top [16], solvent polarity effects can be evaluated by direct inspection of the measured activation energies. These were calculated as usual from Arrhenius plots and are shown in Table 2. As the solvent dielectric constant decreases, the measured activation energy increases on going from methanol to dichloromethane, decreases slightly in chlorobenzene and, finally, drops to a substantially lower value in toluene. The preexponential factor, too, shows a strong decrease in toluene with respect to the other solvents, reaching an atypical value of 6×10^{10} s^{-1}. A similar behaviour of the

Table 2. Preexponential factors (A) and activation energies (E$_a$) for the ground state *cis-trans* back isomerization of BMPC in solvents with different dielectric constants (ε).

Solvent	ε[#]	log [A (s^{-1})]	E$_a$ (kcal mol^{-1})
CH$_3$OH	32.7	12.8 ± 0.7	16.4 ± 0.8
CH$_2$Cl$_2$	8.9	12.8 ± 0.6	17.8 ± 0.3
C$_6$H$_5$Cl	5.6	12.9 ± 0.6	17.4 ± 0.6
C$_6$H$_5$CH$_3$	2.38	10.7 ± 0.7	14.8 ± 0.6

[#] From ref. [65].

Arrhenius parameters has been observed with several cationic cyanine dyes in low-polarity solvents, and has been attributed to the extensive formation of dye-counterion ion pairs on the basis of conductivity measurements [60]. Support for this interpretation comes from some calculations on cyanine model system which indicate a catalytic effect of ion-pairing on the ground state isomerization [67]. In the two most polar solvents, where ion pairs are not formed, a decrease of dielectric constant causes a significant, though not very large, increase of the activation energy. It is not clear whether chlorobenzene should be considered in the analysis of this trend or if the formation of some kind of ion pairs, looser than those formed in toluene, is responsible for the slight decrease of activation energy in this solvent relative to the more polar dichloromethane. However, apart from phenomena traceable to ion pair formation, both the value of the activation energy and its increase with a decrease of the solvent polarity (already observed with DOC and DTC [55]) are in good agreement with those theoretically predicted for the 3-4 *cis→trans* back isomerization in S$_0$. More precisely, the barriers of 20.99 and 22.93 kcal mol^{-1} calculated for this process in solvents of high (ε=40) and low (ε=2) polarities (Fig. 4) match fairly well with the activation energies of 16.4 and 17.8 kcal mol^{-1} measured in CH$_3$OH (ε=32.7) and CH$_2$Cl$_2$ (ε=8.9) solutions (Table 2).

In summary, all the experiments expressly selected to check the theoretical description provided fairly clear evidence in favour of both the basic electronic model proposed for the BMPC photoisomerization (involving a TICT-like state) and the essential characteristics of the intramolecular S$_0$ and S$_1$ potential surfaces as derived from CS INDO CI calculations. Now, combining the results of the present investigation with those of previous studies [24,25] we are in a position to fix the following points about the mechanism and dynamics of BMPC excited-state relaxation: 1) photoexcitation (S$_0$-S$_1$) of the stable (*trans*) form results in the formation of the 3-4 *cis* planar isomer, as well as recovery of the *trans* one, through a perpendicular CT-like S$_1$ minimum of intramolecular origin, 2) a small intramolecular barrier (1.-1.2 kcal mol^{-1}) is interposed between the secondary *trans* and the absolute *perp* minima, 3) the thermal back 3-4 *cis→trans* isomerization requires travelling over a substantial intramolecular barrier ($\approx$18 kcal mol^{-1}) at the *perp* conformation, 4) solvent polarity effects come into play primarily around the *perp* conformation, due to localization of the cationic charge, and result in a slight reduction of the activation energy for the $cis\rightarrow$ $trans$ back isomerization in S_0 (a reduction of the $perp$ S_1 lifetime might also occur due to a polarity induced decrease of the S_1-S_0 energy gap at $\theta=90°$), 5) solvent viscosity essentially affects the dynamics of the $trans\rightarrow perp$ conversion in S_1 because of the combined effects of a flat potential barrier and a rather bulky rotating group.

Detailed studies on this line are in progress in our laboratory in an attempt to reach equally clear conclusions for more complex cyanines characterized by the same (pentamethine) chromophore as BMPC (e.g. DOC and DTC).

4. Conclusion

The aim of the present work was to show that, while awaiting the development of efficient quantum and statistical mechanical procedures able to provide qualitatively and quantitatively satisfactory descriptions of both static and dynamic aspects of photoreactions in condensed phase, at the present time some useful results can be obtained by combining traditional quantum-chemical calculations of potential energy surfaces with specially selected photophysical and photochemical measurements. This simple strategy consists in leading the theoretical description and the experimental analysis to a point where their direct comparison is freed from most arbitrariness factors. For example, with reference to photoreactions where bulky groups perform large amplitude motions combined with substantial changes in electronic distribution (like that reported in section 3), the work should go as far as to obtain kinetic parameters cleared of the solvent viscosity effects and compare them with those deducible from the calculated potential energy surfaces corrected for the solvation effects in a solvent of similar dielectric constant. Procedures of this type can serve a dual purpose: 1) to state to what extent the photoreaction mechanism and dynamics may be controlled by the polarity or the viscosity of the solvent, 2) to test the calculated intramolecular potential surfaces. As regards point 2) the reported study on the $trans$-cis photoisomerism of BMPC gave clear evidence for the soundness of the CS INDO method as well as the reasonableness of the model adopted to estimate the effects of the solvent polarity. On this basis, and the results of several other applications, we can assert that the CS INDO CI technique is a fairly effective and supple tool for dealing with the static (electronic) aspects of photoprocesses, especially those involving large conjugated molecules such as, for example, pigments and dyes having central rôles in biological systems or technological devices.

Acknowledgements

Let us first spare a grateful thought for our dear colleague, Prof. M.C. Bruni, who recently left us prematurely, for her valuable collaboration in the work, both theoretical and experimental, which has led us to results like those presented here. We are greatly indebted to Prof. G. Berthier for having often advised and helped us in several aspects of our recent theoretical work and for inviting us to join in the study of cyanine photophysics and photochemistry. We wish also thank Prof. E.Castelluci for his generous help in the picosecond measurements carried out using the facilities provided by the LENS laboratory of Florence. The sample of BMPC perchlorate used in this work was donated by Prof. Sheves, Rehovot, to whom we are very grateful. This research was supported by the Ministero della Ricerca Scientifica e Tecnologia (Rome), the Consiglio Nazionale delle Ricerche (Rome) and the Centro Interdipartimentale di Calculo Automatico e Informatica Applicata (University of Modena).

References

1.   L. Salem and C. Rowland, *Angew. Chem. Int. Ed. Engl.* **11**, 92 (1972).
2.   J. Michl, *Mol. Photochem.* **4**, 257 (1972).
3.   L. Salem, Electrons in Chemical Reactions : First Principales, Wiley, New York, ch. 3 (1982)
4.   V. Bonacic-Koutecky, J. Koutecky and J. Michl, *Angew. Chem. Int. Ed. Engl.* **26**, 170 (1987).
5.   J. Michl and V. Bonacic-Koutecky, Electronic Aspects of Organic Photochemistry, Wiley, New York (1990).
6.   W. Rettig, *Angew. Chem. Int. Ed. Engl.* **25**, 971 (1986).
7.   J. Michl, in : Modern Theoretical Chemistry, Vol. 8, G.A. Segal Ed. Plenum Press, New York, ch.3 (1977).
8.   A.R. Gregory and D.F. Williams, *J.Phys. Chem.* **83**, 2652 (1979).
9.   F. Momicchioli, I. Baraldi and M.C. Bruni, *Chem. Phys.* **70**, 161 (1982).
10.   F. Momicchioli, I. Baraldi and M.C. Bruni, *Chem. Phys.* **82**, 229 (1983).
11.   L. Salem and W.D. Stohrer, *J.Chem. Soc. Chem. Commun.* **140** (1975).
12.   V. Sundström and T. Gillbro, *Chem. Phys. Lett.* **109**, 538 (1984).
13.   E. Lippert, W.Rettig, V. Bonacic-Koutecky, F. Heisel and J.A. Miehé, *Adv. Chem. Phys.* **68**, 1 (1987).
14.   P.F. Barbara and W. Jarzeba, *Acc. Chem. Res.* **21**, 195 (1988).
15.   D.H. Waldeck, *Chem Rev.* **91**, 415 (1991).
16.   G. Ponterini and M. Caselli, *Ber. Bunsenges. Phys. Chem.* **96**, 564 (1992).
17.   I. Baraldi, F. Momicchioli and G. Ponterini, *J. Mol. Struct.* **110**, 187 (1984).
18.   G. Bartocci, F. Masetti, U. Mazzucato, A. Spalletti, I. Baraldi and F. Momicchioli, *J.Phys. Chem.* **91**, 4733 (1987).
19.   F. Momicchioli, I. Baraldi and E. Fischer, *J. Photochem. Photobiol. A: Chem.* **48**, 95 (1989).
20.   U. Mazzucuto and F. Momicchioli, *Chem. Rev.* **91**, 1679 (1991).
21.   I. Baraldi and G. Ponterini *J. Mol. Struct.* **122**, 287 (1985).
22.   I. Baraldi and G. Ponterini and F.Momicchioli, *J. Chem. Soc. Faraday II* **83**, 2139 (1987)
23.   I. Baraldi, M.C. Bruni, M. Caselli and G. Ponterini, *J. Chem. Soc. Faraday II* **85**, 65 (1989).
24.   F. Momicchioli, I. Baraldi and G. Berthier, *Chem. Phys.* **123**, 103 (1988).
25.   F. Momicchioli, I. Baraldi, G. Ponterini and G. Berthier, *Spectrochim. Acta* **46A**, 775 (1990).
26.   I. Baraldi, A. Carnevali, F. Momicchioli and G. Ponterini, *Chem. Phys.* **160**, 85 (1992).
27.   S. Marguet, J.C. Mialocq, P. Millie, G. Berthier and F. Momicchioli, *Chem. Phys.* **160**, 265 (1992); S. Marguet, D. Sc Thesis, University of Paris Sud, n.1936 (1992)
28.   F. Momicchioli, I. Baraldi, A. Carnevali, M. Caselli and G. Ponterini, *Coord. Chem. Rev.* **125**, 301 (1993).

29. S. Diner, J.P. Malrieu and P. Claverie, *Theoret. Chim. Acta* **13**, 1 (1969).
30. For a general review of the PCILO method see : J.P. Malrieu, in : Modern Theoretical Chemistry, Vol. 7, G.A. Segal Ed. Plenum Press, New York, ch.3 (1977).
31. G. Del Re, *Theoret. Chim. Acta* **1**, 188 (1963).
32. J. Langlet , *Theoret. Chim. Acta* **27**, 223 (1972).
33. J. Langlet and J.P. Malrieu, *Theoret. Chim. Acta* **30** , 59 (1973).
34. D. Perahia and A. Pullman, *Chem. Phys. Lett.* **19**, 73 (1973).
35   B. Huron, J.P. Malrieu and P. Rancurel, *J. Chem. Phys.* **58,** 5745 (1973)
36   M.C. Bruni, F. Momicchioli, I. Baraldi and J. Langlet, *Chem. Phys. Lett.* **36**, 484 (1975)
37   I. Baraldi, M.C. Bruni, F. Momicchioli, J. Langlet and J.P. Malrieu, *Chem. Phys. Lett.* **51**, 493 (1977)
38   M. Martin, R. Carbo, C. Petrongolo and J. Tomasi, *J. Am. Chem. Soc.* **97**, 1938 (1975)
39   J. Douady, V. Barone, Y. Ellinger and R.Subra, *Int. J. Quantum Chem.* **17**, 211 (1980).
40   O. Gropen and H.M. Seip, *Chem. Phys. Lett.* **11**, 445 (1971).
41   J.A. Pople and D.L. Beveridge ,_Appropximate Molecular Orbital Theory_, McGraw-Hill, New York (1970).
42   J. Del Bene and H.H. Jaffé, *J. Chem. Phys.* **48**, 1807 (1968); **48**, 4050 (1968); **49**, 1221 (1968).
43   J. Ridley and M. Zerner, *Theoret. Chim. Acta* **32**, 111 (1973); **42**, 223 (1976).
44   I. Baraldi, E. Gallinella and F. Momicchioli, *J. Chim. Phys.* **83**, 653 (1986).
45   I. Baraldi, A. Carnevali, F. Momicchioli and G. Ponterini, *Spectrochim. Acta* **49A,** 471 (1993)
46   S.A. Pozzoli, A. Rastelli and M.Tedeschi, *J. Chem. Soc. Faraday II* **69**, 256 (1973)
47   S. Evangelisti , J.P. Daudey and J.P. Malrieu, *Chem.Phys.* **75**, 91 (1983)
48   P. Birner, H.J. Köhler and C. Weiss, *Chem. Phys. Lett.* **27**, 347 (1974).
49   R. Benedix, P. Birner, F. Birnstock, H. Henning and H.J. Hofmann, *J.Mol. Struct.* **51**, 99 (1979).
50   J.L. Rivail and D. Rinaldi, *Chem Phys.* **18**, 233 (1976).
51   R. Constanciel and O. Tapia, *Theoret. Chim. Acta* **48**, 75 (1978).
52   S. Miertus, E. Scrocco and J. Tomasi, *Chem. Phys.* **55**, 117 (1981).
53   D. Heidrich, U. Göring, W. Förster and C. Weiss, *Tetrahedron* **35**, 651 (1979).
54   V. Bonacic-Koutecky, P. Bruckmann, P. Hiberty, J. Koutecky, C. Leforestier and L. Salem, *Angew. Chem. Int. Ed. Engl.* **14**, 575 (1975).
55   G. Ponterini and F. Momiccheli, *Chem. Phys.* **151**, 111 (1991)
56   F.Dietz, W. Förster, C.Weiss, A. Tadjer and N. Tyutyulkov, *J.Signalaufz. Mater.* **9**, 177 (1981).
57   F. Dietz and S.K. Rentsch, *Chem. Phys.* **96**, 145 (1985).
58   M. Arvis and J.C. Mialocq, *J. Chem. Soc. Faraday II* **75**, 415 (1979).
59   C. Carre, C. Reichart and D.J. Lougnot, *J.Chim. Phys.* **84,** 577 (1987).
60   A.S. Tatikolov, L.A. Shvedova, N.A. Derevyanko, A.A. Ishchenko and V.A. Kuzmin, *Chem. Phys. Lett.* **190**, 291 (1992)
61   F. Barigelletti, *J. Phys. Chem.* **92**, 3679 (1988)
62   R.A. Velapoldi and K.D. Mielenz, NBS Sp. Publication 260. Washington (1980)
63   D.F. Eaton, *J. Photochem. Photobiol. B: Biology* **2** , 523 (1988)
64   D. Ben-Amotz and C.B. Harris, *J.Chem. Phys.* **86**, 4856 (1987)
65   S.L. Murov, "Handbook of Photochemistry", M.Dekker, New York, 1973.
66   "Landolt-Börnstein Zahlenwerte und Funktionen", Vol. 5a, Springer, Berlin (1969)
67   K.Schöffel, F. Dietz and T. Krossner, *Chem. Phys. Lett.* **172**, 187 (1990)

C_3H_2 : A Puzzling Interstellar Small Molecule

F. PAUZAT and D. TALBI
Laboratoire de Radioastronomie E.N.S. et Observatoire de Paris
24 rue Lhomond, 75005 Paris, France

1. A key molecule and a model compound

The radio detection of a small molecule formed of three carbons and two hydrogens by Thaddeus et al. [1] in 1985 came as a surprize to all astrochemists : cyclopropenylidene C_3H_2, last born to the small world of detected interstellar species was soon to become famous, though competition is high in this world where exotism is common.

First, it is not a common molecule on earth, not being used in laboratory for any synthesis. Second, it is cyclic, which was and still is a rare feature among small interstellar molecules identified up to day; only SiCC and C_3H present the same characteristic.
Third, it seems to be present everywhere in the interstellar space and one of the most abundant after CO.
And over all, as time and studies around this new interstellar component increase, it reveals to be possibly related to the polycyclic aromatic hydocarbon family (PAHs), those controversial molecules of prime interest which could be omnipresent in the interstellar medium and an essential link between simple molecules and grains.

However, there is a critical lack of information on this system, mainly due to insufficient studies of its spectral signatures, which makes it difficult to insert this molecule with confidence in the astrochemical schemes. During these years, only a few experimental and theoretical studies were performed, aiming to the different spectra useful for interstellar identification and chemistry. Still a lot remains to do.

The rotational spectrum has been calculated accurately by ab-initio methods [2], and has been measured in the laboratory with high precision [3,4] , so that the radio detection of C_3H_2 can be done without ambiguity, encouraging its search in different environments as dense dark clouds [5], diffuse interstellar medium [6] or HII regions [7].

The first laboratory IR detection of C_3H_2 is from Reisenauer et al. in 1984 [8], who reported infrared bands at 1279, 1063, 888, and 789 cm^{-1} attributed to C_3H_2 trapped in an Argon matrix. Later, Huang and Graham, in 1990 [9], studied the infrared spectrum of C_3H_2 as part of a systematic investigation of tricarbon hydride transient species in a low temperature Argon matrix. Although they confirmed the assignment of Reisenauer et al. for the band at 1279 cm^{-1}, their studies of the deuterated isotopomers did not support the assignments proposed for the three other bands. Theoretically, both Lee et al. in 1985 [10]

Y. Ellinger and M. Defranceschi (eds.), Strategies and Applications in Quantum Chemistry, 401–419.
© *1996 Kluwer Academic Publishers.*

and Defrees and McLean in 1986 [2], calculated the harmonic IR spectra of C_3H_2, confirming Reisenauer et al. attributions; however, IR spectra for the corresponding deuterated molecules are not available to discuss Huang and Graham experimental measurements.

In the stellar environment where PAHs are supposed to be at the origin of the observed IR emission, satellite bands have been observed around the 3.3 μm feature. In the PAHs model, one of the hypothesis for these bands is what is called the "hot band hypothesis", which states that some of these lines are transitions from upper vibrationally excited levels of the PAHs molecules [11]. The band at 3.3 μm has been identified as the CH stretch (v=1→0). Since the vibrational potential well is anharmonic, the transitions from higher energy levels (v=3→2, v=4→3, ...) do not appear at the same energies and therefore are separated from the v=1→0 transition, so that they can be observed. If we consider C_3H_2 as an aromatic molecule obeying the (4n+2) π electron rule, it is then the smallest PAH existing in space; consequently, the calculation of its anharmonic IR spectra should be helpful for testing the hot band hypothesis.

Concerning the electronic spectra, very little has been done. No experimental work is known on this singlet $^1A_1$ ground state carbene. Theoretical calculations on the lowest two lying triplet states ($^3B_1$ and $^3A_2$) of C_3H_2 have been performed by Lee et al. in 1985 [10]. However, because the transitions towards these triplet states are not allowed, they are of no help for the astrophysical observations and a much more complete vertical spectrum is needed in order to assist in the search of C_3H_2 from its electronic transitions. Till now, the few attempts to find signatures of the molecule in the Visible-UV region have been unsuccessful. But this search has still to be done systematically when data are available, based on the fact that a molecule seen widely in radio and possibly in IR, should necessarily absorb energy at shorter wave length, somewhere in the UV or visible. Considering the real lack of information about this spectrum, we might assume that the observational windows currently chosen for such a search could be erroneous.

From this brief review of the data available, it is obvious that more theoretical work is needed for a better understanding of the C_3H_2 story. First, and even though the rotational spectrum is known with a good precision from experimental work, we found it useful to perform calculations of the rotational constants in order to compare with the observational or experimental values and illustrate the ab-initio approach. Then, we calculated the IR spectrum, vibrations and intensities, for the molecule and its deuterated isomers, allowing to answer the pending questions in the experimental spectra; taking anharmonicity into account showed interesting features for the interpretation of the satellite bands observed at 3.3 μm in space. Finally, in order to decide the window to be used for a search of the molecule in the Visible-UV area, we determined its electronic spectrum, i.e. transition energies and transitions moments at a highly sophisticated level of wave functions.

2. Radio signature

Directly linked to the geometry and dipole moment of a molecule, the rotational spectrum is an unambiguous fingerprint that has enabled the radioastronomers community to identify more than a hundred species. Optimized geometries of C_3H_2 calculated at increasing levels of theory (from RHF to MP4 [12]) are presented in Table 1. The rotational constants obtained for C_3H_2 and its deuterated isomers are presented in Table 2.

Table 1: Optimized geometries of C_3H_2

	RHF 6-31G(d,p)	RHF 6-311++G(d,p)	MP2 6-311++G(d,p)	MP3 6-311++G(d,p)	MP4 6-311++G(d,p)
C=C	1.3096	1.3116	1.3366	1.3292	1.3384
CC	1.4080	1.4035	1.4282	1.4249	1.4363
CH	1.0698	1.0694	1.0799	1.0780	1.0821
∠CCC	55.43	55.71	55.80	55.60	55.54
∠HCC	149.60	150.04	150.12	149.94	149.86

Distances are in angstroms and angles in degrees

The molecule appears close to cyclopropene [4] with geometrical parameters (C=C = 1.296 Å; CC = 1.509 Å; CH = 1.072 Å; ∠CCC = 50°.8; ∠HCC = 149°.9) tending towards aromatic values. We note the lengthening of the double bond opposite to the carbene center and the shortening of the other two bond lengths to a value close to that of aromatic compounds; at the same time, the angles relax to be closer to a regular triangle in order to accomodate the possible conjugation of the two electrons in the π system over the three-membered ring.
These calculations also show a systematic behaviour of the MP3 calculations to provide bond lengths slightly shorter than MP2 due to the correction of an overestimated correlation by third-order terms.

Despite little differences between the geometries, especially those taking correlation effects into account, it can be seen that the rotational constants calculated from the frozen geometries are not accurate enough for a search of the molecule on a radiotelescope.

Table 2: Predicted rotational constants for C_3H_2 and deuterated isomers (GHz)

	RHF 6-31G(d,p) B_e	MP2 6-311++G(d,p) B_e	MP3 6-311++G(d,p) B_e	MP4 6-311++G(d,p) B_e	Obs/Predict B_0
C_3H_2					
A	35.64550	34.85498	34.92945	34.36473	35.093[a]
B	33.01378	31.76857	32.09849	31.70654	32.296
C	17.13960	16.62014	16.72709	16.49108	16.746
C_3HD					
A	35.10359	34.25578	34.36495	33.82845	34.526[b]
B	27.56546	26.61769	26.85424	26.52423	27.019
C	15.44058	14.97877	15.07443	14.86717	15.091
C_3D_2					
A	32.09354	31.43956	31.49072	30.98767	31.638[b]
B	24.87502	23.96281	24.20264	23.92481	24.352
C	14.01347	13.59834	13.68491	13.50101	13.700

(a) observations are from Thaddeus et al. [1]
(b) corrected values from MP3 calculations

At that point, it should also be kept in mind that the values of bond lengths and angles are not directly accessible from experiments but are indirectly determined so as to reproduce the rotational constants which are themselves deduced from microwave experiments. Thus, comparison are always subject to some controversy since there is no biunivoque correspondence between the geometry and the rotational parameters.

At all events, the rotational constants have to be corrected for the electronic correlation still missing in the electronic wave function and for the contribution of the nuclear vibrations. These effects are to be taken into account with a precision depending on the error bar to be admitted. A now classic way to proceed is to perform calculations on model compounds to determine the error in theoretical bond lengths and angles as a function of the level of theory and to use it as a correcting factor for the corresponding parameters in the molecule under consideration [13-15]. It has to be noted that such a strategy is designed to account, not only for the errors inherent to the theoretical model but also for the zero-point vibrational effects as experimental r_0 parameters are used in place of r_e to make the corrections.

In this study where we are interested in isotope substituted systems, that is in systems with the same electronic wave function, a more global approach can be used. From Table 2 it is obvious that MP3 calculations give the best overall results. The compensation of errors that we find here is a general characteristic of this level of wave function, as illustrated by previous calculations on various series of molecules [16]. Thus, we will use the MP3 level of theory together with the formula

$$B_0^D(\text{Est}) = B_0^D(\text{MP3})\frac{B_0^H(\text{MP3})}{B_0^H(\text{Exp})}$$

for the estimation of the rotational constants $B_0^D(\text{Est})$ of the deuterated isomers from the experimental values of the hydrogenated species. The present values should be precise enough to help in the laboratory search of these deuterated isomers.

3. IR signature and interstellar UIR bands

3.1.   HARMONIC IR SPECTRUM

Harmonic IR spectra of C_3H_2 calculated at the RHF/6-311++G(d,p), MP2/6-311++G(d,p) and MP4/6-311++G(d,p) levels are reported in Table 3. The results are nicely converging as electronic correlation is progressively included in the wave function. Excellent agreement between theory and experiment is thus obtained at the MP4 level, which allows for a correct treatment of simultaneous correlation effects in coupled vibrations. The only discrepancies which could show up, would proceed from anharmonicity, as illustrated by the CH stretching vibrations which are found shifted to higher frequencies than anticipated.

For larger systems, where MP4 calculations are no longer tractable, it is necessary to use scaling procedures. The present results make it possible to derive adapted scaling factors to be applied to the force constant matrix for each level of wave function. They can be determined by comparison of the raw calculated values with the few experimental data, each type of vibration considered as an independent vibrator after a normal mode analysis.

A least square fitting leads to the following values:

> CH stretching : 0.80 (RHF) ; 0.87 (MP2) ; 0.89 (MP4)
> CC stretching : 0.85 (RHF) ; 0.95 (MP2) ; 1.0 (MP4)
> CH in-plane bending : 0.80 (RHF) ; 0.94 (MP2) ; 0.96 (MP4)
> CH out-of-plane bending : 0.80 (RHF) ; 0.96 (MP2) ; 1.0 (MP4)

Corrected frequencies are then obtained following Pulay's procedure [17] and the intensities recalculated from the scaled force constants matrix. It can be seen on Table 3 that neither the basis set extension, nor the inclusion of part of the correlation change the results significantly if corrections are adapted to the method of calculation, which is particularly encouraging for an application to larger systems. Frequencies, once corrected by the above scaling procedure or by uniform scaling using an averaged value should then be accurate within a few percent for molecules of the same family, except for the presence of strong coupling between vibrations. An example of such situation can be found here for the asymmetric CH bending and CC stretching vibrations, which, from the composition of the normal coordinate, appear to be strongly mixed. As a consequence, the $v_5(b_2)$ CC stretching estimated at 1078 cm^{-1} at the RHF/6-311++G(d,p) level differs from the experimental value by 2% of its value.

Table 3: IR Harmonic Spectra of C_3H_2

Vibration		Frequencies				Intensities	
		RHF	MP2	MP4	Exp.	RHF	MP2
CH sym str	$v_1(a_1)$	3439	3321	3292		0.9	1.2
(scaled)		(3097)	(3101)	(3111)			
CH asym str.	$v_2(b_2)$	3399	3284	3255		1.3	2.5
(scaled)		(3060)	(3064)	(3070)			
CC sym str	$v_3(a_1)$	1752	1617	1597		1.7	0.7
(scaled)		(1592)	(1571)	(1582)			
CC sym str	$v_4(a_1)$	1413	1322	1291	1278	59	46.
(scaled)		(1274)	(1284)	(1276)			
CC asym str	$v_5(b_2)$	1187	1088	1069	1061	22	6.2
(scaled)		(1078)	(1059)	(1062)			
CH asym wag	$v_6(a_2)$	1093	996	981		0.	0.
(scaled)		(978)	(973)	(980)			
CH asym bend	$v_7(b_2)$	1012	912	894		2.	4.5
(scaled)		(914)	(885)	(881)			
CH sym bend	$v_8(a_1)$	990	916	905	886	20.	22.9
(scaled)		(886)	(888)	(888)			
CH sym wag	$v_9(b_1)$	880	806	787	787	27.	21.7
(scaled)		(787)	(787)	(787)			

Frequencies are reported in cm^{-1}, Intensities in Km/mol

The calculated shifts of the bands for the partially (C_3HD) and totally (C_3D_2) deuterated forms of C_3H_2 are given in Table 4 ; only the vibrations with non-zero intensities have been reported.

- The $v_4(a_1)$ vibration featuring a symmetric CC stretching vibration (at 1278 cm^{-1} experimentally) is shifted by 8-10 cm^{-1} towards lower frequencies when going from C_3H_2 to C_3HD and from C_3HD to C_3D_2, in excellent agreement with experimental measurements by Huang and Graham [9]. As in the experiments, this vibration is found to be the most intense one with an intensity only reduced by 20% with full deuteration.

- The $v_5(b_2)$ vibration featuring the asymmetric CC stretching vibration, which is strongly coupled with the asymmetric CH in-plane bending vibration (at 1061 cm^{-1} experimentally) is shifted by 20-25 cm^{-1} towards lower frequencies, each time deuterium is incorporated in the molecule. The intensity ratio of $1/3$ between the v_5 (b_2) and the v_4 (a_1) dominant feature is almost unchanged with deuteration.

- The $v_7(b_2)$ vibration featuring the asymmetric CH in-plane bending with some blend of asymmetric CC stretching, which is not observed in C_3H_2 because of its weakness has its intensity increased to 7 Km/mol with partial deuteration; it is shifted to 882 cm^{-1} at the MP4 level. The same behavior is seen for the $v_6(a_2)$ shifted to 925 cm^{-1} in C_3HD with an intensity of 6 Km/mol. Could these vibrations be the small features seen by Huang et Graham at 885.5 or 884.5 cm^{-1} ?

- The v_8 (a_1) vibration featuring the symmetric CH in-plane bending (at 886 cm^{-1} experimentally) is drastically shifted after deuteration because it implies mostly the CH bond. At the MP4 level our calculations show that $v_8(a_1)$ is shifted to 678 cm^{-1} in C_3HD and to 648 cm^{-1} in C_3D_2 with weaker intensities of 9 and 7 Km/mol respectively.

Moreover the same authors reported that there is no evidence for the $v_9(b_1)$ 787.4 cm^{-1} band after deuteration. At the MP4 level we found that this band, in C_3HD, is shifted to about 659 cm^{-1} with an intensity of 10 Km/mol and, in C_3D_2, to 603 cm^{-1} with a weaker intensity of 7 Km/mol, about ten times lower than $v_4(a_1)$. Could the weakness of these v_8 and v_9 bands be a possible explanation for their experimental non-observation ? If it is the case, then the attribution of the v_6 and v_7 vibrations to the 885.5 or 884.5 cm^{-1} bands is most problably erroneous. Could these two features simply be an impurity ?

3.2.   ANHARMONICITY OF THE CH STRETCHING MODES

3.2.1. *Electronic and vibrational calculations*

The anharmonic modes for both the a_1 symmetric and b_2 asymmetric CH stretching vibrations have been explored. In order to perform a reasonable anharmonic treatment, we had to take into account the stretching of the bonds to larger elongations than for the harmonic description where displacements can be confined close to the equilibrium geometry. Consequently, correlation effects were included in the determination of the potential surface. The electronic calculations were carried out at the MP2 level, which insures a good description of the CH bond potential towards dissociation. A double zeta basis set extended by diffuse and polarization functions [18,19] was used and proper scaling was applied to the calculated frequencies.

In the vibrational treatment we assumed, as usually done, that the Born-Oppenheimer separation is possible and that the electronic energy as a function of the internuclear variables can be taken as a potential in the equation of the internal motions of the nuclei. The vibrational anharmonic functions are obtained by means of a variational treatment in the basis of the harmonic solutions for the vibration considered (for more details about the theory see Pauzat et al [20]).

Table 5: Anharmonic CH stretching vibration frequencies of C_3H_2

| transition | symmetric CH stretching | | | | asymmetric CH stretching | | | |
| | calculated | | observed[a] | | calculated | | observed[a] | |
	ν (cm⁻¹)	$\Delta\nu$ (cm⁻¹)	ν (cm⁻¹)	$\Delta\nu$ (cm⁻¹)	ν (cm⁻¹)	$\Delta\nu$ (cm⁻¹)	ν (cm⁻¹)	$\Delta\nu$ (cm⁻¹)
$v = 1 \rightarrow 0$	3025		3040		3094		3083	
$v = 2 \rightarrow 1$	2978	-47	2993*	-47	3154	+60	3137	+54
$v = 3 \rightarrow 2$	2930	-48	2944	-49	3213	+59	3194	+57
$v = 4 \rightarrow 3$	2881	-49	2889	-55	3271	+58		
$v = 5 \rightarrow 4$	2833	-46	2846	-43	3328	+57		

(a) from DeMuizon et al. 1986 (IRAS 03035+5819 Fig. 2)
(*) read directly from IRAS spectra (Fig. 2)

Examination of Table 5 immediately reveals that the two anharmonic progressions diverge from the 3.3 µm origin. One is shifted towards larger wavelengths (CH sym. stretch.) while the other is shifted towards smaller ones (CH asym. stretch.). This behavior can be understood from the shape of the potential as a function of the CH normal coordinates for the two vibrations (Figure 1). As illustrated, the two anharmonic surfaces deviate from the harmonic potentials in different ways.

- The symmetric stretching surface behaves like the usual Morse potential with two CH bonds undergoing dissociation simultaneously. The potential surface widens from the harmonic profile and the vibrational levels come closer when the energy increases; frequencies are shifted towards longer wavelengths.

- The asymmetric stretching surface behaves differently. The main reason is the opposite displacements of the nuclei during the vibration ; while one of the CH bonds is going to dissociation, the other undergoes strong compression. Since the potential is much steeper at short distances (repulsion wall) than at larger CH elongations, the potential first narrows from the harmonic profile and the vibrational levels separate when the energy increases; frequencies are shifted towards shorter wavelengths.

Table 4 : Frequency shifts with respect to C_3H_2 and intensities for active vibrations of deuterared isomers

Symmetry	C₃HD						C₃D₂					
	Δν (cm⁻¹)				I (Km/mole)		Δν (cm⁻¹)				I (Km/mole)	
	RHF	MP2	MP4	Exp*	RHF	MP2	RHF	MP2	MP4	Exp*	RHF	MP2
a_1 (1278)*	8	9	10	8	54	40	17	18	20	16	47	34
b_2 (1061)*	22	18	21		19	5	48	38	46		14	3
a_2 (980)£	55	54	55		6	5						
b_2 (881)£	11	1	1		7	12						
a_1 (886)*	199	210	210		9	9	240	240	240		13	15
b_1 (787)*	130	130	128		11	8	186	185	184		7	6

* Experimental frequencies in cm⁻¹ for C_3H_2 from Huang et al. 1990
£ MP4 scaled frequencies
Shifts are scaled with an average value of 0.90 for RHF, 0.975 for MP2, 0.99 for MP4

harm
anh
a_1 symmetric stretching C-H
harm
anh
b_2 asymmetric stretching C-H

Figure 1: Schematic representation of CH stretching potentials
Dotted lines : harmonic potentials and vibrational levels
Full lines : anharmonic potential and vibrational levels

As the main point is the relative positions of these bands with respect to the vibrational origin, the results of the first calculated frequency ($v = 1 \rightarrow 0$) can be directly adjusted to the 3.3μm observed one(see below). Thus, we find rather regular progressions of about 50 cm^{-1} for the a_1 and 60 cm^{-1} for the b_2 stretching vibrations due to anharmonicity.

3.2.2. *Anharmonicity and emission of space carriers*

Many celestial objects show a distinctive set of emission features in the infrared, known as the unidentified infrared emission bands (UIR bands) [21,22]. Since 1981 when Duley and Williams [23] pointed out that a few of the bands fell at the frequencies characteristic of polycyclic aromatic hydrocarbon molecules (PAHs) and suggested that these bands were produced by aromatic units in thermally excited dust grains, a number of arguments coming from observations, experiments and calculations, converge towards the hypothesis that the features eventually arise from free molecular PAHs rather than dust grain [24-27]. Each band of this set, i.e. 3.3, 6.2, 7.7, 8.6 and 11.3 μm, is identified as a fundamental vibrational mode of this class of molecules, respectively the CH stretch, the C=C stretch, C=C deformation modes of the skeleton, the CH in-plane and finally the CH out-of-plane bending vibrations.

More recent observations have revealed that the 3.3 μm emission is often accompanied by a feature at 3.4 μm (2940 cm^{-1}) which is, in fact, part of a rich structure in the 3.3-3.6 μm region as shown for example in Figure 2. The whole structure, including three newly discovered features at 3.46, 3.51, and 3.56 μm (2890, 2850 and 2805 cm^{-1}), was first observed by DeMuizon et al. [28] in two IRAS sources who mentioned also the presence of possible satellites at shorter wavelengths. In the "hot band hypothesis", some of these lines should originate from transitions between upper vibrational levels in PAH molecules[11].

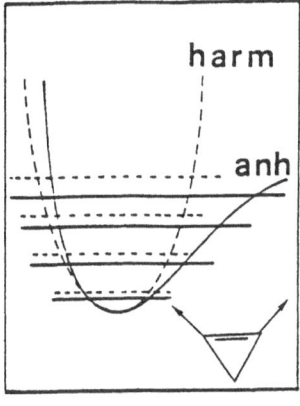

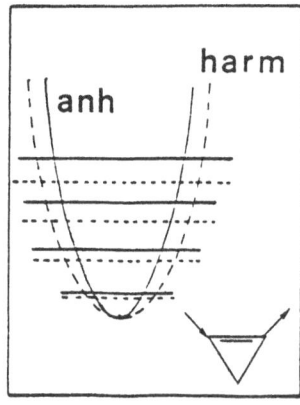

C_3H_2 being the first conjugated ring satisfying the aromaticity concept is in this respect the smallest representative of PAHs. Moreover, there is a strong possibility that C_3H_2 molecules exist in the same environment as PAHs: C_3H_2 most probably comes from the electronic dissociative recombination of $C_3H_3^+$ [1], a molecular ion which would be the result of the coulombic explosion of doubly ionized PAHs [29]. Where PAHs exist, C_3H_2 might exist too. If the anharmonic progression of C_3H_2 can generate satellite bands, then PAHs are most probably responsible for a much larger contribution. For all these reasons, comparing the interstellar observations with our calculated anharmonic progressions is relevant.

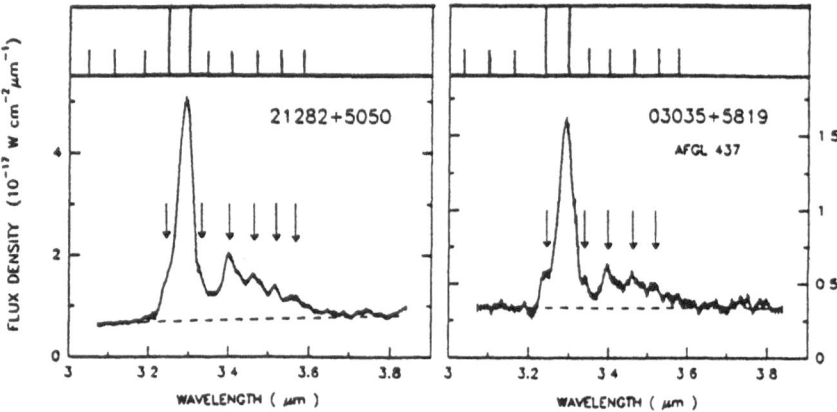

Figure 2: Positions of anharmonic progressions superimposed on the IRAS spectra.

The main conclusion to be drawn is that the calculated IR signatures do match the observed peaks in relative position. They appear with the same frequency interval of about 50 cm^{-1} . The agreement is very satisfying. It is visualized on Fig.2 where we have superimposed our calculated anharmonic progressions on two IRAS spectra. If we now consider that C_3H_2 is more than a spatial curiosity but also a suitable model for PAHs, then the results suggest that a large number of these species should have similar structural characteristics, namely *duo* hydrogens which are essentially found in compact PAHs. This work then provides a structural constraint on the chemistry of the carbonaceous matter.

4. UV signature and observation window

For C_3H_2 to be identified in space from its UV -VUV spectra, one needs to know reliable values of its low vertical electronic excitation energies. For that purpose, a number of experimental studies [30] have been realized in an attempt to observe C_3H_2 electronic transitions between 2000 and 6000 Å. These experiments having failed, computational chemistry is the alternative left to search for stable electronic states, if any, which might have been overlooked in the region between 2 and 6 eV.

Since no experimental work is available to confront the theoretical model designed to describe C$_3$H$_2$ excited states correctly, test calculations had to be done in a preliminary step. For that purpose, we have chosen ethylene, for which extensive calculations of the vertical spectrum as well as experimental measures are available. It is well known indeed that a correct quantitative and even qualitative description of small π-electron systems, is still a challenge for theoretical chemistry. The difficulties are found at each step of the computational approach :
- extended basis sets are needed in order to account for the diffuse character of the valence excited states and for the differential correlation effects;
- dynamic correlation effects, important for s orbitals in excited states, can only be recovered by including high levels of excitations in the configuration interaction (CI), which rapidly leads to untractable expansions;
- strong interactions between excited valence and Rydberg states often result in orbitals which are much too diffuse to properly describe the former states; electronic densities are thus badly described and difficult to correct even with large CI calculations.

4.1. TEST CALCULATIONS ON C$_2$H$_4$

In terms of Lewis orbitals, the electronic configuration of ethylene can be written as
$$.....\sigma_{CC}^2 \; \sigma_{CH}^2 \; \sigma_{CH}^2 \; \sigma_{CH}^2 \; \sigma_{CH}^2 \; \pi^2$$
The extensive calculations of Serrano-Andres et al [31] have shown a spurious valence-Rydberg mixing in the CASSCF wave functions when valence (π,π^*) and Rydberg orbitals are optimized all together in a state average calculation; it was shown that these orbitals loose their diffuse character and instead tend to provide an extra correlation to valence orbitals. To avoid such interaction, the orbitals used for the CI treatment of the electronic spectrum were obtained by a two step procedure :

1- We decided to first optimize C$_2$H$_4$ bonding and antibonding orbitals through an MCSCF procedure where all single and double excitations of the 12 valence electrons from the 6 occupied orbitals to the 6 antibonding orbitals were allowed. This procedure, hereafter referred to as MCSCF/SD, was applied using a triple zeta basis set augmented by polarisation functions (6-31+G*).

2- To account for Rydberg states, the MCSCF/SD set of orbitals was extended by atomic diffuse functions of s,p and d type (3s=0.023, 4s=0.015, 3p=0.021, 3d=0.015) which were Schmidt-orthogonalized.

Correlation effects were recovered in CI calculations in which the wavefunction was expanded in the N-particle space of Table 6. The internal space was partitioned in two subspaces, the first one containing all the valence occupied orbitals and the second one corresponding correlating functions plus the Rydberg orbitals. Such a repartition was designed to give an even-handed treatment for both types of valence and Rydberg states. The orthogonal complement composed the third set of MOs.

The CI space was gradually augmented, each level of calculation being referred to as CI$_{(i,j)}$ according to the classes of configurations i and j incorporated in the CI expansion. The evolution of the vertical excitation energies with regard to this systematic building of the N-particle space has been investigated for the Ag symmetry which is well known to be representative of the difficulties encountered in calculations of electronic spectra.

Table 6: Configuration space of the CI calculations for C_2H_4

symmetry	Internal sets set 1	Internal sets set 2	External set
a_g	$\sigma_{CC}\,\sigma_{CH}$	σ_{CH}^* R R R R R	complement
b_{3u}	σ_{CH}	$\sigma_{CC}^*\,\sigma_{CH}^*$ R R R R	complement
b_{2u}	σ_{CH}	σ_{CH}^* R R	complement
b_{1g}	σ_{CH}	σ_{CH}^* R R	complement
b_{1u}	π	R R	complement
b_{2g}		π^* R R	complement
a_u		R	complement
b_{2g}		R	complement

Configuration classes		Electronic distributions[a]		
	All CSFs	12	0	0
	All CSFs	11	1	0
	All CSFs	10	2	0
1	All CSFs	10	1	1
	All CSFs	11	0	1
	S+D internal excitations from the reference CSFs	9	3	0
		10	4	0
2	All CSFs	8	0	2
3	S internal excitations from the reference CSFs	9	2	1
4	D internal excitations from the reference CSFs	9	2	1
5	No internal excitation from the reference CSFs	9	1	2
6	S internal excitations from the reference CSFs	9	1	2

a)   3 K-shell orbitals are doubly occupied in all CSFs.
   3 reference configurations are selected for the CI expansion.

Table 7 shows the crucial importance of triple excitations. They have to be considered for a balanced and quantitative evaluation of the energies for both ground and excited states. The vertical excitation energy for the first excited state of Ag symmetry, which presents a strong Rydberg character, has converged to 8.26 eV for the $CI_{(1-6)}$ expansion, in excellent agreement with the recent two electron REMPI experimental value of 8.28 eV[32] and compares favorably to the 8.40 eV found in the CASPT2 approach [31] for the same geometry and basis set.

Table 7: Vertical excitation energies for the two lowest excited states of C_2H_4

CI type	Nb of CSFs	$X^1A_g \rightarrow 2\,^1A_g$ 3px	$X^1A_g \rightarrow 3\,^1A_g$ 3pd
$CI_{(1)}$	10998	9.10	10.6
$CI_{(1,2)}$	14074	10.7	12.2
$CI_{(1-3)}$	30256	9.40	10.8
$CI_{(1-3,5)}$	42402	8.10	9.60
$CI_{(1-3,5,6)}$	314810	8.18	9.64
$CI_{(1-6)}$	498032	8.26	9.70
experimental		8.28	

Excitation energies are in eV.
Experimental value from Williams et al [32]

4.2. TEST CALCULATIONS ON C_3H_2

Cyclopropenylidene is a singlet state of A_1 symmetry. In terms of symmetry adapted Lewis orbitals, its electronic configuration can be written as follows :

$$.....\sigma_{CC}^2\ \sigma_{CCC+}^2\ \sigma_{CCC-}^2\ \sigma_{CH+}^2\ \sigma_{CH-}^2\ lp_C^2\ \pi^2$$

where σ_{CC} refers to the CC bond opposite to the lp carbene lone pair orbital, σ_{CCC+}, σ_{CCC-} are the symmetric and antisymmetric linear combinations of the CC single bonds orbitals and σ_{CH+}, σ_{CH-} the corresponding combinations for the CH bonds (see Figure 3).

The first series of calculations was aimed at determining which orbitals should be taken for which excited state to reach convergence in the excitation energies, using the CI space directly transposed from C_2H_4 and reported in Table 8. All calculations were done at the MP4/6-311++G** geometry of the $^1A_1$ ground state using the Alchemy II codes [33].

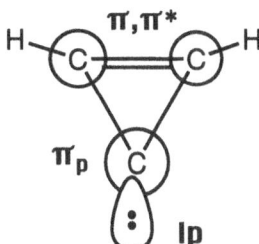

Figure 3: Lewis representation of the valence orbitals used in the MCSCF treatments

Table 8: Configuration space of the CI calculations for C_3H_2

symmetry	Orbital repartition[a]		External set
	Internal sets		
	set 1	set 2	
a_1	$\sigma_{CCC+}\sigma_{CC}\ \sigma_{CH+}$ lp_C	$\sigma_{CCC+}^{*}\ \sigma_{CH}^{*}$ R R R {R R}	complement
b_1	π	π_p R {R}	complement
b_2	$\sigma_{CCC-}\ \sigma_{CH-}$	$\sigma_{CCC-}^{*}\sigma_{CC}^{*}\sigma_{CH-}^{*}$ R R R {R}	complement
a_2		π^{*} R {R}	complement

Configuration classes		Electronic distributions[b]		
	All CSFs	14	0	0
	All CSFs	13	1	0
	All CSFs	12	2	0
1	All CSFs	12	1	1
	All CSFs	13	0	1
	S+D internal excitations from the reference CSFs	11	3	0
		10	4	0
2	All CSFs	12	0	2
3	S internal excitations from the reference CSFs	11	2	1
4	D internal excitations from the reference CSFs	11	2	1
5	No internal excitation from the reference CSFs	11	1	2
6	S internal excitations from the reference CSFs	11	1	2

(a) in brackets : the second series of Rydberg.
5 reference configurations are selected for the CI expansion with the first Rydberg series.
7 reference configurations are selected for the CI expansion with the two Rydberg series.
b) 3 K-shell orbitals are doubly occupied in all CSFs

The atomic basis consists in a double-zeta set expanded with polarization functions (DZP) and augmented by diffuse functions (DZPR). Exponents and contraction coefficient are from McLean and Chandler 1980 [18]; diffuse functions, centered on the heavy atoms with exponents of 0.023 for the s orbitals and 0.021 for the p orbitals are from Dunning and Hay 1977 [34]. Extension of the DZP basis set with two sets of diffuse s (0.0437, 0.0184) and p (0.0399, 0.0168) functions (DZPRR) has also been tested.

Based on the same two step procedure as presented above for C_2H_4 (MCSCF calculations followed by Schmidt orthogonalization of Rydberg functions), a systematic search was conducted by progressively incorporating groups of orbitals in the active space. Two types of wave functions proved well adapted to the problem, one for in-plane excitations, the other for out-of-plane excitations from the carbene orbital. The case of the $^1A_1$ states will serve as an illustration of the general approach done for all symmetries and wave functions.

4.2.1. In-plane excited states

The lowest A_1 excited states of C_3H_2 are of Rydberg type, arising from the promotion of one electron from the carbene lone pair orbital to 3s and 3p Rydberg orbitals, are better represented by orbitals generated by a MCSCF/SD treatment (Table 9).

Table 9: Configuration space of the MCSCF/SD calculations for C_3H_2

symmetry	Orbital repartition[a]	
	set 1	set 2
a_1	$\sigma_{CCC+}\ \sigma_{CC}\ \sigma_{CH+}\ lp_C$	$\sigma_{CCC+}{}^*\ \sigma_{CH+}{}^*$
b_1	π	π_p
b_2	$\sigma_{CCC-}\ \sigma_{CH-}$	$\sigma_{CCC-}{}^*\ \sigma_{CC}{}^*\ \sigma_{CH-}{}^*$
a_2		π^*
Configuration classes	Electronic distributions[b]	
All CSFs	14 {13}	0
All CSFs	13 {12}	1
All CSFs	12 {11}	2

(a) Electronic distributions for the ion in brackets
(b) 3 K-shell orbitals are doubly occupied in all CSFs.
Nb of CSFs: A_1 370 {1761} B_2 344 {1748}

The corresponding vertical excitation energies calculated with the DZPR basis set with 5 references are reported Table 10. The same type of energy convergence as observed for C_2H_4 appears for C_3H_2 when configuration classes, noted by indices, are added to the CI expansion. As energy calculations for excited states of π electron systems are sensitive to the "diffusseness" of the basis set, we have tested the DZPRR basis at the $CI_{(1-3,5)}$ level. The results (in brackets in Table 10) closely resemble those obtained when only one series of diffuse functions (DZPR) is considered, justifying the use of the DZPR basis set for the rest of our calculations.

Because Rydberg states are peculiar states with a core resembling the positive ion and one electron in a diffuse orbital, the $^1A_1$ Rydberg states have been recalculated with orbitals optimized for the ion, with the same MCSCF/SD expansion. An improvement of 0.2 eV is obtained, arguing for the use of this type of MOs for Rydberg $^1A_1$ states.

Table 10: Convergence of excitation energies for the lowest A_1 excited states of C_3H_2

CI type	Nb of CSFs	X^1A_g	$2\,^1A_g$ (3s)	$3\,^1A_g$ (3pz)	$^2A_1$ IP
$CI_{(1)}$	20759	0.0	9.50	10.1	
$CI_{(1,2)}$	34754	0.0	11.5	12.0	
$CI_{(1-3)}$	100942	0.0	10.0	10.6	9.02
$CI_{(1-3,5)}$	212591	0.0	8.80 (8.70)	9.30 (9.20)	9.04
$CI_{(1-5)}$	450264	0.0	8.60	9.12	8.98
$CI_{(1-3,5,6)}$	1113873	0.0	8.44	8.97	
$CI_{(1-6)}$	1383356	0.0	8.01	8.53	

Excitation energies are in eV.
The energies in brackets are calculated with the DZPRR basis set .

The same conclusion, that MCSCF/SD expansions using orbitals optimized for the ion provide a better representation, is reached for the lowest states of B_2 symmetry which are also states of Rydberg type arising from an in-plane excitation from the carbene orbital.

4.2.2. Out-of-plane excited states

The lowest B_1 and A_2 excited states of C_3H_2 correspond to valence excitations from the lp_C carbene lone pair to the π_p and π^* orbitals, the next states being Rydbergs. Because of this mixing, the orbitals have been optimized in the configuration space of Table 11, hereafter referred to as MCSCF/{6422}. Here the orbitals are distributed in four different spaces according to their chemical nature and the electrons assigned so as to define a direct product of CAS subspaces. Because the second $^1B_1$ state is of Rydberg character, orbitals optimized for the ion in an equivalent expansion MCSCF/{6322} have been tested.

Table 11: Configuration space of the MCSCF/{6422} calculations for C_3H_2

symmetry	Orbital repartition[a]			
	set 1	set 2	set 3	set 4
a_1	σ_{CCC+} σ_{CC} σ_{CCC+}^*	lp_C	σ_{CH+} σ_{CH+}^*	
b_1		π π_p		
b_2	σ_{CCC-} σ_{CCC-}^* σ_{CC}^*			σ_{CH-} σ_{CH-}^*
a_2		π^*		
Constraints	Electronic distributions[b]			
All CSFs	6	4 {3}	2	2

(a) Electronic distributions for the ion in brackets
(b) 3 K-shell orbitals are doubly occupied in all CSFs.
Nb of CSFs: B_1 4716 {6354} A_2 4692 {6330}

No significant improvement for the vertical excitation energy of the $2^1B_1(3p)$ state was found. From these results we have decided to describe the lowest states of B$_1$ and A$_2$ symmetries with the same set of molecular orbitals, optimized for the neutral molecule within the MCSCF/{6422} expansion.

4.3. LOW ELECTRONIC EXCITED STATES OF C$_3$H$_2$

Our best estimation for the vertical excitation energies for states of A$_1$ symmetry are reported in Table 12. They correspond to a ground state calculated at CI$_{(1-6)}$ level using orbitals optimized for the neutral molecule with the MCSCF/SD expansion, and excited Rydberg states calculated at the CI$_{(1-6)}$ level using orbitals optimized for the positive ion with the same expansion. The first excited $^1A_1(3s)$ state lies at 7.8 eV above the ground $^1A_1$ state and the second excited $^1A_1(3p)$ state at 8.4 eV. They are all below the first ionization potential which, in our best calculation CI$_{(1-5)}$, is 8.98 eV. Transition moments have been evaluated in a first order treatment, CI$_{(1,3)}$. The very weak value found between the 1^1A_1 and $2^1A_1(3s)$ states leaves little hope for observation and the effort should be concentrated on the 1^1A_1 to $3^1A_1(3p)$ transition.

The vertical excited states of B$_2$ symmetry, calculated at the CI$_{(1-6)}$ level, are very high in energy. The first one, $^1B_2(3p)$ is already at 8.60 eV above the ground state (Table 12) with a transition moment of 0.16 a.u., probably too weak for the transition to be observed.

Table 12: Electronic spectra of C$_3$H$_2$

State	ΔE(a)	μ(b)
X^{1}A$_1$	0.00	
1^1A$_2$	4.66	-
1^1B$_1$	5.20	0.60
2^1B$_1$(3p)	7.50	0.50
2^1A$_1$(3s)	7.80	0.08
3^1A$_1$(3p)	8.40	0.90
2^1A$_2$(3d)	8.59	-
1^1B$_2$(3p)	8.60	0.17
X^{2}A$_1$	8.98 (c)	

a) relative energies to the X^{1}A$_1$ are in eV.
Nb of CSFs: ^{1}A$_1$ {1 383 356}; ^{1}B$_1$ {639 818}; ^{1}B$_2$ {689 119};
^{1}A$_2$ {639 806}
b) transition moments to the X^{1}A$_1$ state are calculated at the first order
level CI$_{(1,3)}$ in a.u.
c) Ionisation potential calculated at the CI$_{(1-5)}$ level
Nb of CSFs : ^{2}A$_1$ {786 143}

Vertical excitation energies to states of B_1 symmetry, calculated at the $CI_{(1-6)}$ level using the orbitals optimized for the neutral molecule with the {MCSCF/6422} expansion, are reported Table 12. The 1^1B_1 valence state and $2^1B_1(3p)$ Rydberg state of C_3H_2 are respectively 5.2 eV and 7.5 eV above the ground state with large transition moments of 0.60 and 0.50 a.u.respectively.

Finaly, the lowest two A_2 states, calculated at the $CI_{(1-6)}$ level using orbitals optimized for the neutral molecule with the MCSCF/{6422} expansion are at 4.66 eV and 8.59 eV. Transitions from the $^1A_1$ to these states are not symmetry-allowed and it is hardly probable that vibronic coupling could make them observable in transient conditions.

The only state which could be seen in the 2000-6000 Å window is the 1^1B_1 valence state. The fact that this state was not seen in spite of its strong transition moment may well be due to the experimental uncertainty of 10% at the limit of the window.

Concluding remarks

The present contribution illustrates the possible role of computational chemistry in supporting astrophysical studies aimed at the detection of new species from their radio, infra-red and electronic signatures. In the case of a very peculiar molecule such as C_3H_2, we have shown that theoretical approaches provide assistance at all levels of spectroscopy.

-The rotational constants, although difficult to establish with the accuracy needed for a direct search on the telescope, should be precise enough to identify the deuterated isomers in the laboratory.
-The IR spectrum of the deuterated isomers is different from what has been estimated by simple extrapolation of the hydrogenated species, which explains why several bands were not recognized in the experiments. In addition, the anharmonic progressions of the CH stretching are found in agreement with the satellites of the 3.3 μm band observed in space and support the "hot band hypothesis" for explaining part of their origin.
-The electronic spectrum reveals at least two states that should be observed, provided the experimental window is enlarged beyond the 2000-6000 Å region.

The results presented here show the adequation of Computational Chemistry to problems of astrophysical interest. They illustrate a promising partnership in a field largely promoted by G. Berthier in the late seventies.

References

1.   P.Thaddeus, J.M.Vritilek and C.A. Gottlieb, *Astrophys J. Lett..* **299**, L63 (1985).
2.   D. J. DeFrees and A.D. McLean, *Astrophys. J. Lett..* **308** L31 (1986).
3.   M. Vrtilek, C.A. Gottelib and P. Thaddeus, *Astroph. J.* **314**, 716 (1987).
4.   M. Bogey, C. Demuynck, J.L. Destombes and H. Dubus *J. Mol. Spectr.* **122**, 313 (1987).
5.   P. Cox, C.M. Walmsley and R. Gusten *Astron. Astrophys.* **209**, 382 (1989).
6.   P. Cox, R. Gusten and C. Henkel, *Astron. Astrophys.* **206**, 108 (1988).
7.   N. Brouillet, A. Baudry and G. Daigne *Astron. Astrophys.* **199**, 312 (1988).

8.	H.P. Reisenauer, G. Maier, A. Riemann and R.W. Hoffmann, *Angew.Chem. Int. Ed. Engl.* **23**, 641 (1984).

9.	J.W. Huang and W.R.M. Graham . *J. Chem. Phys.* **93**, 1583 (1990).

10.	T. J. Lee, A. Bunge and H.F. Schaeffer III, *J. Am. Chem. Soc.* **107**, 137 (1985).

11.	J.R. Baker, L.J. Allamandola and A.G.Tielens, *Astrophys. J. Lett.*. **315**, L61 (1987).

12.	M.J. Frisch, G.W. Trucks, M. Head-Gordon, P.M.W. Gill, M.W. Wong, J.B. Foresman, B.G. Johnson, H.B. Schlegel, M.A. Robb, E.S. Replogle, R. Gomperts, J.L. Andres, K. Raghavachari, J.S. Binkley, C. Gonzales, R.L. Martin, D.J. Fox, D.J. DeFrees, J. Baker,.J.P.P. Stewart and J.A. Pople, Gaussian 92, Gaussian Inc., Pittsburg (1992).

13.	P.R. Taylor and P. Scarlett *Astrophys. J. Lett.* **293**, L49 (1985).

14.	S. Saebo, L. Farnell, N.V. Riggs and L. Radom, *J. Am. Chem. Soc.* **106**, 5047 (1984).

15.	W.J. Bouma and L. Radom, *J. Mol. Spectr.* **43**, 267 (1987).

16.	D.J. DeFrees, K. Raghavachari, H.B. Schlegel and J.A. Pople *J. Am. Chem. Soc.* **104**, 5576 (1982) .

17.	P. Pulay, G. Fogarasi and J.E. Boggs, *J. Chem. Phys.* **74**, 3999 (1981).

18.	A.D. Mc Lean and G.S. Chandler, *J. Chem. Phys.* **72**, 5638 (1980).

19.	H. Gewer and W. Meyer, *Mol. Phys.* **31**, 855 (1976).

20.	F. Pauzat, S. Chekir and Y. Ellinger, *J. Chem. Phys.* **85**, 2861 (1986).

21.	D.K. Aitken in : "Infrared Astronomy", IAU Symposium 96, C.G.Wynn-Williams and D.P.Cruikshank Eds., Reidel, Dordrecht, p.207 (1981).

22.	S.P. Willner in : "Galactic and Extragalactic Infrared Spectroscopy", M.F. Kessler and J.P. Phillips Eds., Reidel, Dordrecht, p. 37 (1984).

23.	W.W. Duley and D.A. Williams, *MNRAS* **196**, 269 (1981).

24.	L.J. Allamendolla, A.G. Tielens and J.R. Barker, *Astrophys. J. Lett.* **290**, L25 (1985); L.J. Allamendolla, A.G. Tielens and J.R. Barker, in : "Physical Processes in Interstellar clouds", G.E. Morfill and M. Scholer Eds., p. 305 (1987). L.J. Allamendolla, A.G. Tielens and J.R. Barker, *Astrophys. J. Suppl.* 71, 733 (1989).

25.	D.J. Defrees, M.D. Miller, D. Talbi, F. Pauzat and Y. Ellinger, *Astrophys. J.* **408**, 530 (1993).

26.	A. Leger and J. L. Puget *Astron. Astrophys.* **137**, L5 (1984).

27.	F. Pauzat, D. Talbi, M.D. Miller, D.J. Defrees and Y.Ellinger,*J. Phys. Chem.* **96**, 7882 (1992)

28.	M. De Muizon, T.R. Geballe, L.B. d'Hendecourt and F. Baas, *Astrophys. J.* **306**, L105 (1986).

29.	S. Leach , J.H. Eland and S.D. Price *J. Phys. Chem.* **93**, 7575 (1989).

30.	P. Thaddeus (Private communication)

31.	L. Serrano-Andres, M. Merchan, I. Nebot-Gil, R. Lindh and B.O. Roos, *J. Chem. Phys.* **98**, 3151 (1993)

32.	B.A. Williams and T.A. Cool, *J. Chem. Phys.* **94**, 6358 (1991)

33.	ALCHEMY II by A.D. McLean, M. Yoshimine, B.H. Lengsfield, P.S. Bagus and B. Liu in MOTECC-90

34.	T.H. Dunning and P.J. Hay in : "Modern Theoretical Chemistry", Vol 3, H.F. Schaefer III Ed., Plenum Press, New York p 1. (1977)

Ab-initio Study of the Intramolecular Hydrogen Shift in Nitromethane and its Acid-dissociated Anion

Y. TAO
Department of Chemistry, Yunnan University, Kunming 650091, People's Republic of China

1. Introduction

The photodecomposition and thermodecomposition of nitromethane have been extensively studied as model systems in combustion, explosion and atmosphere pollution processes[1]. On another hand, nitromethane was selected as a model solvent in experiments aimed at examining non hydrogen-bonded solvent effects in a general acid-base theory of organic molecules [2.3]. This selection is based on the electronic and structural characteristics of nitromethane that has a high dielectric constant, and at the same time cannot form hydrogen bonds with solute molecules.

One might believe that a tautomeric system between nitromethane and acinitromethane could be formed, and that the equilibrium would shift toward aci-nitromethane under the effet of a base. However, even under this assumption, it is not clear whether the hydrogen in the α position to the nitro group of nitromethane is sufficiently active that the tautomeric equilibrium between nitromethane and aci-nitromethane can be established by passing through the intramolecular hydrogen shift, or whether an acid-base equilibrium between nitromethane and its acid-dissociated anion has to take place.beforehand (see Fig.1). In this contribution we present a theoretical study of the 1,3-intramolecular hydrogen shift in the nitromethane and nitromethylene anion undertaken in order to describe the dynamics of these systems and to assess the adequacy of nitromethane as a model for aprotic solvents.

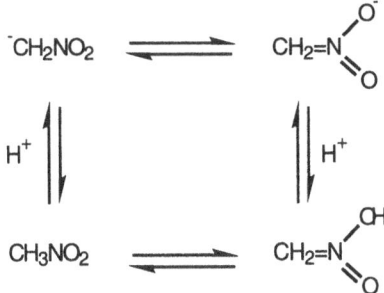

Figure 1. Hydrogen rearrangment and acid-dissociated reactions of nitromethane

421

Y. Ellinger and M. Defranceschi (eds.), Strategies and Applications in Quantum Chemistry, 421–425.
© 1996 *Kluwer Academic Publishers.*

2. Method and Results

The different structures and transitions states of interest in the neutral and negative ion reactions are represented in Fig. 2. A first approach was done at the SCF level, using the split-valence 4-31G basis set. In order to provide a better estimation of the energy differences implied in this reaction schemes, extensive calculations have been performed at the MP2 level of theory using the 6-311++G** basis set which contains the diffuse orbitals necessary to quantitatively describe the negative ions.

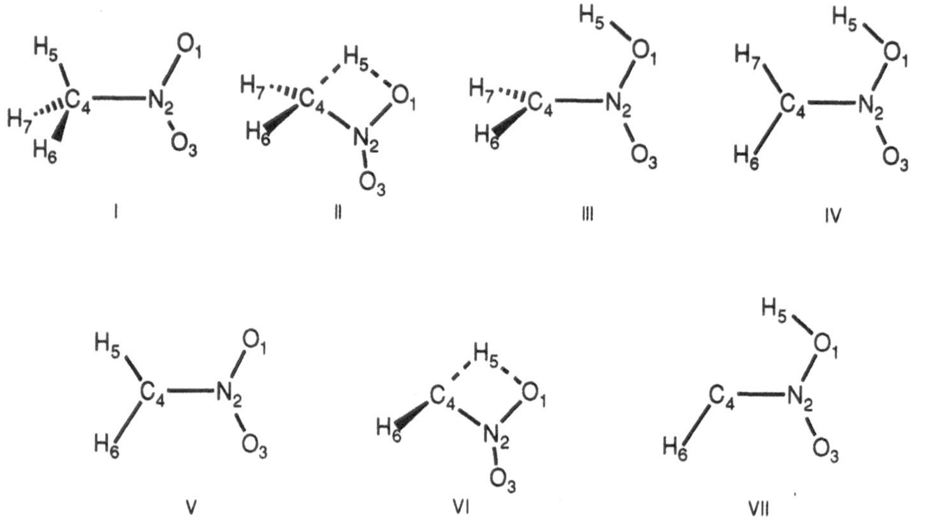

Figure 2. Stationary points on the neutral and anionic reaction pathways

The optimized geometries are reported in Table 1. The total and relative energies of all species illustrated in Figure are presented in Table 2. All calculations have been carried out with the 82 and 90 versions of the GAUSSIAN program system [4].

3. Discussion

3.1. GEOMETRIES

For nitromethane and aci-nitromethane, the optimized structures of the present calculations with the small basis set (4-31G) are very similar to the 3-21G and 6-31G * McKee's optimized structures [1]; the variation of bond lengths and bond angles with the level of theory follows the expected trends with an increase in the bond lengths when correlation effects are taken into account. The results of the present calculations for nitromethane are also in better agreement with the cristal structure [5].

The geometry of nitromethane (**1**) is characterized by the equivalence of the two NO bonds, the single bond character of the CN bond, the coplanarity of the four nonhydrogen atoms, and a value of the ∠ONO angle larger than 120°. The geometry of aci-nitromethane (**4**) is characterized by the nonequivalence of two NO bonds, the double bond character of the

Table 1. Optimized geometries for the 1,3 -intramolecular hydrogen shift of nitromethane and its acid-dissociated anion

	I			II		III		IV		V		VI		VII	
	SCF	MP2	Exp.	SCF	MP2	SCF	MP2	SCF	MP2	SCF	MP2	SCF	MP2	SCF	MP2
N_2O_1	1.218	1.230	1.214	1.316	1.315	1.337	1.333	1.398	1.430	1.315	1.278	1.446	1.407	1.447	1.482
N_2O_3	1.221	1.231	1.231	1.175	1.201	1.172	1.200	1.277	1.213	1.315	1.278	1.325	1.252	1.363	1.257
C_4N_2	1.479	1.493	1.450	1.570	1.482	1.559	1.467	1.270	1.326	1.290	1.354	1.274	1.340	1.276	1.314
H_5O_1				1.122		0.971	1.002	0.957				1.260		0.957	
H_5C_4	1.072	1.087		1.624								1.422			
H_6C_4	1.076	1.089		1.082	1.097	1.088	1.099	1.066	1.078	1.065	1.079	1.065	1.096	1.094	1.104
H_7C_4	1.076	1.089		1.082	1.097	1.088	1.099	1.066	1.078	1.065	1.079				
$C_4N_2O_3$	117.0	116.4		136.7	133.8	133.6	133.1	127.6	129.9			135.2	136.4	134.0	137.5
$O_3N_2O_1$	125.0	125.7	123.4	122.5	120.0	112.2	116.8	112.7	113.2	119.2	120.0	121.9	119.7	111.0	110.3
$C_4N_2O_1$	118.0	117.9		100.8	106.2	110.2	110.1	119.7	116.9	120.4	120.0	102.9	103.9	115.0	112.2
$H_5C_4N_2$	107.9	108.2				102.7	94.5	110.9		118.5	117.3			102.4	
$H_5O_1N_2$															
$H_6C_4N_2$	107.4	107.2		105.6	104.4	101.9	102.8	121.1	119.5	118.5	117.3	118.5	114.0	106.7	103.9
$H_7C_4N_2$	107.4	107.2		105.6	104.4	101.9	102.8	117.2	115.8						
$C_4N_2O_1O_3$	180.0	180.0		180.0	180.0	180.0	180.0	180.0	180.0	180.0	180.0	180.0	180.0	180.0	180.0
$H_5C_4N_2O_1$	0.0	0.0	0.0	0.0	0.0	0.0	0.0			0.0	0.0	0.0	0.0		
$H_5O_1N_2C_4$				0.0	0.0	0.0	0.0	0.0	0.0			0.0	0.0		
$H_6C_4N_2O_1$	120.9	121.0		121.2	123.5	123.7	124.8	180.0	180.0	180.0	180.0	180.0	180.0	180.0	180.0
$H_7C_4N_2O_1$	239.1	239.0		238.8	236.5	236.3	235.2	0.0	0.0	180.0	180.0	180.0	180.0	180.0	180.0

NO_3 bond, the coplanarity of all atoms, and the retention of the ethylene-type double bond character for the CN bond.

Nitromethylene anion (5) and aci-nitromethylene anion (7) are planar molecules. In the nitromethylene anion, there is an equivalence of the two NO bonds and a clear trend of the various bond angles toward the trigonal value of 120°. By contrast, the aci-nitromethylene anion shows the nonequivalence of the two NO bonds, a typical double bond character for the CN bond, and a larger deviation of all bond angles from 120°.

Comparing with the neutral molecules, it can be seen that the presence of the negative charge makes all bond lengths in the anion increase, particulary the bond connected to the atom bearing the negative charge. This suggests that the π conjugation in the anionic systems is reduced, and the trend toward a single bond increased. In addition, the angle variation in the anions shows a smaller steric repulsion and a greater electrostatic attraction between atoms.

The transition state for the 1,3-hydrogen shift of the neutral molecule (2) involves a planar four-membered ring. The requirement for cyclization brings the NO_1 bond distance to a value intermediate between the NO_1 bond lengths of the two tautomers; there is an increase in the CN bond, a shortening of the NO_3 bond, and a closing of the $\angle CNO_1$ angle. The fact that the $H_5 \cdot O_1$ bond (1.123/1.065 Å)is less than the $H_5 \cdot C_4$ bond (1.624/1.725 Å) in the (SCF/MP2) transition structure shows that the shifted hydrogen atom is closer to the oxygen with the larger electronegativity, namely the transition structure resembles aci-nitromethane with an higher energy.

The 1,3-hydrogen shift transition state of the anion (6) is a planar molecule. Similarly, due to the requirement for cyclization, the bond distances between the ring-forming atoms show a tendency to averaged values relative to those of the anions in their equilibrium states, while the $\angle CNO_1$ angle becomes smaller. In the same way, the transition structure in the negative ion is more similar to aci-nitromethylene though of higher energy because of the shifted hydrogen atom being closer to the negatively charged oxygen.

3.2. RELATIVE ENERGIES

It can be seen from Table 2 that the order of the relative energies is identical for the calculations at the SCF/4-31G and MP2/6311++G** levels of theory.

Table 2: Total and relative energies for the stationary points

Structure	4-31G		MP2/6-311++G**	
	E(a.u.)	ΔE(kJ/mol)	E(a.u.)	ΔE(kJ/mol)
I	-243.2745	0	-244.4785	0
II	-243.1380	358	-244.3601	310
III	-243.1469	335	-244.3608	308
IV	-243.2402	90	-244.4370	109
V	-242.6928	1527	-243.8911	1539
VI	-242.5872	1804	-243.8111	1749
VII	-242.6317	1688	-243.8364	1683

The MP2 values are certainly the most reliable; they predict that nitromethane is more stable than planar aci-nitromethane by 109 kJ/mol. and perpendicular aci-nitromethane by 308 kJ/mol.

The rearrangment of nitromethane to aci-nitromethane via the postulated 1,3-intramolecular hydrogen shift is a high barrier reaction (barrier height of 310 kJ/mol), in agreement with the prediction based on the higher tension of four-membered ring and orbital symmetry considerations.

In view of the energy profile it is clear that the equilibrium shifts essentially to the side of the more stable nitromethane in the tautomerism system between nitromethane and aci-nitromethane. These quantum chemistry calculations, although they describe principally the gaseous molecules, provide a theoretical explanation for the fact that nitromethane cannot form any hydrogen bond with the solute molecules. Nitromethylene anion, predicted to be 144 kJ/mol (MP2) more stable than aci-nitromethylene anion, involves a barrier height of 210 kJ/mol for the corresponding 1,3-hydrogen rearrangement. This suggests that for the acid-dissociated anion, as for the neutral system, the nitro-type molecules are more stable, and the 1,3-hydrogen shift can hardly take place.
Finally, we can see that, neutral meolecules, either in nitro-type or in aci-nitro-type, are more stable than acid-dissociated anions; the anion formation is a high endothermic reaction. The energy difference between neutral molecules and acid-dissociated anions calculated at the MP2/6311++G** level is 1539 kJ/mol for nitro-type species, and 1683 kJ/mol for aci-nitro-type species. It is clear that, in these conditions, the acid dissociation of the neutral molecules can hardly occur.in pure nitromethane solutions. It provides another theoretical support for nitromethane as an ideal model of aprotic solvents.

4. Conclusions

1. In either neutral molecules or acid-dissociated anions, the nitro-type species are more stable than the aci-nitro-type species. The 1,3-intramolecular hydrogen rearrangment is a high barrier process. In the tautomeric system formed via the 1,3-hydrogen shift, the equilibrium is therefore strongly displaced to the side of nitro-type species.

2. The acid dissociation of neutral molecules is such a highly endothermic reaction that the acid dissociation of nitromethane can hardly take place. The results of the calculations presented here provide a theoretical support for nitromethane as an ideal model of aprotic solvent in the acid-base theory of organic molecules.

References

1.   M.L. McKee, *J. Amer. Chem. Soc.* **108**, 5784 (1986).
2.   Huang Da-Rong, *Chemistry*, **1**, 44 (1984)
3.   Huang Da-Rong, *Chemistry*, **7**, 55 (1986)
4.   M.J. Frisch, M. Head-Gordon, G.W. Trucks, J.B. Foresman, H.B. Schlegel,
     K. Raghavachari, M. Robb, J.S. Binkley, C. Gonzales, D.F. DeFrees, D.J. Fox,
     R.A. Whiteside, R. Seeger, C.F. Melius, J. Baker, R.L. Martin, L.R. Kahn,
     J.J.P. Stewart, S. Topiol and J.A. Pople, Gaussian 90, Gaussian Inc., Pittsburgh
     (1990).
5.   S.F. Trevino, E. Prince, C.R. Hubbard, *J. Chem. Phys.* **73**, 2996 (1980).

From Cluster to Infinite Solid: a Quantum Study of the Electronic Properties of MoO3

A. RAHMOUNI and C. BARBIER
Laboratoire de Chimie Physique Moléculaire, Université Claude Bernard-LYON 1
43 Boulevard du 11 Novembre 1918, 69622 Villeurbanne Cedex, France

1. Introduction

Transition metal oxides are often selective oxidation catalysts; it is the case of molybdenum trioxide MoO_3, the electronic properties of which will be studied by molecular orbital calculations hereafter. Application of *ab initio* SCF methods to solid clusters requires large calculation times owing to the system size, so the extended Hückel theory (EHT) [1] has been adopted; indeed this method is well adapted for such an approach as proved by previous calculations on oxides and mixed oxides : the electronic structure of some phases of Bi_2O_3, $(MoO_6)^{6-}$ and $(Mo_6O_{24})^{14-}$ have been investigated by Anderson and *al.* [2], some polyanions containing molybdenum by Moffat [3], polyoxomolybdates by Masure [4] and Fournier [5], the propene adsorption on the MoO_3 (100) face by Sylvestre [6] and the electronic band structure of molybdenum bronzes have been studied by Whangbo [7-9] and Canadell [10,11].

After having described molybdenum trioxide, we intend to specify the best finite clusters allowing to represent each of the (010), (001) and (100) faces in order to study surface properties such as energy and electronic distribution. For this purpose, the evolution of the electronic properties will be studied as a function of the cluster size and referred to the results of an EHT - band calculation [12]; all calculations have been made with QCPE programs [13,14] and Hoffmann parameters [15].

2. MoO3 description

Molybdenum trioxide has a layered structure with orthorhombic symetry [16] (a=3.963, b=13.855, c=3.696 Å), this structure consists of double layer sheets parallel to the (010) cleavage plane. The building unit is a distorted MoO_6 octahedron, with Mo-O distances : 1.67, 1.73, 1.95 (twice), 2.25 and 2.33 Å (Fig.1)

Understanding the mechanism of reactions on the catalyst surface requires an adequate description of the surface ; it must modelled either by infinite slab or by clusters having similar properties. The interesting feature of the MoO_3 surface is the existence of three structurally different oxygen atoms, a terminal one O_I coordinated to one molybdenum atom, and two bridge-like oxygen atoms O_{II} and O_{III}, coordinated to two and three Mo atoms, respectively.

427

Y. Ellinger and M. Defranceschi (eds.), Strategies and Applications in Quantum Chemistry, 427–439.
© *1996 Kluwer Academic Publishers.*

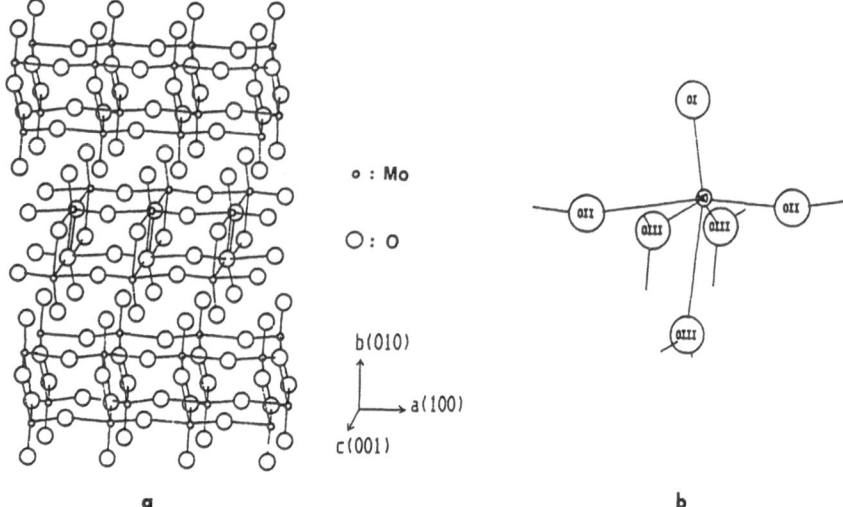

Fig. 1 a) Cristallographic structure of orthorhombic molybdenum trioxide MoO_3
        b) Distorde octahedron $MoO6$ with different types of oxygen O_I, O_{II}, O_{III}

The infinite slab is a monolayer limited by two (010) planes (model 1). It is built with a unit cell Mo_2O_6 and two translation vectors in the **a** and **c** directions, all the atoms having their usual coordination number as in the bulk.

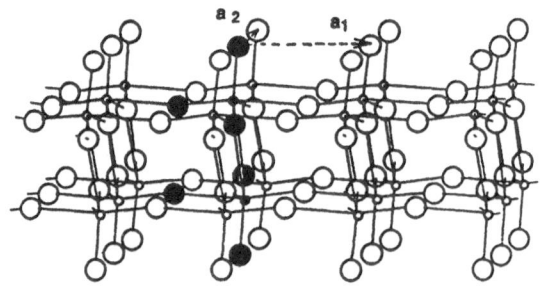

model 1

In the cluster model approach, a finite number of atoms is chosen in order to describe a well defined crystal surface site. Afterwards, the cluster size is increased by adding additional shells of surface atoms until the electronic properties of the active site have reached their convergence value [17]. Depending on the nature of the active site considered and the crystal face it would be better to choose non stoichiometric clusters; on the other hand the ionic character of metal oxides is well known, so it is reasonable to assign a (-2) charge to oxygen and (+6) to molybdenum [18,19], so that non stoichiometric clusters are electronically charged according to the Mo:O ratio.

The (010) face covers a large area of orthorhombic MoO_3 crystallites ; apart from the edge atoms it contains only coordinatively saturated Mo and O, the $Mo-O_I$ being in the perpendicular direction. The selected clusters are obtained from the extension of $(Mo_3O_{12})^{6-}$ along the **a** and **c** directions (model 2).

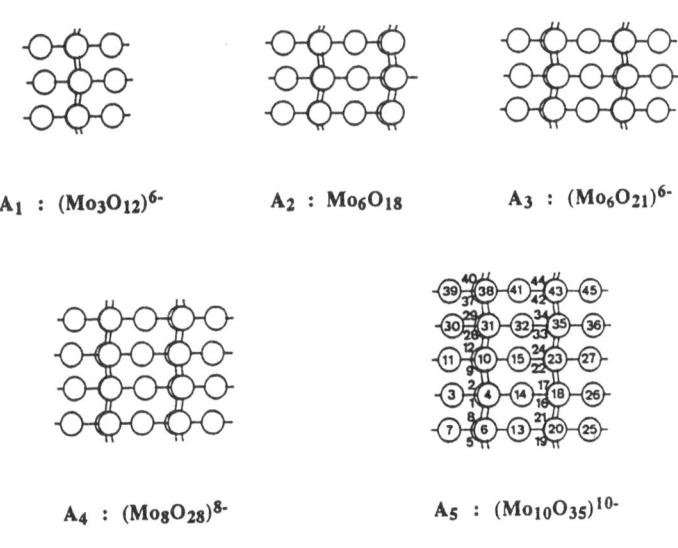

A_1 : $(Mo_3O_{12})^{6-}$ A_2 : Mo_6O_{18} A_3 : $(Mo_6O_{21})^{6-}$

A_4 : $(Mo_8O_{28})^{8-}$ A_5 : $(Mo_{10}O_{35})^{10-}$

model 2

The (100) face is modelled by neutral clusters of general formula $(MoO_3)_n$ (model 3), both containing coordinatively unsaturated Mo and O_{III} with respective coordination of 5 and 2.

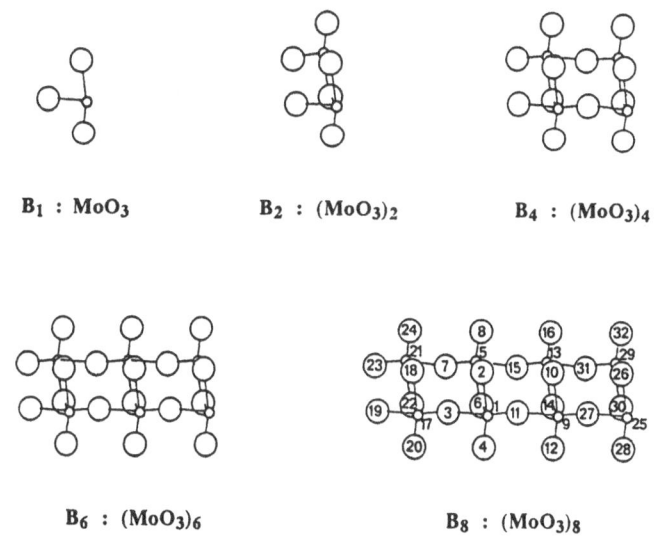

B_1 : MoO_3 B_2 : $(MoO_3)_2$ B_4 : $(MoO_3)_4$

B_6 : $(MoO_3)_6$ B_8 : $(MoO_3)_8$

model 3

The (001) face has the highest density of Mo atoms; depending on the lattice fracture plane, unsaturated Mo and O_{II} can appear. In the **a** direction two types of bonds arise : Mo-O_{II} (1.73Å) and Mo-O_{II}' (2.25Å); several crystal surfaces can then be envisaged (model 4), viz. $(Mo_6O_{24})^{12-}$, $(Mo_6O_{23})^{10-}$, $(Mo_6O_{21})^{6-}$ and Mo_6O_{18}.

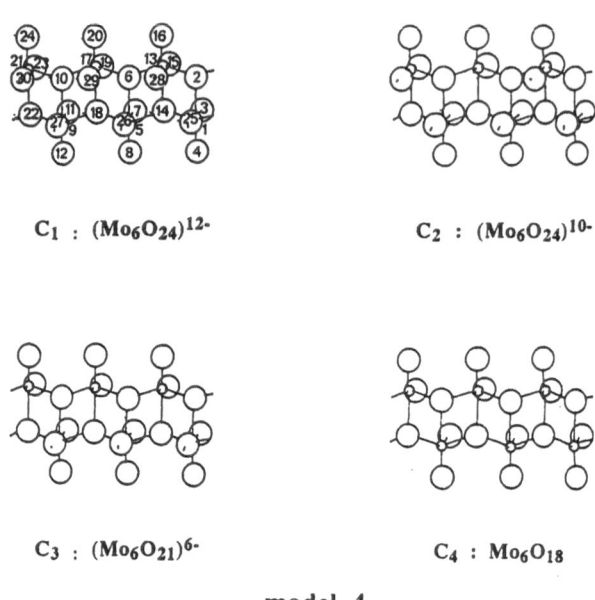

C_1 : $(Mo_6O_{24})^{12-}$ C_2 : $(Mo_6O_{24})^{10-}$

C_3 : $(Mo_6O_{21})^{6-}$ C_4 : Mo_6O_{18}

model 4

3. Results and discussion

3. 1. INFINITE SLAB

The density of states (DOS) of the infinite slab (Fig. 2) presents four blocks, two of them are located under the Fermi level (-14.58 eV); the analysis of orbital contributions to the total DOS (Fig. 3) reveals an occupied O_{2s} band at -33 eV and a second mainly occupied O_{2p} band at -15.3 eV, while Mo t_{2g} (around -9 eV) and Mo e_g (around -6.5 eV) bandes are unoccupied. The t_{2g} orbitals of molybdenum give two bands: the first one at -9 eV is intense and unoccupied, the second at -15.3 eV is weak and occupied; e_g orbitals form three bands: the most intense at -6 eV and two other very weak ones in an occupied region at -15.3 eV and -33 eV. Therefore, only the d bands located below -15 eV participate in a solid bond. The e_g orbitals weakly contributing to O_{2s} and O_{2p} bands give rise to σ bonding levels and the t_{2g} orbitals weakly contributing to the O_{2p} band give rise to π bonding levels. Furthermore, the O_{2p} contribution to Mo t_{2g} and e_g bands is both σ and π antibonding.

The "Crystal Orbital Overlap Population" (COOP) [20] shows (Fig. 4) that all levels arising below the Fermi level are σ and π bonding and the highest energy levels are σ and π antibonding; however the specific COOP curves for each Mo-O distance (Fig. 5) show a large π character at short distance Mo-O_I (1.67 Å). This result confirms the hypothesis of the existence of a double bond between molybdenum and O_I (coordination of 1).

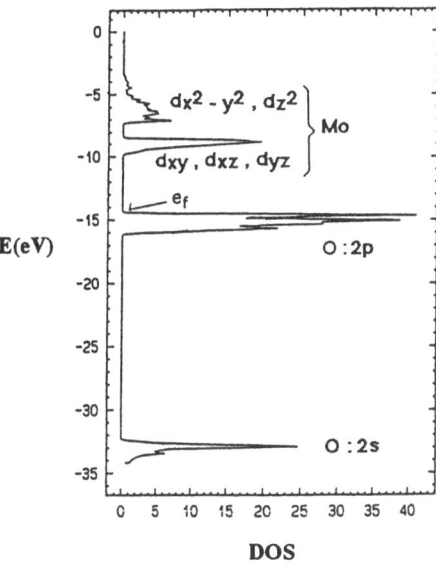

Fig. 2- Density of states of infinite slab of MoO_3

3. 2.   (010) SURFACE CLUSTERS

When the cluster size increases (Fig. 3), the occupied O_{2s} energy levels are concentrated in two blocks around -15.2 and -33.2 eV of widths 1.3 and 1.7 eV respectively; the Fermi level is slightly removed and stabilized at -14.57 eV, the same value as in the band calculation.

In every case, the Mulliken population analysis displays a large electronic transfer from the molybdenum atom towards neighbouring oxygen atoms according to the ionic character of metal oxides. The charges on the molybdenum and oxygen atoms depend on the coordination number and also on the cluster size whether the latter is sufficient to reach the charge convergence or not. Indeed the latter is attained for a cluster containing 6 or 8 molybdenum atoms (Table 1).

Table 1: (010) face evolution of net atomic charges in the octahedra centred on Mo(1)

Atom	$(Mo_3O_{12})^{6-}$	Mo_6O_{18}	$(Mo_6O_{21})^{6-}$	$(Mo_8O_{28})^{8-}$	$(Mo_{10}O_{35})^{10-}$
Mo (1)	3.786	3.770	3.769	3.761	3.761
O_{III} (2)	-1.118	-1.113	-1.113	-1.131	-1.131
O_{II} (3)	-1.818	-1.818	-1.818	-1.817	-1.817
O_{I} (4)	-1.366	-1.364	-1.364	-1.363	-1.363
O_{III} (6)	-1.487	-1.485	-1.485	-1.485	-1.485
O_{III} (10)	-1.487	-1.485	-1.485	-1.135	-1.154
O_{II} (14)	-1.439	-1.201	-1.224	-1.221	-1.221

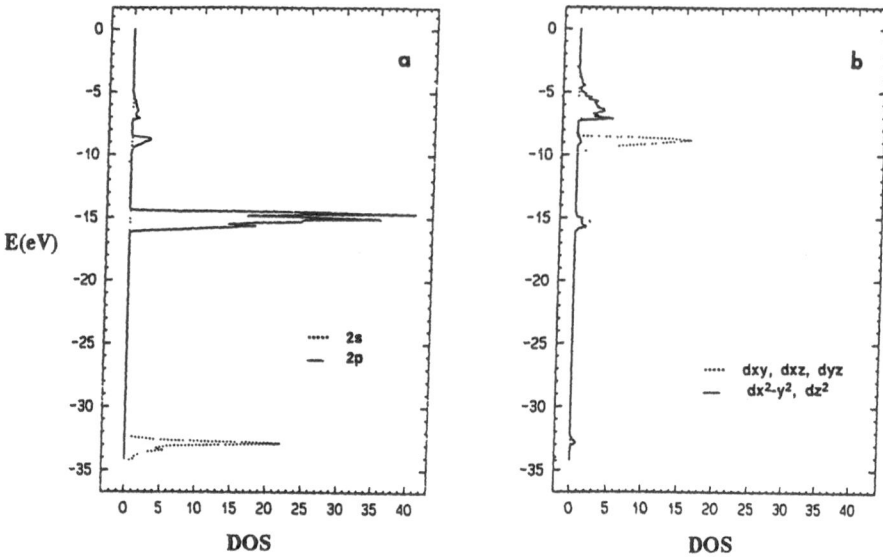

Fig.3- Contributions to the total DOS of MoO_3 : a) O_{2s} and O_{2p} ; b) Mo_{4d}

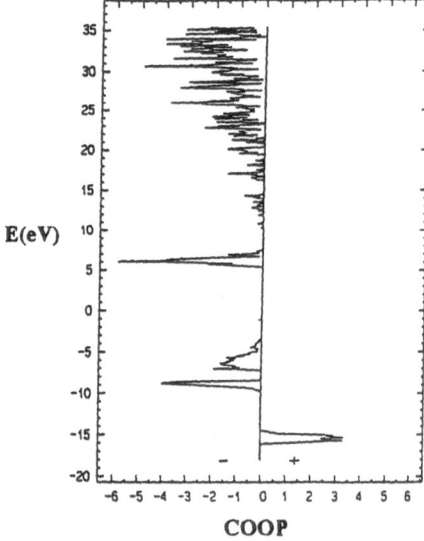

Fig.4- Crystal overlap population of MoO_3
(negative part: antibonding overlap; positive part: bonding overlap)

Table 2 shows that the $Mo(1)$-O_{II} (14) overlap population increases when the coordination number of O_{II} (14) varies from 1 to 2 whereas between $Mo(1)$-O_{III} (10), it decreases when the coordination number varies from 2 to 3 ; in other words the bonding capacity between oxygen and its nearest neighbours is shared between all bonds. Generally, the Mo-O overlap populations converge rapidly.

Table 2: (010) face evolution of the Mo(1)-O overlap populations .

Atom	d_{Mo-O} (Å)	$(Mo_3O_{12})^{6-}$	Mo_6O_{18}	$(Mo_6O_{21})^{6-}$	$(Mo_8O_{28})^{8-}$	$(Mo_{10}O_{35})^{10-}$
O_{III} (2)	2.332	0.1411	0.1414	0.1414	0.1427	0.1426
O_{II} (3)	2.251	0.2341	0.2347	0.2347	0.2359	0.2359
O_I (4)	1.671	0.7230	0.7250	0.7249	0.7287	0.7287
O_{III} (6)	1.948	0.4322	0.4330	0.4330	0.4332	0.4332
O_{III}(10)	1.948	0.4322	0.4330	0.4330	0.4239	0.4243
O_{II} (14)	1.731	0.6563	0.6707	0.6711	0.6739	0.6739

The infinite slab model can be seen as an infinite extension of the (010) clusters along cristallographic directions **a** and **c**, so one can compare limit values of (010) surface clusters to results obtained for infinite slab. The DOS of $(Mo_8O_{28})^{8-}$ does not differ very much (Fig. 6) from that of the infinite slab, especially for the occupied O_{2s} and O_{2p} bands. Table 3 shows that absolute values of the O_{II} oxygen and molybdenum charges are slightly higher in the cluster model whereas the negative charges of O_I and O_{III} are identical in both models. The Mo-O overlap populations are generally in good agreement in both models, except for Mo-O_{III} bonds (1.95 Å).

Table 3: Comparison of limit values of the cluster $(Mo_8O_{28})^{8-}$ with infinite slab.

	Cluster	Infinite slab
	Net charge (e)	
Mo	3.76	3.68
O_{III}	-1.13	-1.14
O_{II}	-1.22	-1.18
O_I	-1.36	-1.36
d(Å)	Mo-O Overlap population	
1.671	0.729	0.736
1.731	0.674	0.682
1.948	0.424	0.320
2.251	0.236	0.245
2.332	0.143	0.146

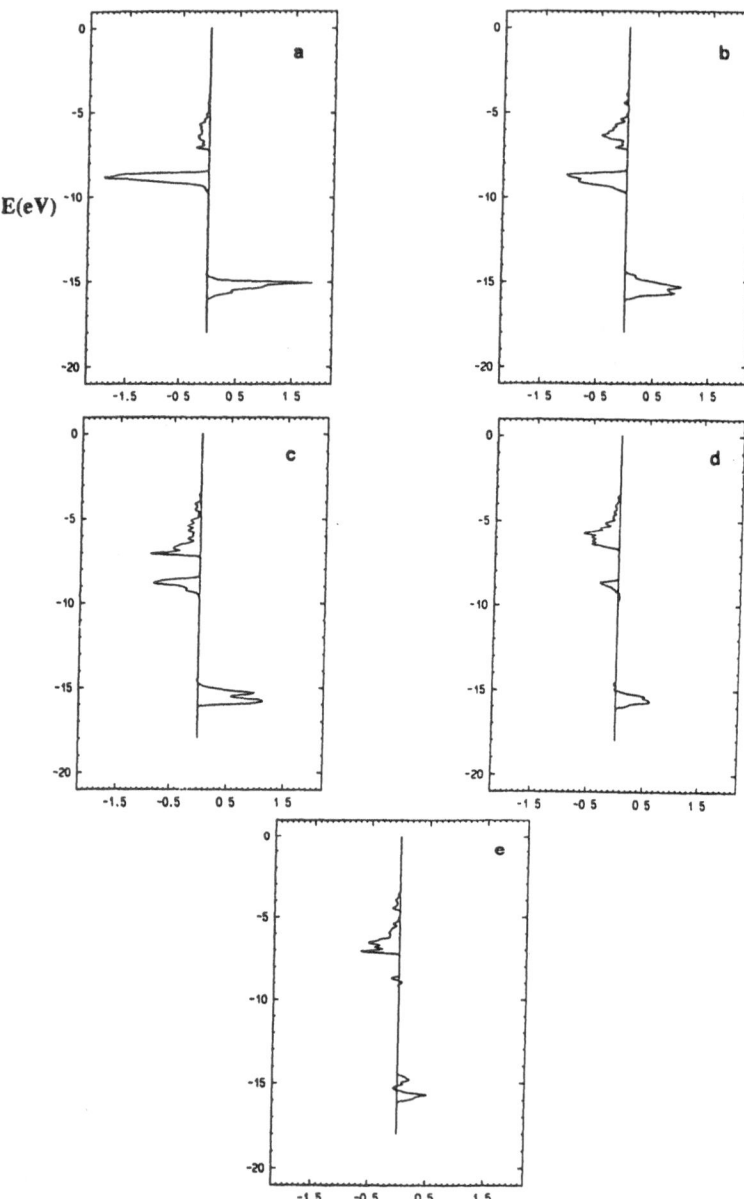

Fig. 5- Crystal orbital overlap population projections for different bond distances;
a : 1.67Å ; b : 1.73Å ; c : 1.95Å ; d : 2.25Å ; e : 2.33Å.

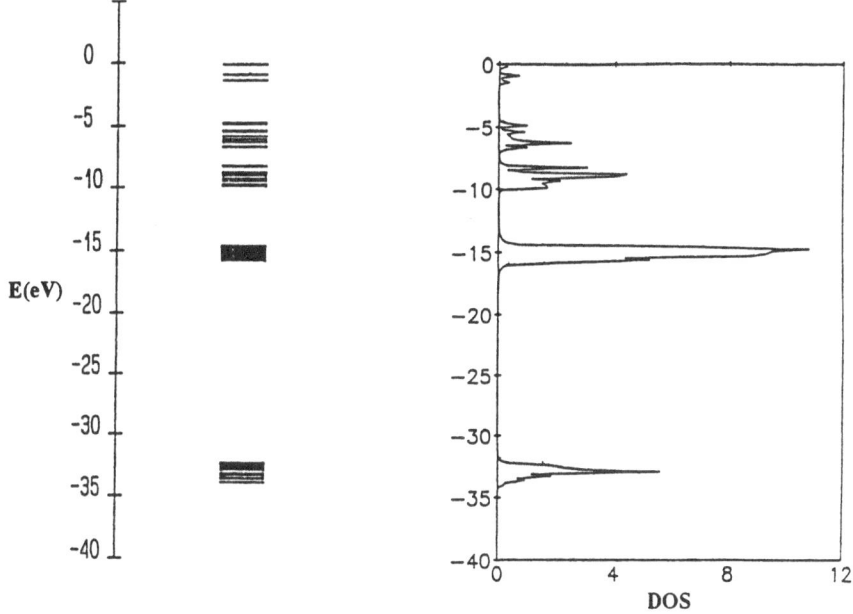

Fig.6- Energy diagram and density of states of $(Mo_8O_{28})^{8-}$

3.3.   (001) AND (100) SURFACE CLUSTERS

Since molybdenum trioxide is built from distorted MoO_6 octahedrons, the atomic arrangement on each face is different. On the (010) cleavage plane each atom has the same coordination number as in the bulk whereas the other faces obtained by lattice breaking contain unsaturated atoms. The type of unsaturated atom depends on the face considered : unsaturated O_{III} oxygen on the (001) face and unsaturated O_{II} oxygen on the (100) face. The evolution of electronic properties with increasing size of (001) clusters shows that whatever the face may be, convergence of most of the electronic properties is reached by a cluster containing 6 or 8 molybdenum atoms (Table 4).

The study of the surface cluster which models the (100) face shows that the creation of oxygen surface vacancies slightly increases the overlap population between unsaturated molybdenum and the nearest oxygen atoms (Table 5). In ionic compounds such as MoO_3, one can consider that electron transfer takes place from O^{2-} to Mo^{6+} so that the creation of the oxygen vacancies involves a decrease in the global electron transfer and consequently the molybdenum charge increases. On the other hand in order to compensate this decrease of electron transfer, the nearest remaining oxygen atoms supply more electrons to unsaturated molybdenum, this explains the decrease in the negative charge on the nearest oxygen and the increase in the overlap population between unsaturated molybdenum and the remaining neighbouring oxygen atoms.

Table 4: Net atomic charge of increasing size of (001) surface clusters.

n°	type	MoO_3	$(MoO_3)_2$	$(MoO_3)_4$	$(MoO_3)_6$	$(MoO_3)_8$
1	Mo	4.912	4.556	4.042	4.015	4.014
2	O_{III}	-1.821	-1.448	-1.465	-1.463	-1.463
3	O_{II}	-1.774	-1.784	-1.807	-1.201	-1.201
4	O_I	-1.316	-1.327	-1.354	-1.351	-1.351
5	Mo		4.192	4.03	4.015	4.015
6	O_{III}		-1.444	-1.466	-1.463	-1.463
7	O_{II}		-1.404	-1.424	-1.200	-1.200
8	O_I		-1.341	-1.352	-1.351	-1.351

In order to compare the electronic properties of the (001), (010) and (100) faces three clusters have been selected each one modelling one specific face of the MoO_3 crystal, these clusters contain the same number of oxygen and molybdenum atoms : B_6 , A_2 and C_4 for the (001), (010) and (100) faces respectively.

The total energy values (Table 6) show that the (010) face is more stable than the (001) and (100) faces, in agreement with the results of Firment[21], also our calculations show that the (001) face is more stable than the (100) one.

Table 5: (100) face overlap population around Mo (17) and Mo (5).

P_{Mo-O}	$(Mo_6O_{24})^{12-}$	$(Mo_6O_{24})^{10-}$	$(Mo_6O_{21})^{6-}$	Mo_6O_{18}
P_{17-6}	0.4237	0.4408	0.4403	0.4398
P_{17-10}	0.4232	0.4402	0.4398	0.4393
P_{17-18}	0.1436	0.1502	0.1503	0.1495
P_{17-20}	0.7306	0.7449	0.7450	0.7451
P_{17-29}	0.2366	-	-	-
P_{17-19}	0.6619	0.6702	0.6704	0.6704
P_{5-6}	0.1436	0.1427	0.1419	0.1486
P_{5-7}	0.2366	0.2366	0.2366	0.2466
P_{5-8}	0.7307	0.7307	0.7307	0.7510
P_{5-14}	0.4232	0.4232	0.4226	0.4338
P_{5-18}	0.4235	0.4229	0.4229	0.4342
P_{5-26}	0.6620	0.6621	0.6621	-

Since the type of oxygen vacancy depends on the (001) or (100) faces, the unsaturated molybdenum charge is highest in B_6; the negative charge of unsaturated oxygen is higher than in the coordinatively saturated atoms. Analysis of the Mo-O overlap population shows that the Mo-O_I bond is stronger on the (100) face whereas the other Mo-O bonds are stronger on the (001) face.

Table 6a: EHT energy and energy of frontier orbitals (eV) of (001), (010) and (100) surface clusters.

	(001)	(010)	(100)
E_{total} (EHT)	-2813.84	-2815.94	-2812.85
HOMO	-14.61	-14.60	-14.61
LUMO	-10.22	-9.99	-10.21

Table 6b: Net atomic charge on the three faces (001), (010) and (100) (* is referred to unsaturated atom).

	(001)	(010)	(100)
Mo	4.02*	3.77	3.88*
O_I	-1.35	-1.36	-1.35
O_{II}	-1.18	-1.20	-1.41*
O_{III}	-1.46*	-1.11	-1.12

Table 6c: Mo-O overlap population in the three studied surfaces (* is referred to unsaturated oxygen).

d_{Mo-O} (Å)	type	(001)	(010)	(100)
1.671	O_I	0.738	0.725	0.745
1.731	O_{II}	0.679	0.671	0.670*
1.948	$O_{III'}$	0.444*	0.433	0.440
2.251	$O_{II'}$	0.247	0.235	-
2.332	O_{III}	0.184*	0.141	0.150

4. Conclusion

The similarity of the results obtained for finite clusters and the infinite slab allows to conclude in favour of the validity of the cluster model of adequate size (6 or 8 molybdenum atoms). In addition to the chemisorption of organic molecules on solid surfaces which is generally considered as a localized phenomenon, the interaction between molybdenum oxide and an adsorbate can also be represented by a local complex formed by a finite cluster and the adsorbed molecule. Indeed, the study of the evolution of the electronic properties as a function of the cluster size shows that, for a cluster containing 6 or 8 molybdenum atoms, most of the electronic properties converge towards limit values. This convergence is sensitive to the direction of the cluster growth. On the other hand, the electronic properties of the (001), (010) and (100) faces are not identical, the type of surface atoms being different ; these results allow to predict that the characteristics of the chemisorption step will depend on the particular face on which it takes place.

Acknowledgement

The authors gratefully acknowledge technical assistance of the "Centre d'Informatique Scientifique et Médical" of the University Claude Bernard Lyon I.

References

1.   R. Hoffmann, *J. Chem. Phys.* **39**, 1397 (1963).
2.   A. B. Anderson, Y. Kim, D. W. Ewing, R. K. Grasselli and M. Tenhover, *Surf. Sci.*, **134**, 237 (1983).
3.   J. B. Moffat, *J. Mol. Catal.* **260**, 385 (1984).
4.   D. Masure, P. Chaquin, C. Louis, M. Che and M. Fournier, *J. Catal.* **119**, 415 (1989).
5.   M. Fournier, C. Louis, M. Che, P. Chaquin and D. Masure, *J. Catal.* **119**, 400 (1989).
6.   J. Sylvestre, *J. Am. Chem. Soc.* **109**, 594 (1987).
7.   M.-H. Whangbo and L. F. Schneemeyer, *Inorg. Chem.* **25**, 2424 (1986).
8.   M.-H. Whangbo and E. Canadell, J. *Am. Chem. Soc.* **110**, 358 (1988).
9.   M.-H. Whangbo, E. Canadell and C. Schlenker, *J. Am. Chem. Soc.* **109**, 6308 (1987).
10.   E. Canadell and M.-H. Whangbo, *Inorg. Chem.* **27**, 228 (1988).
11.   E. Canadell, M.-H. Whangbo, C. Schlenker and C. Escribie-Filippini, *Inorg. Chem.* **28**, 1466 (1989).
12.   M.-H. Whangbo, R. Hoffmann and R. B. Woodward. *Proc. R. Soc. London*, **A366**, 23 (1979).
13.   J. Howell, A. Rossi, D. Wallace, K. Haraki and R. Hoffmann, *QCPE*, **4**, 344 (1984).
14.   M.-H. Whangbo , M. Evain, T. Hugbanks, M. Kertesz, S. Wijeysekera, C. Wilker, C. Zheng and R. Hoffmann, *QCPE*, **9**, 591 (1989).
15.   R. Hoffmann , *J. Am. Chem. Soc.* **112**, 50 (1990).
16.   L. Kihlborg, *Arkiv Kemi*, **21**, 357 (1963).
17.   G. Pacchioni, P. S. Bagus, M. R. Philpott and C. J. Nelin, *Int. J. Quant. Chem. XXXVIII*, 675 (1990).

18. V. D. Sutula , A. P. Zeif, B. I. Popov and P. I. Vadash, *React. Kinet. Catal. Letters* , **9**, 79 (1978).
19. A. B. El Awad, E. A. Hassan, A. A. Said and K. M. Abd El Salaam, *Monatschete für Chemie*, **120**, 199 (1989).
20. R. Hoffmann, *J. Angew, Chem. Int. Ed. Engl.* **26**, 846 (1987).
21. L. E. Firment and A. Ferretti, *Surf. Sci.* **129**, 155 (1983).

Ab initio Calculations on Muonium Adducts of Fullerenes

T.A. CLAXTON
Department of Chemistry, University of Leicester, Leicester LE1 7RH, United Kingdom

1.Introduction

This paper is concerned with the structures of the simplest possible adducts of the C_{60} and C_{70} fullerenes, namely the monohydrides, $C_{60}H$ and $C_{70}H$. These open shell species or radicals may be considered as the product of the addition of one atom of hydrogen or one of its isotopes, among which we include specifically the light pseudo-isotope of hydrogen known as muonium, $Mu = \mu^+e^-$. Although $C_{60}H$ has been observed [1], the stimulus for these calculations arose from the experiments on muon implantation in solid C_{60} [2,3] and C_{70} [4].

C_{60} and higher fullerenes are distinguished from other allotropes of carbon, diamond and graphite, in that they exist as discrete molecules. The spherical or ellipsoidal nature of the monotropes opens up the possibility of intriguing new areas of chemistry. Here we are only interested in the hydrogen (or muonium) adducts, although this study has important implications to the very vigorous and extensive research in fullerene chemistry.

Two types of species have been detected in the μSR spectrum of C_{60}. One shows an unreacted or meta-stable muonium state which may well correspond to an 'internal' state, muonium is trapped *inside* the cage: $Mu@C_{60}$ in the current notation [2]. This may be compared with 'normal' muonium (Mu') in diamond and many other elemental and compound semi-conductors, where the trapping site is in one of the cavities of tetrahedral symmetry. This state of $C_{60}Mu$ is not discussed here, but it does exhibit all the characteristics expected of the 'internal' chemistry of C_{60}. The 'anomalous' muonium state, Mu*, observed in semi-conductors and generally accepted to arise from muonium being trapped within one of the chemical bonds of the crystal, is unknown in molecules [5,6]. The constraints of the crystal lattice are necessary for the bond-centred state to be stable.

The other muonium adduct of C_{60} has very similar μSR hyperfine coupling constants, 326MHz [2], to the the addition compound of muonium and ethene, C_2H_4Mu (329MHz)[7]. This is strongly indicative of a similar local structure and formation mechanism, represented formally in process (1). Muonium attacks and reduces one the short C-C bonds (common to two hexagons), bonds which have formally double bond character. It bonds to and saturates one of the carbon atoms leaving major spin density on the other. The rotational degree of freedom present in the ethyl radical is absent in caged structures and in this respect is simpler theoretically [8]. Nevertheless

441

Y. Ellinger and M. Defranceschi (eds.), Strategies and Applications in Quantum Chemistry, 441–456.
© *1996 Kluwer Academic Publishers.*

the similarity in the coupling constants argues against substantial delocalisation of the unpaired electron.

$$> C = C < \quad + \quad Mu \quad \rightarrow \quad > \dot{C} - CMu < \qquad (1)$$

These observations are supported by the calculations of Estreicher et al. [9], using the the PRDDO method and density functional theory, the semi-empirical calculations of Percival and Wlodek [10] and *ab initio* ROHF calculations [11]. They confirm that the most stable state of $C_{60}Mu$ results from the muon bonding itself to one of the carbon atoms from *outside* the cage.

Here this work is continued and extended to C_{70} where a considerable amount of experimental work is currently in progress. The observation [4] of three electron-muon hyperfine coupling constants in not unexpected since there are five chemically distinct sites for muon to attack. The lower symmetry of C_{70} makes the molecule much more interesting than C_{60}.

2.Methodology

2.1. AB INITIO CALCULATIONS

It is feasible to carry out Hartree-Fock calculations on our available computer resources (an SGI Crimson Elan Workstation) using an STO-3G basis set with full geometry optimization of $C_{60}Mu$ but only partial geometry optimisations of the $C_{70}Mu$ isomers. Fig. 1 shows planar graphs of C_{60} and C_{70} with the carbon atoms suitably labelled for future reference.

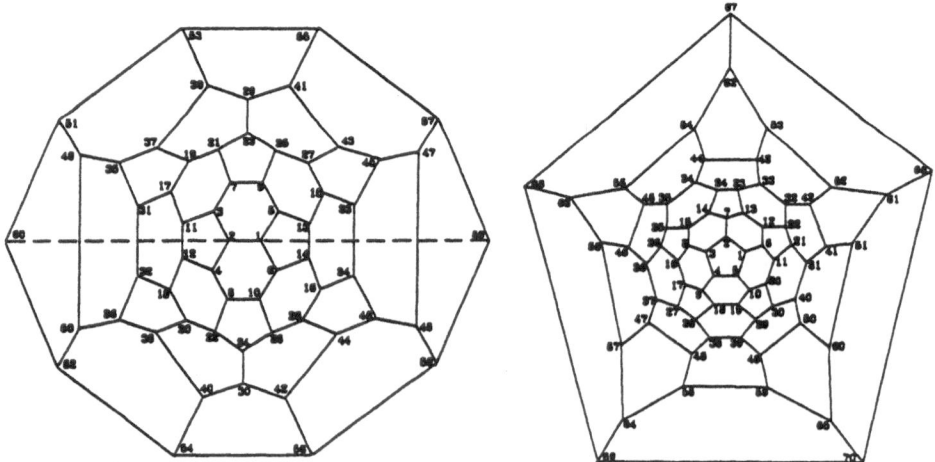

Figure 1: Labelling system used for C_{60} and C_{70}. The dotted line represents the bond between atoms 59 and 60.

Whereas there are only two different bond lengths in C_{60}, short between atoms 1 and 2 and long between atoms 2 and 3, there are seven different bond lengths in C_{70}. The C_{70} bond lengths have been calculated here and previously [12] by the restricted Hartree-Fock method using an STO-3G basis set and are discussed in some detail

later. For the present it is sufficient to split these into three groups as indicated, using an obvious notation, in Fig. 2a.
Previous calculations on $C_{60}Mu$ [11] have indicated that the distortion to the C_{60} cage in the neighbourhood of the point of attachment is so localised that it can be well represented by allowing the positions of only six carbon atoms to relax. It is assumed here that the same will apply to muonium adducts of C_{70} and addition will take place according to process (1) at sites of unsaturation (Fig. 2a).

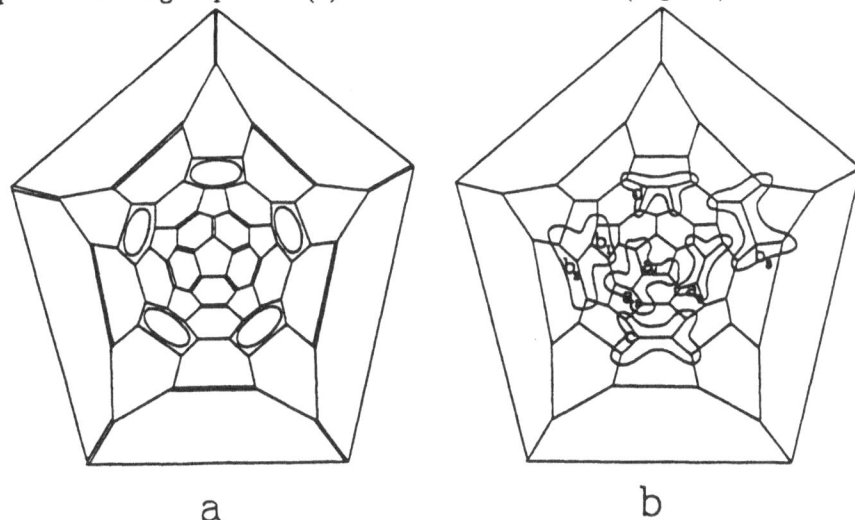

a   b

Figure 2: a) Valence bond structure for C_{70}. b) Definitions of the various types of defect, each enclosed in a closed loop. The muonated centres are labelled. Types a_3, c and b_3 refer only to one muonated centre because the other is symmetrically equivalent. The closed loops enclose the atoms which are allowed to relax in each calculation.

A typical group are atoms 1 through 6 in Fig. 1, forming a type a structure using a terminolgy introduced elsewhere [4]. A type a is identified as involving atoms from two pentagonal arrangements in the fullerene structure, the connecting bond (1–2) being short and presumably unsaturated. Three distinguishable type a structures are illustrated for $C_{70}Mu$ in Fig. 2b, the atoms in closed loops having the points of muonium attachment indicated by a_1, a_2 and a_3.
The groups of atoms within each closed loop is called a *defect*, the a_1 defect being different from a_2 simply by the point of muonium attachment. Defect a_3 is at right angles to defects a_1 and a_2. From previous calculations [11] on $C_{60}Mu$ the central two atoms of the defect are expected to change their hybridisation from sp^2 with an inevitable distortion of the underlying fullerene cage. For example if the a_1 site is the point of muonation, the unpaired electron tends to localise on the a_2 site, each site changing its hybridisation from sp^2 to approximately sp^3, the associated distortion also affecting the four nearest neighbours. Note that all defects of type a contain at their centres one of the double bonds of Fig. 2a enabling the adduct to be formed according to process (1). Although $C_{70}Mu$ can theoretically exist in one of five isomeric forms we have already identified three isomers of type a with an *alkene* type of unsaturation. The rest must be limited to regions of *arene* type unsaturation which form a band around the equator of C_{70}. Even so this gives rise to a further

four defects of which two, at least, must be discarded. All are illustrated in Fig. 2b. Type b structures involve atoms from only one pentagonal arrangement. If type c is considered [4] (involving no pentagons) it should be noted that the point of muonium attachment is the same as for defect b_2. The same applies to defect b_3 which shares the same point of attachment as b_1. Where defects share common muonium points of attachment it is possible that a more extensive relaxation than that considered here will be necessary to describe these isomers accurately. All these defect types have been considered in ref. [4] except type b_3 where the central bond is exceptionally unsymmetric in that it is common to pentagonal and hexagonal arrangements.

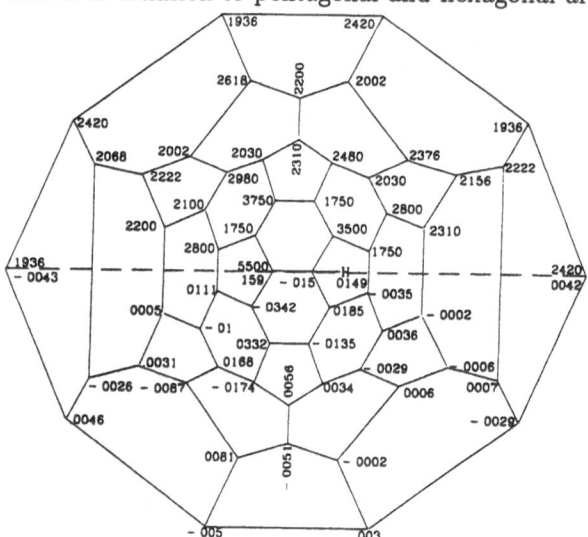

Figure 3: The numbers at each site in the top half (above the dotted line connecting the extreme atoms to the left and right of the diagram) are the numbers of classical structures which can be constructed with hydrogen (muonium) attached to the position indicated and the unpaired electron at the indicated site. The corresponding numbers in the bottom half are the spin densities in atomic units from UHFAA calculations on the fully optimised geometry of $C_{60}Mu$ using an STO-3G basis set within the ROHF method.

Typical structures are specified in Table 1 which uses the labelling of carbon atoms in C_{70} defined in Fig. 1. The restricted open-shell Hartree-Fock (ROHF) method was used in all geometry optimizations using a minimal basis set of orbitals (STO-3G) [13]. These calculations are therefore exploratory in nature. Here we have chosen to use the standard *ab initio* ROHF method since it is well-known that the UHF method (as used in the PRDDO approximation [9]) does not give wave functions which are eigenstates of the total spin operator $\hat{S}^2$. The effect of spin contamination on molecular properties is uncertain, particularly if the contamination is high (the $\hat{S}^2$ value obtained by Estreicher was not reported [9] but values larger than 1 have been reported [11]). The ROHF method is normally unsuitable for investigating spin properties because spin polarisation of the closed shell electrons is not allowed and so some UHF calculations after quartet spin state annihilation (UHFAA)[14] are also reported on the geometries optimised using the ROHF method.

2.2. RESONANCE THEORY

Resonance theory [15] contains essentially three assumptions beyond those of the valence bond method. Perhaps the most serious assumption is the contention that only unexcited canonical forms, non-polar valence bond structures or classical structures need be considered. Less serious, but no more than intuitive, is the proposition that the molecular geometry will take on that expected for the average of the classical structures. This is extended to the measurement of stability being greater the greater the number of classical structures. These concepts are still widely used in chemistry in very qualitative ways.

Molecules of the size of the fullerenes require such approaches to help rationalise the results. Even though the number of classical structures can be very large the same qualitative reasoning can be used as with the smaller molecules, typically benzene. The enumeration of classical structures is easily accomplished using computers. It is simply a problem of determining how many ways a set of points (in our case carbon atoms) can be connected given rules governing their connectivity.

3.Results

All of the *ab initio* results are collected in Tables 2 and 3. The tables differ only in the theoretical method used. Table 2 used the ROHF method and Table 3 used the UHF method using optimised geometries from Table 2.

Table 1: Specification of the carbon atoms (see Fig. 1) allowed to relax in an example of each *type* of defect in C_{70}.

Type	Site of muon	Anticipated site of unpaired electron	Sites of other atoms allowed to relax
a_1	1	6	2,5,11,12
a_2	6	1	2,5,11,12
a_3	11	20	6,7,10,30
b_1	21	31	11,22,40,41
b_2	31	21	11,22,40,41
b_3	21	22	11,12,31,32
c	31	40	21,30,41,50

The numbering system of the carbon atoms used are given in Fig. 1 for $C_{70}Mu$. Further results from the *ab initio* calculations are collected in various figures where they are compared to the results of resonance theory. The $C_{60}Mu$ results are displayed in Figs. 3, 4 and 5. The number of classical structures is compared with the UHFAA spin densities at each centre in Fig. 3. Correlations between the number of classical structures and spin density (Fig. 4) and bond length (Fig. 5) are plotted. The calculated bond lengths in C_{70} are rationalised using resonance theory in Table 4 from which a valence bond type structure of C_{70} is suggested in Fig. 2a. The structures for each isomer of $C_{70}Mu$ are shown in Figs. 6 and 7 and Fig. 8 illustrates the spin delocalisation of each defect suggested by resonance theory.

4.Discussion

4.1. CLASSICAL STRUCTURES

There are at least two ways of describing C_{60}. Its association with the ball in football has led to some unattractive nicknames. It could be regarded as twelve pentagons, each connected to five other pentagons through a connection at each apex. The connection could formally be given a carbon-carbon double bond. The bonds in each pentagon are formally single, emphasising its alkene nature, conjugation between adjacent double bonds taking a secondary role. To illustrate the possible aromatic character of C_{60} it is necessary to describe it as a continuous the sheet of hexagons, each hexagon sharing three of its sides with other hexagons. The extreme view would be to have extended conjugation over the whole ring, although hindered by the curvature imposed by the spherical structure. Whatever view is taken both indicate that they should be good candidates for treatment by resonance theory [15].

Table 2: Restricted Hartree-Fock energies (Hartrees) for C_{60} and C_{70} and their muon adducts. ΔE is the difference in energy between the carbon allotrope and its adduct. In all cases, except where indicated by †, only the six carbon atoms in the immediate vicinity of the muon have had there positions optimised. † means that a full geometry optimisation has been carried out. The *type* specifies the 'defect' and for C_{70} is identified in Table 1. a_μ is the spin density at the muon in atomic units (and the hyperfine coupling constant in MHz). ‡Muon constrained to lie in equatorial plane. * indicates geometry not fully optimized.

Molecule	Type	ROHF Energy		ΔE	a_μ
C_2H_4Mu		-77.659420	†		.01141(160.0)
C_{60}		-2244.221245	†		
$C_{60}Mu$	a	-2244.786084		-.564839	.01070(150.0)
$C_{60}Mu$	a	-2244.790148	†	-.568903	.00978(137.1)
C_{70}		-2618.357387	†		
$C_{70}Mu$	a_1	-2618.912329		-.554942	.01084(152.0)
$C_{70}Mu$	a_2	-2618.911553		-.554166	.01067(149.6)
$C_{70}Mu$	a_3	-2618.916187		-.558800	.01065(149.3)
$C_{70}Mu$	b_1	-2618.904012		-.546625	.00757(106.1)
$C_{70}Mu$	b_3	-2618.900314		-.542927	.00805(112.9)
$C_{70}Mu$	b_2	-2618.8817	*		
$C_{70}Mu$	c	-2618.874827	‡	-.517440	.01848(259.1)

Resonance theory relies on the intuitive generalization of the valence bond method, although its theoretical justification comes from molecular orbital theory [16]. Resonance theory can be used in very qualitative ways. For example the rationalisation of the bond lengths of small aromatic molecules involve very simple enumeration of the classical structures. The intermediate C–C bond length in benzene is interpreted to be the result of both Kekulé structures being equally dominant in the final structure. Since any particular C–C bond is a single bond in one Kekulé structure and a double bond in the other the observed intermediate bond length is understood. It is known that the shortest C–C in naphthalene connects the α- and β- carbon atoms. When this bond is chosen to be double two classical structures can be drawn whereas only one for every other bond selection. This correlation is only expected to be good for molecules where all the carbon atoms involved have similar environments.

Table 3: Unrestricted Hartree-Fock energies (Hartrees) for C_{60} and C_{70} and their muon adducts. muon in atomic units (and MHz) after quartet spin state annihilation. The value of the total spin operator, $< S^2 >$, is also given after quartet spin state annihilation, $< S_{AA}^2 >$.

Molecule	Type	UHF Energy		$< S^2 >$	$< S_{AA}^2 >$	a_μ
C_2H_4Mu		-77.662993	†	.7637	.7501	.01638(229.7)
C_{60}		-2244.221245	†			
$C_{60}Mu$	a	-2244.811621		1.1238	.8937	.01492(209.5)
$C_{60}Mu$	a	-2244.821999	†	1.1500	0.9340	.01495(209.6)
C_{70}		-2618.357387	†			
$C_{70}Mu$	a_1	-2618.951505		1.3066	1.1105	.01501(210.5)
$C_{70}Mu$	a_2	-2618.953237		1.4153	1.3341	.01511(211.9)
$C_{70}Mu$	a_3	-2618.939611		1.0155	.8183	.01362(191.0)
$C_{70}Mu$	b_1	-2618.957123		1.2627	1.0849	.01253(175.7)
$C_{70}Mu$	b_3	-2619.060254		5.3254	19.7503	.03106(435.5)
$C_{70}Mu$	c	-2618.965074	‡	1.8348	2.5429	.02183(306.1)

The fullerenes, notably C_{60}, are probably good examples of such molecules since, for example, there are only two different bond lengths in C_{60}, each carbon atom being symmetrically identical to all other carbon atoms. Using the labelling in Fig. 1 the number of classical structures which can be counted if a double bond is fixed between atoms 1 and 2 is 5500. If the double bond is between atoms 1 and 5 or 1 and 6 the corresponding number is 3500. This immediately explains why in C_{60} the bond length between atoms 1 and 2 (1.3759Å) is shorter than that between atoms 1 and 5 or 1 and 6 (1.4627Å) since resonance theory assumes that the larger the number of classical structures the more stable is the chosen configuration.

It is common in the interpretation of electron spin resonance spectroscopy of organic radicals to draw classical structures to rationalise the observed distribution of spin. In this spirit the number of possible classical structures of $C_{60}Mu$, with the muon at position 1, has been enumerated for the unpaired electron at each other position in turn. In all the discussions using these enumerations two assumptions are made. Firstly the double bonds in each classical structure can exist equally probably in the pentagons as well as the hexagons. Secondly each classical structure is equally probable enabling the spin density distribution in $C_{60}Mu$ to be discussed [16].

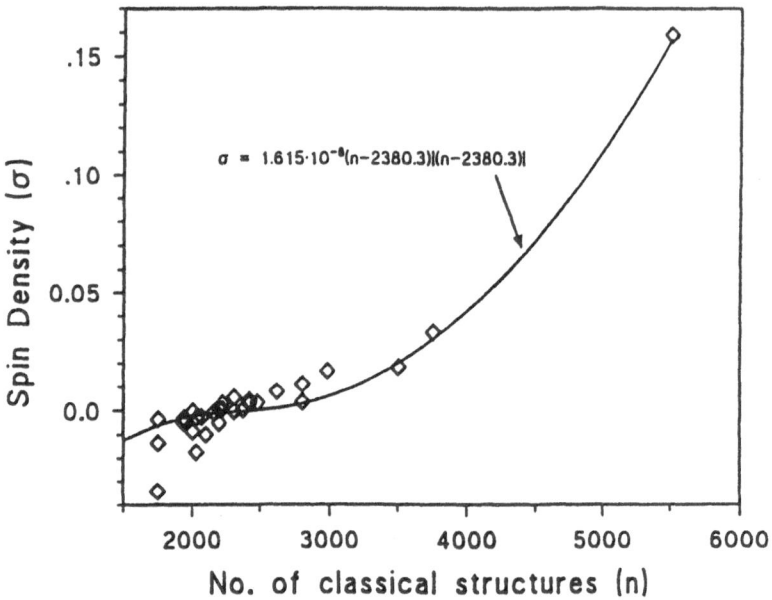

Figure 4: Plot of spin density as a function of the number of classical structures.

In Fig. 3 the number of possible classical structures arising from the spin being localised at each carbon atom (top half) is compared to the UHFAA spin density results (lower half). Note that the number of classical structures when the unpaired electron is at sites 2, 5 and 6 is the same as for the double bonds involving atoms 1 and 2, 1 and 5 or 1 and 6 in C_{60}. The correlation between the number of classical structures and the spin density is excellent. With only one exception all centres with the number of classical structures larger than 2200 show positive spin density and all those less than 2200 show negative spin density. This anticipated correlation can be further quantified.

The expectation that the spin density should be proportional to n^2 [16], should be tempered with the knowledge that an explanation must include negative spin densities. The mechanism which give rise to negative spin densities is called spin polarization. Very qualitatively the unpaired electron 'attracts' electrons of like spin in its immediate neighbourhood imposing electrons of the opposite on neighbouring atoms (or negative spin density). In the case of classical structures the spin density on each centre i will be a function of the number of classical structures, n_i^2, and *reduced* by the spin density of each of its neighbours. For example if the average of the n's of the three nearest neighbours is subtracted from the n of any centre the result has the same sign as the spin density without exception. This suggests that we should look for a relation of the type

$$\sigma_i = a((n_i - bn_j)|n_i - bn_j| + (n_i - bn_k)|n_i - bn_k| + (n_i - bn_l)|n_i - bn_l|$$

where a and b are adjustable constants, centres j, k and l are adjacent to centre i and where the modulus sign is necessary to allow for negative spin densities. A simpler, and more easily manipulated, relation is $\sigma_i = a(n_i - b)|n_i - b|$, The equation has been plotted on Fig. 4 to show how well it correlates the data. The oscillation of sign of the spin density distribution over the C_{60} cage clearly has an explanation from resonance theory.

This relationship has been found in spite of the fact that the spherical cage of C_{60} has been severely distorted in the region of the defect which could have mitigated against correlations of this sort. This comment also applies to the $C_{70}Mu$ isomers below.

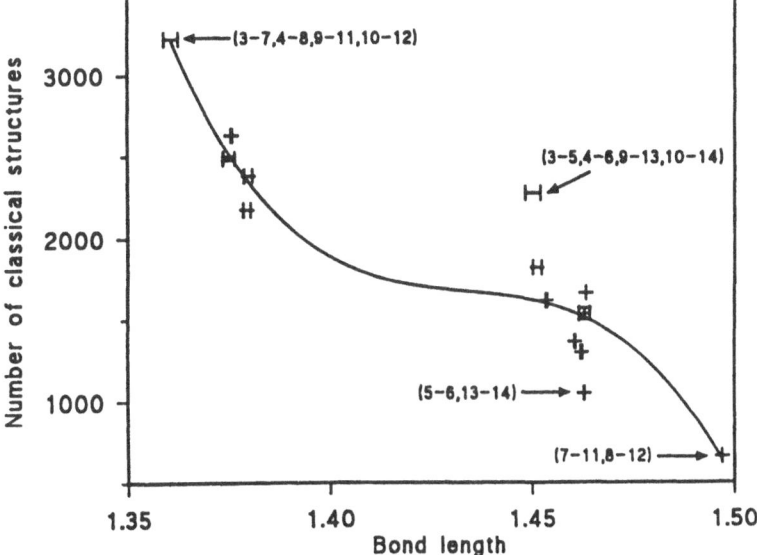

Figure 5: Plot of calculated bond lengths of C_{60}Mu as a function of the number of classical structures. Some points are labelled by (i–j) which identify the bond between atom i and atom j.

It is also possible to try and rationalise the variation in bond length of $C_{60}Mu$ over the surface using resonance theory. This is rather more difficult than the similar calculation made above for C_{60}. To do this we assume that the muon is attached to atom 1 (Fig. 1) and the unpaired electron is localised at position 2. Certainly this structure will dominate. Two other adjacent atoms are selected to form a bond and the number of classical structures which can be drawn from this configuration is enumerated and the number is a measure of the double bond character. These have been directly plotted (Fig. 5) against the optimised bond lengths calculated for $C_{60}Mu$. Biaxial error bars are used to include a range of values which would otherwise have overlapped. A number of points on the graph have been indicated for further comment. These points refer to bonds which are the immediate vicinity of the defect and conceivably still liable to the large distortion.

Nevertheless the unexpectedly long bonds, (7-11,8-12), are well predicted as are the very short bonds (3-7,4-8,9-11,10-12). All of these bonds are part of the hexagonal carbon atom structure. On the other hand the bonds of the neighbouring pentagons, 3-5,5-6,etc., are certainly out of place although still qualitatively in the right order. Presumably these bonds have absorbed most of the residual distortion of the defect for geometrical reasons.

Apart from these all other bond lengths correlate well with the corresponding number of classical structures giving the two bond lengths of C_{60}. The bond lengths close to 1.46Å correspond closely to the central carbon-carbon bond length in butadiene, associated by Dewar and Schmeising [17] to a measure of the normal single bond length for sp^2-hybridised carbon atoms, (1.48Å). [1] The other large group of bond lengths are near 1.38Å, a little larger than the 1.35Å associated with a double bond between sp^2 carbon atoms [16]. This would suggest that the bond lengths are well predicted by resonance theory, the bond lengths being either slightly longer than a typical double bond or slightly shorter than a single bond. This would seem to rule out significant conjugation as expected in an arene. The alkene properties of C_{60} are apparently dominant since cluster calculations on $C_{18}H_{12}Mu$ and $C_{30}H_{12}Mu$ [11] closely reproduce the results of $C_{60}Mu$.

However C_{70} should provide a much clearer insight into the importance of conjugation since there are 5 different types of carbon atom leading to eight different bond lengths. These have been obtained from a calculation on C_{70} fully optimising the geometry using the ROHF method and an STO-3G basis set (see Table 4, the bond lengths are within .0001Å in ref.[12]).

The corresponding numbers of classical structures for each bond are also given. The column marked † in Table 4 is $1+(n-n_0)/(n_t-3n_0)$ where n is the number of classical structures for the bond in question, $n_0 = 14196$, the number of classical structures for the typical bond 31-40, and n_t the total number of classical structures for C_{70}. We are, without justification, assuming that n_0 classical structures are unimportant to the bond order and are subtracted from all classical structures. The ratios of the residuals are taken as a measure of the mobile bond order [16]. This fixes the typical bond 31-40 to be single. The bond order for the aromatic bond is then 1.5 (the same as benzene in resonance theory). At present this is no more than a useful correlation. The almost identical numbers of classical structures for the ajacent typical bonds 21–22 and 21–31 strongly indicates that these form an aromatic type system because the symmetry of the molecule makes these typical bonds form complete hexagonal arrangements. The arene properties are also demonstrated from the almost identical bond lengths which arise even though the bond environments of 21–22 and 21–31 are different. These aromatic rings are apparently not directly conjugated since the bond which connects them, typically 31–40, is the longest bond with the smallest number of classical structures. The other single bonds have similar immediate environments. Their bond lengths are typical for single bonds connecting sp^2 carbon atoms [17]. One

---

[1]The point is emphasised from STO-3G RHF calculations on butadiene. The optimised central bond is 1.4893Å and the other bonds are 1.3134Å. Of more significance are the corresponding calculated bond lengths from the doubly positive ion of butadiene, that is, two π-electrons are removed. The central bond is now reduced to 1.3553Å, the double bond as expected, but the outer bonds are the single bond lengths 1.4953Å.

of the double bonds (1.3643Å) is significantly shorter than the short bond (1.3759Å) in C_{60} suggesting that the alkene characteristics could be enhanced in C_{70}. The other double bond is only marginally longer than the C_{60} short bond. The calculated long bond (1.4627Å) in C_{60} is just about the average of the single bonds of C_{70} which have the same bond environment (pentagon/hexagon). It was therefore with some confidence that the valence-bond type structure in Fig. 2a was presented as representing the structure which describes the chemistry of C_{70}. If muonation occurs more readily at alkene bonds the expected sites are typically 1, 6 and 11, that is, all type a structures (Fig. 2b). Type b and c structures involve bonds with arene character.

Table 4: Correlation of the number of classical structures with calculated bond lengths for C_{60} and C_{70}. The bond environment column describes the arrangements of the carbon atoms which have the bond in common. The column labelled with a † is the bond order calculated using resonance theory as described in the text.

Typical bond	Calculated bond length	No. of classical structures	Bond order [12]	†	Interpretation of bond	Bond Environment
31-40	1.4858	14196	1.20	1.000	Single	Hex/Hex
11-21	1.4701	14239	1.26	1.004	Single	Pent/Hex
1-2	1.4609	14528	1.30	1.035	Single	Pent/Hex
6-12	1.4554	14528	1.32	1.035	Single	Pent/Hex
21-22	1.4173	18943	1.48	1.496	Aromatic	Pent/Hex
21-31	1.4135	18986	1.50	1.500	Aromatic	Hex/Hex
1-6	1.3787	23112	1.66	1.931	Double	Hex/Hex
11-20	1.3643	23401	1.74	1.961	Double	Hex/Hex

4.2. AB INITIO CALCULATIONS

Our interest is two-fold, we wish to know whether the defect is firstly *structurally* localised and secondly *electronically* localised. Our second interest extends to know how important is the 'continuous surface' arising from the spherical shape of C_{60} or the ellipsoidal shape of C_{70}.

This prompted us [11] to try to represent $C_{60}Mu$ by clusters of carbon atoms, $C_{18}H_{12}Mu$ and $C_{30}H_{12}Mu$, the external atoms being constrained to lie on a part of a spherical surface with the same radius as C_{60}. The results were very similar to the $C_{60}Mu$ calculations with partial geometry optimisation to suggest that this adduct did not depend on the full structure but corresponded to a localised 'defect', both structurally and electronically.

The $C_{60}Mu$ calculations reported here are for a fully optimised geometry. The difference between the calculation using the fully optimised geometry and that which allowed only six of the carbon atoms to relax from the underlying C_{60} structure is not

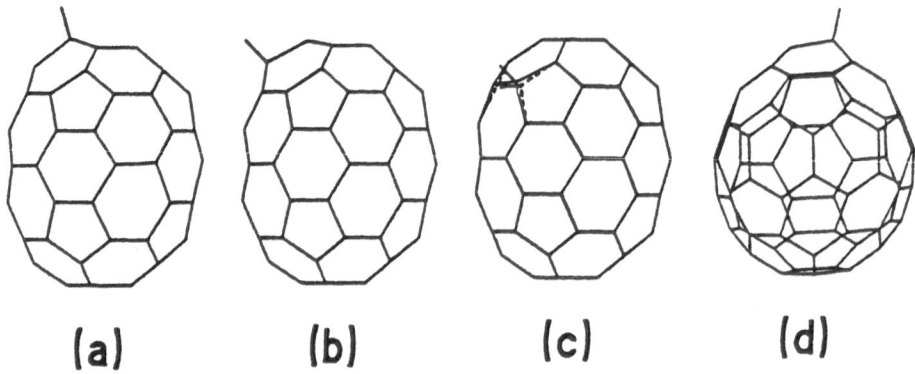

Figure 6: Polyhedral view (to scale) of structure types a_1(a), a_2(b) and two orientations of a_3(c and d) for $C_{70}Mu$. The muon is at the end of the dangling bond and in views (a) and (b) lies in the plane of the paper. For views (a), (b) and (c) four edge carbon atoms are also in the plane of the paper. The other visible atoms are above the paper. Each atom above the paper hides a corresponding atom below the paper except for type a_3 where in the region of the muon the undistorted structure below the plane is shown with dashed lines. This is useful since it clearly shows the nature of the distortion. View (d) is an orientation of type a_3 to illustrate that the distortion is similar to the other type a structures

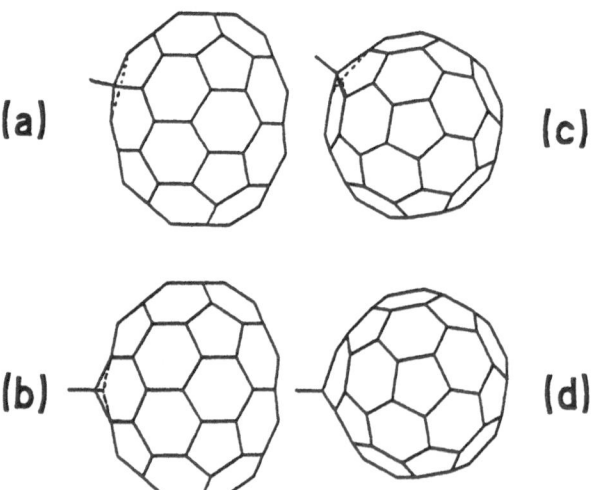

Figure 7: Polyhedral view (to scale) of type b_1 (a and c) and c (b and d) structures for $C_{70}Mu$. The views (c) and (d) look down the 5-fold axis. The views (a) and (b) are perpendicular to the 5-fold axis (similar to Fig. 8). The dashed lines correspond to the undistorted structure.

significant (Table 2) and it is confirmed that the muonium adduct of C_{60} produces only a localised distortion.

The ROHF calculations (Table 2) show a clear progression of decreasing stability the closer the muon is placed to the equator of C_{70}. This is shown by ΔE, which should be less than -.5 Hartrees for a stable adduct. The most stable structures are of type a, characteristic of the $C_{60}Mu$ adduct, and analogous to muonium adding to an alkene (compare Figs. 2a and 2b). The corresponding UHF calculations (Table 3) do not show the same behaviour although many of the differences can be associated with the problems expected from large values of $< S^2 >$, in other words, the exceptionally large spin contamination makes these UHF calculations particularly uncertain. Perhaps this has arisen due to the restrictions imposed by the partial geometry optimisation procedure. In any case the results will not be discussed further.

Apart from type b_2, which is only slowly convergent to the optimised geometry, the other centres are well described by the ROHF method. Polyhedral views of the three type a structures are shown in Fig. 6. These all illustrate the change of hybridisation at the point of muonium attachment and at the adjacent carbon atom where the unpaired electron is effectively localised as expected from addition to an alkene. The b_1 and c defects (Fig. 7) are quite different. The expected hybridisation change to sp^3 is clearly present for the atom bonded to muonium, but other significant distortions are not obvious. This is consistent with the prediction from resonance theory (Fig. 8) that the unpaired electron for these structures is delocalised over a large number of centres.

This seems to imply that the associated distortion of the cage is only large for one atom (not two as for the type a structures) and therefore possibly better accomodated by the defects defined in Table 4. It must be concluded that either structures b and c are chemically unrealistic or there is a subtle, but significant, stabilisation provided by a complete geometry optimisation.

In Fig. 8 the preferred sites for the unpaired electron are estimated from the numbers of resonance structures counted when the electron is fixed at that site. The structural differences between Figs. 6 and 7 can now be understood. In addition the similarity between type a_1 and a_2 is explained and also its marginally more stable centre a_3. Also the similar properties of b_1 and b_3 are rationalised as is the expectancy that b_2 is a more likely structure than c.

Overall Tables 2 and 3 suggest that only type a has a firm foundation. If we accept that the other structures may be possible only types b_1 and b_3 show smaller muon spin density. Experimentally [4] three sites for muonium attachment have been detected with muon hyperfine coupling constants 278.2MHz, 342.8MHz and 364.6MHz with amplitudes 17.1%, 13.1% and 11.2% respectively. The smallest coupling constant of the three is the strongest. The most obvious explanation is that only type a structures have been observed, type a_3, with the smallest of the type a coupling constants, has twice the number of sites in the molecule than the other type a structures. This would fit both the amplitude and coupling constants results qualitatively. Since structures type a_1 and a_2 are possibly too similar to be detected individually it is possible that the three resonances are (i) types a_1 and a_2 (ii) type a_3 (iii) type b_1 or b_3 in order of decreasing hyperfine coupling constant. It is not adequate to suggest that the number of sites of each should be proportional to the amplitude in this case since type a structures are 'alkene' and type b structures are 'arene'. However since alkene sites are expected to form adducts more readily than arene sites [18] the predicted amplitudes are qualitatively incorrect.

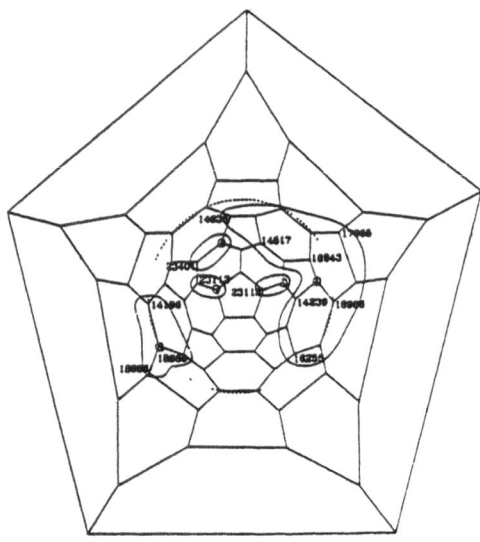

Figure 8: Measure of delocalisation of each defect type predicted by resonance theory. The loops enclose centres which have numbers of classical structures larger than .74 times the greatest number in the type. The cut-off point for type b_1 (or type b_3) centres is particularly arbitrary since the delocalisation is spread around the equator. The small circles are the point of muonium attachment. The dotted circle is coincident with the equator of C_{70}.

5.Conclusions

The application of resonance theory to adducts of C_{60} and C_{70} has proved very successful in reproducing the trends obtained from *ab initio* calculations. Not only has this helped the intepretation of the results but has contributed to testing the reliability of these *exploratory* calculations. The assignment of the isomers of $C_{70}Mu$ to the μSR results is still uncertain the current preference is that they arise from the (type *a*) alkene structures. It does seem unlikely that the type *c* structure will be observed. Although partial geometry optimisations seem satisfactory for the type *a* defects, they are not for the other structures and full geometry optimisations seem to be necessary. It is possible that only type *a* structures are chemically feasible. Further work is necessary to confirm these conjectures but it is clear that the interpretation of the experimental data in terms of isomers of $C_{70}Mu$ is correct.

References

1. J.R. Morton and K.F. Preston *private communication*

2. E.J. Ansaldo, C. Niedermeyer and C.E. Stronach, *Nature* **353**, 129 (1991); E.J. Ansaldo, J. Boyle, C.H. Niedermeyer, G.D. Morris, J.H. Brewer, C.E. Stronach and R.S. Cary: *Z. Phys. B, Condensed Matter* **86**, 317 (1992).

3. R.F. Kiefl, J.W. Schneider, A. MacFarlane, K. Chow, T.L. Duty, T.L. Estle, B. Hitti, R.L. Lichti, E.J. Ansaldo, C. Schwab, P.W. Percival, G. Wei, S. Wlodek, K. Kojima, W.J. Romanow, J.P. McCauley Jr., N. Coustel, J.E. Fischer and A.B. Smith III. *Phys. Rev. Letters* **68**, 1347 (1992).

4. Ch.Niedermayer, I.D. Reid, E. Roduner, E.J. Ansaldo, C. Bernhard, U. Binninger, J.I. Budnick, H. Glückler, E. Recknagel, A. Weidinger, *Phys. Rev. Letters* to be published

5. S.F.J. Cox and M.C.R. Symons, *Chem. Phys. Letters* **126**, 516 (1986).

6. T.A. Claxton, A. Evans and M.C.R. Symons, *J. Chem. Soc. Faraday Trans. II*, **82**, 2031 (1986).

7. P.W. Percival, J.-C. Brodovitch, S.-K. Leung, D. Yu, R.F. Kiefl, D.M. Garner, D.J. Arseneau, D.G. Fleming, A. Gonzalez, J.R. Kempton, M. Senba, K. Venkataswaran and S.F.J. Cox, *Chem. Phys. Letters* , **163** (1989) 241.

8. T.A. Claxton, A.M. Graham, S.F.J. Cox, Dj.M. Maric, P.F. Meier and S.Vogel, *Hyperfine Interactions* ,**65** 913 (1990).

9. S.K. Estreicher, C.D. Latham, M.I. Heggie, R. Jones, S. Öberg, *Chem. Phys. Letters*, **196**, 311 (1992).

10. P.W. Percival and S. Wlodek, *Chem. Phys. Letters* , **196**, 317 (1992).

11. T.A. Claxton and S.F.J. Cox, *Chem. Phys. Letters* , in press.

12. J. Baker, P.W. Fowler, P. Lazzeretti, M. Malagoli and R. Zanasi *Chem. Phys. Letters*, **184**, 182 (1991).

13. *Gaussian 92*, Revision A, M.J. Frisch, G.W. Trucks, M. Head-Gordon, P.M.W. Gill, M.W. Wong, J.B. Foresman, B.G. Johnson, H.B. Schlegel, M.A. Robb, E.S. Repogle, R. Gomperts, J.L. Andres, K. Raghavachari, J.S. Binkley, C. Gonzalez, R.L. Martin, D.J. Fox, D.J. Defrees, J. Baker, J.J.P. Stewart, and J.A. Pople, Gaussian, Inc., Pittsburgh PA, 1992

14. A.T. Amos and G.G. Hall, *Proc. Roy. Soc.*, **A263**, 483 (1961).

15. G.W. Wheland The theory of resonance 1944, Wiley, New York.

16. M.J.S. Dewar and H.C. Longuet-Higgins, *Proc. Roy. Soc.*, **A214**, 482 (1952).

17. M.J.S. Dewar and H.N. Schmeising, *Tetrahedron* **5**, 166 (1959).

18. E. Roduner, *Lecture Notes in Chemistry No. 49*, Springer-Verlag, Berlin, 1988.

List of Authors

Subject Index

TOPICS IN
MOLECULAR ORGANIZATION AND ENGINEERING

Honorary Chief Editor: W. N. Lipscomb, Harvard, U.S.A.
Executive Editor: Jean Maruani, Paris, France

KLUWER ACADEMIC PUBLISHERS – DORDRECHT / BOSTON / LONDON